Multidimensional Databases: Problems and Solutions

Maurizio Rafanelli
Istituto di Analisi dei Sistemi ed Informatica – C.N.R., Italy

IDEA GROUP PUBLISHING
Hershey • London • Melbourne • Singapore • Beijing

Acquisition Editor: Mehdi Khosrow-Pour
Senior Managing Editor: Jan Travers
Managing Editor   Amanda Appicello
Development Editor: Michele Rossi
Copy Editor: Maria Boyer
Typesetter: Amanda Lutz
Cover Design: Weston Pritts
Printed at: Integrated Book Technology

Published in the United States of America by
        Idea Group Publishing (an imprint of Idea Group Inc.)
        701 E. Chocolate Avenue, Suite 200
        Hershey PA 17033
        Tel: 717-533-8845
        Fax: 717-533-8661
        E-mail: cust@idea-group.com
        Web site: http://www.idea-group.com

and in the United Kingdom by
        Idea Group Publishing (an imprint of Idea Group Inc.)
        3 Henrietta Street
        Covent Garden
        London WC2E 8LU
        Tel: 44 20 7240 0856
        Fax: 44 20 7379 3313
        Web site: http://www.eurospan.co.uk

Library of Congress Cataloging-in-Publication Data

Rafanelli, M. (Maurizio)
   Multidimensional databases : problems and solutions / Maurizio
Rafanelli.
        p. cm.
   ISBN 1-59140-053-8 (hard cover) -- ISBN 1-59140-086-4 (ebook)
   1. Multidimensional databases. I. Title.
   QA76.9.D3 R219 2003
   005.74--dc21
                                                        2002153246

British Cataloguing in Publication Data
A Cataloguing in Publication record for this book is available from the British Library.

NEW from Idea Group Publishing

- **Digital Bridges: Developing Countries in the Knowledge Economy**, John Senyo Afele/ ISBN:1-59140-039-2; eISBN 1-59140-067-8, © 2003
- **Integrative Document & Content Management: Strategies for Exploiting Enterprise Knowledge**, Len Asprey and Michael Middleton/ ISBN: 1-59140-055-4; eISBN 1-59140-068-6, © 2003
- **Critical Reflections on Information Systems: A Systemic Approach**, Jeimy Cano/ ISBN: 1-59140-040-6; eISBN 1-59140-069-4, © 2003
- **Web-Enabled Systems Integration: Practices and Challenges**, Ajantha Dahanayake and Waltraud Gerhardt ISBN: 1-59140-041-4; eISBN 1-59140-070-8, © 2003
- **Public Information Technology: Policy and Management Issues**, G. David Garson/ ISBN: 1-59140-060-0; eISBN 1-59140-071-6, © 2003
- **Knowledge and Information Technology Management: Human and Social Perspectives**, Angappa Gunasekaran, Omar Khalil and Syed Mahbubur Rahman/ ISBN: 1-59140-032-5; eISBN 1-59140-072-4, © 2003
- **Building Knowledge Economies: Opportunities and Challenges**, Liaquat Hossain and Virginia Gibson/ ISBN: 1-59140-059-7; eISBN 1-59140-073-2, © 2003
- **Knowledge and Business Process Management**, Vlatka Hlupic/ISBN: 1-59140-036-8; eISBN 1-59140-074-0, © 2003
- **IT-Based Management: Challenges and Solutions**, Luiz Antonio Joia/ISBN: 1-59140-033-3; eISBN 1-59140-075-9, © 2003
- **Geographic Information Systems and Health Applications**, Omar Khan/ ISBN: 1-59140-042-2; eISBN 1-59140-076-7, © 2003
- **The Economic and Social Impacts of E-Commerce**, Sam Lubbe/ ISBN: 1-59140-043-0; eISBN 1-59140-077-5, © 2003
- **Computational Intelligence in Control,** Masoud Mohammadian, Ruhul Amin Sarker and Xin Yao/ISBN: 1-59140-037-6; eISBN 1-59140-079-1, © 2003
- **Decision-Making Support Systems: Achievements and Challenges for the New Decade**, M.C. Manuel Mora, Guisseppi Forgionne and Jatinder N.D. Gupta/ISBN: 1-59140-045-7; eISBN 1-59140-080-5, © 2003
- **Architectural Issues of Web-Enabled Electronic Business**, Nansi Shi and V.K. Murthy/ ISBN: 1-59140-049-X; eISBN 1-59140-081-3, © 2003
- **Adaptive Evolutionary Information Systems**, Nandish V. Patel/ISBN: 1-59140-034-1; eISBN 1-59140-082-1, © 2003
- **Managing Data Mining Technologies in Organizations: Techniques and Applications**, Parag Pendharkar/ ISBN: 1-59140-057-0; eISBN 1-59140-083-X, © 2003
- **Intelligent Agent Software Engineering**, Valentina Plekhanova/ ISBN: 1-59140-046-5; eISBN 1-59140-084-8, © 2003
- **Advances in Software Maintenance Management: Technologies and Solutions**, Macario Polo, Mario Piattini and Francisco Ruiz/ ISBN: 1-59140-047-3; eISBN 1-59140-085-6, © 2003
- **Multidimensional Databases: Problems and Solutions**, Maurizio Rafanelli/ISBN: 1-59140-053-8; eISBN 1-59140-086-4, © 2003
- **Information Technology Enabled Global Customer Service**, Tapio Reponen/ISBN: 1-59140-048-1; eISBN 1-59140-087-2, © 2003
- **Creating Business Value with Information Technology: Challenges and Solutions**, Namchul Shin/ISBN: 1-59140-038-4; eISBN 1-59140-088-0, © 2003
- **Advances in Mobile Commerce Technologies**, Ee-Peng Lim and Keng Siau/ ISBN: 1-59140-052-X; eISBN 1-59140-089-9, © 2003
- **Mobile Commerce: Technology, Theory and Applications**, Brian Mennecke and Troy Strader/ ISBN: 1-59140-044-9; eISBN 1-59140-090-2, © 2003
- **Managing Multimedia-Enabled Technologies in Organizations**, S.R. Subramanya/ISBN: 1-59140-054-6; eISBN 1-59140-091-0, © 2003
- **Web-Powered Databases**, David Taniar and Johanna Wenny Rahayu/ISBN: 1-59140-035-X; eISBN 1-59140-092-9, © 2003
- **E-Commerce and Cultural Values**, Theerasak Thanasankit/ISBN: 1-59140-056-2; eISBN 1-59140-093-7, © 2003
- **Information Modeling for Internet Applications**, Patrick van Bommel/ISBN: 1-59140-050-3; eISBN 1-59140-094-5, © 2003
- **Data Mining: Opportunities and Challenges,** John Wang/ISBN: 1-59140-051-1; eISBN 1-59140-095-3, © 2003
- **Annals of Cases on Information Technology** – vol 5, Mehdi Khosrowpour/ ISBN: 1-59140-061-9; eISBN 1-59140-096-1, © 2003
- **Advanced Topics in Database Research** – vol 2, Keng Siau/ISBN: 1-59140-063-5; eISBN 1-59140-098-8, © 2003
- **Advanced Topics in End User Computing** – vol 2, Mo Adam Mahmood/ISBN: 1-59140-065-1; eISBN 1-59140-100-3, © 2003
- **Advanced Topics in Global Information Management** – vol 2, Felix Tan/ ISBN: 1-59140-064-3; eISBN 1-59140-101-1, © 2003
- **Advanced Topics in Information Resources Management** – vol 2, Mehdi Khosrowpour/ ISBN: 1-59140-062-7; eISBN 1-59140-099-6, © 2003

Multidimensional Databases: Problems and Solutions

Table of Contents

Foreword

Some months ago, I was invited by Idea Group Publishing to write a book on multidimensional databases because of my long experience in this area (more than 20 years). I accepted because I felt there was a need to create a point of reference to the most important issues in this particular field of research.

This presented me with the opportunity of writing a brief history of this field and of distinguishing between what is new in recent research and what is merely a renaming of old concepts. As many authors have rightly pointed out, a large number of concepts, operations, etc. had already been proposed in the '80s, appearing in papers, journals, conferences, workshops, and books. Important papers by Chin; Denning; Klug; Malvestuto; McCharty; Ghosh; Olken; Meral and Gultekin Ozsoyoglu; Rafanelli; Sato; Schlorer; Shoshani; Su; Tansel; and Wong document the history of this field.

Many of the most important and well-known researchers have contributed to this book, and each of them has written about their own specialist field. I hope, after many months of work, that this book will represent a milestone.

It is not only a review of past papers, but also of current research projects, and I hope it will encourage the search for new solutions to the many problems that are still open. There have been incredible advances in technology and ever-increasing demands from users in the most diverse applicative areas (finance, medicine, statistics, business, etc.). This presents a challenge for researchers to discover new methodological tools for investigating both old and new research areas in the database field and, in particular, that of multidimensional databases.

I am very proud to have had the cooperation of such distinguished scientists, who have given prestige and significance to this work, and I would like to thank them with all my heart for their valuable contribution.

Preface

The term "multidimensional data" generally refers to data in which a given *fact* is quantified by a set of *measures*, obtained by applying one more or less complex aggregative function (from count or sum to average or percent, and so on) to raw data. Such measures are characterized by a set of variables, called *dimensions*. In reality, a dimension often consists of a more complex structure than a simple variable, as we will see in the following chapter. Multidimensional data can be modeled by different representations, depending on the application field which uses them. For example, some years ago the term "multidimensional data" referred essentially to statistical data, that is, data whose use was (and is) basically for socio-economic analysis. The visual representation used most was (and is) the table (even if histograms, cakes, graphics, etc. are used too). Recently, the metaphor of the data cube, already proposed at the beginning of the '80s, was taken up again and used for new applications—such as On-Line Analytical Processing (OLAP)—which refer to business analysis.

Many of the problems concerning statistical multidimensional databases and many of the concepts defined in this context, especially if referring to models, operators, and algebras in general, have been up again and enlarged for new management system types and new applications, such as OLAP. Studies on this data type started at the beginning of the '80s. In this introduction we would like to give a brief history of the various topics in this research area, covering the period of the last 20 or more years.

At the beginning of the 1980s, a group of researchers began to look at some of the problems which arose when they considered data obtained by applying simple aggregation functions (count or sum) to row disaggregate data. This was the main reason that prompted some researchers to organize a workshop in California (1st SDBM (1981)) on the main topics of statistical databases (the term used, at that time, for multidimensional aggregate data). This name was coined because such data, organized in a database as flat files or multidimensional "tables," were mainly used to carry out statistical analysis or socio-economic type applications, such as census data on national production and consumption patterns, etc., as discussed in Shoshani (1982), Brown, Navathe, & Su (1983), and Shoshani & Wong (1985). However, they were also used for business applications, such as financial summary reports, sales forecasting, etc., as described in Wong (1984). Generally, these tables are represented by using two dimensions (for this reason they are also called "flat tables"), but, in reality, often a row and/or a column each consists of two or more dimensions.

As mentioned above, since 1981 a series of conferences have been organized: 1st SDBM (1981), 2nd SDBM (1983), 3rd SSDBM (1986), 4th SSDBM (1988), 5th SSDBM (1990), 6th SSDBM (1992), 7th SSDBM (1994), 8th SSDBM (1996), 9th SSDBM (1997), 10th SSDBM (1998), 11th SSDBM (1999), 12th SSDBM (2000), 13th SSDBM (2001), and 14th SSDBM (2002). Their purpose was to discuss issues of interest regarding statistical (and scientific) database management and to propose original solutions to the problems which arose in this area, both from the theoretical and from the application point of view. Recently, these conferences have also covered other new applications, in particular, on-line analytical processing and, in general, data warehousing. The papers presented at the conferences mentioned above cover practically all the topics of this area, from data models to new operators and relative query languages, from temporal summary table management to data privacy, from physical storage to metadata management, from graphic and visual interfaces to query optimization. Important issues on statistical databases are discussed in Shoshani (1982), Denning, Nichelson, Sande, & Shoshani (1983), Shoshani & Wong (1985), Rafanelli (1989), Rafanelli (1990), and in a book edited by Michalewicz (1990).

The first problem studied in the literature and regarding statistical multidimensional databases was that of privacy. At the end of the '70s, a few papers appeared in conferences and journals, for example, Yu & Chin (1977), Chin (1978), Denning D.E., Denning P.J., & Schwartz (1979). In them the authors began to study "how" to protect the individual privacy and, moreover, the privacy of companies, organizations, etc., which can be violated by clever manipulation of summary data. This manipulation could lead to the exact or approximate disclosure of confidential data or individuals. In the subsequent years other papers researched this topic further: Denning (1980), Schlorer (1980), Beck (1980), Schlorer (1981), Chin & Ozsoyoglu (1982), Denning (1983), Denning & Schlorer (1983), Chin (1986), Gusfield (1988), Adams & Wortmann (1989), Malvestuto, Moscarini, & Rafanelli (1991), and Michalewicz (1991). Recent proposals, such as Malvestuto & Moscarini. (1996b), Kao (1997), Chu (1997), Adam, Gangopadyay, & Holowczak. (1999), and Kleinberg, Papadimitriou, & Raghavan (2000), have been made to guarantee the privacy of individual records. These proposals include mechanisms of inference control, which can be said to be effective only if sensitive summary data are neither explicitly nor implicitly released. There have also been proposals on how to answer complex queries (in the sense of structured data, such as a table). The privacy problem is widely discussed in Chapter 11 of this book.

During the '80s it was immediately evident that the above-mentioned (statistical) multidimensional tables had a more complex data structure than the conventional disaggregate data which were, in general, represented by relations (bi-dimensional tables in which the row, called "tuple," had the meaning of "instance of relationship" and the column represented an attribute of the same relation). From a given point of view, each attribute of the relation represented one of its "dimensions," even if in the literature the classic relations are considered as corresponding to 0-dimensional tables, all of whose attributes are essentially measure attributes, as affirmed in Gyssens & Lakshmanan (1997).

Instead, a statistical, multi-dimensional table was characterized by different elements. First of all, the concept of "summary attribute" was introduced; see, for example, the 1st SDBM (1981), the 2nd SDBM (1983), Shoshani (1982), Rafanelli & Ricci (1983), and Shoshani & Wong (1985). The summary attribute represents the result (summary) of aggregation operations performed on disaggregate data, as defined in Shoshani (1982), and in Bezenchek, Rafanelli, & Tininini (1996a), i.e., the "measure" of the aggregation is carried out. This point will be described in more detail in the following chapter. The previous attributes, called "category attributes" in order to distinguish them from the others, were enriched with a descriptive characteristic, inheriting and reinforcing the concept of table dimension.

The multidimensionality of the statistical tables was emphasized many years ago. For example, Rafanelli & Ricci (1983) wrote: "Category attributes represent a cross-product of an *n-dimensional space*...." Later, in Shoshani & Wong (1985), speaking on statistical databases, we find the following phrases: "In addition to statistical operators, such as sampling and aggregation, the access of the data is of a different nature. For example, it is quite common to access a region in *multidimensional space*, or to find materials with certain approximate properties, or cases that fit a statistical pattern. For such cases *multidimensional data structures* and search methods are desirable."

And in Fortunato, Rafanelli, Ricci, & Sebastio (1986), the authors affirm: "Every complex table can always be decomposed into different simple tables...The logical description of a simple table takes its *multidimensionality* into account." Again, Rafanelli (1991) wrote: "Multidimensionality is a dominant feature in SSDBs. Many of the characteristics and requirements of SSDBs can be traced to this feature...The difficulty of dealing with multidimensional spaces is further compounded by the fact that each dimension can itself have a complex (usually hierarchical) structure."

Finally, in Rafanelli and Shoshani (1990a) and, subsequently, in Shoshani & Rafanelli (1991), we find the following phrases:" *Multi-dimensionality*; typically a multidimensional space defined by the category attributes, is associated with a single summary attribute (for example, the three-dimensional space defined by "state," "race," and "year" can be associated with "population." The implication is that a combination of values from "state," "race," and "year" (e.g., Alabama, Black, 1989) is necessary to characterize a single population value (e.g., 21,373)). *Classification hierarchies*—a classification relationship often exists for the categories. For example specific items: "fruit," "vegetable," "grain," etc. can be classified as "agricultural products"; "tables," "chairs," "sofas," etc. as "furniture." Moreover, "agricultural products," "furniture," etc. can themselves be classified as "products," thus forming a three-level classification hierarchy."

Recently studies have been made to discuss different problems regarding hierarchy. This topic will be discussed in more detail in Chapter 4. For example, the possibility of browsing along different levels of the same dimension, where each level represents the different granularity of the dimension instances, will be discussed. At the beginning, the possibility of describing the dimension of a table at a different level of aggregation (the above-mentioned granularity) and consequently the summary data described by this dimension, was seen as reclassifying a category attribute as another category attribute which represented the same more or less aggregate dimension, as written in Rafanelli & Ricci (1985). For example, the category attribute "months" of the dimension "time" could be *reclassified* as a new category attribute "quarter" by a function which groups the first three months into the first quarter, the second three months into the second quarter, and so on. Linked to the issue of the different dimensions which characterize and describe an aggregate data, there is the problem of the different classifications (see Malvestuto, Rafanelli, & Zuffada, 1988; Malvestuto & Zuffada, 1989) of the same dimension which appears, as a descriptive variable, in different databases. This topic will also be briefly discussed in Chapter 4.

Another issue which attracted the interest of scientific researchers from the beginning of the '80s was the conceptual data models for multidimensional databases and the operators to manipulate these data. The first graphical model, called SUBJECT, was proposed by Chan & Shoshani at the 7th VLDB Conference (Chan & Shoshani, 1981). Subsequently, different other graphical models were proposed: GRASS (see Rafanelli & Ricci, 1983; Rafanelli, 1987), SAM* (see Su, 1983), STORM (see Rafanelli & Shoshani, 1990a; Rafanelli & Shoshani, 1990b, 1990c; Shoshani & Rafanelli, 1991), ADAMO (see Bezenchek, Massari, & Rafanelli, 1994; Bezenchek, Rafanelli, & Tininini, 1996b; Tininini, Bezenchek,

& Rafanelli, 1996). This last model is briefly discussed in the following chapter. At the same time, models for summary data were proposed in Johnson (1981), Rafanelli & Ricci (1988), Chen, McNamee, & Melkanoff (1988), Sato (1991), Rafanelli & Ricci (1991), Malvestuto (1993), and Rafanelli & Ricci (1993).

More recently, a variety of formal multidimensional data models have been proposed, such as in Li & Wang (1996), Tininini, Bezenchek, & Rafanelli (1996), Agrawal, Gupta, & Sarawagi (1997), Datta & Thomas (1997), Gyssens & Lakshmanan (1997), Cabibbo & Torlone (1997), Lehner (1998), Vassiliadis (1998), Gingras & Lakshmanan (1998), Franconi & Sattler (1999), Pedersen & Jensen (1999), Nguyen, Tjoa, & Wagner (2000), and Pedersen, Jensen, & Dyreson (2001), both by academics and by industrial communities, even if there is not yet a broad consensus on a common terminology and formalism, as discussed in Chapter 3. At the same time, different proposals were presented regarding the algebras (operators with different characteristics from the classic relational operators, as discussed in Chapter 5, and the relative query languages with aggregate functions), for example, in Klug (1982), Fortunato, Rafanelli, Ricci, & Sebastio (1986), Ozsoyoglu, Ozsoyoglu, & Mata (1985), and Ozsoyoglu, Ozsoyoglu, & Matos (1987). Visual query languages were also proposed, such as in Wong & Kuo (1982), Rafanelli & Ricci (1990), and Meo Evoli, Rafanelli, & Ricci (1994). New elaborations or redefinitions of many of these operators have been recently proposed (see, for example, Gray, Bosworth, Layman, & Pirahesh, 1996; Gray et al., 1997; Gyssens & Lakshmanan, 1997; Cabibbo & Torlone, 1998; Lehner, 1998; Pedersen, Jensen, & Dyreson, 2001) with regard to OLAP applications, sometimes simply changing their name, other times extending the original operator, but, in particular, giving a rigorous formal definition of them. Similarities and differences between these two kinds of operators will be briefly discussed in Chapter 5.

Multidimensional querying is often based on the metaphor of the data cube and the concepts of facts, measures, and dimensions. It is often an exploratory process, performed by navigating along dimensions and measures, increasing/decreasing the level of detail, and focusing on specific subparts of the cube that appear "promising" for the required information. More recently, important results on query languages for bags (e.g., those in Albert, 1991; Libkin & Wong, 1993; Grumbach & Milo, 1996) have led to a more "natural" characterization of aggregate functions. Several approaches to the problem of querying multidimensional data have been based on extensions of the relational algebra and calculus and/or of SQL, the most common relational query language. In Chapter 9, techniques to retrieve multidimensional (aggregate) data and problems of evaluation, related to the efficient data retrieval and calculation (known as the problem of *rewriting a query using views*), are discussed.

Another problem which arose at the beginning of the '80s was the study of metadata. One of the first papers in which this issue was discussed and this important concept defined was McCarthy (1982). In it the author provides an excellent overview of metadata, including statistical metadata. As it is well known, metadata are data about data, i.e., systematic descriptive information about data content and organization. Metadata are necessary to specify information about multidimensional aggregate statistical data. They can provide the definition of logical models, as well as a more detailed documentation. Well-defined and differentiated metadata are necessary to allow software links between different logical and physical representations; between multidimensional aggregate statistical databases, application programs, and user interfaces; as well as between multiple distributed and heterogeneous systems (see 2nd SDBM, 1983; Rafanelli, 1991). Studies on metadata for multidimensional aggregate statistical data can be found in McCarthy (1982) and Wong (1984).

A problem which seriously impedes the integrated use of databases, and, in particular, the manipulation of multidimensional aggregate databases, is the use of different units of measure, for example, dollars or pesetas for currency, pounds or Kilograms for fruit production, and so on. There have been few studies on this issue. It was discussed in general in Karr & Loveman (1978) and Gehani (1982). Other authors, like Sparr (1981) and Rafanelli, Bezenchek, & Tininini (1996), studied this problem, focusing on statistical multidimensional aggregate databases.

Also the problem regarding the counting unit (i.e., if the counting unit is 106 and a summary value "10" actually represents "10 million") was briefly discussed in Rafanelli, Bezenchek, & Tininini (1996) and will be discussed in Chapter 1.

The management of the time in MDDB is a topic which was initially studied in Rafanelli (1990), and subsequently, in Yang & Widom (1998, 2000), and in Mendelzon & Vaisman (2000, 2001). One problem which arises is the changing of the semantics of a given term at two different moments in time. For instance, suppose we have two states, S1 and S2, that, at time t1, have a given boundary. If at time t2 state S1 gives state S2 part of its territory, then the boundary, the area, the population, etc. of the two states change, but their names remain the same. This means that we call objects, which are different in time, by the same name.

In recent years there has been growing interest in multidimensional database systems, essentially to perform on-line analysis of transaction-based business data, such as retail store transactions, and to consolidate and summarize data in order to put enterprise data into multidimensional perspectives. OLAP is considered a tool for developing business analysis and decision support applications. As explained in many papers, OLAP is an area of active commercial and research interest which developed in recent years, especially with the advances in hardware for on-line mass storage which have made the warehousing of large amounts of data possible. One of the goals of OLAP tools is to provide fast answers to queries which aggregate the warehouse data. The term OLAP was coined in Codd, Codd, & Salley (1993), in order to characterize the dynamic aspects (with particular reference to performance issues) of such data. Subsequently there were other papers on OLAP applications, for example, Codd, Codd, & Salley (1993), Finkelstein (1995), and Priebe & Pernul (2000). Traditional data models, such as the ER model (Chen, 1976) and the relational model (Codd, 1970), do not provide good support for OLAP applications. As a result, new data models based on a multi-dimensional view of data have emerged. In all the chapters in this book, OLAP is referred to from different points of view: basic concepts, models, operators, time management, dynamic data cubes, materialized views, querying, incomplete information , etc. are some of the topics discussed.

Another issue to which some researchers have focused their attention on is the *interoperability* between multidimensional (aggregate) data and other data types, essentially geographic data. In fact, at present, neither models or algebras which include the possibility to represent, store, and manage complex data of this type exist. Consequently, it is necessary to study how to answer queries which have both multidimensional operators, and topological operators, i.e., queries of the type:

1. "Select the regions which border on the Tuscany region and in which the oil production was greater than 10,000 hectolitres in the years from 1990 to 1999," or
2. "Select the regions crossed by the Danube River and in which the average fruit production in a given year was greater than the previous year, in the period 1990-1999."

This problem is widely discussed in Chapter 13.

The aim of this book is to provide the first text on multidimensional databases and on the main topics of this area, discussing properties and peculiarities of such databases. Various well-known researchers, who are experts in one or more fields of multidimensional databases, were personally invited to write a chapter on their research area for this volume. As a result, this book presents the history, the current state, and future trends of each of the above-mentioned areas.

In particular, in Chapter 1, Maurizio Rafanelli presents the basic notions regarding multidimensional (aggregate) databases by referring to different definitions given for them in the literature. He discusses the differences which exist between the disaggregate, microdata and the aggregate, macrodata, and the importance, in this context, of the macrodata. He presents the process which makes it possible to obtain aggregate data (macrodata), from a very large set of raw data (microdata) (see Rafanelli, Bezenchek, & Tininini, 1996). He also introduces definitions of the different data structures proposed in the literature for multidimensional aggregate data (statistical object, table, cube, multidimensional aggregate data [MAD]). In this context he also introduces the concept of dimension and of the possible hierarchies which could be present in each dimension and which characterize the measure of the multidimensional aggregate data. Finally, he describes the different approaches used in the varying fields of application, such as the statistical environment and On-Line Analytical Processing (OLAP). Both of them use these data, but the approaches mentioned above depend on the different aspects emphasized in them. For example, most of the work they carry out is characterized by their change of use (from statistical and socio-economic type applications to the analysis of transaction-based business data). At the end of the chapter, the author explains the necessity to use a graph model to represent multidimensional aggregate data and discusses, in particular, the ADAMO model. Finally, he discusses tabular models, used both in SDB and OLAP, and gives a set of definitions regarding OLAP terminology.

Arie Shoshani, in Chapter 2, discusses the multidimensionality in statistical, OLAP, and scientific databases. He affirms that the term "multidimensional database" typically refers to a collection of objects, each represented as a point in a multidimensional space. Even data that is represented in a tabular form, such as relations, can be thought of as multidimensional data, if each row (tuple) is considered an object, and the columns (attributes) are considered the dimensions. The problem of viewing high-dimensional data to identify clusters, outliers, and various patterns has been the subject of several research projects. An important reason for viewing data in the multidimensional space is summarization. This need is most obvious in databases that represent statistical data or in databases used for decision support. These are referred to as "Statistical Databases" and "On-Line Analytical Processing" (OLAP), respectively. In the OLAP literature the multidimensional space is referred to as a "cube." Therefore, the Author affirms that another important aspect which links Statistical and OLAP databases is that each dimension can have a category hierarchy associated with it. Dimension hierarchies can become fairly complex depending on the type of dimension.

In Chapter 3, Riccardo Torlone presents the requirements that an ideal conceptual multidimensional data model should fulfill. Because it is widely accepted that traditional conceptual data models, like the entity-relationship model, are not well suited to describe the multidimensional and aggregative nature of OLAP applications, a variety of multidimensional data models have been recently proposed (both by the academic and industrial communities), but it should be said that a consensus on a common terminology and formalism has not emerged yet. Then, he presents a first version of the above-mentioned ideal conceptual multidimensional data model from a general point of view. Far from being complete, this model aims to capture the core of the various multidimensional data models proposed in the scientific literature and the means adopted by OLAP systems to represent and manipulate data. The model relies on two main and agreed concepts: dimension and cube. A dimension represents a business perspective under which data analysis is to be performed and is organized in a hierarchy of levels. The levels of a dimension correspond to data domains at different granularity. A cube represents factual data on which the analysis is focused, and associated measures with coordinates, defined over a set of dimension levels. Therefore, a number of extensions of this model and a survey on various multidimensional models proposed in the literature are discussed.

In Chapter 4, Elaheh Pourabbas and Maurizio Rafanelli discuss the different types of hierarchies which can appear in a dimension of a multidimensional (aggregate) data structure. They study their behavior when the user browses along them or when he manipulates them by using operators, such as roll-up, drill-down, etc. Moreover, they classify four different basic types of hierarchies and three forms of data abstraction, and discuss their characteristics. These hierarchies divide a single dimension into different levels of aggregation. Depending on them, the authors discuss the characteristics of some OLAP operators which refer to hierarchies in order to maintain the data cube consistency. In particular, they discuss the important concept of a summarizable function and different types of mapping between two category attributes (which represent the instance of a dimension in a multidimensional data structure). Then, they propose a set of operators for changing the hierarchy structure. The issues discussed provide modeling flexibility during the schema design phase and correct data analysis. In particular, they deal more with multiple hierarchies, and introduce the multiplicity of a hierarchy as a semantic variant of a simple one.

In Chapter 5, Maurizio Rafanelli discusses the set of operators proposed by different authors in the literature. He draws distinctions between operators for statistical aggregate data (and between the relational approach and the multidimensional approach) and the operators for OLAP. A comparison between the two sets is also made, highlighting their similarities and differences. In particular, he discusses the operators for multidimensional aggregate data which extend relational algebra and relational calculus, then the operators for multidimensional aggregate data defined in a tabular environment, and finally the operators for OLAP applications. A comparison between the OLAP terminology and the multidimensional aggregate (statistical) database terminology is also made. Finally, the author resumes giving particular emphasis to the operators which form the skeleton of the basic algebra.

In Chapter 6, Alberto Mendelzon and Alejandro Vaisman show the need for a temporal approach to OLAP, giving a review of temporal database concepts and work regarding time in multidimensional databases. They argue that, in the presence of dimension updates which trigger changes on the granularity of the facts, a temporal model supporting schema versioning is required. They show that existing data models are not suitable for these requirements. Therefore, they present a new temporal multidimensional model which makes it possible to keep track of the history of a multidimensional database, and introduce a temporal query language, called *TOLAP*, supporting this temporal model. A preliminary *TOLAP* implementation is described, and the results of experiments are commented on. At the end of the chapter, the authors suggest possible research directions.

In Chapter 7, Mirek Riedewald, Divyakant Agrawal, and Amr El Abbadi discuss several techniques for update-efficient aggregation on MOLAP data cubes. First they give an introduction and a background, focusing on data cubes and aggregate query types. Then they define an invertible aggregate operator and affirm that the existence of inverse operations enables the construction of elegant techniques for speeding up queries on MOLAP data cubes that do not require additional storage as opposed to materializing the original cube. Then very sparse data set problems are briefly discussed and the Progressive Data Cube (pCube) is proposed and described, in order to incorporate any hierarchical multidimensional index structure, e.g., R-tree, quad tree, etc. Following these are techniques that explicitly address the sparseness issue are presented. As the authors say, "the similarity between all approaches is that they try to find an appropriate balance between query, update, and storage cost. While earlier proposals mostly focused on query and storage aspects, large data sets with frequent updates created a need for more dynamic solutions."

In Chapter 9, Stefano Paraboschi, Giuseppe Sindoni, Elena Baralis and Ernest Teniente present the view selection problem, first showing how a materialized view can be a query optimization solution applicable to generic databases. A general approach can be developed as an extension of multi-query optimization techniques, where it is important to detect the query parts that are common to separate query plans, in order to compute them only once but reuse their results many times. View materializations have proven to be an important query optimization solution for multidimensional databases. The reasons for the success of view materialization in multidimensional databases are twofold: first, the extremely high ratio between queries and updates, typical of all data analysis systems, makes the task to maintain all the materialized views considerably easier; also, the limited variety of queries typically supported by multidimensional databases permits the design of sophisticated models that are able to represent well the contribution that a view materialization can offer to view computation.

In particular, this chapter presents a model where a star schema is enriched with functional dependencies that identify hierarchies among the dimensional attributes. Queries, and potential view materializations, can be represented as nodes on a lattice. Ordering on lattice nodes means that a query can be computed using another query. Typical multidimensional database operations like roll-up and drill-down correspond nicely to typical lattice operations: meet and join. In the second part of the chapter the authors present the important characteristics of the solutions for materialized views that have been devised for commercial DBMS-based analysis tools. The chapter will base this analysis on the papers by research groups of DBMS vendors that have appeared in technical conferences and on the analysis of the technical documentation accompanying these systems.

In Chapter 9, Leonardo Tininini discusses techniques to retrieve multidimensional (aggregate) data proposed in the literature and which are based on the idea of determining the cube of interest and then navigating along the dimensions, increasing or decreasing the level of detail or selecting specific subparts of the cube. He gives a brief presentation of query languages based on an extension of the relational algebra and those based on a calculus (again an extension of the relational one), where queries are expressed in a more declarative way. He also presents visual languages, which usually rely on an underling algebra or calculus, and are based on a more interactive and iconic querying paradigm. In addition, he focuses on the problem of query evaluation, i.e., issues related to efficient data retrieval and calculation, possibly (often necessarily) using precomputed data, known in the literature as the problem of *rewriting a query using views*. He also presents some results on the equivalence and rewriting of aggregate queries that can be used to optimize the evaluation of queries on multidimensional data. Finally, he briefly outlines how specific techniques of indexing can significantly improve query evaluation in the multidimensional context.

In Chapter 10, Curtis E. Dyreson, Torben B. Pedersen, and Christian S. Jensen, after a brief introduction and a background of the topic, explain uncertainty in the data and in the aggregated data/metadata, and then discuss future directions of research. In particular, they present techniques for handling incompleteness in base data, derived data, and metadata. With regard to the incomplete measures, the authors outline the problems posed by three of the kinds of incomplete information that could appear in a measure attribute: unknown, imprecise, and probabilistic values. Then they discuss incompleteness in a grouping of attributes, focusing on problems and techniques for the grouping of exclusive and inclusive disjunctive values only. The study of incomplete information in a hierarchy follows. In it the incomplete derived data are considered and some additional strategies for improving the responsiveness of queries on the incomplete data are briefly discussed. Finally, incomplete information in metadata is discussed, with regard to non-covering, non-onto, and non-strict hierarchies.

In Chapter 11, Marina Moscarini and Francesco M. Malvestuto discuss the compromise of individual privacy through statistical queries on summary tables, reviewing some recent results on the inference problem which only refer to simple count and sum queries with data of real and non-negative real types. Attacks on confidentiality come from sensitive statistical queries whose answers allow a knowledgeable user to determine exactly or estimate accurately the value of a confidential field in some individual records. To guarantee privacy of individual records, a mechanism of inference control must be embodied in the (statistical) multidimensional aggregate database interface according to a security policy, which can be said to be effective only if sensitive summary statistics are neither explicitly nor implicitly released. The different control mechanisms proposed are analyzed as to their effectiveness and efficiency. In this chapter, the authors also review some recent results on the computational complexity of the inference problem for additive queries on a statistical database, with regard to simple count and sum queries, which will be referred to as simple "additive queries," with data of real and non-negative real types. They met with some computationally hard problems (on strong safety and strong p-safety), which are likely to have no efficient solutions, and conclude that there still remains much work to do to solve the problem of the security of statistical databases in a satisfactory way. Future and emerging research trends are listed to answer all the complexity questions raised by the inference problem discussed in this chapter, as well as the case of data of integer and non-negative integer types. Other lines of research will have to cover multidimensional tables and special classes of hypergraphs for which safety tests can be efficiently worked out.

In Chapter 12, Andrea Calí, Domenico Lembo, Maurizio Lenzerini, and Riccardo Rosati revise the various approaches for dealing with schema and data integration in data warehousing. They explain why data integration is important in data warehousing, present a comprehensive framework for data integration in data warehousing, and then survey existing approaches both from an academic and from an industrial perspective. They affirm that a fundamental aspect in the design of a data warehouse system is the process of acquiring the raw data from a set of relevant information sources. They will call the component of a data warehouse system that deals with this process a "source integration system." The main goal of a source integration system is to deal with the transfer of data, from the set of sources constituting the application-oriented operational environment, to the data warehouse. Since sources are typically autonomous, distributed, and heterogeneous, this task has to deal with the problem of cleaning, reconciling, and integrating data coming from the sources. The design of a source integration system is a very complex task, which comprises several different issues. The authors highlight the fact that the main purpose of this chapter is to discuss the most important problems arising in the design of a source integration system, with special emphasis on schema integration, processing queries for data integration, and data cleaning and reconciliation. They conclude by affirming that, at present, these still need to be studied and investigated.

In Chapter 13, Elaheh Pourabbas focuses on the common key elements between geographic and multidimensional databases which allow effective support to data federation. She points out that the enormous increase in data and its sources, due to the growing number of independent databases widely accessible through computer networks, has motivated cooperation between database systems, creating systems that are sometimes referred to as multi-database or federated database systems. These databases may represent business information (such as transaction data), medical information (such as patient treatment and results), scientific data (such as large sets of experimental measurements), or spatial information (such as geographic data and its visualization as maps). Currently Multidimensional Databases (MDDBs) and Geographic Information Systems (GISs) are seen as the most promising and efficient information technologies for supporting decision making.

In this chapter the author gives some basic definitions for describing geographic data to support multidimensional data and proposes an approach which extends the geographic data structure through special attributes, called *binding attributes*, in order to describe all phenomena represented by MDDB. This extension will make it possible to answer more specific "OLAP-based" queries within GDB without modifying the physical organization of data in both environments. The chapter concludes by proposing some future lines of research.

Maurizio Rafanelli
Instituto di Analisi dei Sistemi ed Informatica – C.N.R., Italy

REFERENCES

1st SDBM. (1981). *Proceedings of the 1st LBL Workshop on Statistical Database Management,* Menlo Park, California. Berkeley, CA: LBL, University of California.

2nd SDBM. (1983). *Proceedings of the 2nd International Workshop on Statistical Database Management,* Los Altos, California. Berkeley, CA: LBL, University of California.

3rd SSDBM. (1986). *Proceedings of the 3rd International Workshop on Statistical and Scientific Database Management,* Grand Duchy of Luxembourg. Luxombourg: Eurostat.

4th SSDBM. (1988). *Proceedings of the 4th International Conference on Statistical and Scientific Database Management (SSDBM'88),* Rome, Italy. Lecture Notes in Computer Science, No. 339. Berlin Heidelberg, Germany: Springer Verlag.

5th SSDBM. (1990). *Proceedings of the 5th International Conference on Statistical and Scientific Database Management (SSDBM'90),* Charlotte, North Carolina. Lecture Notes in Computer Science, No. 420. Berlin Heidelberg, Germany: Springer Verlag.

6th SSDBM. (1992). *Proceedings of the 6th International Conference on Statistical and Scientific Database Management (SSDBM'92),* Ascona, Switzerland. Zurich, Switzerland: ETH, University of Zurich.

7th SSDBM. (1994). *Proceedings of the 7th International Conference on Scientific and Statistical Database Management (SSDBM'94),* Charlottesville, Virginia. Los Alamitos, CA: IEEE Computer Society Press.

8th SSDBM. (1996). *Proceedings of the 8th International Conference on Scientific and Statistical Database Management (SSDBM'96),* Stockholm, Sweden. Los Alamitos, CA: IEEE Computer Society Press.

9th SSDBM. (1997). *Proceedings of the 9th International Conference on Scientific and Statistical Database Management (SSDBM'97),* Olympia, Washington. Los Alamitos, CA: IEEE Computer Society Press.

10th SSDBM. (1998). *Proceedings of the 10th International Conference on Scientific and Statistical Database Management (SSDBM'98),* Capri, Italy. Los Alamitos, CA: IEEE Computer Society Press.

11th SSDBM. (1999). *Proceedings of the 11th International Conference on Scientific and Statistical Database Management (SSDBM'99),* Cleveland, Ohio. Los Alamitos, CA: IEEE Computer Society Press.

12th SSDBM. (2000). *Proceedings of the 12th International Conference on Scientific and Statistical Database Management (SSDBM'00),* Berlin, Germany. Los Alamitos, CA: IEEE Computer Society Press.

13th SSDBM. (2001). *Proceedings of the 13th International Conference on Scientific and Statistical Database Management (SSDBM'01),* Fairfax, Virginia. Los Alamitos, CA: IEEE Computer Society Press.

14th SSDBM. (2002). *Proceedings of the 14th International Conference on Scientific and Statistical Database Management (SSDBM'02),* Edimbourg, H.K. Los Alamitos, CA: IEEE Computer Society Press.

Adam, N.R., & Wortmann, J.C. (1989). Security control methods for statistical databases: A comparative study. *Journal of the ACM Computing Surveys,* 21, 515-556.

Adam, N.R., Gangopadyay, A., & Holowczak, R.D. (1999). A survey of research on database protection. *Statistical Data Protection.* Luxombourg: Eurostat, European Community, 29-43.

Agrawal R., Gupta A., & Sarawagi S. (1997). Modeling multidimensional databases. *Proceedings of the 13th International Conference on Data Engineering (ICDE'97),* Birmingham, UK, 232-243.

Albert, J. (1991). Algebraic properties of bag data types. *Proceedings of the 17th International Conference on Very Large Data Bases (VLDB'91),* Barcelona, Spain, 211-219.

Beck, L.L. (1980). A security mechanism for statistical databases. *Journal of ACM Transactions on Database Systems,* 5(3), 316-338.

Bezenchek, A., Massari, F., & Rafanelli, M. (1994). Storm+: Statistical data storage and manipulation system. *Proceedings of the 11th International Symposium on Computational Statistics (Compstat'94),* Vienna, Austria. Wien, Austria: Physica-Verlag, 351-356.

Bezenchek, A., Rafanelli, M., & Tininini L. (1996a). A data structure for representing aggregate data. *Proceedings of the 8th International Conference on Scientific and Statistical Database Management,* Stockholm, Sweden. Los Alamitos, CA: IEEE Computer Society Press, 22-31.

Bezenchek, A., Rafanelli, M., & Tininini, L. (1996b). ADAMO: A conceptual model to describe and to represent aggregate data. *12th International Symposium on Computational Statistics (Compstat'96)*, Barcelona, Spain. Short papers, 97-98. Also: ADAMO: A conceptual model for aggregate data. Technical Report. IASI R.440, September 1996.

Brown, V.A., Navathe, S.B., & Su, S.Y.W. (1983). Complex data types and operators for statistical data and metadata. *Proceedings of the 2nd International Workshop on Statistical Database Management,* Los Altos, California, 188-195.

Cabibbo, L., & Torlone, R. (1997). Querying multidimensional databases. *Proceedings of the 6th International Workshop on Databases and Programming Languages (DBPL'97),* Estes Park, Colorado. Lecture Notes in Computer Science, No. 1369. Berlin Heidelberg, Germany: Springer-Verlag, 319-335.

Cabibbo, L., & Torlone, R. (1998). From a procedural to a visual query language for OLAP. *Proceedings of the 10th International Conference on Scientific and Statistical Database Management (SSDBM'98),* Capri, Italy. Los Alamitos, CA: IEEE Computer Society Press, 74-83.

Chan, P., & Shoshani, A. (1981). SUBJECT: A directory-driven system for organizing and accessing large statistical databases. *Proceedings of the 7th International Conference on Very Large Data Bases (VLDB'81),* Cannes, France, 553-563.

Chen, P.P. (1976). The entity-relationship model—Toward a unified view of data. *Journal of ACM Transactions on Database Systems,* Volume(1), 9-36.

Chen, M.C., McNamee, L., & Melkanoff, M. (1988). A model of summary data and its applications in statistical databases. *Proceedings of the 4th International Conference on Statistical and Scientific Database Management (SSDBM'88), Rome, Italy.* Lecture Notes in Computer Science, No. 339. Berlin Heidelberg, Germany: Springer Verlag, 356-387.

Chin, F.Y. (1978). Security in statistical databases for queries with small counts. *Journal of ACM Transactions on Database Systems*, 3(1), 92-104.

Chin, F.Y. (1986). Security problems on inference control for sum, min, and max queries. *Journal of the ACM*, 33, 451-464.

Chin, F.Y., & Ozsoyoglu, G. (1982). Auditing and inference control in statistical databases. *IEEE Transactions on Software Engineering*, 8, 574-582.

Chu, P.C. (1997). Cell suppression methodology: The importance of suppressing marginal totals. *IEEE Transactions on Knowledge and Data Engineering*, 9, 513-523.

Codd, E.F. (1970). A relational model of data for large shared data banks. *Journal of Communications of the ACM*, 13(6), 377-387.

Codd, E.F., Codd, S.B., & Salley, C.T. (1993). *Providing OLAP (On-Line Analytical Processing) to User-Analysts: An IT Mandate.* Technical Report, E.F. Codd and Associates.

Datta, A., & Thomas, H. (1997). A conceptual model and algebra for On-Line Analytical Processing in decision support databases. *Proceedings of the Seventh Annual International Workshop on Information Technologies and Systems (WITS'97)*, 91-100.

Denning, D.E. (1980). Secure statistical databases with random sample queries. *Journal of ACM Transactions on Database Systems*, 5(3), 291-315.

Denning, D.E. (1983). A security model for the statistical database problem. *Proceedings of the 2nd International Workshop on Statistical Database Management,* Los Altos, California, 368-390.

Denning, D.E., & Schlorer, J. (1983). Inference controls for statistical databases. *Journal of IEEE Computer*, 16(7), 69-82.

Denning, D.E., Denning, P.J., & Schwartz, M.D. (1979). The tracker: A threat to statistical database security. *Journal of ACM Transactions on Database Systems*, 4(1), 76-96.

Denning, D.E., Nichelson, W., Sande, G., & Shoshani, A. (1983). Topics in statistical database management. *Proceedings of the 2nd International Workshop on Statistical Database Management,* Los Altos, California, 46-51.

Espil, M.M., & Vaisman, A.A. (2001). Efficient intentional redefinition of aggregation hierarchies in multidimensional databases. *Proceedings of the Fourth ACM International Workshop on Data Warehousing and OLAP (DOLAP'01),* Atlanta, Georgia.

Finkelstein, R. (1995). MDD: Database reaches the next dimension. *Journal of Database Programming and Design*, (April), 27-28.

Fortunato, E., Rafanelli, M., Ricci, F.L., & Sebastio, A. (1986). An algebra for statistical data. *Proceedings of the 3rd International Workshop on Statistical and Scientific Database Management.* Grand Duchy of Luxembourg. Luxombourg: Eurostat, European Community, 122-134.

Franconi, E., & Sattler, U. (1999). A data warehouse conceptual data model for multi-dimensional aggregation. *Proceedings of the International Workshop on Design and Management of Data Warehouses (DWDM'99),* 13.1-13.10.

Gehani, N.H. (1982). Databases and units of measure. *Journal of IEEE Transactions on Software Engineering,* 8(6), 605-611.

Gray, J., Bosworth, A., Layman, A., & Pirahesh, H. (1996). Data cube: A relational operator generalizing group-by, cross-tab, and roll-up. *Proceedings of the International Conference on Data Engineering (ICDE'96),* New Orleans, Louisiana. Los Alamitos, CA: IEEE Computer Society Press, 152-159.

Gray, J., Chaudhuri, S., Bosworth, A., Layman, A., Reichart, D., Venkatrao, M., Pellow, F., & Pirahesh, H. (1997). Data cube: A relational aggregation operator generalizing group-by, cross-tab and sub-totals. *Journal on Data Mining and Knowledge Discovery*, 1(1), 29-54.

Grumbach, S., & Milo, T. (1996). Towards tractable algebras for bags. *Journal of Computer and System Sciences,* 52(3), 570-588.

Gusfield, D. (1988), A graph theoretic approach to statistical data security. *Computing*, 17, 552-571.

Gyssens, M., & Lakshmanan, L.V.S. (1997). A foundation for multi-dimensional databases. *Proceedings of the 23rd International Conference on Very Large Data Bases (VLDB'97),* Athens, Greece, 106-115.

Johnson, R.R. (1981). Modeling summary data. *ACM SIGMOD International Conference on Management of Data*, Ann Arbor, Michigan, 93-97.

Kao, M.-Y. (1997). Efficient detection and protection of information in cross-tabulated tables II: Minimal linear invariants. *Journal of Combinatorial Optimization*, 1, 187-202.

Karr, M., & Loveman, D.B. III. (1978). Incorporation of units into programming languages. *Communications of the ACM*, 21(5), 385-391.

Kleinberg, J.M., Papadimitriou, C.H., & Raghavan, P. (2000). Auditing Boolean attributes. *Proceedings of the Nineteenth ACM SIGMOD-SIGACT-SIGART Symposium on Principles of Database Systems*, Dallas, Texas, 86-91.

Klug, A. (1982). Equivalence of relational algebra and relational calculus query languages having aggregate functions. *Journal of Association for Computing Machinery*, 29(3), 699-717.

Lehner, W. (1998). Modeling large-scale OLAP scenarios. *Proceedings of the 6th International Conference on Extending Database Technology (EDBT),*

Valencia, Spain. Lecture Notes in Computer Science, No. 1377. Berlin Heidelberg, Germany: Springer-Verlag, 153-167.

Li, C., & Wang, X.S. (1996). A data model for supporting On-Line Analytical Processing. *Proceedings of the Fifth International Conference on Information and Knowledge Management (CIKM'96)*, Rockville, Maryland, 81-88.

Libkin, L., & Wong, L. (1993). aggregate functions, conservative extensions, and linear orders. *Proceedings of the Fourth International Workshop on Database Programming Languages—Object Models and Languages (DBPL'93)*, New York, 282-294.

Malvestuto, F.M. (1993). A universal-scheme approach to statistical databases containing homogeneous summary tables. *Journal of ACM Transactions on Database Systems*, 18(4), 678-708.

Malvestuto, F.M., Moscarini, M., & Rafanelli, M. (1991). Suppressing marginal cells to protect sensitive information in a two-dimensional statistical table. *Proceedings of the 10th International ACM Symposium on Principles of Database Systems (PODS'91)*, Denver, Colorado, 252-258.

Malvestuto, F.M., Rafanelli, M., & Zuffada, C. (1988). Many-source databases: Some problems and solutions. *Technical Report IASI*, R.218, June.

Malvestuto, F.M., & Zuffada, C. (1989). The classification problem with semantically heterogeneous data. *Proceedings of the 4th International Conference on Statistical and Scientific Database Management (SSDBM'88)*, Rome, Italy. Lecture Notes in Computer Science, No, 339. Berlin Heidelberg, Germany: Springer-Verlag, 157-176.

McCarthy, J. (1982). Metadata management for large statistical databases. *Proceedings of the 8th International Conference on Very Large Data Bases (VLDB'82)*, Mexico City, Mexico, 234-243.

Meo Evoli, L., Rafanelli, M., & Ricci F.L. (1994). An interface for the direct manipulation of statistical data. *Journal of Visual Languages and Computing*, 5, 175-202.

Mendelzon, A.O., & Vaisman, A.A. (2000). Temporal queries in OLAP. *Proceedings of 26th International Conference on Very Large Data Bases (VLDB'00)*, Cairo, Egypt, 242-253.

Michalewicz, Z. (Ed.). (1990). *Statistical and Scientific Databases*. New York: Ellis Horwood.

Nguyen, T.B., Tjoa, A.M., & Wagner, R. (2000). Conceptual multidimensional data model based on metacube. *Proceedings of the 1st International Conference on Advances in Information Systems (ADVIS'00)*, Izmir, Turkey. Lecture Notes in Computer Science, No. 1909. Berlin Heidelberg, Germany: Springer-Verlag., 24-33.

Ozsoyoglu, G., Ozsoyoglu, Z.M., & Mata, F. (1985). A language and a physical organization technique for summary tables. *Proceedings of the 4th ACM International Conference on Management of Data (SIGMOD '85),* Austin, Texas. New York, NY, 3-16.

Ozsoyoglu G., Ozsoyoglu Z.M., & Matos (1987). Extending relational algebra and relational calculus with set-valued attributes and aggregate functions. *Journal of ACM Transaction on Database Systems,* 12(4), 566-592.

Pedersen, T.B., & Jensen, C.S. (1999). Multidimensional data modeling for complex data. *Proceedings of the 15th International Conference on Data Engineering (ICDE '99),* Sydney, Australia. Los Alamitos, CA: IEEE Computer Society Press, 336-345.

Pedersen, T.B., Jensen, C.S., & Dyreson, C.E. (2001). A foundation for capturing and querying complex multidimensional data. *Journal of Information Systems,* 26(5), 383-423.

Priebe, T., & Pernul, G. (2000). Towards OLAP security design—Survey and research issues. *Proceedings of the Third ACM International Workshop on Data Warehousing and OLAP (DOLAP '00),* Washington, DC, 33-40.

Rafanelli, M. (1987). A graphical approach for statistical summaries: The GRASS model. *Proceedings of the International Symposium on Microcomputers and Their Applications,* Cairo, Egypt. Anaheim, CA: Acta Press, 78-81.

Rafanelli, M. (1989). Research topics in statistical and scientific database management: IV SSDBM. *Proceedings of the 4th International Conference on Statistical and Scientific Database Management (SSDBM '88),* Rome, Italy. Lecture Notes in Computer Science, No. 339. Berlin Heidelberg, Germany: Springer Verlag, 1-18.

Rafanelli, M. (1990). Statistical and scientific database management systems. In Kent, A., & Williams, J.G. (Eds.), *Encyclopedia of Computer Science and Technology,* 23(8). Pittsburgh, PA: Dekker Inc. Publications.

Rafanelli M. (1991). Data models. Chapter 6 in Michalewicz, Z. (Ed.), *Statistical and Scientific Databases.* London: E. Horwood Lim.

Rafanelli, M., Bezenchek, A., & Tininini, L. (1996) The aggregate data problem: A system for their definition and management. *Journal of ACM-Sigmod Record,* 25(4), 8-13.

Rafanelli, M., & Ricci, F.L. (1983). Proposal of a logical model for statistical databases. *Proceedings of the 2nd International Workshop on Statistical Database Management.* Los Altos, California, 264-272.

Rafanelli, M., & Ricci, F.L. (1985). STAQUEL: A query language for statistical macro-database management systems. *Proceedings of the International Conference, Convention Informatique Latine (CIL '85),* Barcelona, Spain, 625-637.

Rafanelli, M., & Ricci, F.L. (1988). A statistical functional model for statistical tables. *Proceedings of the International Symposium on Modeling, Identification and Control*, Grindelwald, Switzerland. Anaheim, CA: Acta Press, 136-139.

Rafanelli, M., & Ricci, F.L. (1990). A visual interface for browsing and manipulating statistical entities. *Proceedings of the 5th International Conference on Statistical and Scientific Database Management (SSDBM'90)*, Charlotte, North Carolina. Lecture Notes in Computer Science, No. 420. Berlin Heidelberg, Germany: Springer-Verlag, 163-182.

Rafanelli, M., & Ricci, F.L. (1991). A functional model for macro-databases. *Journal of ACM-Sigmod Record*, 20(1), 3-8.

Rafanelli, M., & Ricci, F.L. (1993). Mefisto: A functional model for statistical entities. *Journal of IEEE Transactions on Knowledge and Data Engineering*, 5(4), 1-12.

Rafanelli, M., & Shoshani, A. (1990a). STORM: A statistical object representation model. *IEEE Data Engineering Bulletin*, 13(3), 12-18.

Rafanelli, M., & Shoshani, A. (1990b). STORM: A statistical object representation model. *Proceedings of the 5th International Conference on Statistical and Scientific Database Management (SSDBM'90)*, Charlotte, North Carolina. Lecture Notes in Computer Science, No. 420. Berlin Heidelberg, Germany: Springer-Verlag, 14-29.

Rafanelli, M., & Shoshani, A. (1990c). On the representation problems of statistical object. *International Conference on Database Theory and Application (TechnoData'90)*, Berlin, Germany, 122-132.

Sato, H. (1991). Statistical data models: From a statistical table to a conceptual approach. Chapter 7 in Michalewicz, Z. (Ed.), *Statistical and Scientific Databases*. London: E. Horwood Lim.

Schlorer, J. (1980). Disclosure from statistical databases: Quantitative aspects of trackers. *Journal of ACM Transactions on Database Systems*, 5(4), 467-492.

Schlorer, J. (1981). Security of statistical databases: Multidimensional transformation. *Journal of ACM Transactions on Database Systems,* 6(1), 95-112.

Shoshani, A. (1982). Statistical databases: Characteristics, problems and solutions. *Proceedings of the 8th International Conference on Very Large Data Bases*, Mexico City, Mexico, 208-222.

Shoshani, A., & Rafanelli, M. (1991). A model for representing statistical objects. Advances in data management. In. Sadanandan, P., & Vijayaraman, T.M. (Eds.), *Proceedings of the 3rd International Conference on Management of Data (COMAD'91)*, Bombay, India, 161-177.

Shoshani, A., & Wong, H.K.T. (1985). Statistical and scientific database issues. *IEEE Transactions on Software Engineering*, 11(10), 1040-1047.

Sparr, T.M. (1981). Units and accuracy in statistical databases. *Proceedings of the 1st LBL Workshop on Statistical Database Management,* Menlo Park, California, 59-60.

Su, S.Y.W. (1983). SAM*: A semantic association model for corporate and scientific statistical databases. *Journal of Information Sciences,* 29(2-3), 151-199.

Tininini, L., Bezenchek, A., & Rafanelli, M. (1996). A system for the management of aggregate data. *Proceedings of the 7th International Conference on Database and Expert Systems Applications (DEXA'96),* Zurich, Switzerland. Lecture Notes in Computer Science, No. 1134. Berlin Heidelberg, Germany: Springer-Verlag, 533-543.

Vassiliadis, P. (1998). Modeling multidimensional databases, cubes and cube operations. *Proceedings of the 10th International Conference on Scientific and Statistical Database Management (SSDBM'98),* Capri, Italy. Los Alamitos, CA: IEEE Computer Society Press, 53-62.

Yang, J., & Widom, J. (1998). Maintaining temporal views over non-temporal information sources for data. *Proceedings of the 6th International Conference on Extending Database Technology (EDBT'98),* Valencia, Spain. Lecture Notes in Computer Science, No. 1377, Berlin Heidelberg, Germany: Springer-Verlag, 389-403.

Yang, J., & Widom, J. (2000). Temporal view self-maintenance. *Proceedings of the 7th International Conference on Extending Database Technology (EDBT'00),* Konstanz, Germany. Lecture Notes in Computer Science, No. 1777, Berlin Heidelberg, Germany: Springer-Verlag, 395-412.

Yu, C.T., & Chin, F. (1977). Study on the protection of statistical databases. *Proceedings of the 1977 ACM SIGMOD International Conference on Management of Data,* Toronto, Canada, 169-181.

Wong, H.K.T. (1984). Micro and macro statistical/scientific database management. *1st International Conference on Data Engineering (ICDE'84),* Los Angeles, California, 104-106.

Wong, H.K.T., & Kuo, I. (1982). GUIDE: Graphical user interface for database exploration. *Proceedings of the 8th International Conference on Very Large Data Bases (VLDB'82),* Mexico City, Mexico, 22-32.

Chapter I

Basic Notions

Maurizio Rafanelli
Istituto di Analisi dei Sistemi ed Informatica – C.N.R., Italy

ABSTRACT

This chapter presents the basic notions regarding multidimensional (aggregate) databases by referring to different definitions given for them in the literature. It illustrates the important concepts of micro, macro, and metadata; presents a formal definition of the aggregation process, discussing the concepts of dimension and dimension hierarchies; describes the multidimensional aggregate data structure, distinguishing between simple, complex, and composite structure; illustrates the different types of null values; and discusses differences and similarities which exist between multidimensional aggregate data (generally called statistical data because they are used mainly by statisticians) and the On-Line-Analytic Processing (OLAP) of multidimensional data represented by different data cubes, also discussing the different (symmetric and non-symmetric) treatment of dimensions and measures required by OLAP and aggregate multidimensional databases. Finally it discusses a graph model and a tabular model for this kind of data, and gives a set of definitions regarding the OLAP terminology.

INTRODUCTION

In this chapter we present the basic notions regarding multidimensional (aggregate) databases by referring to different definitions given for them in the literature. In particular, we illustrate the differences which exist between disaggregate (micro) and aggregate (macro) data structures and the importance, in this context, of metadata. The role they play in a multidimensional aggregate database is also discussed. Then, we present the process which makes it possible to obtain aggregate data (macrodata) from a very large set of raw data (microdata). In this context we introduce the concept of dimension, which will be discussed in Chapter 4, specifically the possible hierarchies which could be present in each dimension. We distinguish between two different user activities performed on these macrodata. Such activities are orthogonal to each other, and are characterized by:

1   *Multidimensional Data Manipulating*, i.e., the manipulation of the descriptive part of macrodata, which is metadata; this kind of activity does not change the summary types (count, percent, average, etc.) of the macrodata measure, and the fact described by the single object.
2   *Data Analysis*, i.e., the processing and elaboration of the summary values, which can change the summary value type (for example, from *sum* to *percent*).

Finally, we describe the different approaches used in the varying fields of application, such as the statistical environment and On-Line Analytical Processing (OLAP). Both of them use these data, but the approaches mentioned above depend on the different aspects emphasized in them. For example, most of the work they carry out is characterized by their change of use (from statistical and socio-economic type applications to the analysis of transaction-based business data).

This chapter is outlined as follows: the first section illustrates the important concepts of micro, macro, and metadata. A formal definition of the aggregation process follows, and we also briefly discuss the concepts of dimension and dimension hierarchies. We then introduce definitions of the different data structures proposed in the literature for multidimensional aggregate data (statistical object, table, cube, MAD). The different types of null values are illustrated, along with how they are treated in a multidimensional database. The next section discusses differences and similarities which exist between multidimensional aggregate data (generally called statistical data because they are used mainly by statisticians) and the On-Line-Analytic Processing (OLAP) of multidimensional data represented by different data cubes. It also discusses the different (symmetric and non-symmetric) treatment of dimensions and measures required by OLAP and aggregate multidimensional databases.

Microdata, Macrodata, and Metadata

During the 1980s there was a lot of activity in the area of multidimensional aggregate databases, called statistical databases, focusing mostly on socio-economic-type applications. The fundamental data structure of a multidimensional database is what is known in the literature as an n-dimensional table (referred to by Gyssens & Lakshmanan, 1997). During the 2nd Statistical Data Base Management workshop, held in 1983 in California (see 2nd SDBM, 1983), the multidimensionality of statistical and scientific data was underlined as a/the storage structure in order to organize disk files into blocks. In particular, the majority of scientific data models had a rectangular three-dimensional grid, i.e., a parallelepiped whose individual cells were minicubes. Each rectangular three-dimensional grid was subdivided into "cubes," as proposed in Bell (1983). This paper was the first example of multidimensional representation of data by cubes. In the 1990s the On-Line Analytic Processing (OLAP) area was introduced for the analysis of transaction-based business data, such as retail store transactions (see Codd, 1993). In this area the cube metaphor was proposed again in order to underline the representation and also to support business data in a multi-dimensional space.

As discussed in more detail later in this chapter, both OLAP and multidimensional statistical databases (SDBs) deal with multidimensional datasets, and both are concerned with applications of aggregate functions over the dimensions of the data sets. Much of the work on SDBs took place in the 1980s and still continues today, as does OLAP database work, which started mainly in the 1990s.

Some of the main topics in the literature on MDDBs are: metadata, conceptual and logical data models, data structure, data selection, data manipulation, and data querying and visualization (output display).

In McCarthy (1982) the author gives a definition of *metadata for large statistical databases* (in this book, by *statistical databases* we mean a class of databases which allow the definition, manipulation, elaboration, and storing of aggregate multidimensional data, obtained by applying aggregative functions to raw data, and which are used to obtain statistical analyses):

"Metadata is data about data, that is, systematic descriptive information about data content and organization that can be retrieved, manipulated, and displayed in various ways. Metadata may be simple and unstructured, such as a typewritten narrative describing a data tape, or structured and complex, such as an active machine-readable DBMS dictionary used to control multiple databases."

Generally there are two broad classes of multidimensional statistical data and micro and macrodata, as described by Wong (1984). The former refers to SDBs containing microdata (sometimes called "elementary" or raw data), that is, records of individual entities or events, such as mortality data of individual people or population census. The latter refers to databases containing multidimensional aggregate data (MAD), often shown as statistical tables, that result from the application of aggregate functions (for example, count, sum, or average) on raw data. Examples of MAD are tables of "energy consumption" or charts of "mortality by disease." These two classes of databases differ from various points of view. For example, from a descriptive point of view, there is an intentional and an extensional level of metadata for the macrodata, unlike for the microdata; from the querying point of view, due to the different (and more complex) structure of the data, the relational algebra operators are suitably modified so as to correctly carry out queries on the relations which represent the structures mentioned above.

The distinction between metadata and data is not always a clear one. For example, consider a table of population counts by age, race, and sex. In one sense, age, race, and sex categories are data values that characterize the individuals summarized in the table. In another sense, they are metadata that serve as labels for cross-product cells of a three-dimensional structure. From a relational perspective, they may be viewed as composite keys of a single *population* attribute. As noted in Smith & Smith (1978), "relationship, entity, component, *category*, attribute, and instance are just different interpretations of the same abstract objects."

In addition to these concepts, multidimensional aggregate (statistical) databases require special types of metadata to describe statistical characteristics, and to provide information for data manipulation and analysis software. At the same time, the problem of representing conceptually aggregate data, as well as defining a suitable data structure for them arose. As it is known, a relational database consists of a large set of relations, each of them with their own relation schema, their own set of attributes (each with an instance domain), and so on. In these databases different aspects of reality can be stored, from a population census to data regarding car production of a given industry, from data on epidemiological studies to data on cars sold by a given organization, and so on. All these data are called *microdata*, i.e., raw data which represent different peculiarities of the reality which they describe and which refer to information of different individual entities, such as persons, objects, events, etc.

These microdata are generally stored as relations in a relational database and when an aggregation function, like a sum or a count is applied, the result is a complex data, called *macrodata*, which consists of a descriptive part and a summary part. The latter is called a/the summary attribute or measure, and it is characterized by the descriptive part mentioned above. This descriptive part is called *metadata* and its simplest definition is "data describing data." Many researchers proposed using the relational model also for this kind of data. Regarding this point, we observe that the modeling of aggregate data by means of relations has several drawbacks. In particular:

i)   Each multidimensional aggregate table has its own relation scheme which corresponds to a distinct DB file at the physical level. As a result information within the database is highly fragmented and the integrated access to such information may become difficult or even unfeasible. For this reason Malvestuto (1993) proposes *universal schemes* to collect, in a single relation, all the multidimensional aggregate tables regarding the same population of units of observation, i.e., obtained from the same set of microdata. Similarly, Sato (1991) introduces *conceptual files* to unify all the aggregate data corresponding to the same abstract concept, i.e., which describe the same phenomenon.

ii)   As already stressed in Chan & Shoshani (1981) and, successively, in many other papers, conventional database organizations are not adequate to exploit the regularity induced by the cross-product of the category attribute domains and therefore this results in large redundancies. The introduction of universal schemes and indexing techniques increases such redundancies further.

iii) The operators of the relational algebra are inadequate at manipulating aggregate data correctly. Let us consider for example a relation scheme R (Year, Nation, Sex, Population, AvgIncome), where the underlined words represent the category attributes or descriptive parameters which form a key of R, and Population and AvgIncome represent the measures of the fact studied, classified by year, nation, and sex. If only the population and average income statistics, classified by year and nation, are required (such values are known as *marginal values*, as defined in Malvestuto, Moscarini, & Rafanelli, 1991), one has to "remove" the (category) attribute sex. However, a projection would obviously yield an incorrect value. On the other hand, the operation where a simple projection would work correctly is the removal of the (summary) attribute AvgIncome from the same scheme R. Consequently, several extensions of the relational model have been proposed by Ghosh (1985), who sometimes distinguishes category and summary attributes in *generalized relations*, such as in Su (1983), or by allowing set-valued relations and introducing new specific operators, as in Ozsoyoglu, Ozsoyoglu, & Mata (1985), and in Ozsoyoglu, Ozsoyoglu, & Matos (1987). However, as noted by different researchers, multidimensional aggregate data is fairly complex. It needs to support semantics, operations, and physical structures of multidimensional space, as well as classification structures. Supporting classification structures implies the storage and management of all the metadata of the category values, and their hierarchical associations. For this reason a different approach was studied. It was based on different data structures and operators, which were able to consider the complexity of the data.

The Aggregation Process

In this section we present a formal description of the aggregation process, in order to achieve a more precise modeling of aggregate data. We will define these data from both a conceptual and a logical point of view.

The main difference between them is that in the case of conceptual data, which we will call *multidimensional aggregate data* (MAD), we do not consider their physical storage, while in the case of logical data, which we will call the *multidimensional aggregate data structure* (MADS), we refer explicitly to their physical storage. In the following we will give their formal definitions and show that this structure corresponds exactly with the aggregation process.

Before speaking about the aggregation process, we describe the concept of the *aggregation function* as presented in Klug (1982). This concept is quite simple. An aggregation function takes a set of tuples (a relation) as an argument and produces a single simple value (usually a number) as a result. Many relational query languages require that aggregate functions are able to accept arguments with duplicates. For example, to sum the salaries in an employee relation, the relation would be projected on the salary column, duplicates would be retained, and the projection would be sent to the sum function. Besides being unnecessary, as we will see, using the notion of "duplicates" has a number of disadvantages. For example, a disadvantage is the following: the usual algebraic identities for relational algebra fail to hold when duplicates must be retained. The author gives one example of this. Consider the following two "relations" with duplicates:

$$R \quad \begin{array}{|cc|} \hline 1 & 2 \\ 1 & 2 \\ \hline \end{array} \quad , \quad S \quad \begin{array}{|cc|} \hline 1 & 2 \\ 1 & 2 \\ \hline \end{array}$$

Intersection should be related to join and projection by the equation,
$$R \cap S = (R\,[1, 2 = 1, 2\,]\,S)\,[1, 2].$$
The right-hand part evaluates the relation,

$$\begin{array}{|cc|} \hline 1 & 2 \\ 1 & 2 \\ 1 & 2 \\ 1 & 2 \\ \hline \end{array}$$

Now, intersection should also satisfy the property,
$$R \cap S \subseteq R$$
Clearly, this is impossible.

The solution proposed by the author is quite simple. Instead of providing one sum function (or average, max, etc.), he provides a parameterized family of sum functions:
$$\text{sum}_1, \text{sum}_2, \text{sum}_3, \ldots, \text{sum}_i, \ldots$$
The function sum_i sums the numbers in the ith column of its input. Now there is no need for the notion of "duplicates." For example, to determine the sum of salaries (column 3) in the relation

$$R \quad \begin{array}{|lll|} \hline \text{Joe} & 25 & 12000 \\ \text{Nancy} & 21 & 13000 \\ \text{Sue} & 28 & 13000 \\ \text{Pete} & 35 & 14000 \\ \text{John} & 30 & 12000 \\ \hline \end{array}$$

we would write $\text{sum}_3(\text{R})$.

Formally, we hypothesize a countable set
$$\text{Agg} = (f_1, f_2, f_3, \dots)$$
of *aggregate functions*. Each $f_i \in \text{Agg}$ is a function
$$f_i: R \to N,$$
where R is the set of all relations. Thus, an aggregate function, given a set of tuples, produces a single number as its value.

Most of the authors describe multidimensional aggregate data as a mapping from the domains of the category attributes (independent variable) to the (numerical) domains of the summary attributes (dependent variable). Each category attribute often represents a level of a hierarchy present in that dimension of that MAD. Since the independent variable ranges over an n-dimensional space (the space of the n-tuples of category attribute instances), in Shoshani & Wong (1985) the concept of *multidimensionality* of aggregate data was introduced. Consequently, many of the authors mentioned above proposed modeling aggregate data by means of particular relations where the category attributes form the key. In other words, the category attributes C and the summary attributes S are the building blocks of a relation scheme $R = C, S$, where the functional dependency $C \to S$ holds. Such data can assume different graphical representations: generalized relations (as proposed in Su, 1983), statistical tables, histograms, pie-charts, graphics, cubes, etc. For the sake of simplicity, unless differently specified, we will consider a table as the base structure. A multidimensional aggregate table can be represented by a relation r over R such that each tuple in r corresponds to an entry in the table.

Now we give some definitions useful in describing the aggregation process.

Let Θ be the database universe, i.e., the set of all the relations which form the very large relational database in which raw data (microdata) are stored (for example, a census made in a given year and in a given country, data on production of a given industry in the last 20 years, and so on).

Let R be the subset of Θ relative to all the relations used in the definition of the multidimensional aggregate (macro) database and which, therefore, refers to all the phenomena studied. Note that each phenomenon consists of one or more *facts* which are the physical objects stored in the database. For example, a phenomenon can be "Production of cars" and the set of facts which refer to this phenomenon could be: "Production of cars in Italy classified by year of production, manufacturer, and model"; "Production of cars in Italy classified by year of production, builder, and color"; "Production of cars in France classified by year of production, color, and model"; and so on.

Let $R_x = \{ \mathcal{R}_i \}_{i=1,\dots,h}$ be the set of all the relations $\mathcal{R}_1, \mathcal{R}_2, \dots, \mathcal{R}_h$ (each of them with attributes different in number and names), which refer to the x-th phenomenon. Let $A_1^1, A_2^1, \dots, A_{k1}^1$ be the set of attributes of the relation $\mathcal{R}_1$, where the apex refers to the index which characterizes the considered relation, k_1 is the number of attributes of this relation (i.e., its cardinality), each of which has a definition domain Δ^i_1, Δ^i_2, ..., Δ^i_{k1}, and likewise for the other relations. Note that, in general, not all the attributes are used in the definition of one fact.

In order to clarify how the subsets of R to be aggregated are characterized, let us analyze the well-known concept of the *category attribute* of a multidimensional aggregate database. A category attribute is the result of an abstraction on one or more attributes of the microdata; analogously its *instances* are the result of an abstraction on the (numerical, Boolean, string, etc.) values actually associated with the single microdata. Abstraction plays an important role in the aggregation process, because in general many attributes with different structures and names may correspond to the same category attribute. In order to be aggregated, the microdata must "talk about the same thing," but this by no means implies that the *way* of doing it is the same in all microdata. Thus, the same category attribute "town" of a given statistical object may correspond to a 30-character string field named "TOWN" in a group of microdata, to a 40-character string field named "city" in a second group of microdata, and even to a numerical field named "town_code" in another group.

Let Ω be the set of all the attributes which appear in R, and let $A_x \in \Omega$ be the generic attribute of this set.

Definition 1 Let R be the set of all the relations used in the definition of a multidimensional aggregate database, let $A_x \in \Omega$ be a generic attribute of this database, and let a_{xy} be one of its instances (with y = 1, ..., k, where k is the cardinality of the definition domain of A_x). The logical predicate ($A_x = a_{xy}$), defined on the microdata of R, is called *base predicate*.

For example, if A_x is the generic attribute "Year" and if its instance domain is <1990, 1991, 1992, 1993, 1994, 1995>, the base predicate $A_x = a_{x1}$ is Year = 1990.

Definition 2 The *base set* of the base predicate ($A_x = a_{xy}$) is the subset of Θ consisting of all microdata which satisfy the base predicate. In the following such a subset will be denoted by $B_{Ax = axy}$.

Let $\mathcal{A}$ be the subset of all the attributes of Ω that will become descriptive (or category) attributes or measures of all the MAD which will form the multidimensional aggregate database at the end of the aggregation process. Then $\mathcal{A}$ is the set of all and only the attributes which describe all the facts which appear in the multidimensional aggregate database. Many of these attributes appear in different relations of R. Different attributes can contribute to form one *hierarchy*. Different hierarchies can belong to the same *dimension*, on the condition that pairs of hierarchies have at least one attribute in common. Note that parallel hierarchies, called *specialization hierarchies*, can exist.

For example, country → state → city can have, as a specialization hierarchy, country_of_residence → state_of_residence → city_of_residence. Moreover, other attributes, which do not appear in $\mathcal{A}$, can complete the hierarchies mentioned above (on the condition that the relationship between them and the other attributes of the same hierarchy is defined). We call $\mathcal{A}$ * the set of these last attributes plus the attributes of $\mathcal{A}$.

For example, supposing, in the microdatabase, we have the attributes city, zone, state, region, and country which are all linked in this order by *a part-of* relationship. Also supposing the aggregation process created a set of fact tables in the multidimensional aggregate database which only has the category attributes city, state, and country forming the above-mentioned hierarchy of the dimension "space." Then, set $\mathcal{A}$ consists of the attributes "city, state, and country," while set $\mathcal{A}$ * consists of the attributes "city, zone, state, region, and country."

We call the latter hierarchies *primitive hierarchies* because all the hierarchies which refer to one of them are included in it. Analogously, we call the dimension which includes all its primitive hierarchies the *primitive dimension*.

Let $\mathcal{H}$ be the set of all the hierarchies (including the specialized hierarchies) defined in $\mathcal{A}$ *. Let D be the set of all the dimensions defined in $\mathcal{A}$ * (which can consist of different hierarchies).

Note that the users often give the name of a dimension to descriptive variables of a MAD which are, in reality, levels of a hierarchy relative to this dimension. For example, they say "Production of Cars, classified by the dimensions *model, year of production*, and *manufacturer*," where the dimension *year of production* is, in reality, the level *year of production* of the (specialized) hierarchy "year of production → month of production → day of production" of the dimension "time." In fact, this last is a conceptual structure which can have different category attributes referring to different hierarchies of the same dimension in the same MAD. As another example, we can consider a given MAD which represents the fact "Work situation in USA" classified by "birth state" and by "state of residence," which are two specialization hierarchies of the same hierarchy in the same (primitive) dimension, but which define two different dimensions in this MAD.

Let Δ be the set of all the definition domains (i.e., of all the instances) of the attributes of $\mathcal{A}$, and let Δ* be the set of all the definition domains of the attributes of $\mathcal{A}$ * which also include all the possible instances that each attribute can assume (therefore, also including the instances not present in the relations of Θ). We call these definition domains "primitive domains." This means that all the attributes (and all the relative instances) which appear in the multidimensional aggregate database are part of $\mathcal{A}$ * and Δ* respectively.

Category attributes are not the only metadata of multidimensional aggregate data: several other properties may provide a semantic description of the summary data. Among them we consider, in particular the following:

- the *aggregation type*, which is the function type applied to microdata (e.g., count, sum, average, etc.) to obtain the macrodata (i.e., a MAD, see Tansel, 1987; Rafanelli & Ricci, 1993) and which defines the *summary type* of the measure. This property must always be specified;

- the *data type*, which is the type of summary attribute (e.g., real, integer, non-negative real, non-negative integer) (see Malvestuto, 1993);
- the *fact F_j* described by the multidimensional aggregate table considered (e.g., production, population, income, life expectancy) (see Bezenchek, Massari, & Rafanelli, 1994);
- other properties may be missing, for example "data source" (which may be unknown), "unit of measure," and "unit of count," as defined in the following.

Let Γ be the set of the *functional dependencies* which are possibly present in the multidimensional aggregate database and which, therefore, exist among groups of attributes. Functional dependency is another important concept which it is necessary to consider in order to manipulate the MADS of the multidimensional aggregate database correctly. For example, if one fact stored in the MDDB is "Car production in Italy," described by year of production, model, color, etc., and another fact stored is "Car production in France," described by cylinder, color, etc., after having defined the functional dependency "year of production, model → cylinder," we can join the two MADS by aggregating the first MADS to "cylinder." Obviously, the result will be at a minor level of descriptive granularity.

Given a phenomenon x and given the set of relations $R_x \subset R$, we consider the subset of R_x formed only by the relations involved in the building of the fact $\mathcal{F}_j$. We call this subset an *aggregation relation*, and denote it by R_j^x, where $R_j^x = \{ \mathcal{R}_{j,1}, ..., \mathcal{R}_{j,s} \}^x$. Every fact $\mathcal{F}_j$ has its own descriptive space formed by s category attributes (where s is the cardinality of the j-th fact), which are a subset of all the attributes in the relations R_j^x. We denote the set of the above-mentioned s category attributes by $\mathbb{A}_j^x = \{A_{j,ks}\}^x = \{A_{j1}, ..., A_{js}\}^x$ We call the relation B_j^x, formed by these attributes, a *base relation* of the fact $\mathcal{F}_j$.

The measure values are the result of the aggregation process, i.e., of the application of the aggregation function to the base relation of the fact. The fact obtained by this aggregation process is called *base fact*, because its representation cannot even be disaggregated (i.e., only more aggregate views can be obtained). Each fact consists of a set of materialized views, obtained by applying different operators of aggregation (roll-up, group-by), or of reduction of the definition domains of its category attributes (dice). For a discussion on operators, see Chapter 5. This set of materialized views defines the *lattice* of this fact. The source of this lattice is formed by the total of all the summary category instances of the base fact (called *grand total*), and the sink formed by all the summary category instances at the lowest level of disaggregation.

Let $\mathcal{F} = \{ \mathcal{F}_j \}$ be the set of all the *fact names* described by the MAD of the multidimensional aggregate database.

Let $S = \{S_j\}$ be the set of all the subjects described in the facts, in other words, the "what is" of the summary attributes (cars, people, fruit, workers, dollars, etc.).

Let $R_j^x = \{\mathcal{R}_{j,1}, ..., \mathcal{R}_{j,s}\}^x$ be the subset of the relations in the microdatabase which are involved in the x-th fact. Let $\mathbb{A}_j^x = \{A_{j,ks}\}^x = \{A_{j1}, ..., A_{js}\}^x$ be the set of attributes of $\mathcal{R}_j^x$ which are the only ones considered in the building of this MAD.

Definition 3 Let $\{\mathcal{R}_1, ..., \mathcal{R}_s\}_x^j$ be the set of all the relations involved in the building of the generic j-th fact $\mathcal{F}_j$ of the x-th phenomenon P_x. These relations are included in R_x and have *all* the category attributes $\mathbb{A}_x^j = \{A_{j,ks}\}^x$ of the fact $\mathcal{F}_j$ simultaneously present in it (and possibly other attributes which are not used in this aggregation process). 'j' characterizes the different category attributes of $\mathcal{F}_j$. We call the relation (in non-normal form) formed by all the tuples of the previous set $\{\mathcal{R}_1, ..., \mathcal{R}_s\}_x^j$ *aggregation relation* R_j^x of the j-th MAD which describes the fact $\mathcal{F}_j$.

For example, let $\mathcal{R} = \{\mathcal{R}_1, \mathcal{R}_2, \mathcal{R}_3, \mathcal{R}_4, ..., \mathcal{R}_{521}\}$ be the set of all the relations which refer to the set of facts F studied in the multidimensional aggregate database. Let $\mathcal{R}_1 = \{A_1, A_2, A_3, A_4, A_5\}$, $\mathcal{R}_2 = \{A_1, A_2, A_3, A_6\}$, $\mathcal{R}_3 = \{A_1, A_2, A_4, A_7, A_8\}$, and $\mathcal{R}_4 = \{A_2, A_3, A_4, A_7, A_9\}$ be the set of relations which refer to the j-th fact $\mathcal{F}_j$ described, for example, by the category attributes A_1, A_2, A_4. Let $\{\mathcal{R}_1, \mathcal{R}_3\}$ be the set of relations which have the set of all the attributes $\mathbb{A}_j^x = \{A_{j,ks}\}^x$ (with $k_s = 1,2,4$) of the fact $\mathcal{F}_j$ simultaneously present. The descriptive space of the MAD which describes the fact $\mathcal{F}_j$, the relations $\mathcal{R}_1, \mathcal{R}_2, \mathcal{R}_3$, and $\mathcal{R}_4$ which refer to the fact $\mathcal{F}_j$, and the aggregation relation $\mathcal{R}_j$ of a MAD which describes the fact $\mathcal{F}_j$, are shown in Figure 1.

Definition 4 *The classification set* ζ_{Ax} of the category attribute A_x with the domain Δ_{Ax} is the set whose elements are the base sets defined on Θ by A_x:
$$\zeta_{Ax} = \{B_{Ax = a_{xy}} : a_{xy} \in \Delta_{xy}\}$$
For example, if $\Delta year = \{1991, 1992, 1993\}$, then:
$$\zeta year = \{B_{year} = 1991, B_{year} = 1992, B_{year} = 1993\}$$

Definition 5 By υ_{BA_x} we denote the *union of all base sets* defined on Θ by the instances of the category attribute A_x:
$$\upsilon_{BA_x} = \bigcup_{a_{xy} \in \Delta_{Ax}} B_{Ax = a_{xy}}$$

and we call the set $\{\upsilon_{A_x}\}$ *union set* of A_x.

Figure 1: Phases of the Aggregation Process

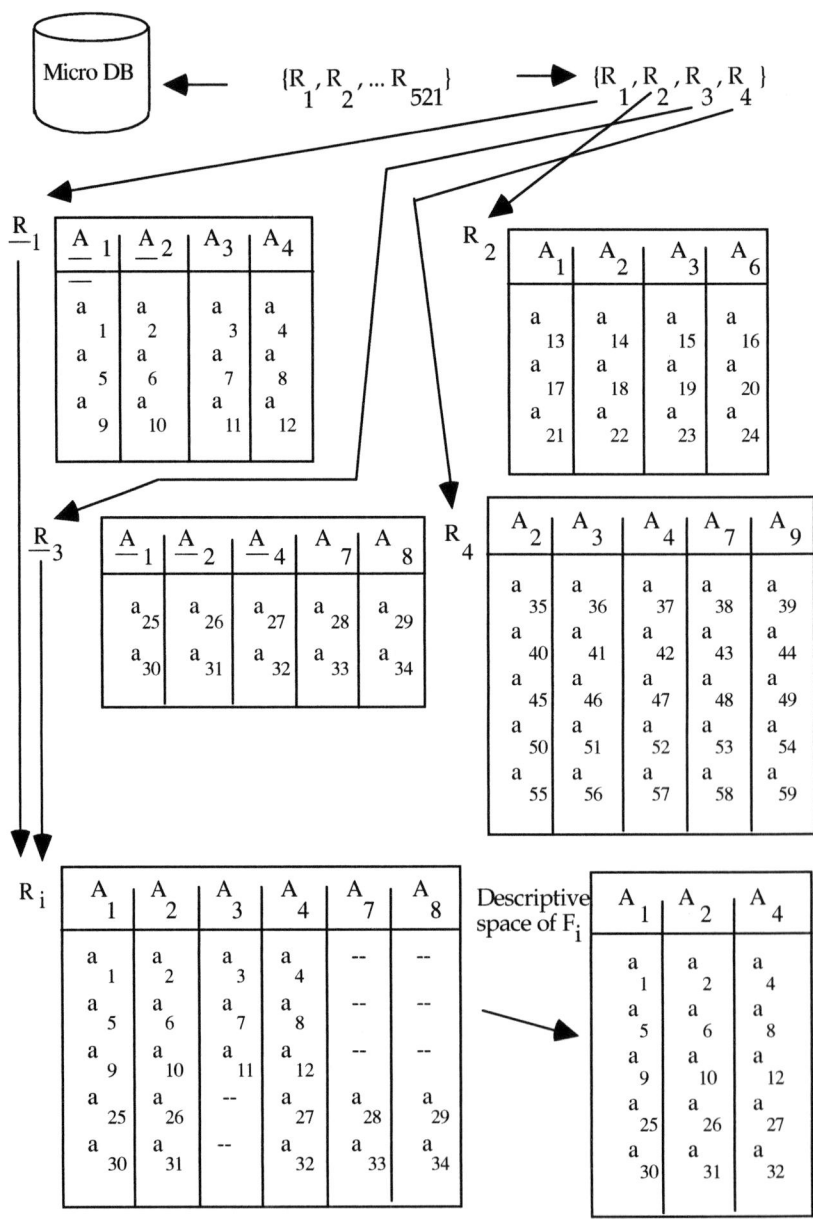

For example, if the attribute year has the same domain indicated above, then:

$$\upsilon_{year} = B_{year = 1991} \cup B_{year = 1992} \cup B_{year = 1993}.$$

Definition 6 We call *aggregation schema* $\mathbb{A}_i = \{A_{ij}\}$ of the i-th MAD, with the aggregation relation $\mathcal{R}_i$ which describes the fact $\mathcal{F}_i$, the set of all the names of the category attributes A_{ij} of the fact $\mathcal{F}_i$. This set defines the dimensions of the MAD. The number of these category attributes defines the *cardinality* of the MAD.

Definition 7 The *base relation* $\mathcal{R}_{Bi}$ of the i-th MAD (with its predefined descriptive space), which describes fact $\mathcal{F}_i$, is the subset of the aggregation relation $\mathcal{R}_i$ of this fact which has all and only the descriptive attributes A_{Bij} of the fact $\mathcal{F}_i$.

For example, the base relation $\mathcal{R}_{Bi}$ of the i-th MAD of the previous example is its descriptive space formed by the attributes A_{ij} (with j = 1,2,4) of the i-th fact $\mathcal{F}_i$.

Definition 8 The *base schema* A_{ij} of the i-th MAD is the tuple of the base relation attribute names of the MAD.

Now we can give a more precise definition of the base set.

Definition 9 The *base set* $\mathcal{B}_{A_x} = a_{xy}$ of the base predicate $A_x = a_{xy}$ relative to the base relation $\mathcal{R}_i$ of the i-th MAD is the subset of the base relation tuples which satisfy the base predicate.

Now we can describe the aggregation process, starting from a relational database of raw data Θ. Supposing, for example, we refer to "Data on production in Italy." This production refers to transportation, food, building trade, clothing trade, etc.

The *first step* in building an aggregate database is to define set R of all possible microdata relations to which the aggregation process must be applied, and, within it, to choose the set R_x of relations which refer to a given phenomenon P_x. Suppose that R_x refers to the phenomenon P_x = "Data on transportation" and that the relations involved $\{\mathcal{R}_i\}$ i = 1,2,3 are *Car* (make, model, cubic capacity, color, air-conditioning, automatic gears, number of airbags, month-year of production), *Sale* (make, model, cubic capacity, arrival date, sale date, name of the seller, branch office city, zone of sale), and *Manufactures* (manufacturer's name, sex, age, region of residence, make, model, cubic capacity, color).

Then, having defined set Ω of all the attributes which appear in R (i.e., make, model, cubic capacity, color, air-conditioning, automatic gear, number of airbag, month-year of production, arrival date, sale date, seller name, city of the branch, sale zone, buyer's name, sex, age, region of residence), the *second step* consists of choosing attributes which will become descriptive (or category) attributes in the multidimensional aggregate database, that is, set $\mathcal{A}$ (in this case, for example, {model, cubic capacity, color, month-year of production, arrival date, sale date, city of the branch, sale zone, sex, age, region of residence}.) In this step we also define the possible functional dependencies, as well as the attributes which possibly complete the hierarchies, and the specialization hierarchies. For example, in this case, a functional dependency is (model, month-year of production → cubic capacity).

The *third step* consists of recognizing all the attributes which belong to the same dimension, and the hierarchies (and their specializations) within this dimension, i.e., sets $\mathcal{H}$ and D. With reference to the example mentioned above, we have different hierarchies (time, space, etc.), which form set $\mathcal{H}$. In this step, all the attributes which do not appear explicitly but which complete a given hierarchy, as well as the possible specialization hierarchies (explained in Chapter 4), have to be inserted into it. In our example, we define the first hierarchy of the dimension "space," i.e., region → province → city, then the connected specialization hierarchies, i.e., region (of residence) → province (of residence) → city (of residence), and region (of branch office) → province (of branch office) → city (of branch office). Note that the instances of specialization attributes are exactly the same instances as the reference attributes, even if the semantics are different. At this point we define the second hierarchy of the dimension "space," i.e., zone (of sale) → city (of sale). Note that the two hierarchies have (at least) one attribute in common (in this case, city). In this step the complete definition domains of all the above-mentioned attributes (then the primitive domains) will also be defined, i.e., set Δ^*.

At this point, the part of the multidimensional aggregate database which refers to the phenomenon "Data on transportation" can be built repeatedly by performing the following steps.

The *fourth step* is selection of the subset $\mathcal{R}_i$ of relations which are involved in the i-th fact, and of the set A^i_{kn} of attributes of $\mathcal{R}_i$ which are the only ones considered in the building of the MAD. In this way we have defined the general characteristics of this object. At the end of the aggregation process, all the sets A_j of attributes will form the descriptive space of the multidimensional aggregate database. In this step we also define the subsets in D and in $\mathcal{H}$ which characterize respectively the dimensions and the possible hierarchies in each dimension of the fact. Therefore, a hierarchy represents the same dimension at different granularity levels.

One of the innovative features of multidimensional aggregate data has been the introduction of a third type of attribute, namely the *implicit category attribute*, which can considerably enlarge and enhance the manipulation capabilities of an aggregate database. In fact, if among the attributes which define the dimensions of a MAD, one or more of them have a definition domain which consists of only one instance, we transform each of them into part of the "fact name" and call it an implicit attribute or an *implicit dimension* (because it does not explicitly appear in the MAD dimensions). Note that also implicit attributes contribute to MAD cardinality (i.e., the number of dimensions present in this MAD).

The *fifth step* involves application of the aggregation function to the attributes of the relations in A. The result of this operation is the numeric data which represents the measure carried out on microdata, called simply *measure* (or summary attribute). Depending on the type of function applied, the parameters which characterize each fact have to be defined, i.e.:

- *summary type*—defined by the aggregate function type applied to microdata (a count, a sum, etc.);
- the *count unit*—suppose that the result instances are 100, 213, 47, etc. and suppose also that the subject is "fruit"; the count unit defines if, for example, 100 really means 100 (100 *x 1*), or 100,000 (100 *x 1,000*) or 100,000,000 (100 *x 100,000*), and so on;
- *measure unit*—in the previous example, the number could be "number of pieces of fruit," or "kilograms," or "tons," etc.;
- *data source* (this information is not always available).

In general, the aggregation function applied is initially a *count* or a *sum*. Subsequent aggregations can be obtained by applying algebraic operations and statistical-mathematic operations to these aggregate data, for example, to obtain averages, percentages, maxims, minimums, index functions, and so on.

The *sixth step* is definition of the fact name (name of the MAD), for example Production of vegetables, Sale of cars, Number of births in a given state, Incidence of a given disease in a fixed country, etc. At the end of the aggregation process, all the names of the *facts* defined will form the set $\mathcal{F}$.

The *seventh step* then is definition of the subject described in the fact, by choosing it from among the attributes of the relations in $\mathcal{A}$ * (for example, *cars*). At the end of the aggregation process, all the subjects described in the facts will form set S .

The *eighth step* is definition of possible *Notes* which characterize possible anomalies of the MAD (this step can be lacking). For example, if car production (name of the fact described by the MAD) is described by year of production, region, and model, we could have the summary data for the years 1985, 1990, 1995, and 2000 for all the regions, apart from the region Piedemont, where car production refers to the years 1986, 1991, 1996, and 2001. In the MAD these last data are classified as if they were relative to the years 1985, 1990, 1995, and 2000, but, with a suitable *Note*, we would know the real years.

When we apply the aggregation process, all the MADs produced are conceptual structures, in the sense that, for each of them, we have not yet defined how to store them in the database, or any order among the attributes of a fact or among the domain instances of each attribute. To store them in the multidimensional aggregate database, we have to define these orders.

Figure 2: Simple MADS

Cereal production during the period 1980-81 (all values are in thousands of tons)	wheat	rice
Italy	437	214
France	319	42
Germany	116	6

The result of this further step (the *ninth step*) is the definition, for each MAD, of a corresponding logical structure, called *Multidimensional Aggregate Data Structure* (MADS). In it each instance of the measured data is characterized exactly by a tuple, whose elements are defined by one instance for each dimension (category attribute) of the MAD.

Definition 10 An *aggregation process* is formally described by the six-tuple

$$< P, \; \mathcal{R}_i, A_i, S_p, N, f_{agg}>,$$

where:

P is a fact name which identifies a fact universe R through a mapping φ; this mapping is defined from the fact set F to the powerset of Θ.

$\mathcal{R}_i$ is a set of relations involved in the building of the i-th MAD.

A_p is a set of the descriptive (category) attributes $\{A_{p,1}, A_{p,2}, ..., A_{p,k}\}$ of $\mathcal{R}_p$ (with its own instance domains $\Delta (A_{p,j})$, with j = 1, ..., k) on which to apply an aggregation function.

S_p is the *subject* which characterizes the measure of the MAD.

N is a numerical domain.

$f_{agg}: \mathcal{R}_p \rightarrow S_p$ is an aggregation function (initially, in general, a sum or a count).

A MAD is, therefore, a concise representation of the aggregation process result.

Definition 11 Let H and K be two sets whose elements are sets; we define the *cross intersection* of H and K, and denote it by $H \otimes K$, as the set of all the possible intersections between each element in H and each element in K:

$H \otimes K = \{ j : j = h \cap k, h \in H, k \in K \}$

In order to clarify the meaning of these definitions, let us consider a simple MADS (represented as a table), shown in Figure 2, whose fact studied is cereal production in some European countries during the period 1980-81. In it the summary type is *sum*, the count unit is *thousands*, the measure unit is *ton*, and the data source of the MADS is not specified.

The described fact is "cereal production." The category attributes "country" and "cereal" (directly obtainable from the rows and columns of the table) are not sufficient to fully characterize the aggregation sets. For instance, the cross intersection of the *base set* $B_{country = Italy}$ of the base predicate "country = Italy" with the base set $B_{cereal = wheat}$ of the base predicate "cereal = wheat" determines the set of all microdata regarding wheat production in Italy and such a set also includes data relative to the periods before 1980 and after 1981.

To characterize the *base relation* $\mathcal{R}_{Bi}$ of the i-th MAD (with its predefined descriptive space) correctly, which describes fact $\mathcal{F}_i$, we must introduce a third category attribute "years," implicitly present in the textual description of the table. Then, into the base relation (and, therefore, into the base schema) of this fact we have to insert all the category attributes which do not appear explicitly in the relation because they always assume the same (unique) value (in our case, the category attribute "years" with its definition domains formed by the set-value {1980, 1981}). We define the attributes like "years" as *implicit*, because they do not explicitly express a classification in the MAD, but are necessary for a correct definition of the base relations. The use of implicit attributes takes account of the full dimensionality of the MAD and as a consequence makes their manipulation more effective.

THE MULTIDIMENSIONAL AGGREGATE DATA STRUCTURE

In this section we will describe the data structure (simple, complex, and composite) of multidimensional aggregate data. First of all, we will discuss the important concept of *dimension* and the hierarchy concept, which is implicit in it.

Figure 3: The "Geography" Primitive Dimension

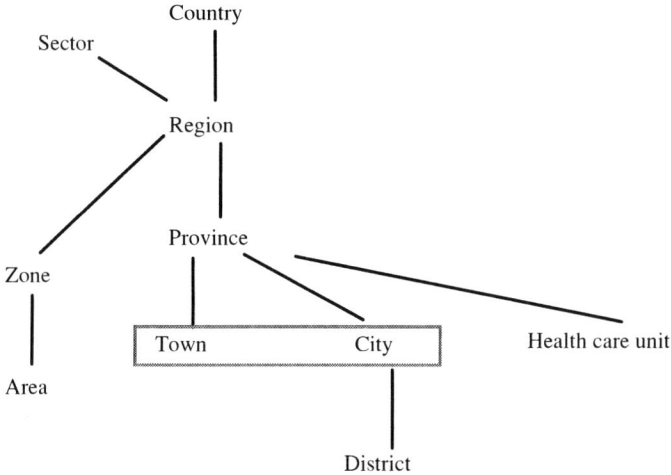

As previously said, a measure of a MAD is described by a set of *dimensions*. These are, in reality, a set of category attributes (see, for example, Rafanelli, 1990); each of them can be a level of one of the possible hierarchies which form the primitive dimension. Each category attribute also has a *primitive category attribute* (several category attributes can have the same primitive category attribute), to which all the attributes with the same semantics are linked and whose domain consists of the union of all the domains of the category attributes previously linked to it. This is quite important when, for example, different category attributes linked to the primitive attribute have different names. For example, in one MADS the category attribute *year* appears and in another MADS the category attribute *years* appears. Therefore, they will be linked to the same primitive category attribute *year*, which semantically expresses the same concept. Thus, the primitive dimension is the union of all the primitive category attributes which refer to this dimension. These attributes are part of one or more hierarchies. For example, "province" could be one level of the hierarchy "political classification," which, together with the other hierarchies "health care zone classification" and "energy zone classification," form the *unique* dimension "geography," as shown in Figure 3. Note that the political classification considers province as partitioned into two different kinds of cities: small cities (town) and big cities (city). These last are also sub-divided into districts.

Note also that this representation can assume the so-called "spider" configuration, in the sense that transversal hierarchies can generate different roots, so that the "all" term, as formulated in Gray et al. (1997), can also be lacking. With the term "primitive" we mean to express the concept of completeness regarding that dimension (in our case, geography), i.e., that such a dimension cannot assume other values (from the intentional point of view, other category attributes, from the extensional point of view, other instances of definition domain of each category attribute which defines every level) than those which appear in it. These and other concepts will be discussed in Chapter 4.

The complexity of aggregate data is also due to the fact that the same operator may require different recomputation functions for the summary values, depending on the summary type of the multidimensional aggregate table, as illustrated in Fortunato et al. (1986), and in Rafanelli & Ricci (1993). Summary data are always fixed in time, in the sense that every instance of the time dimension *statically* characterizes the measure described by it (and by the other category attributes which describe the numerical values that are in the table cells). It is always (implicitly or explicitly) present in every MAD. Another dimension always implicitly or explicitly present in every MAD is space. Shoshani (1982) wrote: "…the partitioning of geography can change with the application. In addition, the classification of the parameters can change over time." There followed an example in which problems without strict geographical hierarchies (for example, descendent elements not always fully contained within a single parent element) and changes over time (for example, definitions of regions changed over time due to legislative action or political needs or

Figure 4: The 2-D Representation of a Multidimensional MADS

Employment in California			Professional Category				
			Engineer		Secretary		Teacher
			Profession		Profession		Profession
			Chemical Engineer	Civil Engineer	Junior Secretary	Executive Secretary	Elementary Teacher
Sex	Male	**Year** 81	197,700	241,100	534,300	154,100	212,943
		82	209,900	278,000	542,100	169,8 00	213,521
		..	......	....	....	......	....
		88	237,800	439,200	550,100	293,700	216,230
	Female	**Year** 81	25,800	112,000	667,300	162,,300	216,071
		82	28,900	127,600	692,500	174,400	217,520
		..	.....	......	......	......	......
		88	39,800	194,100	811,500	242,600	221,067

changes which evolve over time like the classification of diseases) underlay the research over the following years.

Subsequently, it was modeled on the fact that both time and space can be either implicit, in the sense that they might appear in the name of the table but not as descriptive variables, or explicit, as claimed by Bezenchek, Rafanelli, & Tininini (1996b). Virtually, using a 2-D representation requires squeezing the multi-dimensional space into two dimensions. This is usually done by choosing several of the dimensions to be represented as rows and several as columns. In general, the 2-D representation of multi-dimensional aggregate data forces a (possibly arbitrary) choice of two hierarchies for the rows and columns. The apparent conclusion is that a proper model should retain the concept of multi-dimensionality and represent it explicitly, as observed in Rafanelli & Shoshani (1990). In the 2-D representation, classification hierarchies are represented in the same manner as the multi-dimensional categories, while actually the classification hierarchy represents a single dimension. Consider, for example, the table "Employment in California" classified by "sex," "year," and by "professional categories"—"Professions" (the numbers are fictitious), taken from Rafanelli & Shoshani (1990) and shown in Figure 4.

It is obvious from this example that the values of average income are given for specific combinations of "sex," "year," and "profession" only. Thus, "professional category" is not part of the multi-dimensional space of this statistical object, but part of a "hierarchical" classification relationship, "professional categories" → "professions" (where the notation → means a one-to-many relationship). This means that a hierarchical relationship exists between the instances of "professional class" (e.g., "engineer") and the instances of the "profession" (e.g., "civil engineer"), i.e., there is a fundamental difference between category structure and multi-dimensionality. Usually, only the low-level elements of the classification relationship participate in the multi-dimensional space. This fundamental difference should be explicitly represented in a semantically correct multidimensional aggregate data model. Moreover, by necessity, more than one dimension must be represented by the rows and the columns if more than two dimensions exist in the dataset. This is accomplished by selecting an arbitrary order of the dimensions for the rows and the columns, as noted in Shoshani (1997). The label "Employment in California" represents the summary measure for this table, but it also says that this table has an additional dimension "state" where the instance value selected is a singleton "California." Finally, there is a summary function implied (in this case it is "sum") with this table for further summarization to be done.

Another important characteristic of aggregate data, in contrast to conventional disaggregate data, is that they are essentially *static*. Only recently, with the increasing interest in developing business analysis and decision support applications, the on-line analysis of data (using sophisticated mathematical and statistical formulas, and consolidating summary data), such data have also assumed a dynamic connotation. But, even if the values change over time, it is usually necessary to record the evolution, rather than just the current version of the database, i.e., to add the new values without changing the old ones. For these reasons, the dimensions "time" and "space" (*when* and *where* the fact happens) have a particular importance in this kind of data. Different data structures, in the context of a data model, were proposed. For example, in Ozsoyoglu & Ozsoyoglu (1983) and, subsequently, in Ozsoyoglu, Ozsoyoglu, & Mata (1985), a summary table schema and a summary table instance were formally defined. In particular, a *summary table schema* S (F_r, F_c, C) is a three-tuple where F_r and F_c denote a row and column category attribute forest, and C is an ordered multi-set of cell attributes. In a summary table, a category attribute may be elementary or set-valued, but the cell attributes are always elementary. A *summary table instance* is a collection of cell instances structured as specified by the summary table schema.

In Rafanelli & Shoshani (1990), the Storm model is proposed. The authors speak explicitly of aggregate databases calling them statistical databases. Moreover, independently of the different, possible ways of representation (tables, relations, vectors, pies, bar-charts, graphs, and so on) of the structure stored in this database, such a structure is called a statistical object (SO). Each statistical object is characterized by having two different types of attributes: (a) a single *summary attribute* (that is, the result of the application of aggregate functions on microdata). A summary attribute has a *summary type*, which depends on the type of function applied to the microdata; and (b) a set of *category attributes*, which describe the summary attribute. therefore, the authors formally define a statistical object as in the following:

A *Statistical Object* is a data structure defined by a quadruple

$< N, C, S, f >$, where:

N is the name of the SO, which describes the universe of the phenomenon of interest (for example, "Average income in California").

C is a finite set of category attributes; each category attribute has a domain associated with it, and a "domain cardinality" which corresponds to the number of values (sometimes called "modalities" by statisticians) of its domain.

S is a single *summary attribute* associated with the SO. The summary attribute has a domain and a domain cardinality associated with it.

f is a *function* which maps, from the cross-product of the category attribute values, to the summary attribute values of the SO.

By *cross-product* the authors mean the Cartesian product in which its member order has no importance (i.e., A x B = B x A). Each category attribute also has a *primitive category attribute* (several category attributes can have the same primitive category attribute), to which all the attributes with the same semantics are linked. Its domain also consists of the union of all the domains of the previous category attributes linked to it.

In the following we will propose the multidimensional aggregate data (MAD) and the multidimensional aggregate data structure (MADS), respectively the conceptual and logical structure of aggregate data.

DIFFERENT USE OF MULTIDIMENSIONAL DATA: OLAP APPLICATIONS

Recently other proposals for modeling multidimensional databases have been presented. These proposals are based on the cube metaphor, which represents the concept of multidimensionality by a three-dimensional representation (see Agrawal, Gupta, & Sarawagi, 1997; Gyssens & Lakshmanan, 1997; Lehner, 1998; Pedersen & Jensen, 1999; Nguyen, Tjoa, & Wagner, 2000). In fact, there has been significant interest in multidimensional database systems for developing business analysis and decision support applications. E.F. Codd (1993: Codd, Codd, & Salley, 1993) proposed the term On-Line Analytical Processing (OLAP) for rendering enterprise data in multidimensional perspectives, performing on-line analysis of data using mathematical formulas or more sophisticated statistical analyses, and consolidating and summarizing data according to multiple dimension.

The new data models for OLAP, based on a multidimensional view of data, typically categorize data as being "measurable business facts (measures) or dimensions." These are mostly textual and characterize the facts. In reality, in OLAP research, most work has concentrated on performance issues. Techniques have been developed for computing the data cube (see Agarwal et al., 1996), for deciding what subset of a data cube to pre-compute (see Harinarayan, Rajaraman, & Ullman, 1996; Gupta, Harinarayan, Rajaraman, & Ullman, 1997), for estimating the size of multidimensional aggregates (see Shukla, Deshpande, Naughton, & Ramasamy, 1996), and for indexing pre-computed summaries. It has been suggested that researchers who study the conceptual model should try to combine the OLAP models with the more advanced and consolidated data model concepts from the field of statistical multidimensional aggregate databases.

For OLAP databases, we have the following conceptual structure: summary measure; summary function; dimensions; classification hierarchies, exactly like the statistical databases, which have the same components, as claimed by Shoshani (1997). In spite of the different origins of these two areas (application types, socio-economic motivation, and business applications to collect and analyze information for decision making), there are many similarities in the problems they tackle.

Apart from the different emphasis given to the use of data made in the two environments, and to the research done (modeling and privacy the former, efficiency of access and of data analysis the latter), another distinction which can be made between OLAP and SDB is that, while statistical databases are usually derived (summarized) from other base data, OLAP databases often represent the base data directly.

Similarities and differences between OLAP and statistical databases (SDBs), that is, the unobvious connection between analyzing business data and socio-economic data, are discussed in Shoshani (1997). It is important to underline the fact that both of them deal with multidimensional data sets, and both are concerned with statistical summarizations over the dimensions of the data sets.

The data about individuals or original objects from which statistical databases are derived is referred to in statistical database literature as "microdata," and the summarized dataset as "macrodata." In addition, the data associated with classification structures is referred to as "metadata." Metadata can be quite extensive, and are often managed by specialized systems, or general purpose database systems, such as relational systems. Statistical databases mostly present macrodata either for reasons of privacy, or because the original dataset is of no interest (i.e., only the summaries are needed for statistical analysis). In OLAP, summaries may obscure the phenomena we wish to discover, thus we start with the original dataset. These generalizations, of course, do not always hold, but, as observed in Shoshani (1997), by and large most examples bear this observation. In any case, more or less formal models were proposed in the literature. The more significant proposals are the following.

Agrawal, Gupta, & Sarawagi (1997) propose a data model (and a few algebraic operators) that provide a semantic foundation to multidimensional databases. The distinguishing feature of the proposed model (similar to other authors') is the symmetric treatment of dimensions and measures. The model also provides support for multiple hierarchies along each dimension. Its data model is a multidimensional cube with a set of basic operations defined on it and which produce a new cube (closed operators) as output. The proposed model is a logical model, so that it does not force any storage mechanism.

Gyssens & Lakshmanan (1997) propose an n-dimensional table as a fundamental data structure of the multi-dimensional database. Drawing on the terminology of statistical databases, the authors classify the attribute set associated with the schema of a table into two kinds: *parameters* and *measures*. Analogous to Agrawal, Gupta, & Sarawagi (1997), there is no a priori distinction between parameters and measures in that any attribute can play either role. The actual *contents* of a table are essentially orthogonal to the associated structure, i.e., the distribution of attributes over dimensions and measure. Separating both features leads to a *relational* view of a table. The cells of an n-dimensional table can have more than one value.

Lehner (1998) proposes a modeling approach, declaring explicitly that the nested multi-dimensional data model proposed is not yet really another data model, but provides necessary extensions in different directions. The multidimensional context of the proposed model uses dimensional structures, which model the business terms of the user's world in a very complex and powerful way, for gaining analytical access to the measures or facts. The author gives the formal definition of "Primary and Secondary Multidimensional Objects," reflecting the multidimensional view of classification and dimensional attributes, in order to represent a consistent and intuitive view of nested multidimensional data cubes. The author also explains that, as introduced in Rafanelli & Ricci (1983), the aggregation type describes the aggregation operators which are applicable to the modeled data (Σ: data can be summarized, Φ: data may be used for average calculations, c: constant data implies no application of aggregation operators). The cardinality of the context descriptor reflects the dimensionality of the corresponding data cube.

In Pedersen & Jensen (1999) and, subsequently, in Pedersen, Jensen, & Dyreson (2001), multidimensional data modeling for complex data is proposed. For every part of the model, the authors define the *intention* and the *extension*. An *n-dimensional fact schema* is defined as a two-tuple S = (Y, D), where Y is a *fact type* and $D = \{T, i = 1, .., n\}$ is its corresponding *dimension*. A dimension type τ is a four-tuple $(C, \leq_\tau, T_\tau, \perp_\tau)$, where $C = \{C_j, j = 1, ..., k\}$ are the *category types* of τ, $\leq \tau$ is a partial order on the C_js, with $T_\tau \in C$ and $\perp_\tau \in C$ being the top and bottom element of the ordering, respectively. Thus, they deduce that the category types form a lattice. The intuition is that one category type is "greater than" another category type if members of the former's extension logically contain members of the latter's extension, i.e., they have a larger element size. The top element of the ordering corresponds to the largest possible element size, that is, there is only one element in its extension, which logically contains all other elements. C_j is a category type of T, written $C_j \in \tau$ if $C_j \in C$. The authors assume a function $Pred : C \to 2^c$ that gives the set of immediate predecessors of a category type C_j. The authors observe that many types of data, e.g., ages or sales amounts, can be added together to produce meaningful results. This data has an ordering on it, so computing the average, minimum, and maximum values makes sense. For other types of data, e.g., dates of birth or inventory levels, the user may not find it meaningful in the given context to add them together. However, the data has an ordering on it, so taking the average or computing the maximum or minimum values do make sense. Some types of data do not have an ordering on them, and so it does not make sense to compute the average, etc. Instead, the only meaningful aggregation is to count the number of occurrences. Therefore, they affirm that it is possible to support correct aggregation of data by keeping track of what types of aggregate functions can be applied to what data. This information can then be used to either prevent users from doing "illegal" calculations on the data completely, or to warn the users that the result might be "wrong." Following this line of reasoning and previous works (Rafanelli & Ricci. 1983; Lehner, 1998), they distinguish between three types of aggregate functions: Σ, applicable to data that can be added together; Φ, applicable to data that can be used for average calculations; and c, applicable to data that is constant, i.e., it can only be counted. Considering only the standard SQL aggregation functions, they deduce that $\Sigma = \{\text{SUM, COUNT, AVG, MIN, MAX}\}$, $\Phi = \{\text{COUNT, AVG, MIN, MAX}\}$, and $c = \{\text{COUNT}\}$. The aggregation types are ordered, $c \subset \Phi \subset \Sigma$, so data with a higher aggregation type, e.g., Σ, also possess the characteristics of the lower aggregation types.

In Nguyen, Tjoa, & Wagner (2000), the authors introduce a conceptual multidimensional data model that facilitates a precise, rigorous conceptualization for OLAP. OLAP systems organize data using the multidimensional paradigm in the form of data cubes, each of which is a combination of multiple dimensions with multiple levels per dimension. Summarized data is preaggregated and stored with the main purpose of exploring the relationship between independent, static variables—*dimensions*—and dependent, dynamic variables—*measures*. Moreover, dimensions always have structures and contain one or more natural hierarchies, together with other attributes that do not have hierarchy relationship to any of the attributes in the dimensions. The first goal of the paper is to propose a model able to represent and capture natural hierarchical relationships among members within a dimension. The model allows the handling of dimensions with complex structures, such as unbalanced and multi-hierarchical structures. Moreover, the proposed data model permits the representation of relationships between dimension members and measure data values by means of cube cells. The second goal of the paper is the modeling of the conceptual multidimensional data model in terms of classes by using UML.

Based on the formal representation of the class specifications in UML, the design and implementation of the data model for object-oriented databases are straightforward.

MULTIDIMENSIONAL AGGREGATE DATA

Most of the existing models for aggregate data represent a MAD as a mapping between category and summary attributes. The analysis of the aggregation process shows aggregate data in a new perspective. A MAD represents a functional link between *aggregation sets* and *summary values*, rather than between tuples of category attribute instances and summary values. In this framework the category attributes and the corresponding instances express the *constraints* which univocally characterize the single aggregation sets.

Now we can formally define multidimensional aggregate data (MAD) and the multidimensional aggregate data structure (MADS). In this three different data structures for multidimensional aggregate data were proposed: simple, complex, and composite MAD (referred to as statistical object, SO, in that paper). The aggregation sets of an aggregation process can often be represented simply by cross-intersecting the classification and union sets of the explicit and implicit attributes respectively. In such cases the aggregation process is represented by a *simple* MAD.

Definition 12 A *simple MAD* is a conceptual multidimensional aggregate data defined by the six-tuple $< P, N, f, \mathcal{A}, S, s >$, where:

- P is the fact name described by the MAD. Through the mapping φ, P identifies a fact universe $\mathcal{F}$.

- N is the numerical domain where the summary attribute is defined (e.g., $\mathbb{R}$, $\mathbb{R}+$, $\mathbb{Z}$, $\mathbb{Z}+$). Note that the summary attribute is actually defined on an extension of N, namely on $N \cup \{\text{'N.A.'}\} \cup \{\text{'S.Z.'}\}$, where 'N.A.' is for not available and 'S.Z.' for structural zero, as defined in Malvestuto (1993).

- f is the type of aggregation function (e.g., count, sum, average).

- $\mathcal{A}$ is the set formed by the two subsets ε and I, where:
    - ε is the set of *explicit category attributes*, each of them with its corresponding ordered instance domain:
        $$\varepsilon = \{ E_1, E_2, ..., E_M \}$$
        $$\text{with } \Delta (E_1) = \{i_{E_1,1}, i_{E_1,2}, ..., i_{E_1,P_1}\}$$
        $$...$$
        $$\Delta (E_M) = \{i_{E_M,1}, i_{E_M,2}, ..., i_{E_M,P_M}\}$$
        (M represents the number of explicit category attributes, while P_j ($j = 1, ..., M$) the number of instances of the j-th attribute).
    - I is the set of *implicit attributes*, each of them with its corresponding unique instance of the definition domain:

$I = \{ I_1, I_2, ..., I_N \}$

(N represents the number of implicit attributes).

- S is the subject of the MAD, i.e., the "what is" of the cell value, the instance of the measure.
- s is a set of $(P_{(j=1, M)} P_j)$ summary values in bijective correspondence with the aggregation schema $\mathbb{A}_i = \{A_{ij}\}$ of the i-th simple MAD.

Definition 13 A *simple MADS* is a logical multidimensional aggregate data structure defined by the six-tuple $< P, N, f, \underline{\mathcal{A}}, S, s >$, where:

- P,
- N, and
- f have the same meaning as the previous definition.
- $\underline{\mathcal{A}}$ is the *ordered* set formed by the two subsets $\underline{\varepsilon}$ and $\underline{\vartheta}$, where:
  - $\underline{\varepsilon}$ is the *ordered* set of *explicit category attributes*, each of them with its corresponding ordered instance domain:

$\underline{\varepsilon} = \{ E_1, E_2,..., E_M \}$

$\Delta (E_1) = \{ i_{E_1,1}, i_{E_1,2},..., i_{E_1,P_1} \}$

...

$\Delta (E_M) = \{ i_{E_M,1}, i_{E_M,2},..., i_{E_M,P_M} \}$

Figure 5: A Complex MADS

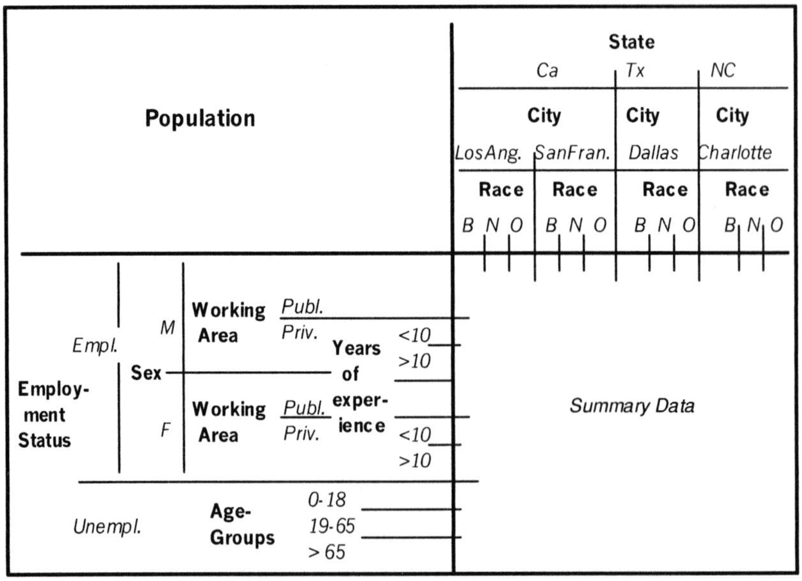

(M represents the number of explicit category attributes, while P_j (j = 1,..., M) the number of instances of the j-th attribute).

- *I* has the same meaning as the previous definition.

- *s* is an *ordered* set of $\Pi_{(j=1,M)}$ P_j) summary values in bijective correspondence with the aggregation schema $\mathcal{A}_{ij}$ of the simple MADS.

For example, a simple MADS is the table shown in Figure 4.

Definition 14 A *Complex MAD* is a conceptual multidimensional aggregate data in which one or more dimensions is partitioned into two or more subsets, each of which is "classified by" different attributes. This situation can produce a union of subMADS, each of them with a possible different cardinality.

Analogously, a *Complex MADS* has the same definition, with the obvious difference of the *ordered* sets *e* and *s*.

In Figure 5 an example of this situation is shown. In this case the MADS has to be split into two (or more) different MADS.

A subtle distinction has to be made between the concept of multidimensionality and that of *polidimensionality*. The former refers to a data structure in which measured data is described by different (two or more) parameters which define a multidimensional descriptive space. The latter refers to a descriptive space in which a category attribute has, for example, one part of its definition domain classified by one or more other category attributes, A, B, …C, while the other part is classified by one or more different category attributes, E, F, …H. Note that the number of the

Figure 6: A Composite MADS

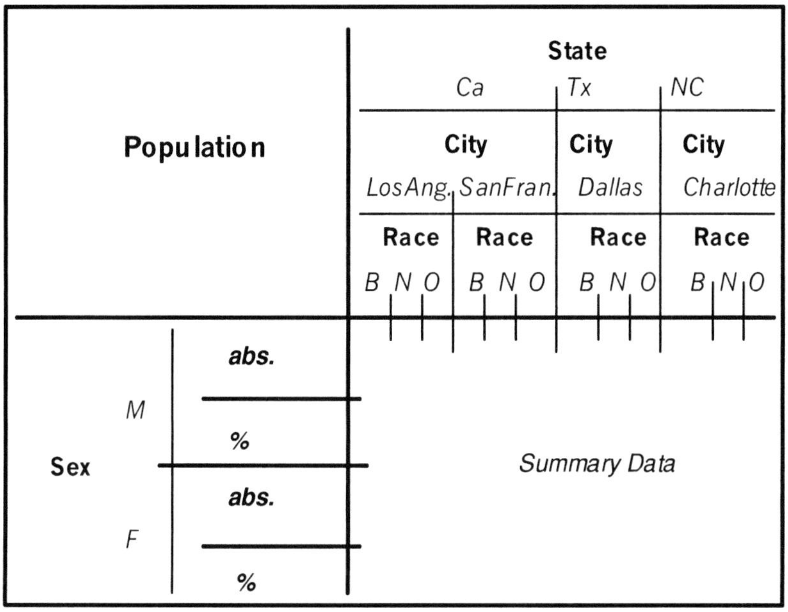

Figure 7: A Complex-Composite MADS

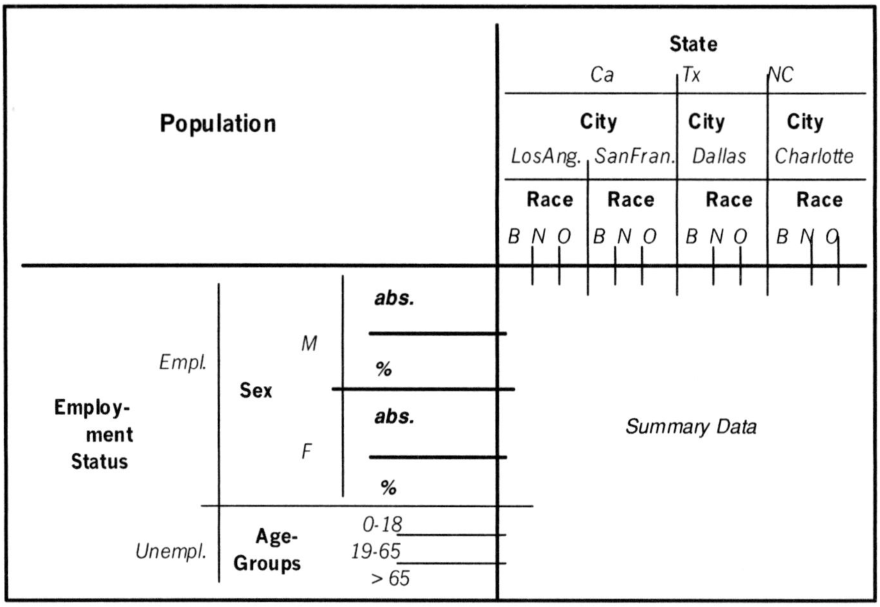

first and the second group of category attributes can be different. For example, with regard to the MADS of Figure 5, which is a typical example of polidimensional MADS, the category attribute "Employment Status" has its instance "Employed" classified by "Sex," "Working Area," and "Years of Experience," while the other instance "Unemployed" is classified by "Age Groups." This means that such a MADS will have one part with the three common dimensions "State-City," "Race," and "Employment Status" (this last with the only instance "Employed"), plus the three other dimensions "Sex," "Working Area," and "Years of Experience" (i.e., with cardinality = 6), while the other part has the above-mentioned three common dimensions (with the only instance "Unemployed" of the dimension "Employment Status"), plus the other dimension "Age Groups" (i.e., with cardinality = 4).

A MAD (or a MADS) may collect data from two or more aggregation processes (composite MAD or MADS). Then:

Definition 15 A composite MAD is a conceptual multidimensional aggregate data defined by the set { $a_1, a_2,..., a_S$ }, where each element a_j is a complex (possibly simple) MADS, and in which different summary types generally appear.

Analogously, a composite MADS is a MADS defined by the *ordered* set { a_1, $a_2,..., a_S$ } as defined in the previous definition.

This means that this type of MAD (or MADS) is obtained by the union of two (or more) MAD (MADS) in which the fact described was the same, but the summary type was different. This union is performed only if a common dimension, along which such a union is made, exists.

For example, in Figure 6 an example of composite MADSs (represented by a table) is shown. Also in this case the MADS can be split into two (or more) MADSs, each of them with a unique summary type. Obviously, we can have the combination of the two situations, i.e., a complex-composite MADS, as shown in Figure 7.

In Malvestuto (1993) two statistical objects are defined as being "homogeneous" if they refer to the same summary variable and to exactly the same *population*, i.e., if they have been obtained by applying the same aggregation function, and the collection of units of observation (microdata) involved is exactly the same. Malvestuto makes the assumptions that 1) the category attributes partition the population and that 2) the aggregation function is additive. He shows that, under these hypotheses, a collection of homogeneous statistical objects can be queried as if they were actually a single, higher dimensional object and that the set of answerable queries can be larger, compared to the one resulting from querying the single statistical objects separately. Moreover, important results on the answerability and evaluability of a query are obtained. Unfortunately, determining homogeneity is generally rather difficult, since there may be thousands of populations in an aggregate database, and each time we want to insert a new statistical object or query the database, we need to know the name of the population of interest or to carefully browse a long list of population names.

REPRESENTING AND INTERPRETING NULLS IN MULTIDIMENSIONAL AGGREGATE DATA

The presence of scarce data or null values is frequent in multidimensional data. In the case of multidimensional aggregate databases, the refined subdivision made in the theory of null values, as mentioned in ANSI/X3/SPARC Report (1975) and Vassiliou (1980), is not necessary. In fact, only two types of null values are basically used: unknown (or non-available, NA) and non-existing (the latter are often called "structural zero entries," see Bishop, 1978).

An example of the null value "unknown" arises when in a MAD only the data relative to the production of fruit in the USA are reported; in this case, if the data relative to the state of California for the years "80, 81, 82" are missing, they will be "unknown" (presumably not recorded after a survey and, in any case, not available).

An example of "non-existing" data, instead, can be found in a MAD where the data relative to the reports of illness subdivided into illness and sex are illustrated; in this case the missing data regard "cancer of the prostate gland" for "sex = female" or "cancer of the uterus" for "sex = male" (the value will be zero, but it will be a structural zero, in that it can never assume a value which is different from zero). It should be noted that there is an important difference between the two types of null values, especially with reference to the relative marginal value (the total regarding the category attribute).

In particular, in the former (unknown) case, the total regarding the category attribute, for which there are unknown values, is not the total of the attribute itself. In the latter (non-existing) case, although it is true that, for example, the total of cancer of the prostate cases is the total reported in the marginal (summarizing with regard to "sex"), in this way the information that these cases are the total of males alone and not the entire population is lost. For this reason, we distinguish, for the summary data of the tables in output, two null values: non-available value (symbol "NA") and structural zero (symbol "-"). The structural zeros are important also in the case in which the cross-product among the different category attributes which describe the summary data is incomplete (for example, in the case of a relation).

GRAPH MODELS

The necessity to use a graph model to represent multidimensional aggregate data was a natural consequence of the fact that the use of 2-D table representation had several deficiencies. 2-D table representation existed, and still continues to be practiced by statisticians today, because statistical data have been historically presented on paper. Using a 2-D representation requires squeezing the multidimensional space into two dimensions. This is typically done by arbitrarily choosing several of the dimensions to be represented as rows and several as columns. In general, 2-D representation of multidimensional data forces a (possibly arbitrary) choice of two hierarchies for the rows and columns (Shoshani, 1997). The apparent conclusion is that a proper graphical model should retain the concept of multi-dimensionality and represent it explicitly, as mentioned in Rafanelli & Shoshani (1990), where the authors referred this concept to the aggregate statistical data and proposed the STORM graph model.

2-D table representation changes some important semantic concepts. For example, *classification hierarchies* are represented in the same manner as multidimensional categories, while actually the classification hierarchy represents a single dimension. Instead, there is a fundamental difference between category attribute structure and multidimensionality. A *category attribute* represents a dimension of the aggregate data structure or an element (level) of the classification hierarchy which forms the dimension mentioned above. Usually, only the low-level elements of the classification relationship participate in the multidimensional space. Moreover, by necessity, more than one dimension must be represented by rows and columns if more than two dimensions exist in the dataset. This is accomplished by selecting an arbitrary order of the dimensions for rows and columns, as mentioned above. Again, the label of each multidimensional aggregate data structure (i.e., the title of the 2-D table) represents the summary measure for this dataset and can express other dimensions implicitly. For example, "Production of vegetables in California in 1995" says that this dataset has two additional dimensions: "state" and "year," where the instance values selected are the singletons "California" and

"1995." In reality, the dimension names should really be "State of vegetable production" and "Year of vegetable production," because the phenomenon studied is "vegetable production." In this case, as we will see in the following, a primitive domain of the dimension (and of the different descriptive attributes in a classification hierarchy of a dimension) is defined as a common reference of the primary concept (in this case, "state" and "year"). Indeed, dataset may be only one "page" of a collection of pages, each representing another state and/or another year. There is also a mixing between categories and category instances. Depending on their use by the user, the same name can be used in both cases. For example, consider the classification hierarchy "professional category"-"profession" of the table "Employment in California" shown in Figure 1, and suppose that the instance domains have, respectively, <engineer, secretary, teacher> for professional category, and <chemical, civil> for "engineer" profession, <junior, executive> for "secretary" profession, and <elementary, intermediate> for "teacher" profession. In such cases, the category attributes and their values do not fit on a page or a screen, so that the representation spreads into multiple pages or screens. This confuses the global understanding of the multidimensional aggregate data. The problem is that the metadata on the multidimensional aggregate data has no separate representation. Furthermore, there should be a separation between structural metadata (the meta-schema) and instance values (the meta-database).

Different graph models have been proposed in the literature: SUBJECT, described in Chan & Shoshani (1981); GRASS, described in Rafanelli & Ricci (1983); SAM*, described in Su (1983), in which there are seven types of associations, each with its own properties (i.e., structure, operations, and constraints), a G-relation

Figure 8: Spaces and Levels of ADAMO

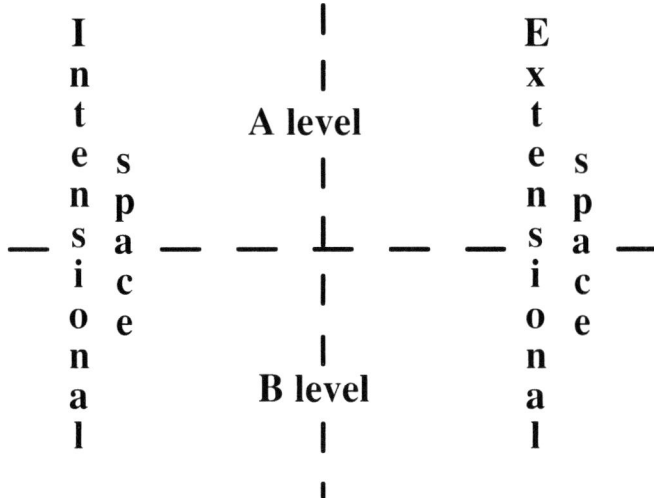

and an algebra are formally defined; STORM, described in Rafanelli & Shoshani (1990); and ADAMO, described in Bezenchek, Rafanelli, & Tininini (1996a) and in Tininini, Bezenchek, & Rafanelli (1996). The distinctive feature of ADAMO in comparison to the other models for aggregate data is in general a major adherence to the aggregation process, and particularly the capability of:

- representing and distinguishing the implicit and explicit category attributes as well as the complex MAD;
- expressing, by simple and intuitive graphical notations, the summarizability characteristics of a category attribute and the marginal values of a MAD;
- expressing the hierarchies in the dimensions of a MAD along with the corresponding summarizability characteristics and marginal values;
- expressing the identification dependencies in the hierarchies of a MAD;
- representing the composite MAD in a compact form by introducing the alias nodes;
- handling the problem of integration among data supplied by different information sources.

The model utilizes two *representation spaces,* called, respectively, *intentional* and *extensional space*. Orthogonally to these spaces, two distinct *representation levels* are defined, called, respectively, *A* (for Aggregate data) and *B* (for Base concept) *level*, as shown in Figure 8.

The intentional side of the A level allows a conceptual description of the MAD

Figure 9: Graphical Representation of a MAD in ADAMO

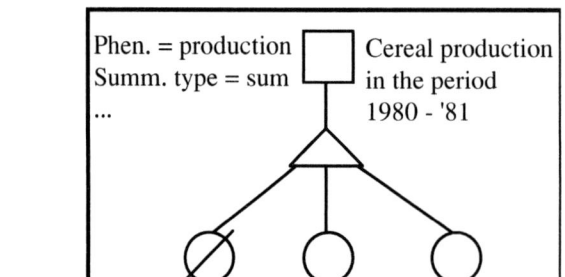

by means of tree-structures, whereas in the extensional side both the category attribute instances and the summary values are listed. The B level contains *primitive concepts,* namely *primitive circle nodes, primitive phenomena,* and *primitive instances* to which all the circle nodes, square nodes, and instances of the A level are linked. The B level allows two concepts sharing the same name as the A level to be distinguished when they are linked to distinct primitive concepts. Also, two

Figure 10: Graphical Notations of Different Summarizability Characteristics

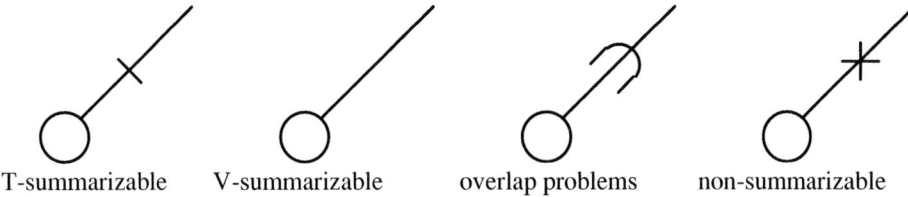

T-summarizable   V-summarizable   overlap problems   non-summarizable

concepts with different names at the A level (synonyms) are unified when they are both linked to the same primitive concept. In the A level a conceptual description of the MAD is given. The main advantage of the graphical approach is that it allows "at a glance" understanding of even very semantically complex multidimensional aggregate data. The conceptual description of a MAD is achieved by means of a tree-structure with variously shaped nodes, each of them in strict correspondence with one of the following concepts:

(1)   a *simple circle node* (O) represents a classification set (i.e., an explicit category attribute);
(2)   a *slashed circle node* (Ø) represents a union set (i.e., an implicit category attribute);
(3)   a *triangle node* (Δ) represents a cross-intersection among its child-nodes (note that the associativity of the operator Ä allows a triangle node to have more than two child-nodes);
(4)   a *butterfly node* (⋈) represents the union set among its child-nodes;
(5)   a *square node* (□) represents the mapping f between the aggregation relations and the corresponding summary values, and also several other properties regarding the MAD such as the described fact, the aggregate function type, the summary type, the information source, the measure unit, and the counting unit.

For example, let us reconsider the table of Figure 2. The corresponding graphical representation by the ADAMO model is shown in Figure 9.

ADAMO provides four graphical notations, shown in Figure 10, to express the

Figure 11: Table Example

Population in Italy in 1991	F	M	Total
0 - 18 years	3,845,000	3,325,000	(1)
19 - 65 years	22,367,000	21,856,000	(1)
> 65 years	3,534,000	3,027,000	(1)
Total	(2)	(2)	(3)

Figure 12: Graphical Representation of Marginals and Implicit Dimensions

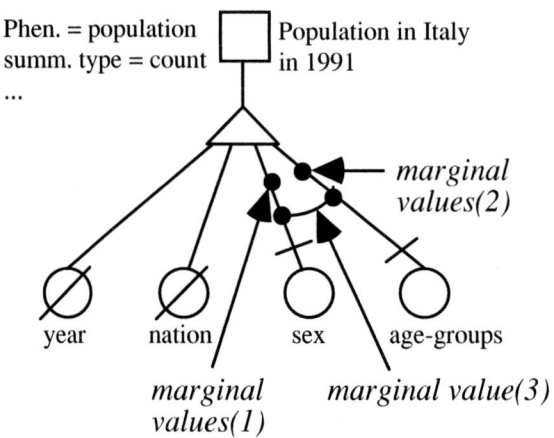

summarizability characteristics of each MAD, namely to express that the category attribute A is (i) T-summarizable (totally summarizable); (ii) I-summarizable (Implicitly summarizable); (iii) non-summarizable due to *overlaps* among its base sets; and (iv) non-summarizable because the operation is meaningless.

A MAD with marginal values can always be represented by a complex tree-structure with (at least) one *union* node. However, a concise notation has been introduced that allows the user to represent the marginal values by means of the structure corresponding to the non-marginal summary values. Let us consider for example the table outlined in Figure 11.

Figure 13: Graphical Representation of the ID-Dependency

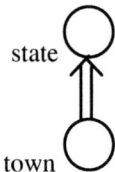

Figure 14: Composite Table

Population and Average Income in 1993	Population		Avg
	F	M	Income
California	val_1	val_2	val_7
Texas	val_3	val_4	val_8
Nevada	val_5	val_6	val_9

This MAD is conceptually represented by the ADAMO tree-structure in Figure 12. Each set of marginal values conceptually obtainable by summarizing a single category attribute is represented by a little black circle on the edge above the correspondent circle node.

The set of marginal values obtainable by summarizing two or more category attributes is represented by an equal number of little black circles connected by an edge.

It often happens that the category attribute instances in a MAD are not uniquely defined due to *homonymy*. For example, in an aggregation hierarchy "state → town," if the domain of state consists of the values New York, Wyoming, Texas, California, Arizona, Nevada, Oregon, and the domain of town includes the value Buffalo, this value is not uniquely defined because a city with such a name exists both in New York and in Wyoming. On the other hand, the pair of instances (state, town) is sufficient to univocally identify each instance of town, since it is known that no two towns with the same name exist in the same state. We say that between the category attributes town and state, an *identification dependency* exists. This is a very important piece of information, since it can greatly simplify the process of linking between the instances of the A level and the corresponding primitive instances of the B level. Figure 3 shows the ADAMO graphical representation of an identification (ID-) dependency (see Rafanelli & Shoshani, 1990).

We have already considered the use of butterfly nodes to define complex MADs. The introduction of a butterfly node is also required when a single MAD collects data obtained by two or more different aggregation processes, that is when it represents a *composite* MAD. The corresponding aggregation functions and/or facts are obviously distinct, and each needs to be represented by a distinct square

Figure 15: The Graphical Representation of an Alias Node

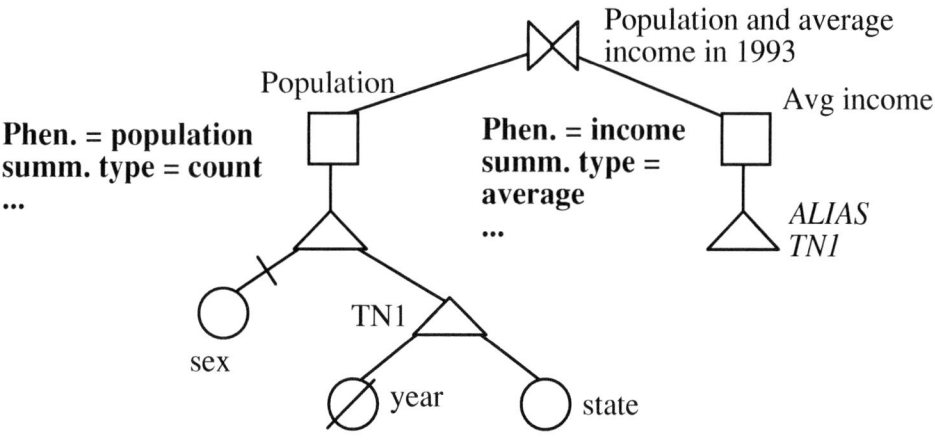

Figure 16: Different Total Levels

Population in 1981		F	M
California	Los Angeles	val_1	val_2
	San Diego	val_3	val_4
	San Francisco	val_5	val_6
	Total towns	val_7	val_8
	Total California	val_9	val_10
...	...	...	...

node. Several square nodes are linked to the same upper butterfly node, opportunely labeled.

Let us consider for example the table of Figure 14, where both the population (classified by state and sex) and the average income (classified by state) are displayed.

The ADAMO representation is shown in Figure 15.

Note that in this case the node ⋈ represents the *union of the underlying structures* (MADs), rather than the union of aggregation sets. Another important observation refers to the *alias node*: an alias node is a shorthand and represents a reference to the entire structure, having as its root the *original node*, or simply to the entire domain of the original node, if it is a circle node.

Alias references are allowed only within the same composite MAD. Butterfly nodes, aliases, and marginals allow a very compact representation of the MAD (in Figure 16) to be obtained. This table fragment is taken from Bezenchek, Rafanelli, & Tininini (1996a) and represents population data in some U.S. towns. In this fragment the towns are grouped by state and two subtotals are also displayed, one representing the aggregate population of the listed towns, and the other representing the whole population of the state. The corresponding ADAMO tree-structure is that of Figure 17. The circle node town refers to the first term in brackets; the marginal symbol refers to the second term (as it represents the summarization of town, that is its implicitation with the explicitation of state), and the circle node *ALIAS state* refers to the last term.

As stressed above, state is a virtual node needed to express both the fact that the towns are grouped by state, and also that the marginal values corresponding to the summarization of town are directly available in exact form.

Notice that, since state is needed both in virtual and in explicit mode (to express the entire state values), the ALIAS mechanism allows there to be two state circle nodes (one in the hierarchy and the other in leaf-position), again to define the instances of the corresponding attribute only once.

Figure 17: Graphical Representation of Butterfly Node with Alias and Marginals

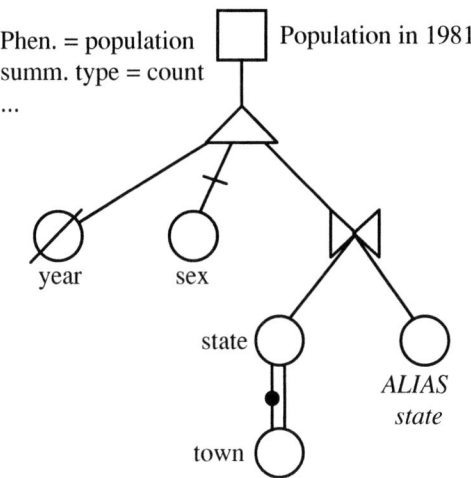

Tabular Models

Tabular models are used both in SDB and OLAP. A 2D statistical table (as in Figure 4), with the *adding* of columns and rows that show summarization over rows and columns respectively, is an example of these models. These totals are often referred to by statisticians as "marginals" (because they appear on the "margins" of tables). The marginals are usually not included in the database if they can be derived. However, in situations where summarizability does not hold, it is necessary to store these values in the database as well. It is generally not efficient to compute the marginals for very large datasets. This is one of the most important problems addressed by OLAP research and it will be discussed in Chapter 9.

Another tabular representation stems from the popularity of the relational model. The advantage of this representation is its familiarity and the fact that it can be readily implemented on a relational system, as well as the possibility of integrating it with other data in the same model.

However, there are several problems with this representation:
i)   There are no semantics associated with a relational table to distinguish between category and summary attributes.
ii)   There is no distinction between the attributes associated with the classification hierarchy and the dimensions.
iii) The repetition of values in the columns relative to the descriptive attributes make it difficult to see the categories of each of these category attributes.

Furthermore, if the relation is implemented in a straightforward way, this layout is very wasteful of space since it stores the entire cross product. Consequently, summarization and query processing is slowed down.

Figure 18: Example of Star Model Representation

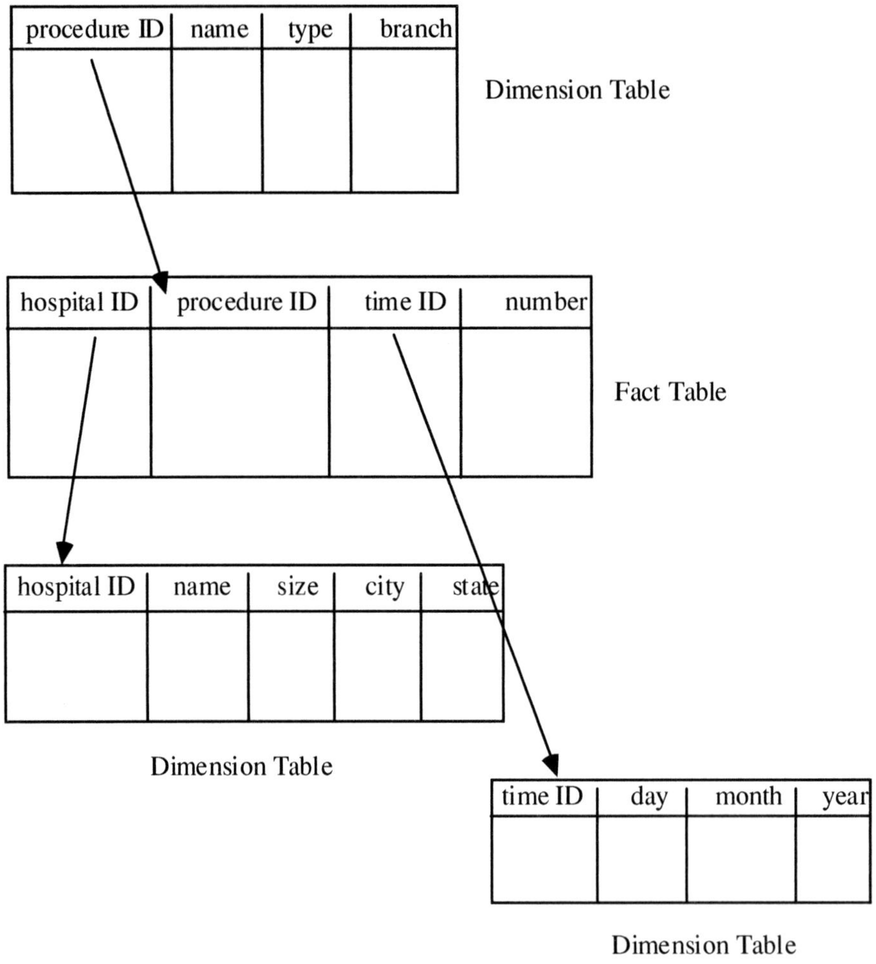

Another issue for using this model is how to represent marginals. In Gray et al. (1996), the authors suggest using the reserved keyword value "ALL" to achieve this purpose. However, one would have to know which of the columns represent the dimension and which the summary attribute, since "ALL" cannot be applied to summary attributes. In addition, this solution does not deal with classification structures which exist in other relations if normalization is applied. With this construct, the authors have defined a "data cube" operator, which we will discuss in Chapter 5.

A relational representation that partially overcomes the above problems is used by an OLAP company (see MicroStrategy). An example of this representation is shown in Figure 20 for a hospital and patient procedures database (the figure is taken from Shoshani, 1997). They call it the "star model," since it has one relation that represents multidimensional space in the center and the other relations that represent dimensions around it.

As can be seen from Figure 18, this is an attempt to introduce the semantics of a MAD in the context of the relational model. By labeling relations as the "fact table" and "dimension tables," they make a distinction between summary data and the data associated with the dimensions. By placing the "fact table" visually at the center of the "star," they convey the multidimensional nature of this database. Furthermore, each dimension table can hold the category attributes of the classification structure for that dimension. For example, the "hospital" dimension table has attributes "city" and "state" that make up the classification structure. Many of the difficulties described previously still exist. Specifically:

i)   The "fact table" still requires the storage of the entire cross-product.

ii)   It is not possible to know whether an attribute in the dimension tables is a "category attribute" (such as "type" in the "procedure" table) or a regular *descriptive attribute* (such as "name" in the "procedure" table).

iii) It is not possible to structurally assess the hierarchical relationship of the attributes that make up the classification structure. For example, there is no structural information on whether "procedures" group into "types" and "type" into "branches," or whether "procedures" group into "types" and also "branches" independently of "types."

The data cube model is the most recent graphical representation of multidimensional data, and it is used in an OLAP environment. This representation is intended to be used only to convey concepts, but it cannot be used for real databases, for which one needs to resort to other representations such as the graphical or tabular representation described above. While the multidimensional space is clearly represented in this data cube model, the structure of the dimensions is not well presented. In spite of the above deficiencies, the data cube concept is useful in visualizing certain operations, such as "slice" and "dice."

In conclusion, we give a set of definitions regarding the OLAP terminology taken from The OLAP Council (1995):

Aggregate (or Consolidate): Multidimensional databases generally have hierarchies or formula-based relationships of data within each dimension. Consolidation involves computing all of these data relationships for one or more dimensions. Such relationships are normally summations, but any type of computation relationship or formula might be defined.

Multidimensional Array (cube): This is a group of data cells arranged by the dimensions of the data. An example of a two-dimensional array is a spreadsheet, with the data cells arranged in rows and columns, each being a dimension. An example of a three-dimensional array is a cube, a physical metaphor of a multidimensional visualization of the data. Each dimension of the cube forms a side. A higher dimensional array has no physical metaphor, but they organize the data in the same way that users think of their enterprise.

Calculated Member: This is a member of a dimension whose value is determined by other members' values (e.g., by the application of a mathematical or logical operation.). It is any member that is not an input member.

Cell: It is a single datapoint that occurs at the intersection defined by selecting one member from each dimension in a multidimensional array.

Children: They are the members of a dimension that are included in a calculation to produce a consolidated total for a parent member. Children may themselves be consolidated levels, which require that they have children. A member may be a child for more than one parent, and a child's multiple parents may not necessarily be at the same hierarchical level, thereby allowing complex, multiple hierarchical aggregations within any dimension.

Derived Data: Derived data is produced by applying calculations to input data at the time the request for that data is made, i.e., the data has not been pre-computed and stored on the database. The purpose of using them is to save storage space and calculation time, particularly for calculated data that may be infrequently called for or that is susceptible to a high degree of interactive personalization by the user.

Detail Member: A detail member of a dimension is the lowest level number in its hierarchy.

Dimension: This is a structural attribute of a cube that is a list of members, all of which are of a similar type in the user's perception of the data. A dimension acts as an index for identifying values within a multidimensional array. If one member of the dimension is selected, the remaining dimensions in which a range of members (or all members) are selected define a sub-cube. If all but two dimensions have a single member selected, the remaining two dimensions define a spreadsheet (or a "slice" or a "page"). If all dimensions have a single member selected, then a single cell is defined.

Drill-Down/Up: Drilling-down or up is a specific analytical technique whereby the user navigates among levels of data ranging from the most summarized (up) to the most detailed (down). The drilling paths may be defined by the hierarchies within dimensions or other relationships that may be dynamic within or between dimensions.

Formula: A formula is a database object, which is a calculation, rule, or other expression for manipulating the data within a multidimensional database Formulae define relationships among members. They are used by OLAP database builders to provide great richness of content to the server database. Formulae are used by end-users to model enterprise relationships. They are also used to personalize the data for greater visualization and insight.

Hierarchical Generation: Two members of a hierarchy have the same generation if they have the same number of ancestors leading to the top. Notice that the term generation and level are both necessary to describe sub-groups of dimension members, since, for example, although two siblings share the same parent and are, therefore, of the same generation, they won't be from the same level if one of the siblings has a child and the other doesn't.

Hierarchical Level: Members of a dimension with hierarchies are at the same level if, within their hierarchy, they have the same maximum number of descendants in any single path below.

Hierarchical Relationships: Any dimension's members may be organized based on parent-child relationships, typically where a parent member represents the consolidation of the members which are its children. The result is a hierarchy, and the parent-child relationships are hierarchical relationships.

Dimension Member: A dimension member is a discrete name or identifier used to identify a data item's position and description within a dimension.

Member Combination: A member combination is an exact description of a unique cell in a multidimensional array, consisting of a specific member selection in each dimension of the array.

Missing Data (Value): This is a special data item which indicates that the data in this cell does not exist. This may be because the member combination is not meaningful, or has never been entered. It is similar to a null value, but is not the same as a zero value.

Navigation: This is a term used to describe the processes employed by users to explore a cube interactively by drilling, rotating, and screening, usually using a graphical OLAP client connected to an OLAP server.

Page Dimension: This is generally used to describe a dimension which is not one of the two dimensions of the page being displayed, but for which a member has been selected to define the specific page requested for display. All page dimensions must have a specific member chosen in order to define the appropriate page for display.

Page Display: The page display is the current orientation for viewing a multidimensional slice. The horizontal dimension(s) run across the display, defining the column dimension(s). The vertical dimension(s) run down the display, defining the contents of the row dimension(s). The page dimension-member selections define which page is currently displayed. A page is much like a spreadsheet, and may in fact have been delivered to a spreadsheet product where each cell can be further modified by the user.

Parent: This is the member that is one level up in a hierarchy from another member. The parent value is usually a consolidation of all of its children's values.

Pivot (Rotate): This represents the changing of the dimensional orientation of a report or page display.

Reach Through: Reach through is a means of extending the data accessible to the end user beyond that which is stored in the OLAP server. A reach through is performed when the OLAP server recognizes that it needs additional data, and automatically queries and retrieves the data from a data warehouse or OLTP system.

Scoping: This term means to restrict the view of database objects to a specified subset. Further operations, such as update or retrieve, will affect only the cells in the specified subset.

Selection: This is a process whereby a criterion is evaluated against the data of members of a dimension in order to restrict the set of data retrieved.

Slice: A slice is a subset of a multidimensional array corresponding to a single value for one or more members of the dimensions not in the subset. From an end-user perspective, the term slice most often refers to a two-dimensional page selected from a cube.

REFERENCES

2nd SDBM. (1983). *Proceedings of the 2nd International LBL Workshop on Statistical Database Management,* Los Altos, California, September 27-29.

Agarwal, S., Agrawal, R., Deshpande, P.M., Gupta, A., Naughton, J.F., Ramakrishnan, R., & Sarawagi, S. (1996). On the computation of multidimensional aggregates. *Proceedings of the 22nd International Conference on Very Large Databases (22nd VLDB)*, Mumbai (Bombay), India,. 506-521.

Agrawal, R., Gupta, A., & Sarawagi, S. (1997). Modeling multidimensional databases. *Proceedings of the 13th International Conference on Data Engineering.* Birmingham, UK. IEEE Computer Society Press, 232-243.

ANSI/X3/SPARC Report. (1975). Study Group on Data Base Management Systems' Interim Report 75-02-08, EDT-Bulletin. *ACM Sigmod*, 7(2).

Bell, J. (1983). Data structures for scientific simulation programs. *2nd International Workshop on Statistical Database Management*, Los Gatos, California, 196-201.

Bezenchek, A., Rafanelli, M., & Tininini, L. (1996a). ADAMO: A conceptual model for the graphical definition and manipulation of aggregate data. *Technical Report IASI*, No. 440, September.

Bezenchek, A., Rafanelli, M., & Tininini, L. (1996b). A data structure for representing aggregate data. *Proceedings of the 8th International Conference on Scientific and Statistical Database Management (8th SSDBM)*, Stockholm, Sweden. Los Alamitos, CA: IEEE Computer Society Press, 22-31.

Chan, P., & Shoshani, A. (1981). SUBJECT: A directory-driven system for organizing and accessing large statistical databases. *Proceedings of the 7th International Conference on Very Large Data Bases*, Cannes, France, 553-563.

Codd, E.F. (1993). Providing OLAP (On-Line Analytical Processing) to user-analysts: An IT mandate. *Technical Report,* E.F. Codd and Associates.

Codd, E.F., Codd, S.B., & Salley, C.T. (1993). Beyond decision support. *Computerworld*, 27(30).

Fortunato, E., Rafanelli, M., Ricci, F.L., & Sebastio, A. (1986). An algebra for statistical data. *Proceedings of the 3rd International Workshop on Statistical and Scientific Database Management (3rd SSDBM),* Luxembourg, 122-134.

Ghosh, S.P. (1986). Statistical relational tables for statistical database management. *IEEE Transactions on Software Engineering,* 12(12), 1106-1116.

Gray, J., Chaudhuri, S., Bosworth, A., Layman, A., Reichart, D., Venkatrao, M., Pellow, F., & Pirahesh, H. (1997). Data cube: A relational aggregation operator generalizing group-by, cross-tab and sub-totals. *Journal of Data Mining and Knowledge Discovery,* 1(1), 29-54.

Gupta, H., Harinarayan, V., Rajaraman, A., & Ullman, J.D. (1997). Index selection for OLAP. *Proceedings of the 13th International Conference on Data Engineering,* Birmingham, UK. IEEE Computer Society Press, 208-219.

Gyssens, M., & Lakshmanan, L.V.S. (1997). A foundation for multi-dimensional databases. *Proceedings of the 23rd International Conference on Very Large Data Bases (VLDB'97),* Athens, Greece, August 25-29, 106-115.

Harinarayan, V., Rajaraman, A., & Ullman, J.D. (1996). Implementing data cube efficiently. *Proceedings of the International Conference on Management of Data (SIGMOD'99),* Montreal, Canada, 205-216.

Klug, A. (1982). Equivalence of relational algebra and relational calculus query languages having aggregate functions. *Journal of the ACM,* 29(3).

Lehner, W. (1998). Modeling large-scale OLAP scenarios. *Proceedings of the 6th International Conference on Extending Database Technology,* (EDBT'98), Valencia, Spain. Lecture Notes in Computer Science, No. 1377. Springer-Verlag, 153-167.

Malvestuto, F.M. (1993). A universal-scheme approach to statistical databases containing homogeneous summary tables. *ACM Transactions on Database Systems,* 18(4), 678-708.

Malvestuto, F.M., Moscarini, M., & Rafanelli, M. (1991). Suppressing marginal cells to protect sensitive information in a two-dimensional statistical table. *Proceedings of the 10th ACM Symposium on Principles of Database Systems (PODS'91),* Denver, Colorado. ACM Press, 252-258.

McCarthy, J.L. (1982). Metadata management for large statistical databases. *Proceedings of the 8th International Conference on Very Large Data Bases (VLDB'82),* Mexico City, Mexico, 234-243.

MicroStrategy. *The Case for Relational OLAP, A White Paper.* Prepared by MicroStrategy, Incorporated, http://www.strategy.com/dwf/wp_b_al.htm.

Nguyen, T.B., Tjoa, A.M., & Wagner, R. (2000). Conceptual multidimensional data model based on metacube. *Proceedings of the 1st International Conference on Advances in Information Systems (ADVIS'00),* Izmir, Turkey. Lecture Notes in Computer Science, No. 1909. Berlin Heidelberg, Germany: Springer-Verlag, 24-33.

OLAP Council. (1995). *OLAP and OLAP Server Definitions: OLAP Glossary.* http://www.olapcouncil.org/research/glossaryly.htm.

Ozsoyoglu, G., & Ozsoyoglu, Z.M. (1983). An extension of relational algebra for summary tables. *Proceedings of the 2nd International Workshop on Statistical Database Management (SDBM'83)*, Los Gatos, California, 202-211.

Ozsoyoglu, G., Ozsoyoglu, Z.M., & Mata, F. (1985). Extending Relational Algebra and Relational Calculus with Set-Valued Attributes and Aggregate Functions. *ACM Transactions on Database Systems, 12*(4), December, pp. 566-592.

Ozsoyoglu, G., Ozsoyoglu, M., & Matos, F. (1987). Ext. Rel. Algebra and Rel. with Set-Valued Attributes and Agg. Functions, *ACM Trans. on Database Systems,* 12(4), December, 566-592.

Pedersen, T.B., & Jensen, C.S. (1999). Multidimensional data modeling for complex data. *Proceedings of the 15th International Conference on Data Engineering (ICDE'99)*, Sydney, Australia. Los Alamitos, CA: IEEE Computer Society Press, 336-345.

Pedersen, T.B., Jensen, C.S., & Dyreson, C.E. (2001). A foundation for capturing and querying complex multidimensional data. *Journal of Information Systems*, 26(5), 383-423.

Rafanelli, M. (1990). Statistical and scientific database management systems. In Kent, A., & Williams, J.G. (Eds.), *Encyclopedia of Computer Science and Technology*, 23(Suppl.8), 369-409. Dekker Inc. Publication, Pittsburgh University, Pittsburgh, Pennsylvania.

Rafanelli, M., & Ricci, F.L. (1983). Proposal of a logical model for statistical databases. *Proceedings of the 2nd International Workshop on Statistical Databases*, (SDB'83), Los Altos, California, 264-272.

Rafanelli, M., & Ricci, F.L. (1993). Mefisto: A functional model for statistical entities. *IEEE Transactions on Knowledge and Data Engineering*, 5(4), 670-681.

Rafanelli, M., & Shoshani, A. (1990). STORM: A statistical object representation model. *Proceedings of the 5th International Conference on Statistical and Scientific Database Management (5th SSDBM)*, Charlotte, North Carolina. Lecture Notes in Computer Science, No. 420. Berlin Heidelberg, Germany: Springer-Verlag, 14-29. Also, (1990). *Bulletin of the IEEE Computer Society Technology Committee on Data Engineering*, 13(3), 12-18.

Sato, H. (1991). Statistical data models: From a statistical table to a conceptual approach. In Michalewicz, Z. (Ed.), *Statistical and Scientific Databases*. Horwood, 167-199.

Shoshani, A. (1982). Statistical databases: Characteristics, problems and solutions. *Proceedings of the 8th International Conference on Very Large Data Bases (VLDB'82)*, Mexico City, Mexico, 208-222.

Shoshani, A. (1997). OLAP and statistical databases: Similarities and differences. *Proceedings of the 16th ACM Symposium on Principles of Database Systems (PODS'97),* Tucson, Arizona, 185-196.

Shoshani, A., & Wong, H.K.T. (1985). Statistical and scientific database issues. *IEEE Transactions on Software Engineering,* 11(10), 1040-1047.

Shukla, A., Deshpande, P., Naughton, J.F., & Ramasamy, K. (1996). Storage estimation for multidimensional aggregates in the presence of hierarchies. *Proceedings of 22nd International Conference on Very Large Data Bases (VLDB'96),* Mumbai (Bombay), India, 522-531.

Smith, J.M., & Smith, D.C.P. (1978). Database abstractions: Aggregation and generalization. *ACM Transaction on Database Systems,* 2(2), 105-133.

Su, S.Y.W. (1983). SAM*: A semantic association model for corporate and scientific/ statistical databases. *Journal of Information Sciences,* 29(2&3), 151-199.

Tininini, L., Bezenchek, A., & Rafanelli, M. (1996). A system for the management of aggregate data. *Proceedings of the 7th International Conference on Database and Expert Systems Applications (DEXA'96),* Zurich, Switzerland. Lecture Notes in Computer Sciences, No. 1134. Springer-Verlag, 531-543.

Vassiliou, Y. (1980). Functional dependencies and incomplete information. *Proceedings of the 6th International Conference on Very Large Data Bases (VLDB'80),* Montreal, Canada, 260-269.

Wong, H.K.T. (1984). Micro and macro statistical/scientific database management. *Proceedings of the 1st International Conference on Data Engineering (ICDE'84),* Los Angeles, California, 104-106.

Chapter II

Multidimensionality in Statistical, OLAP, and Scientific Databases

Arie Shoshani
Lawrence Berkeley National Laboratory, USA

ABSTRACT

The term "multidimensional databases" refers to data that can be viewed conceptually in a multidimensional space, where each dimension represents some attributes of the data. Viewing data in this form is natural for many applications, yet the concepts are not treated in a uniform way in the database literature. In this chapter, we show the commonality of concepts between three database areas: statistical, OLAP, and scientific databases. We show that these domains have two main structural concepts: the cross-product space of the dimensions, and the classification hierarchy structure associated with each dimension. In the first part of this chapter we describe how these structures are sed to represent data in statistical and OLAP databases and how summarization operators can be applied to them. Further, we discuss how these structures can be extended to represent related information using federated database concepts. In the second part of the chapter we show that these concepts are common to many scientific database applications. In particular, we discuss the importance of supporting classification structures and the difficulty in representing them as tables in relational databases. We also discuss data structures to support multidimensional databases, emphasizing space-time representation, clustering in multidimensional space, indexing in multidimensional space, and supporting classification structures. We conclude by arguing that the concepts of multidimensionality and classification structures as well as the operation over them should be elevated to "first class" object types. These object types should be visible by the application user explicitly in the conceptual schemas as well as exposing them in the user interfaces.

INTRODUCTION AND BACKGROUND

There is a lot of data that can be viewed as multidimensional data. The term multidimensional databases typically refers to a collection of objects, each represented as a point in a multidimensional space. Even data that is represented in a tabular form, such as relations, can be thought of as multidimensional data, if each row (tuple) is thought of as an object, and the columns (attributes) are thought of as the dimensions. For example, consider the following table: employee (personID, age, sex, salary) shown in Figure 1a. If each person is represented as a point in the multidimensional space of (age, sex, salary), then that table can be represented as in Figure 1b.

The utility of representing data in the multidimensional space is that it is more natural to view certain features of the data in this way. For example, it is natural to view clusters in the multidimensional space. In Figure 1b, one can easily see that there is a small cluster of highly paid people (perhaps representing managers who are generally older) and a larger cluster of lower paid people. We can also see "outliers" as is the case with the younger person with a high salary. Of course, these concepts extends to data in more than three dimensions, but cannot be viewed as easily. The

Figure 1a: An "Employee" Table

personID	age	sex	salary
1234	28	F	110,000
2345	56	F	150,000
3456	27	M	60,000
4567	30	M	70,000
5678	61	M	130,000
6789	28	F	80,000
7890	25	M	50,000
8901	22	M	65,000
9012	34	F	70,000
.	.	.	.
.	.	.	.
.	.	.	.

Figure 1b: A 3-D View of the Table

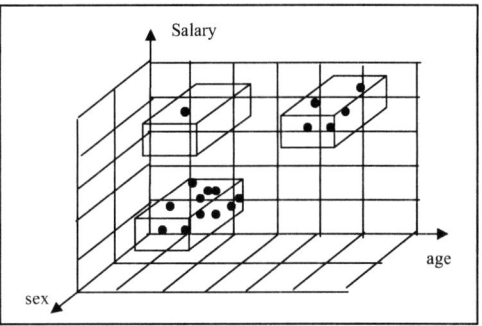

problem of viewing high-dimensional data to identify clusters, outliers, and various patterns has been the subject of several research projects. An extensive review of such methods is provided in Keim & Kriegel (1996) and will not be discussed further here.

Some data is naturally multidimensional such as two-dimensional or three-dimensional spatial data. For example, climate modelers prefer to view their observed or simulated data in a multidimensional structure representing space (two or three dimensions), time, and variables being measured (temperature, wind velocity, etc.) In this case, certain operations, such a selecting spatial regions or performing the operation of "monthly means" on the data, are very common and need to be supported.

Another reason for viewing data in the multidimensional space is summarization. This need is most obvious in databases that represent statistical data or in databases used for decision support. These are referred to as "Statistical Databases" and "On-Line Analytical Processing" (OLAP), respectively. In the OLAP literature the multidimensional space is referred to as a "cube," and by selecting sub-ranges or summarizing over sub-ranges of the multidimensional space, one generates "sub-cubes."

In general, one can summarize over an entire dimension or over a region of the dimension. To illustrate a summarization over an entire dimension, consider again the database of Figure 1. One can summarize (using the operation COUNT) over the dimension "sex" to produce the lower dimensional database: "number of employees by age by salary." This is shown in Figure 2a. Note that this summarization produced a new "summary measure." Each point in this 2-D space now represents the measure: "number_of_employees." This is typical of statistical databases where the base data, called "microdata," is summarized to form "macrodata." When only part of a range is selected or summarized, the dimensionality of the product does not

Figure 2a: A Summary Database in 2D Space

Figure 2b: The Database Further Summarized on Each Dimension

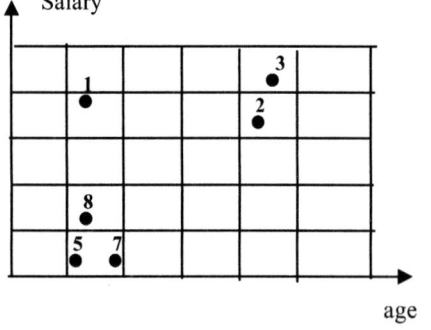

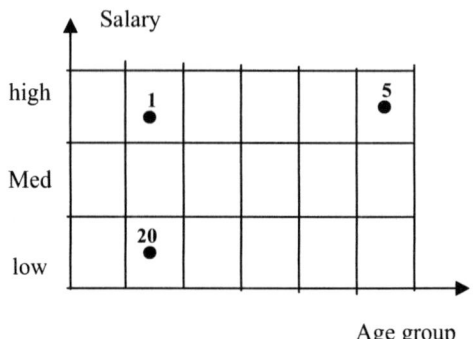

change. For example, selecting only lower paid younger people in the example of Figure 1 still produces a three-dimensional sub-cube.

Another aspect of statistical and OLAP databases is that each dimension can have a category hierarchy associated with it. For example, "age" can be organized as "age groups" of 1-10, 11-20, etc., and "salary" can be organized as "salary level" of "low," "medium," and "high." In this case, summarization can take place over any one of the dimensions. This action does not reduce the dimensionality of the "cube." Figure 2b shows the result of summarization of the dimensions of Figure 2a. Dimension hierarchies can get fairly complex depending on the type of dimension. For example if one of the dimensions is "products" sold in a department store, then it can have a large number of levels in the category hierarchy.

In a previous paper, we identified multidimensionality as a common aspect of both scientific and statistical databases (Shoshani & Wong, 1985). In this document, we elaborate on the concepts multidimensionality as well as category hierarchies, and discuss how they are used in summary and scientific databases in the next two sub-sections.

SUMMARY DATABASES

Because both statistical databases and OLAP databases are mainly designed for summarization operations, we refer to both with the generic term "summary databases." The concept of "statistical databases" was introduced in the 1980's (Chan & Shoshani, 1981; Shoshani, 1982) and was followed by much activity in this area. The requirements of OLAP were introduced in a white paper by Codd & Associates (1993). This was followed by a paper on the Data Cube (Grey et al., 1996), which is an extension to the relational model to support OLAP databases. It was not apparent initially that both the statistical database and OLAP areas are addressing a similar problem. However, in the article (Shoshani, 1997), we compared statistical databases and OLAP databases in terms of their data models, data structures, and operators. We found the fundamental concepts to be similar, but the work performed in these two domains tended to emphasize different aspects. While much of the work in the statistical database area emphasized conceptual modeling and formalization of operators, the OLAP area emphasized data structures and operational efficiency. However, more recently, more attention is paid to the formal definition of the multidimensional aspects of OLAP databases and their query language (see, for example, Tsois, Karayannidis, & Sellis, 2001; Cabibbo & Torlone, 1997). In this section, we discuss some of the main concepts of summary databases, showing parallels of the statistical database terminology to those of OLAP.

Conceptual Modeling

As mentioned in the introduction, there are two structural features to summary databases: 1) the multidimensional space, and 2) the category hierarchy associated with each dimension. Each dimension (such as "sex" or "age") has a set of categorical values (or categories) associated with it (such as the "sex" categories: "male," "female," and the "age" categories: 1, 2, ..., 100). The multidimensional space is simply a cross-product of dimension categories. The categories of a dimension can be further grouped into a category hierarchy (such as grouping "age" into "age groups" of "1-10, ..., 91-100).

Initially, each point in the multidimensional space is associated with some object identifier (such as "personID"), as was shown in Figure 1a. As mentioned above, this is referred to as the "base data" or the "microdata." Summary databases are constructed/derived from the microdata by applying a summary operator (such as "count," "sum," "average," etc.) to produce a "summary measure." This operation transforms a "micro database" into a "summary database."

Consider again the example of Figure 1, where we add "projectID" to the employee table. In the customary "relational" notation, this will be represented as: [Employee (personID, age, sex, projectID, salary)]. Note that this notation alone does not represent the fact that there are functional dependencies between Person_ID and each of the attributes age, sex, projectID, and salary. If we use ":" to denote "functional dependency," this may be represented as [personID: age, sex, projectID, salary]. Constructing/deriving a summary database from this micro database amounts to creating a new "summary measure" (or "variable") on the multidimensional space. For example, we can choose to derive a summary database with a summary measure of "average-salary" associated with the multidimensional space (age, sex, projectID). We may use the notation for that as [(age, sex, projectID): average-salary]. That is, there is a functional dependency from the multidimensional space defined by (age, sex, projectID) to average-salary. Thus, every point in this multidimensional space has an average-salary associated with it.

Of course, the summary database can be represented as a table (a relation), such as [salary-info (age, sex, projectID, average-salary)], but the semantics are then lost. The main concept of summary databases is to capture the multidimensional space and the dependency of the summary measure on that space. Further, the concept of a category hierarchy needs to be explicitly captured. This can be done by a "→" notation, such as [age → age-group] or [city → state → region]. For the example above we might group "projectID" into "project-type." Our summary database would then be represented as: [(age → age-group, sex, projectID → project-type): average-salary].

The above notation captures the two main structural concepts as well as the summary measure. In the area of statistical databases, a graphical notation was used to show these concepts explicitly (Shoshani, 1982). Figure 3a represents the example above in that graphical notation, where the "S" node represents a "summary measure," the "X" node represents a "cross-product" of the dimensions underneath it, and the "C" nodes represent category classes. The categories of a "C" node may be further grouped into categories of the "C" node below it. Another graphical

Figure 3a: A Graphical Representation of a Summary Database

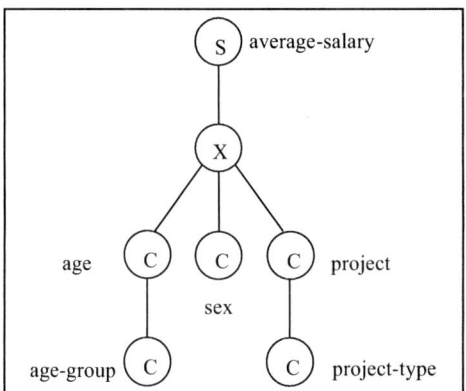

Figure 3b: An Inverted Graphical Representation Using UML Notation

Figure 3c: A Star Schema Representation of the Summary Database

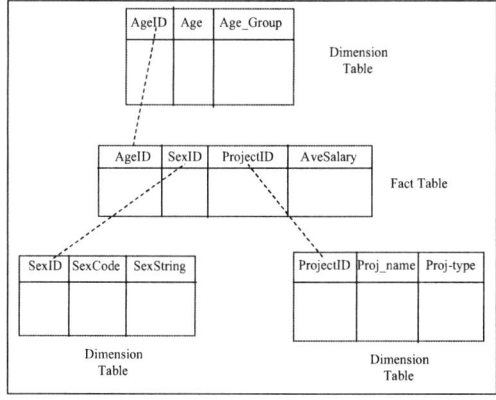

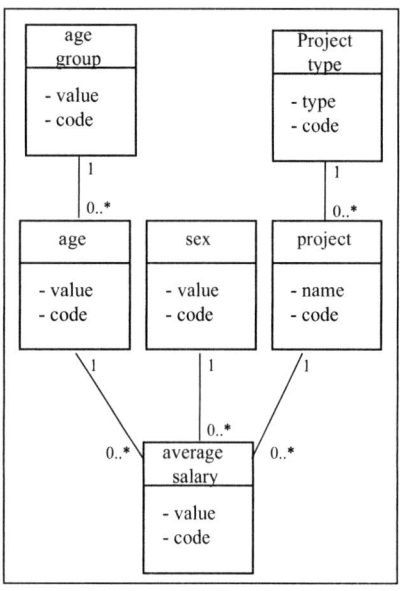

notation preferred by some is the Universal Markup Language Notation (UML) where the summary measure is at the bottom of the graph and the categories are at the top (see, for example, Pedersen, Jensen & Dyreson, 1999). This is shown for the same example in Figure 3b.

Our purpose in showing the above graphical notations is to emphasize that these are various forms that represent the same concepts. In the OLAP area, relational table structures are used, but additional structure was necessary to capture the concept of a multidimensional "cube" and the category hierarchy structure. In order to represent the semantics of a "cube," a "star schema" is used, where the multidimensional cube and the summary measure are represented in a "fact table" that is placed in the center of the "star," and "dimension tables" represent dimension categories as well as the hierarchical structure of the categories. We show this tabular notation for the example above in Figure 3c. Note that without external labels to the tables, there is no way of telling which is the "fact table" and which are the "dimension tables." Also note that in the fact table there is no way of telling which is the summary measure, except from the meaning of the names of the columns. Similarly, there is no explicit way of telling in the dimension tables that there is a classification hierarchy, except by guessing from the labels of the columns. The reason for this situation is that the tables are devoid of any semantics. Other graph representations used in the literature follow the entity-relationship notation (e.g., Tsois, Karayannidis, & Sellis, 2001; Cabibbo & Torlone, 1998) or other data cube notations (e.g., Lehner, Ruf, & Teschke, 1996).

Supporting Related Information

In practice, the categories involved in the summary databases may have secondary data associated with them. In the above example, a "project-ID" may have "project-description" or "project-deadline" associated with it. Furthermore, projects may belong to city departments, and the cities may have attributes such as "size," "total-budget," and "mayor."

Where in the summary model does such information fit? One solution taken by many authors is to extend the summary model to include additional structures that support associations and even generalization/classification (e.g., Lehner, Ruf, & Teschke, 1996; Sellis, 2000). Another approach is to provide for "links" between an "object" data schema and a "summary" data schema. This approach can also be used in cases where the "object" databases and the "summary" databases can reside on different systems. In this case the links can be supported as a federation of databases using the well-known concepts of "wrappers" and a "mediator." We have investigated this approach in Pedersen, Shoshani, Gu, & Jensen (2000), and built a prototype system on top of the ORACLE relational system and Microsoft's "OLE DB for OLAP" to illustrate its usefulness. This approach can be illustrated in the example shown in Figure 4.

In this example we show a single link between the category class "project" in the summary database and the "project" object class in the object database. This is a one-to-one link. In general, one can have multiple links set between the database schemas. Also, links can represent other association cardinalities (one-to-many, etc.), or even have attributes defined for the link instances. Once this federation is defined and the links instantiated, it is now possible to ask queries that span both databases. For the example in Figure 4, suppose that the summary database is for all the employees in a particular state. This database is maintained by the state's financial office. Suppose that another "public works" object database shown on the

Figure 4: Using Links to Federate "Object" and "Summary" Databases

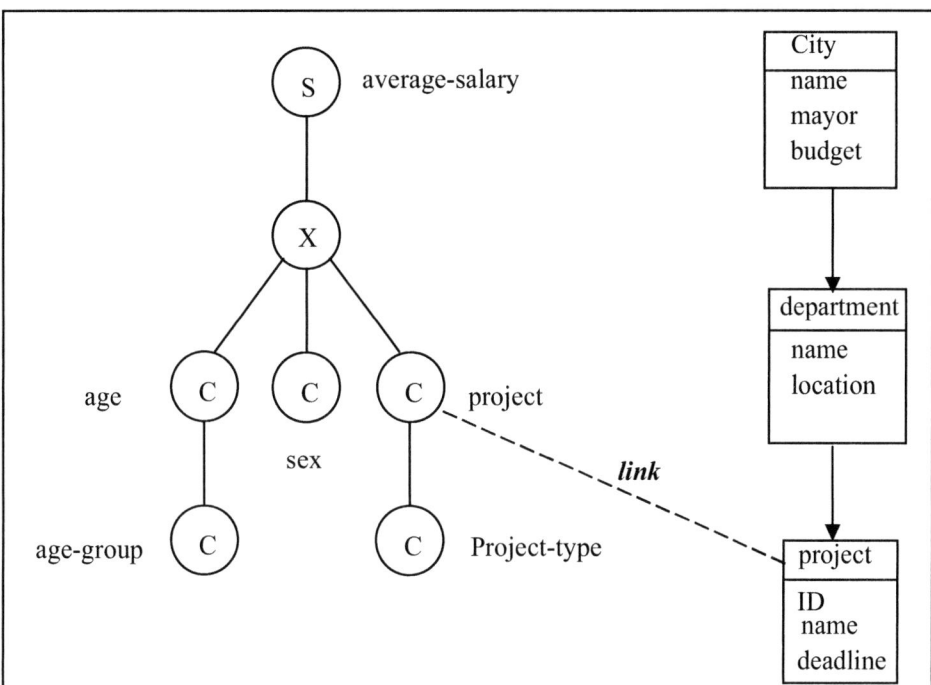

right is maintained by the state's public projects office. One can formulate the query "find Average-Salary by age for all projects in the cities that have a budget greater than $10 million. This query implies that only the projects in departments that belong to cities with a budget greater than $10 million will be selected, and the results summarized over them, as well as over sex.

The linking of databases allows the separation of summary data from other related information. This can be done in a single database system or across a federation. In the example of Figure 4, the separation allows each office to manage its own database, and the federation allows joint queries. For systems that implement OLAP databases as relational tables (such as Microsoft's "OLE DB for OLAP"), "links" are implied when additional tables for the related information are added, but this requires that all the databases are managed by a single system. Also, the semantics of the summary part of the tables and the other information is not explicit. The treatment of summary database semantics explicitly as part of an "object" model is still missing from commercial products. The use of links can facilitate this capability, even if they are applied in a single database system, let alone as a way to define federations between summary and object database systems.

Implied Aggregation

One can take advantage of the structural semantics of summary databases to imply the aggregate operations that need to be performed. This can greatly simplify the query expressions for specifying operations on summary databases. The implied aggregation is based on the multidimensional property, the category hierarchy property, and the "aggregate operator" associated with the summary measure. We discuss first the semantics of the "aggregate measure."

Once a summary database is created, the "summary measure" has an "aggregate operator" (also referred to as "summary operator") associated with it. Typical aggregate operators are "count," "sum," "maximum," "minimum," and "average." For example, the summary measure "population" has the aggregate operator "count" associated with it, and the summary measure "average-income" naturally has the aggregate operator "average" associated with it. We note that in order to support further aggregation with the "average" operator, the system has to carry the "sum" and "count" of the base items involved. More complex operators, such as "mean" and "standard-deviation," can also be calculated by carrying the appropriate base values.

We illustrate the concept of implied aggregation with a simple example. Consider again the summary database of Figure 3. Let's assume the database represents "average-salary-in-California." Recall that it has the following schema: *[avg-sal-Cal (age → age-group , sex, projectID → project-type): average-salary]*, where "avg-sal-Cal" is the name of the database.

Consider the following query applied to this database: "get average-salary for women by project-type." While this seems perfectly understandable to us, there are three implied instructions for performing this query:

1. In calculating the result, aggregate over all ages for each (sex = female, project-type) combination. This is implied because the "age" dimension was NOT mentioned at all in the query.
2. In calculating the result, aggregate over all projects that are in the same project-type. This was implied because of the category hierarchy between "projectID" and "project-type."
3. In calculating the result, use the aggregate operator "average." This is implied because of the default operator associated with the summary measure "average-salary."

As can be seen from this example, the implied aggregation is based on the three structural properties of summary databases: the multidimensionality, the category hierarchy, and the default aggregation operator. As mentioned above this can simplify the query language. We used such a language in Pedersen, Shoshani, Gu, & Jensen (2000). The language is called SumQL, has an SQL format, but takes advantage of the implied aggregation to simplify the query. As an example, the above query is given below as a SumQL query.

SELECT average-salary
BY_CATEGORY sex, project-type
FROM avg-sal-Cal
WHERE (sex = female)

Note that joins are not necessary, and the implied aggregation operations ("sum over ages" and "sum over projects") are omitted.

Summarizing category hierarchies requires that these hierarchies are well formed. For example, aggregating "population" from cities to "states" is not well formed, since states contain regions other that cities, such as farmlands, or small towns. However, if we added one fictitious city called "other" to the list of cities, and associated with it all the non-cities population, then we can summarize correctly to the state level. We call the property of being able to correctly aggregate to the next higher category level "summarizability." We studied the summarizability conditions of summary database in Lenz & Shoshani (1997).

Efficiency Considerations

The main efficiency issue in summary databases is how to efficiently compute aggregate operations over the cells of the multidimensional space, and how to extract efficiently subsets of the multidimensional space. There are basically two types of aggregation operations, corresponding to the multidimensional property and the category hierarchy property. These are:

A.1. Aggregating over an entire dimension. This is referred to in the OLAP literature as a *"consolidate"* operation. For example, in the summary database [population_in_US (state, sex, age): population], aggregating over age to generate [population_in_US (state, sex): population] is a "consolidate" operation over the dimension "age."

A.2. Aggregating over a dimension to a higher level of the category hierarchy. This is referred to in the OLAP literature as a *"roll-up"* operation. For example, suppose that we have the mapping of "states" to "regions" (such as "west," "mid-west," "south," and "east"), then the summary database [population_in_US (state, sex, age): population] can be rolled-up to produce [population_in_US (region, sex, age): population].

As for selecting efficiently subsets of the multidimensional space, most of the attention in the literature was given to two types of "subset selection" operators:

S.1. Selecting a single value out of a dimension. This is referred to in the OLAP literature as a *"slice"* operation. For example, selecting only a single state (say "Alabama") for the summary database [population_in_US (state, sex, age): population] will produce [population_in_Alabama (sex, age): population]. Notice that this reduced the database from a three-dimensional database to a two-dimensional database, which is the intuition for the term "slice."

S.2. Selecting a range of values out of a dimension. This is referred to in the OLAP literature as a "*dice*" operation. For example, selecting only the ages 1-18 (to get only children's population) from the summary database [population_in_US (state, sex, age): population] will produce the multidimensional subset (also referred to as a "sub-cube") [population_of_children_in_US (state, sex, age): population], where age is limited to 1-18. In this case the dimensionality of the sub-cube is the same as the original summary database.

Note that although "slice" can be considered a special case of "dice," they may require support of different data structures and indexes, since they are analogous to performing a single value selection (where hash indexes are most effective for a single attribute) vs. a "range" selection (where tree indexes are most effective for a single attribute). However, in summary databases, multi-attribute indexes are needed for these operations.

There are other operations of interest, such as taking the union of two cubes or disaggregating to lower levels of the category hierarchies, but most of the works published about processing queries efficiently are primarily concerned with the above four operations. As an example for such a cube-query-language (CQL) that supports the above operations and extension to them, see Bauer & Lehner (1997).

Summary query languages combine these four basic operations in various ways. The problem is how to efficiency compute and generate the results for ad-hoc queries involving these operators. The main idea for dealing with this problem is to materialize (pre-compute) sub-cubes. However, just considering the "consolidated" sub-cubes (the operation referred to as operation A.1) is problematic since there are 2^n-1 possible combinations of sub-cubes to consider. For example, a summary database with three dimensions (x,y,z) has sub-cubes with the following dimensions: (x,y,z), (x,y), (x,z), (y,z), (x), (y), (z). A seminal paper addressing this problem was by Harinarayan, Rajaraman, & Ullman (1996). Essentially, the problem they addressed can be formulated as follows: given a limited space (say 20% of the original database size), and the cardinality of each dimension, which sub-cubes should be materialized, given equal probability of all possible ad-hoc queries. They developed an analytic algorithm to determine the optimal selection of sub-cubes to materialize.

Following this work, there were a large number of works that dealt with algorithms that are sensitive to the query patterns or that are better suited for performing update operations. For example, see the paper on dynamic data cubes (Geffner, Agrawal, & El Abbadi, 2000) which also summarizes previous work in this domain.

SCIENTIFIC MULTIDIMENSIONAL DATABASES

In the area of scientific databases, the aspects of multidimensionality that are of interest depend on the application areas and the purpose of organizing or viewing the data as a multidimensional dataset. Scientific databases are growing in numbers and in size at a fast pace as a result of superior instrument automation and faster parallel computers. Instruments are now capable of measuring physical phenomena with higher precision and in shorter time intervals. For example, satellites are now sending large quantities of measurements in the upper atmosphere. Similarly, runs on parallel supercomputers with thousands of processors generate large simulated datasets. These advances require better data management techniques, especially for multidimensional data. In a previous paper (Shoshani, Olken, & Wong, 1984), we identified multiple problem areas common to various scientific databases. In this sub-section, we discuss in some detail four main aspects of dealing with large multidimensional data: space-time representation, clustering in multidimensional space, indexing in multidimensional space, and supporting classification structures.

Space-time Representation

Since many scientific databases represent physical phenomena, the need to represent space-time is common to many applications. A few examples are climate, aerodynamics, combustion, and fluid dynamics. In such applications, the typical method of conducting research is to develop theoretical models, to simulate these models, and to compare the simulation results to real observed data. Simulation models typically partition space into some cell structure, the most common of which is a rectangular cell structure. For example, in climate simulations a point in the cell structure represents a certain longitude, latitude, and height position (X,Y,Z). The simulations are typically run in time steps, where each time step contains a collection of X,Y,Z points, and each point has a set of tens of "variables" (such as "tempera-ture," "wind velocity," "pressure," etc.) associated with it. These types of datasets are so common, that specialized file formats have been developed for them, such as HDF5 [NCSA/HDF5] and NetCDF (UCAR/Unidata/NetCDF).

The file organization of these datasets typically follows the way they are generated. It is a linearized representation of the multidimensional dataset, in the order of time-space variables (i.e., the multidimensional space is linearized in the order of T,X,Y,Z, and for each point in this space the values for (V1,V2,...,Vn) are stored). Each file format representation has an access library associated with it to extract desired subsets of the data for subsequent analysis. The main problem associated with this format organization is that for large simulations that span multiple files, the access requires reading small pieces from many files, making the reading of a subset of data quite inefficient. For example, if a climate analyst wishes to visualize the temperature pattern around the equator for the last 10 years, this implies that all the files containing the time steps for the last 10 years have to be accessed, and for each file the equator points have to be accessed, and for each of these points, only the "temperature" has to be extracted. Consequently, organizing the data according to the intended access pattern is desirable.

The problem of matching the file organization to the predicted access pattern has been studied (see, for example, Chen, et al., 1995), but in practice is not easily applied. There are two reasons for that: 1) the access patterns may suggest physical organizations that conflict with each other, or can change over time, and this requires automated reorganization of the data; and 2) for large datasets (some simulation datasets can be in the order of a few terabytes), it is very expensive to re-organize the data. Another approach is to cluster the data that is being accessed (referred to as "hot clustering"), but this approach is not in wide use, since it requires managing duplication of data. The most useful technique used currently is to partition the simulation dataset into one variable at a time (sometimes referred to as "vertical partitioning"), so that access to a particular variable does not require accessing the other variables. This simple solution is effective for tasks that access only a few variables at a time, such as visualization.

Many physical phenomena have regions of high activity and regions of little activity. For example, an ignition of fuel in a combustion cavity has regions of high turbulence close to where the fuel is injected. Similarly, a stress model of an airplane wing has more stress activity in the joints, and climate models have more activity in certain storm regions. For this reason, there are models that use "Adaptive Mesh Refinement" (AMR) models. The main idea is to use meshes that are more and more refined in the active regions, thus saving a lot of computation and space in the non-active regions. Such representation requires specialized data structures. This is a very important and active field of research in the mathematics area, but is not addressed at all by the database research community. See Plewa (2001) for a comprehensive set of pointers to such work by mathematicians around the world.

Accessing space-time datasets requires the support for specialized operators. Temporal operators include the specification of time spans (e.g., summer months), partial overlap of time periods, etc. Similarly, spatial operators may include distance functions (e.g., find all the points within a given radius) and spatial containment or overlap. The support for such operations requires specialized data structures that depend on the intended application. These have been covered extensively in the literature. There are several survey papers on the subject (Güting, 1994; Abraham & Roddick, 1999; Boehm, Berchtold, & Keim, 2001).

We note that space-time databases also have a natural dimensionality hierarchy. The time dimension is usually used to summarize the data in order to reduce them to a degree that can be handled by an analysis or visualization tool. For example, in climate simulation datasets, it is common to summarize the variables to the level of "monthly means." Such summarization is also useful for a high-level visualization of the data, which can lead to "zooming into" regions of interest. Similarly, the summarization over the spatial regions (which requires support of methods for spatial averaging) is also used in order to reduce the amount of data that the user has to deal with.

Clustering in Multidimensional Space

Given that objects are represented as points in a multidimensional space, it is natural to think about the object clusters as an indication of some common phenomenon. Similarly, an outlier (a point by itself) in the multidimensional space indicates an unusual phenomenon. Various data mining techniques are based on these observations, and numerous methods for data clustering have been proposed (see surveys by Keim & Hinneburg, 1999; Jain, Murty, & Flynn, 1999). In this sub-section, we will describe an example that illustrates the importance of high-dimensional clustering for very large scientific databases, and the reasons that most of the proposed clustering methods are inadequate.

Our example draws from our experience with large databases in High Energy Physics (see Shoshani et al., 1999, for more details). High Energy Physics experiments consist of accelerating sub-atomic particles to nearly the speed of light and forcing their collision. The particles that collide produce a large number of additional particles. Each such collision (called an "event") generates in the order of 1-10 MBs of raw data collected by a detector. A typical rate of data collection is about 10^8-10^9 events/year. Thus, the total amount of data collected is very large, in the order of 300 terabytes to 1 Peatbyte per year. This is why it is important to have efficient indexing and clustering analysis techniques.

In order to be able to find interesting clusters, summary data is extracted for each event. Each event is analyzed to determine the particles it produced, and summary properties for each event are generated (such as the total energy of the event, its momentum, and number of particles of each type). The number of summary elements extracted per event is typically quite large (100-200). Thus, the problem is one of finding clusters over a billion elements, each having 100 descriptors or more. This problem can be thought of as finding clusters in the 100-dimensional space over a billion points. The total amount of space required assuming 4 bytes per value is 400 GBs.

There are two aspects to this difficult problem: size and high dimensionality. The large size implies that the clustering analysis method must be linear to be practical. Since clusters cannot be detected in very high-dimensional space, the high-dimensionality aspect requires dimensionality-reduction methods, that is, methods that select fewer dimensions that represent the clustering properties the best. However, the known techniques are either non-linear and/or they are effective for small memory datasets. Similarly, the known linear clustering methods on large datasets are inadequate because they are based on cell-partitioning methods where the dimensions to be analyzed and the binning of each dimension are pre-selected. Therefore, hybrid techniques have been proposed: running dimension-reduction techniques on a sample of the data, and applying the results of this step to the linear clustering methods. We describe such a methodology in a recent paper (Otoo, Shoshani, & Hwang, 2001). Another effective approach reported recently is based on using different cutting planes for each dimension and finding the optimal grid partitioning (Hinneburg & Keim, 1999).

Indexing in Multidimensional Space

Indexing multidimensional data has been recognized as a specialized area for a long time now. If each dimension is indexed separately, then searching for objects in the multidimensional space will require the intersection of the results of all the index searches, which is an inefficient operation. Therefore, specialized multidimensional indexing methods have been developed. A recent survey covers such indexing methods (Boehm, Berchtold, & Keim, 2001). The specialized methods include such well-known indexing structures as the R-tree (Guttman, 1984), and various improvements to it, such as the R+ tree (Sellis, Roussopoulos, & Faloutsos, 1987), Grid Files (Nievergelt, Hinterberger, & Sevcik, 1984), Quad-trees (Samet, 1984), and Pyramid Indexes (Berchtold, Böhm, & Kriegel, 1998), just to name a few. Such indexing methods work especially well for low-dimensional applications (below six to seven dimensions) and for queries that involve all the dimensions. However, if we need to access part of the dimensions form a high-dimensional dataset, such methods become inefficient because too many branches of the index structures are involved in the search.

As a case in point, consider the high-dimensional space described in the High Energy Physics application in the previous sub-section. Recall that we described the problem as having a very large number of objects (many millions to a billion), and the number of dimensions (or attributes) in the hundreds. The typical search query is a range query over a few of the dimensions. This is referred to as a "partial range multidimensional query." For example, a query might be "find all objects (events) that have energy between 5 and 7 GEV and the total number of particle produced was more that 10,000." There are two problems with using the conventional multidimensional indexing methods: the high dimensionality (100 or more dimensions), and the fact that the query is a "partial range query."

The solution to this problem requires another approach. We can take advantage of the fact that scientific databases are typically "append only," and organize the index into "vertical partitions" where each partition represents a single dimension over all the objects. Further, each vertical partition can be organized as a set of compressed bitmaps. The operation of a partial range query can then be transformed into logical operations over the compressed bitmaps. We describe this technique in Shoshani et al. (1999) and further develope it in Wu, Otoo, & Shoshani (2001). This discussion is illustrative of the special indexing needs of scientific high-dimensional databases, especially as such databases are growing in size.

Supporting Classification Structures

It is a common practice to use specialized terms to describe concepts or objects in scientific domains. If the number of such terms is large, then they are organized as "classification structures," usually hierarchies of terms. For example, in biology organisms are classified by Kingdom, Phylum, Class, Order, Family, Genus, and Species. In medicine, there is a long history of international classification of diseases, contained in a large book and updated every 10-20 years (the latest is ICD-10). Similar large classifications exist in pharmacology, chemistry, material science, etc. These are often referred to as "ontologies." In fact, there are general products that are designed to allow the user to develop his/her classification. An example, in the area of clinical information, is an open source system, called openGALEN (http://www.opengalen.org). Hierarchical organization of concepts is not unique to scientific data. It is a natural way of classifying and categorizing concepts in any domain. For example, in the business world, products are naturally organized into hierarchical categories. This requirement is the same as the category hierarchies we discussed in statistical and OLAP databases, but can be more complex in scientific databases, since the classification structures can be many levels deep.

Scientists who need classification structures in their work find it extremely difficult to use commercial relational databases. This is because the tabular organization of information is not convenient for representing these classification hierarchies. To illustrate this difficulty, consider a simple example of organizing products as shown in Figure 5a. In this figure, each row represents a single product as well as the higher-level categories it belongs to. The problem with this represen-

Figure 5a: Representing a Category Hierarchy in a Tabular Form

Figure 5b: An Alternative, More Compact Representation

product	product-group	product-class	product-category
banana	fruit	food	agricultural
orange	fruit	food	agricultural
apple	fruit	food	agricultural
...	...	...	...
cheese	Milk-product	food	agricultural
yogurt	Milk-product	food	agricultural
...	...	...	...
tomato	vegetable	food	agricultural
broccoli	vegetable	food	agricultural
...	...	...	...
chair	furniture	wood	manufactured
table	furniture	wood	manufactured
desk	furniture	wood	manufactured
...	...	...	...

Product-ID	product-name	product-type	product-parent
1	agricultural	category	--
2	food	class	1
10	fruit	group	2
11	vegetable	class	2
...	...	...	...
100	banana	product	10
101	orange	product	10
...	...	...	...
200	tamato	product	11
201	broccoli	product	11
...	...	...	...

tation is that all the higher-level categories have to be repeated for all the products. This causes several difficulties: 1) For deep hierarchies, the number of columns is as large as the hierarchy depth. Even for the simple example used in Figure 5a, one can imagine refinement of products to additional levels, such as "milk-products" organized into "cheese," "milk," etc. sub-categories, and each of these into "non-fat milk," "low-fat milk," etc. Thus, a 10-deep hierarchy will require 10 columns, where the values in nine of the columns repeat for each instance of the "leaf" column. 2) This repetition is problematic not only because of space considerations, but also for data entry, and the potential for errors. 3) Categories often have a "description" item or a "code" associated with them. In the representation of Figure 5a, these have to be repeated as well.

As a result of the above difficulties, an alternative, more "normalized" representation is often used, as shown in Figure 5b. In this representation each category instance at any level is represented only once, eliminating the repetition. To support the category hierarchy, each row points to its parent category, i.e., pointing to the appropriate row ID. The problem with this representation is that while the table is more compact, operations over it are not supported by relational systems and languages, such as SQL. This is referred to as the "transitive closure" operation. For example, suppose that we wish to summarize sales for all food products in some organization. Using the table in Figure 5b, one would have to start with row 2, then find for it all the "children" by performing a "join" with the same table. This will produce rows 10, 11, ..., etc. For each of these, the next level of "children" has to be found. This will then produce rows 100, 101,..., 200, 201,..., etc. Writing SQL to generate that is too difficult for the user, so the work is delegated to programmers to incorporate into special-purpose user interfaces.

Figure 5c: A Normalized Version of the Category Hierarchy

product-Category ID	product-category	product-Category description
PCAT-1	agricultural	farm products
PCAT-2	manufactured	Factory products
...	...	...

product-Class ID	product-Class	product-Class description	product-Category ID
PCLA-1	food	retail	PCAT-1
PCLA-2	animal feed	wholesale	PCAT-1
PCLA-3	wood	mahogany	PCAT-2
...	...	...	...

product-group ID	product-group	product-group description	product-Class ID
PGRP-1	fruit	organic	PCLA-1
PGRP-2	vegetable	organic	PCLA-1
...	...	...	...
PGRP-100	furniture	Hand-made	PCLA-3
...	...	...	...

product-ID	product-	product-description	product-group ID
PRD-1	banana	large	PGRP-1
PRD-2	orange	large	PGRP-1
...	...	...	...
PRD-200	chair	kitchen	PGRP-100
...	...	...	...

We note that the "transitive closure" operation is straightforward in the case of Figure 5a. For the above example, one has only to specify "select products where product-class=food." This is the reason that many implementations end up using this representation in spite of its repetitiveness.

Finally, we should also consider the "fully normalized" alternative shown in Figure 5c. In this representation each category level is represented in its own table. An entry in one table has a pointer to an entry in its parent category table. This representation is similar to that of Figure 5b, but by splitting the category types into separate tables, it eliminates the duplication of the "product-type" column. We added in Figure 5c a "description" column for each of the category levels to show that this "normalized" representation eliminates their repetition as well.

The advantage of this representation is that a "transitive closure" is expressible by specifying "joins" between the tables. The disadvantage is the need to deal with multiple tables, and the cost of performing a large number of "joins."

Our purpose in discussing these alternatives of tabular representations of category hierarchies is to show why relational databases are rarely used for such structures, or used in a limited way. It points to a need to support such structures as specialized objects. Actually, this problem is more complex in real systems, since the hierarchy structures are not always as well organized as our example implies. In reality, there are cases where a category of one level may point to a parent category two or more levels above it. A recent paper that addresses this problem was written by Pedersen, Jensen, & Dyreson (1999).

Another aspect of classification structures that makes them even more complex to support in a tabular form is that a collection of objects can be classified in more than one category hierarchy. For example, the product category hierarchy of Figure 5 could be classified with a category hierarchy "manufacturer" where manufacturer could be further organized by location (city, state). Thus, the same base objects (in our case, "products") can be classified according several properties they may have. These properties that can be used to classify a collection of objects are referred to as "facets" (Batty, 1998). A database system that supports category hierarchies should permit the search of the objects with category specifications in multiple facets. This is equivalent to having a multidimensional space of facets, where each facet can be represented as a category hierarchy.

Future Trends

It should be evident from the above discussion that the two concepts of multidimensionality and category hierarchies are fundamental to summary and scientific data. In fact, it is our belief that they are fundamental to any domain that requires classification of objects or concepts into manageable collections. Further, since any collection of objects can have multiple "facets" (or attributes), these facets define a multidimensional space. There are two implications to these observations: the need for explicit conceptual modeling of these concepts, and efficient data structure to support these two fundamental concepts.

As far as conceptual modeling is concerned, it is necessary to support multidimensional classes and category-hierarchy classes as first-class concepts. By "first-class concepts" we mean that they are visible to the user. In the area of object modeling, the two dominant concepts are "object classes," which represent a collection of objects, such as "products," and "associations," which represent the associations between object classes, such as the association between "products" and "manufacturers." Object classes have attributes associated with them (such as "product-cost" or "product-weight"). We argue here that these have to be complemented with the concepts of multidimensional classes and category-hierarchy classes. For example, it should be possible to define a "product-type" as a category-hierarchy class and associate the object class "products" with it. This is shown schematically in Figure 6a.

Furthermore, it should be possible to define a multidimensional class with two dimensions, one being "product-type" and one being "manufacturer," and associate this multidimensional class with the "product" object class. This is shown schematically in Figure 6b.

The technology to support efficient data structures for the multidimensional class and the category-hierarchy class may vary depending on the application. For high-dimensional classes a special index may be necessary; for summary databases an efficient summarization technique may be necessary; and for very rich and deep category hierarchies, a specialized software package may be necessary. The main issue here is what is the infrastructure that can support such diversity. Extending the relational paradigm, the "object-relational" approach permits plugging "data blades"

Figure 6a: A Schematic of an Object Model for Supporting Category-Hierarchy Class

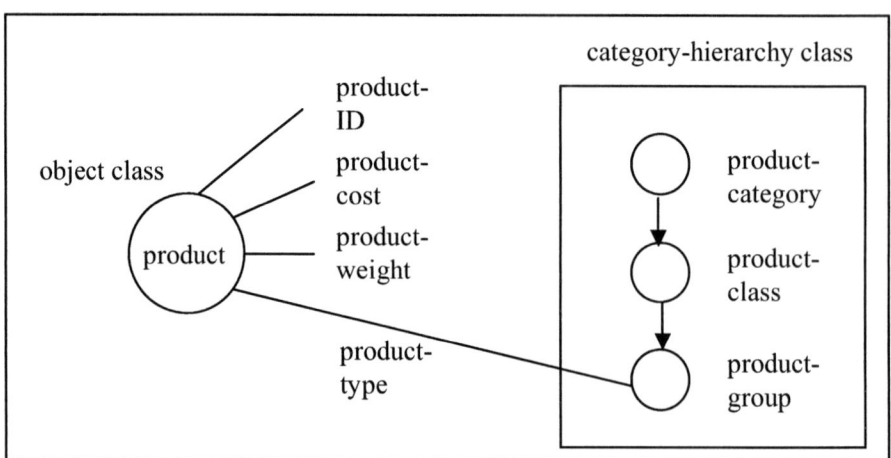

Figure 6b: A Schematic of an Object Model to Support a Multidimensional Class Whose Dimensions Are Two Category-Hierarchy Classes

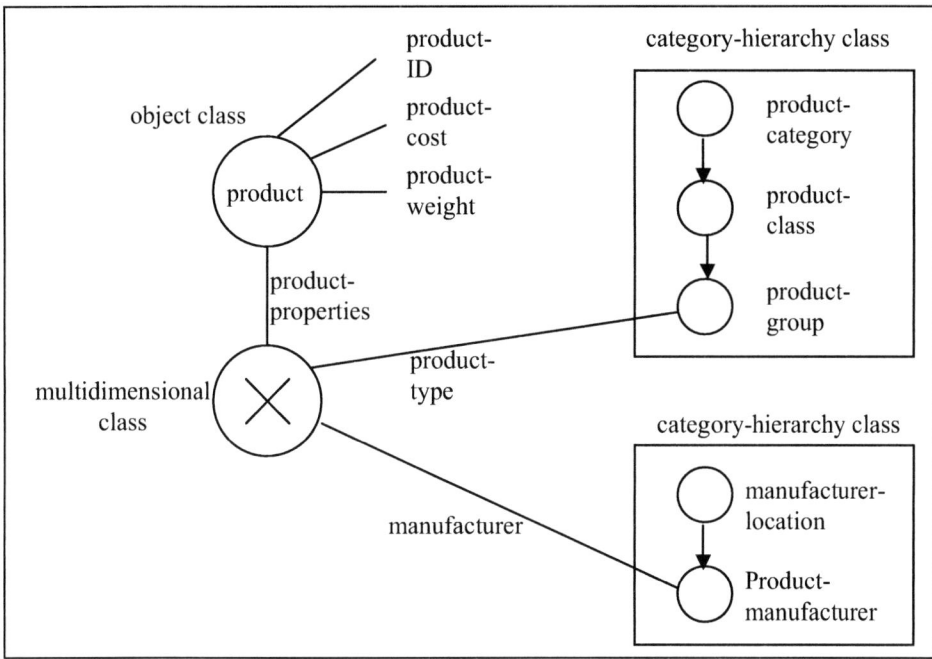

or "data cartridges" to support specialized data types. This approach was adopted by commercial vendors, especially Informix and ORACLE. However, this approach provides a table-driven view of the conceptual model. Furthermore, it does not permit one data type to be built from other data types. This will require a "data blade" that was other "data blades" building blocks.

We believe that the concepts of federation technology and the ability to associate object classes using "links" is a more flexible approach. Federated databases should permit putting together sub-systems that support the object-based data model with various other sub-systems as needed, so that the external conceptual model can support richer classes of objects and operators over them. In particular, we advocate a federated architecture to support sub-systems for object, multidimensional, and category-hierarchy classes. This architecture could exist under a single database system, or could be applied to federate multiple database sub-systems.

REFERENCES

Abraham, A., & Roddick, J.F. (1999). Survey of spatio-temporal databases. *International Journal on Advances of Computer Science for Geographic Information Systems (GeoInformatica),* 3(1), 61-99.

Batty, D. (1998). WWW—Wealth, weariness or waste: Controlled vocabulary and thesauri in support of online information access. *D-Lib Magazine, The Magazine of Digital Library Research,* volume.

Bauer, A., & Lehner, W. (1997). The Cube-Query-Languages (CQL) for multidimensional statistical and scientific database systems. *Proceedings of the International Conference on Database Systems for Advanced Applications (DASFAA),* 263-272.

Berchtold, S., Böhm, C., & Kriegel, H-P. (1998). The pyramid technique: Towards breaking the curse of dimensionality. *Proceedings of the International Conference on Management of Data (SIGMOD),* 142-153.

Boehm, C., Berchtold, S., & Keim, D.A. (2001). Searching in high-dimensional spaces: Index structures for improving the performance of multimedia databases. *ACM Computing Surveys,* 33.

Cabibbo, L., & Torlone, R. (1997). Querying multidimensional databases. *Proceedings of the International Workshop on Database Programming Languages (DBPL).* Lecture Notes in Computer Science, volume. City: Springer-Verlag, 319-335.

Cabibbo, L., & Torlone, R. (1998). A Logical Approach to Multidimensional Databases, *Proceedings of the International Conference on Extending Database Technology (EDBT),* 183-197.

Chan, P., & Shoshani, A.(1981). Subject: A directory-driven system for organizing and accessing large statistical databases. *Proceedings of the International Conference on Very Large Data Bases (VLDB),* 553-563.

Chen, L.T., Drach, R., Keating, M., Louis, S., Rotem, D., & Shoshani, A. (1995). Efficient organization and access of multidimensional datasets on tertiary storage systems. *Information Systems Journal,* 20(2), 155-183.

Codd, E.F. & Associates. (1993). *Providing OLAP (On-Line Analytical Processing) to User-Analysts: An IT Mandate.* White Paper, Commissioned by Arbor Software (now Hyperion Solutions).

Gray, J., Bosworth, A., Layman, A., & Pirahesh, H. (1996). Data cube: A relational aggregation operator generalizing group-by, cross-tab, and sub-total. *Proceedings of the International Conference on Data Engineering (ICDE),* 152-159.

Geffner, S., Agrawal, A., & Abbadi, A. (2000). The dynamic data cube. *Proceedings of the International Conference on Extending Database Technology (EDBT),* 237-253.

Güting, R.H. (1994). An introduction to spatial database systems. *VLDB Journal,* 3(4), 357-399.

Guttman, A. (1984). R-trees: A dynamic index structure for spatial searching. *Proceedings of the International Conference on Management of Data (SIGMOD),* 47-57.

Harinarayan, V., Rajaraman, A., & Ullman, J.D. (1996). Implementing data cubes efficiently. *Proceedings of the International Conference on Management of Data (SIGMOD)*, 205-216.

Hinneburg, A., & Keim, D.A. (1999). Optimal grid-clustering: Towards breaking the curse of dimensionality in high-dimensional clustering. *Proceedings of the International Conference on Very Large Data Bases (VLDB)*, 506-517.

Jain, A.K., Murty, M.N., & Flynn, P.J. (1999). Data clustering: A review. *ACM Computing Surveys*, 31(3), 264-323.

Keim, D.A., & Hinneburg, A. (1999). Clustering techniques for large data sets— From the past to the future. *Proceedings of the International Conference on Knowledge Discovery and Data Mining (KDD)*, Tutorial Notes, 141-181.

Keim, D.A., & Kriegel, H-P. (1996). Visualization techniques for mining large databases: A comparison. *IEEE Transactions on Knowledge and Data Engineering (TKDE)*, 8(6), 923-938.

Lehner, W., Ruf, T., & Teschke, M. (1996). CROSS-DB: A feature-extended multidimensional data model for statistical and scientific databases. *Proceedings of the International Conference on Information and Knowledge Management (CIKM)*, 253-260.

Lenz, H-J., & Shoshani, A. (1997). Summarizability in OLAP and statistical data bases. *Proceedings of the International Conference on Scientific and Statistical Database Management (SSDBM'00)*, 132-143. Available on-line at: http://www.lbl.gov/~arie/papers/summarizability.SSDBM97.ps.

Nievergelt, J., Hinterberger, H., & Sevcik, H.C. (1984). The grid file: An adaptable, symmetric multikey file structure. *ACM Transactions on Database Systems (TODS)*, 9(1), 38-71.

National Center for Supercomputing Applications (NCSA)'s HDF5 homepage. Available on-line at: http://hdf.ncsa.uiuc.edu/HDF5.

Otoo, E.J., Shoshani, A., & Hwang, S-W. (2001). Clustering high dimensional massive scientific datasets. *Proceedings of the International Conference on Scientific and Statistical Database Management (SSDBM)*, 147-157.

Pedersen, T.B., Jensen, C.S., & Dyreson, C.E. (1999). Extending practical pre-aggregation in On-Line Analytical Processing. *Proceedings of the International Conference on Very Large Databases (VLDB'99)*, 663-674.

Pedersen, T.B., Shoshani, A., Gu, J., & Jensen, C.J. (2000). Extending OLAP querying to external object databases. *Proceedings of the International Conference on Information and Knowledge Management (CIKM)*, 405-413.

Plewa, T. (2001). *Compilation of Sources for Adaptive Mesh Refinement for Structured Grids*. Available on-line at: http://www.camk.edu.pl/~tomek/AMRA/amr.html.

Samet, H. (1984). The quadtree and related hierarchical data structures. *ACM Computing Surveys,* 16(2), 187-260.

Sellis, T., Roussopoulos, N., & Faloutsos, C. (1987). The R+-tree: A dynamic index for multidimensional objects. *Proceedings of the International Conference on Very Large Data Bases (VLDB)*, 507-518.

Shoshani, A. (1982), Statistical databases: Characteristics, problems, and some solutions. *Proceedings of the International Conference on Very Large Data Bases (VLDB),* 208-222.

Shoshani, A., & Wong, K.T. (1985). Statistical and scientific database issues. *IEEE Transactions on Software Engineering (TSE),* 11(10), 1040-1047.

Shoshani, A., Olken, F., & Wong, K.T. (1984). Characteristics of scientific databases. *Proceedings of the International Conference on Very Large Data Bases (VLDB),* 147-160.

Shoshani, A. (1997). OLAP and statistical databases: Similarities and differences. *Proceedings of the Symposium on Principles of Database Systems (PODS),* 185-196.

Shoshani, A. Bernardo, L.M., Nordberg, H., Rotem, D., & Sim, A. (1999). Multidimensional indexing and query coordination for tertiary storage management. *Proceedings of the International Conference on Scientific and Statistical Database Management (SSDBM),* 214-225.

Tsois, A., Karayannidis, N., & Sellis, T.K. (2001). MAC: Conceptual data modeling for OLAP. *Proceedings of the International Workshop on Design and Management of Data Warehouses (DMDW).*

University Corporation for Atmospheric Research (UCAR). Unidata's NetCDF homepage. Available on-line at: http://www.unidata.ucar.edu/packages/netcdf.

Wu, K., Otoo, E.J., & Shoshani, A. (2001). A performance comparison of bitmap indexes. *Proceedings of the International Conference on Information and Knowledge Management (CIKM),* 559-561.

Chapter III

Conceptual Multidimensional Models

Riccardo Torlone
Università Roma Tre, Italy

ABSTRACT

A variety of multidimensional data models have recently been proposed by both academic and industry communities. but consensus on formalism or even a common terminology has not yet emerged. In this chapter, we first discuss the requirements that an ideal conceptual multidimensional model should fulfill. These requirements are suggested by general information system modeling principles and the specific characteristics of OLAP applications. Building on these requirements, we then present a general conceptual multidimensional data model and show how it can be used to describe the basic aspects of a business application in a way that is easy to understand and independent of the criteria for actual data organization in the various systems. Starting from the characteristics of the model proposed, we summarize the general features that a multidimensional conceptual model should support. We then survey various multidimensional models proposed and relate their characteristics to these general features. Finally, we discuss the main points raised in the chapter and some problems that remain to be solved in this context.

INTRODUCTION

The ability to represent information in an abstract and implementation-independent way is crucial in the lifecycle of every information system application—not only in its design but also in its operational phase. This is particularly true in the context of data warehousing and OLAP where, because of the level of complexity, application development and management are usually difficult and error-prone tasks.

In spite of this, conceptual data models for data warehousing have received little attention for a long period in the applicative area. Traditionally, multidimensional applications are modeled in a way that strictly depends on the corresponding implementation. One of the most used formalisms for data representation in this context is the relational model, which is clearly well suited in the case of a ROLAP (Relational OLAP) implementation. In general however, using a logical data model has a number of negative consequences. First, a logical representation is conceived to describe, at the appropriate level of abstraction, how data is stored in a specific DBMS, but it is usually not expressive enough to capture in an effective way the essential, multidimensional aspects of a data warehousing application. Second, it is difficult to define a design methodology that includes a general, conceptual step, independent of any specific system but suitable for all. Finally, in specifying aggregations of data, analysts often need to take care of tedious details that refer to the distribution of the information along the various structures used for its storage. For these reasons, data warehouse developers today understand that conceptual data models and methodologies are fundamental ingredients for the realization of good-quality products and for effective employment of their content.

It is now widely accepted that traditional conceptual data models, such as the Entity-Relationship model, are not appropriate for description of the multidimensional and aggregative nature of OLAP applications. For this reason, a variety of multidimensional data models have recently been proposed by both academic and industry communities, although it should be noted that a consensus on formalism or even a common terminology has not yet emerged.

In this chapter, we first discuss the requirements that an ideal conceptual multidimensional model should fulfill. These requirements are suggested by general information system modeling principles and the specific characteristics of OLAP applications. Building on these requirements, we then present a general conceptual multidimensional data model and show how it can be used to describe the basic aspects of a business application in a way that is easy to understand and independent of the criteria for actual data organization in the various systems. Far from being complete, this model aims at capturing the core of the various proposals of multidimensional data models and the conceptual means adopted by OLAP systems for data representation and manipulation. The model relies on a few agreed-upon concepts. The basic notions are the *dimension* and the *data cube*. A dimension represents a business perspective under which data analysis is to be performed and is organized in a hierarchy of *levels*, which correspond to different ways to group its elements. A *data cube* represents factual data on which the analysis is focused and associates *measures* with *coordinates*, defined over a set of dimension levels.

Starting from the characteristics of the model proposed, we summarize the general features that a multidimensional conceptual model should support. We then survey various multidimensional models proposed and relate their characteristics to these general features Finally, we discuss the main points raised in the chapter and some problems that remain to be solved in this context.

We do not address query languages, which are clearly strictly related to the subject of data models, as they are described in Chapter 10.

BACKGROUND AND TERMINOLOGY
Conceptual Data Models and Data Warehousing

A data model is for a database designer what a box of colors is for a painter: it provides a means for drawing representations of reality. Indeed, it has been claimed that "data modeling is an art" (Hull, 1997), even if the product of this activity has the prosaic name of *database scheme*.

When a data model allows the designer to devise schemes that are easy to understand and can be used to build a physical database with any actual software system, it is called *conceptual* (Batini, Ceri, & Navathe, 1992). This name comes from the fact that a conceptual model tends to describe *concepts* of the real world, rather than the modalities for representing them in a computer.

Many conceptual data models exist with different features and expressive powers, mainly depending on the application domain for which they are conceived. As we have said in the introduction, in the context of data warehousing, it was soon realized that traditional conceptual models for database modelling, such as the entity-relationship model, do not provide a suitable means to describe the fundamental aspects of such applications. The crucial point is that in designing a data warehouse, there is the need to represent explicitly certain important characteristics of the information contained therein, which are not related to the abstract representation of real-world concepts, but rather to the final goal of the data warehouse: supporting data analysis oriented to decision making. More specifically, it is widely recognized that there are at least two specific notions that any conceptual data model for data warehousing should include in some form: the *fact* (or its usual representation, the *data cube*) and the *dimension*. A fact is an entity of an application that is the subject of decision-oriented analysis and is usually represented graphically by means of a data cube. A dimension corresponds to a perspective under which facts can be fruitfully analyzed. Thus, for instance, in a retail business, a fact is a sale and possible dimensions are the location of the sale, the type of product sold, and the time of the sale.

Practitioners usually tend to model these notions using structures that refer to the practical implementation of the application. Indeed, a widespread notation used in this context is the "star schema" (and variants thereof) (Kimball, 1996) in which facts and dimensions are simply relational tables connected in a specific way. An example is given in Figure 1. Clearly, this low-level point of view barely captures the essential aspects of the application. Conversely, in a conceptual model these concepts would be represented in abstract terms which is fundamental for concentration on the basic, multidimensional aspects that can be employed in data analysis, as opposed to getting distracted by the implementation details.

Before tackling in more detail the characteristics of conceptual models for multidimensional applications, it is worth making two general observations. First, we note that in contrast to other application domains, in this context not only at the physical (and logical) but also at the conceptual level, data representation is largely influenced by the way in which final users need to view the information. Second, we recall that conceptual data models are usually used in the preliminary phase of the design process to analyze the application in the best possible way, without implementation "contaminations." There are however further possible uses of multidimensional conceptual representations. First of all, they can be used for documentation purposes, as they are easily understood by non-specialists. They can also be used to describe in abstract terms the content of a data warehousing application already in existence. Finally, a conceptual scheme provides a description of the contents of the

Figure 1: An Example of Star Schema

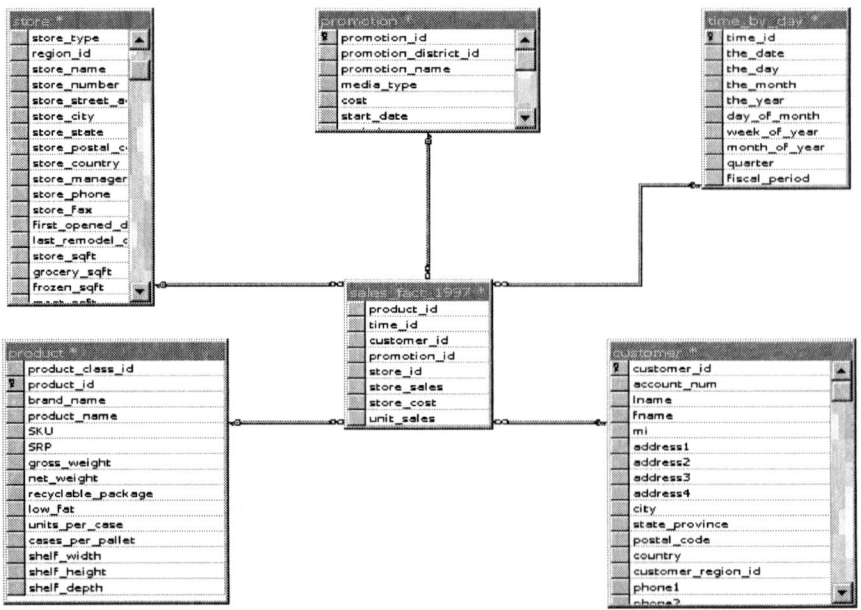

data warehouse which, leaving aside the implementation aspects, is useful as a reference for devising complex analytical queries.

Modelling Multidimensional Applications

Let us now investigate in more detail, but still informally, the fundamental ingredients of a conceptual data model for data warehousing. We start from the observation made above that the effectiveness of data warehousing modeling strictly depends on the ability to describe factual data according to appropriate *dimensions*, that is, "perspectives" under which data can be analyzed. For instance, in a data warehousing application for a retail company, it is useful to organize data along dimensions such as products commercialized by the company, stores selling these products, and days on which sales occur. To better support data analysis, it is useful to organize a dimension into a hierarchy of *levels*, obtained by grouping elements of the dimension according to the analysis needs. For instance, we might be interested in grouping products into brands and categories, and days into months and years. When the members of a level *l* can be grouped to members of another level *l'*, it is often said that *l rolls-up* to *l'*. For instance, the level "product" rolls-up to the level "brand." A level usually has *descriptive attributes* (or simply *descriptions*) associated with it. For instance, descriptions of a store include its name, manager, and address.

Let us consider a more concrete example, which will be used as a simple case study throughout this chapter.

Example 1: *The* Toys4All *company produces and sells a large number of products (mainly toys) in a chain of stores, over a wide territory.*

A main business goal for this company could be to understand the impact of promotions on sales, that is, how promotions influence product sales and to what extent promotions are profitable. Another important business goal could be the analysis of the warehouse process, where inventory levels should be measured monthly, for each product and warehouse controlled by the company. It follows that possible dimensions of the Toys4All data warehouse application are Product, Store, Warehouse, Time, *and* Promotion. *The Product dimension may be organized into levels such as* item *(whose members are products such as Disney's* Dinosaur *and* Duplo Pooh), product-line *(containing members like Mattel's* Disney *and* Lego Duplo), brand *(*Mattel *and* Lego), category *(*Popular Characters *and* Blocks), *and* department *(*Action Figures *and* Blocks). The elements of the* Time *dimension describe days over a period of time; this dimension may be organized into the levels* day, month, quarter, year, *and* season. *A member of the level* day *might be February 27, 2001. Members of the level* day *can be grouped to members of the level* month, *but also to members of the level* season *(e.g., Carnival). Descriptions of the* item *level might be its name and code.*

Traditionally, the entities of an application subject to decision-oriented analysis are called *facts,* and the specific and measurable aspects of a fact relevant for the analysis are known as *measures.* A collection of measures for the same fact can be nicely represented by means of a *data cube* (or hypercube) having a "physical" dimension for each "conceptual" dimension of measurement: a coordinate of the data cube specifies a combination of level members, and the corresponding cell contains the measure associated with such a combination.

Example 2: *For the Toys4All company, a possible fact is the daily sale. This fact can be analyzed with respect to the day of the sale, the product sold, the store of the sale, and the promotion applied to the daily sale. The measurements made for each daily sale could include the number of units sold, the income, and the cost. Thus, a data cube* Sales *can be used to describe daily information about the items sold by the stores of the chain. An instance of this data cube can state the fact that on February 27, 2001 the store Colosseum has sold two pieces of Duplo Pooh, applying a Carnival 2001 Promotion, for a corresponding gross income of 19.98 Euros against a cost of 14.98 Euros.*

In the warehouse process, measurable facts are the inventory levels, to be measured, for instance, monthly, for each product and warehouse. They can be modeled by means of a data cube Inventory. *The measurements made for each monthly inventory could include the inventory level (the quantity in stock at the end of the month), the quantity shipped during the month, and the value at cost of the quantity in stock.*

In the next section we will try to formalize the general notions discussed in this section.

A CONCEPTUAL MULTIDIMENSIONAL MODEL

We now present a simple multidimensional data model "MD" that provides a number of constructs to describe, in an abstract but natural way, the basic notions involved in multidimensional analysis. As is customary in database models, we make a clear distinction between the *scheme* (which specifies the structure of a concept) and the *instance* (that is, the actual values associated with a concept).

Formal Definition of MD

We assume the existence of a finite set of *base types* such as text, integer, decimal, and date. Each base type *t* is associated with a domain of *base values* of that type. We also assume the existence of a countable set of *names* and a countable set of *identifiers* (*ids*). Such ids are values, distinct from base values, that are used to uniquely identify real-life objects.

A dimension has three main components: a set of levels, a set of level descriptions, and a hierarchy over the levels.

Definition 1 [Dimension scheme] *An* MD dimension scheme *D consists of:*
- *a finite set L of names called* levels;
- *a finite set Δ of names called* level descriptions, *for each level in L; each description is associated with a base type t;*
- *a partial order ≤ called* roll-up relation *on the levels in L; if $l_1 \le l_2$ we say that l_1 rolls-up to l_2.*

There is a natural graphical representation of an MD dimension. Some examples are reported in Figure 2. In this representation, levels are depicted by means of round-cornered boxes and there is a direct arc between the two levels l_1 and l_2 if $l_1 \le l_2$. Small diamonds depict the descriptions of a level.

Example 3: *Figure 2 reports the dimensions for the Toys4All company, as described in Example 1: Time, Product, Store, Promotion, and Warehouse.*

As an example, let us consider in more detail the Time dimension. Its levels are day, month, quarter, year, and season. The roll-up relation of Time is the reflexive and transitive closure of the sets of pairs (day, month), (month, quarter), (quarter, year), and (day, season). Thus, for instance, the level day rolls-up to the level month, but also to the level year. Descriptions of the level

Figure 2: Dimension Scheme in the MD Model

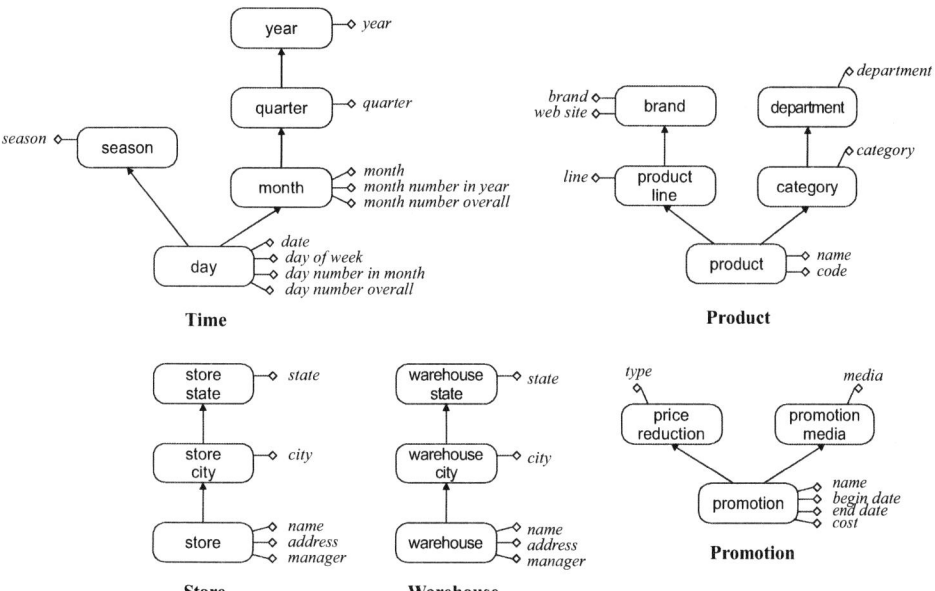

day are date, day-of-week *(mapping each day to the name of the corresponding day),* day-number-in-month *(mapping each day to the number of the day within its month), and* day-number-overall *(coding days in consecutive day numbers).*

Let us now state precisely what is an instance of a dimension scheme.

Definition 2 [Dimension instance]: *An instance of a dimension $D=(L, \Delta, \leq)$ consists of:*

- *a finite set of (real-world) objects, each of which has a unique id associated with it, for each level l in L, called* members *of l;*
- *a function from the members of l to the domain of base type t associated with l, for each level description in Δ;*
- *a roll-up function* ROLL-UP$_{l_1 \to l_2}$ *from the members of l_1 to the members of l_2, for each pair of levels l_1 and l_2 in L such that $l_1 \leq l_2$; if $m_2 =$ ROLL-UP$_{l_1 \to l_2}(m_1)$, we say that m_1 rolls-up to m_2.*

The roll-up functions of a dimension instance must satisfy the following *consistency conditions.*

Condition 1 [Consistency of roll-up]: *The family of roll-up functions of a dimension are* consistent *if:*

1. *for each level l, the function* ROLL-UP$_{l \to l}$ *is the identity on the members of l; and*
2. *if a level l_1 rolls-up to l_2 in different ways (e.g., rolling-up through either l' or l"), then the members of l_1 roll-up to elements of l_2 in a consistent way, that is:*

$$\text{ROLL-UP}_{l \to l}{}' \left(\text{ROLL-UP}'_{l \to l_2} (m) \right) = \text{ROLL-UP}_{l \to l}{}'' \left(\text{ROLL-UP}\, l''_{\to l_2} (m) \right)$$

for each member m of l_1.

Note that, as is customary in conceptual models, a member of a dimension level is not a value but is the object itself (e.g., a member of the store level is the actual building, not its name and address). In fact, although this object has an id and a number of values (the descriptions) associated with it, its existence and identity are clearly independent of them.

We are now ready to introduce the general notion of *multidimensional database scheme*. This has two main components: a collection of dimensions and a number of *data cube schemes*, which are defined over levels of the dimensions.

Definition 3 [Multidimensional Scheme]: *A* multidimensional scheme *consists of:*

- a finite set D of dimension schemes;
- *a finite set F of* data cube schemes *of the form:*

$$f [A1 : l1, \dots , An : ln] \rightarrow [M1 : m , \dots , Mk : mk],$$

where f is a name, each Ai $(1 \le i \le n)$ *is a distinct name called* attribute *of f, each li is a level of D, each Mj* $(1 \le j \le k)$ *is a distinct name called* measure *of f, and each mj is either a base type or a level of D.*

Note that in MD there is a uniform treatment of measures and dimensions, as a measure can be not only a simple value but also a level of a dimension. This allows the analyst to transform measures into attributes and vice versa (Cabibbo & Torlone, 1998b), an important functionality that any OLAP system should have (Pedersen, 2000).

Data cube schemes can also be naturally represented by means of diagrams. An example that refers to the dimensions in Figure 2 is given in Figure 3: facts are represented by boxes and measures by circles.

Example 4: *A multidimensional scheme for the business processes of the Toys4All Company described in Example 1: and Example 2: can be defined using the dimension schemes of Example 3: Specifically, two data cubes,* Sales *and* Inventory, *can be used to model the sale process and the warehouse process respectively. The schemes of these data cubes are represented graphically in Figure 3.*

The data cube Sales *describes daily sales, detailed by item, store, and promotion. Its attributes are* time *(at the day level of the time dimension, describing the day in which the sale occurred),* item *(the product sold),* store *(the store having sold the product), and* promotion *(the promotion applied to the sale). Its measures are* unit-sales *(the number of items sold),* euro-sales *(the income of the sale, in Euros), and* euro-cost *(the cost price of the items sold).*

The data cube Inventory *is instead used to represent the inventory levels of the various products, detailed by warehouse and month. Specifically,*

Figure 3: Two Data Cube Schemes Over the Dimensions in Figure 2

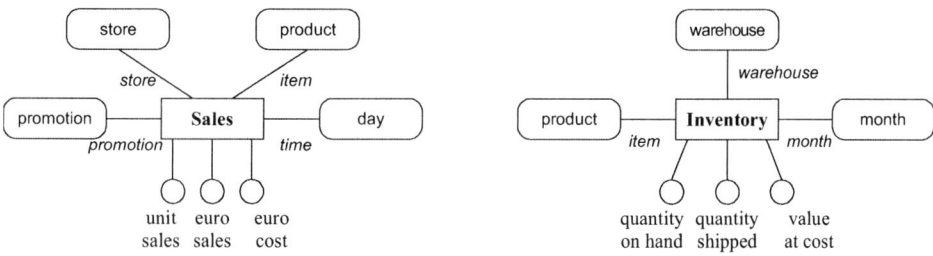

inventory levels are measured at the end of each month. The measures of this data cube are quantity-on-hand *(the quantity in stock of a product at the end of the month),* quantity-shipped *(the quantity shipped from the warehouse during the month), and* value-at-cost *(the value of the quantity in stock, at cost price).*

Before introducing the notion of instance of a data cube scheme, two preliminary notions are needed.

Let $D = (D,F)$ be a multidimensional scheme, $f[A1 : l1,..., An : ln] \rightarrow [M1 : m ,..., Mk : mk]$ be a data cube scheme in F, and d be an instance of D.

Definition 4 [Conceptual coordinate]: *A (conceptual) coordinate for f over d is a tuple over the attributes of f, that is, a function mapping each attribute Ai to a member of the level li occurring in d.*

Definition 5 [Fact]: *A fact for f over d is a tuple over the measures of f, that is, a function mapping each measure name Mj to either a value (if mj is a base type) or a member in d (if mj is a level).*

We are now ready to introduce the notion of instance of a multidimensional scheme.

Definition 6 [Instance of multidimensional scheme]: *An instance of a multidimensional database scheme (D,F) is composed of:*
- a dimension instance d for each dimension scheme in D;
- *a partial function called* data cube *mapping coordinates for f over d to facts for f over d, for each data cube scheme f in F.*

An *entry* of a data cube c is a coordinate over which the instance of c is defined.

Example 3: *A possible instance for the multidimensional scheme of Example 4: is shown in Figure 4. In this example, level members are represented by their ids.*

A coordinate over the data cube scheme Sales is, for example,

$$[time : d423, item : p98, store : s12, promotion : pr111]$$

where d423 is, for instance, the id associated with the physical item at hand.

The actual instance associates with this entry the value 2 for the measure unit-sales, the value 19.98 for the measure euro-sales, and the value 14.98 for the measure euro-cost.

In Figure 4, data cubes are graphically represented as a table. This representation suggests how data cubes can be implemented using the relational model: a data

Figure 4: A Sample Instance Over the Multidimensional Scheme of Example 4:

SALES

time	item	store	promotion	unit-sales	euro-sales	euro-cost
d_{423}	p_{98}	s_{12}	pr_{111}	2	19.98	14.98
d_{423}	p_{41}	s_{12}	pr_1	3	44.94	28.20
d_{423}	p_{56}	s_{21}	pr_{111}	1	2.99	1.10
d_{424}	p_{98}	s_{12}	pr_1	1	11.99	7.49
...	...	...	...	...	...	...

INVENTORY

time	item	warehouse	quantity-on-hand	quantity-shipped	value-at-cost
m_{13}	p_{98}	w_2	100	60	749.00
m_{13}	p_{41}	w_3	80	100	752.00
m_{14}	p_{98}	w_2	50	70	374.50
...	...	...	...	...	...

cube over a scheme f can be represented by a relation over the attributes of f, with additional columns for the measures. The attributes of f form the key of the relation. In practice, a data cube having n attributes and m measures can also be represented by means of an n-dimensional array in which each (non-null) entry corresponds to an entry of f and is associated with an m-tuple of measures. This representation recalls the way in which multidimensional systems usually store data, thus confirming that the MD is a conceptual model which describes multidimensional data independently of any specific (logical) implementation.

It is apparent that the notation we have used for coordinates resembles subscripting into a multi-dimensional array (although in a non-positional way). However, there is an important difference between data cubes and multi-dimensional arrays. Specifically, in arrays, "physical" coordinates vary over intervals within (linearly ordered) domains of *values*, whereas domains over which coordinates range in the MD model are *conceptual entities*. In this sense, our notion of coordinate is "conceptual."

Roll-up functions are a distinctive feature of the model proposed: they describe *intentionally* how members of different levels are related. This description is independent of any effective implementation: roll-up functions can be implemented by means of materialized relations, built-in functions, or external procedures. Moreover, roll-up functions provide a powerful tool for querying multidimensional data, as they can be used to specify how data can be grouped, and how data cubes involving data at different levels of granularity can be joined (Cabibbo & Torlone, 1997, 1998b).

Basic Features of a Multidimensional Model

The MD data model presented in the previous section exhibits those fundamental features that any multidimensional model should include in some form in order to be suitable for OLAP applications. According to Pederson (2000) and Blaschka et al. (1998), these "mandatory" features can be summarized as follows.

- *Explicit separation of structure and contents.* This is indeed a basic requirement of database models that make a clear distinction between the *schema*, which describes the structure of data, and the *instances*, which correspond to the actual contents.
- *Explicit notions of dimension and data cube.* These are the basic concepts of multidimensional data representation, as we discussed earlier.
- *Explicit hierarchies in dimensions.* A dimension should be structured into a hierarchy of levels to suggest the modalities in which data can be grouped along dimensions.
- *Multiple hierarchies in each dimension.* In one dimension, there can be more than one path along which to aggregate data. This is captured in MD by having a partial order relationship between the levels of a dimension.
- *Level attributes.* Other descriptive properties of the analysis dimensions, independent of the hierarchy relationship among levels, should also be representable. Level descriptions are used in MD for this purpose.
- *Measure sets.* This refers to the possibility of defining complex cell structures (grouping more than one measure) related to the same fact. In MD this is implemented by associating several measures to the same cube coordinate.
- *Symmetrical treatment of dimensions and measures.* The data model should allow measures to be treated as dimensions and vice versa. This is important because there are concepts (for instance, the age of customers) that can be measured (for instance, the average age of customers can be of interest) but which can also be used to group facts. This aspect is implemented in MD by allowing measures to be defined over dimension levels. This solution also makes it possible to register factual data at different granularities.

Advanced Features of a Multidimensional Model

There are a number of further advisable features that a conceptual multidimensional model should support. We have classified these features as "advanced" because they model concepts that either: i) are difficult to represent in a simple way (such as the notion of "summarizability"), or ii) serve to capture specific application cases. Adopting once more a terminology inherited from Pedersen (2000) and Blaschka et al. (1998), these basic features can be summarized as follows.

- *Support for aggregation semantics.* The data model should provide a support for the identification of aggregations whose result is *incorrect*, that is, meaningless to the user. This undesirable situation may occur for two main reasons.
  – A single fact can be counted more than once. Let us consider for instance the data cube Sales of our case study, whose scheme is described in Example

4: and reported in Figure 3. If we need the number of sales with respect to a specific media used for their promotion, we should only count a given sale once, even if several promotions have been applied to the sale.

– Some types of aggregation along certain paths of a dimension can be meaningless for a specific type of measure. For example, it may not be meaningful to add inventory levels of different products together, but calculating their average may make sense. This concept is strictly related to the notion of *summarizability* studied in the context of statistical databases (Lenz & Shoshani, 1997; Rafanelli & Shoshani, 1990), which defines when an aggregation, for instance, total sales, can be calculated by directly combining results from lower-level aggregations, for instance, the sales for each store. This problem has been recently investigated by various authors (Hurtado & Mendelzon, 2001; Lehner, Albrecht, & Wedekind, 1998).

- *Support for non-standard aggregations of facts.* There are various possible cases.

– *Non-strict hierarchies.* The hierarchy of levels in a dimension is non-strict if some of the mappings between the members of one level to the members of a higher level are many-to-many rather than one-to-many relationships. In our example, the Product dimension, described in Example 1 and represented in Figure 2, becomes non-strict if, for instance, a product can be classified according to different categories. The MD model can be extended to include non-strict hierarchies by assuming that the mappings $ROLL\text{-}UP_{l_1 \to l_2}$ are simple binary relations over members of levels l_1 and l_2 such that l_1 rolls-up to l_2, rather than functions.

– *Non-onto hierarchies.* A hierarchy in a dimension is "onto" if, for each member m of a level, there is a member m' of a lower level (if any) such that m' rolls-up to m. This property is not satisfied in our case study if, for example, there is a brand in an instance of the Product dimension (see Figure 2) with no associated product. In MD non-onto hierarchies are allowed as no restrictions are posed on the functions $ROLL\text{-}UP_{l_1 \text{-} l_2}$, which can be therefore non-onto.

– *Non-covering hierarchies.* A hierarchy in a dimension is non-covering if the member of a level rolls-up to a member of a higher level in the hierarchy by "skipping" one or more intermediate levels. In the Toys4All example, this may happen if, for example, in an instance of the Store dimension (see again Figure 2), there is a member of the Store level that rolls-up to a member of the State level, without rolling-up to any members of the City level. This would occur if the corresponding store is located not in a city but in a rural area. In MD non-covering hierarchies can be supported by allowing the roll-up functions to be *partial*.

– *Many-to-many relationships between facts and dimensions.* It may happen that the relationship between a fact and its corresponding dimensions is not a many-to-one mapping. In our case study, it may be the case that a specific sale (a row in the fact cube reported in Figure 4) is actually associated with a combination of promotions rather than just one. This is not strictly forbidden in the model (new rows can be added for this purpose) but can lead to incorrect aggregations (see above). This problem can be solved in many cases with an appropriate instantiation of the dimensions (Pedersen, Jensen, & Dyreson, 2001).

- *Handling change and time.* Schemes and data change over time, and there may sometimes be an interest in performing analysis across changes. In our example, a category of products might be moved from one department to another; we would then analyze the impact of this change on the number of sales. The problem of the management of slowly changing dimensions (Kimball, 1996) is related to this aspect. The maintenance of data cubes under dimension updates is also a relevant problem and has been recently investigated (Hurtado, Mendelzon, & Vaisman, 1999). Temporal analysis can also be of interest, for instance, the variations in inventory levels over time. Approaches taken in temporal data models (Tansel et al., 1993) could be applied to deal with these cases.

- *Handling imprecision.* Any real application must deal with the intrinsic problem of imprecision in representing and managing information. This problem has been widely studied in conceptual modeling. However, few studies have addressed this interesting and important problem in the context of multidimensional analysis, where imprecise data (for instance, the presence of missing values) can lead to incorrect results in calculating aggregations (Dyreson, 1996; Pedersen, Jensen, & Dyreson, 1999). A simple way to include a notion of imprecision in the measurement of facts in MD is to allow the presence of null values in data cubes. Conversely, incomplete knowledge of the dimension hierarchies can be taken into account by assuming that the roll-up functions are partial.

AN OVERVIEW OF MULTIDIMENSIONAL DATA MODELS

In this section we briefly report on data models that have been proposed for multidimensional databases, in relation to the requirements reported in the previous section. A more thorough examination and comparison of many such models can be found in several survey papers appearing in the literature (Blaschka et al., 1998; Pedersen, 2000; Rafanelli, 1995; Vassiliadis & Sellis, 1999). General discussion on OLAP, multidimensional analysis, and data warehousing can be found in Chaudhuri & Dayal (1997), Codd, Codd, & Salley (n.d.), Colliat (1996), Inmon (1996), and Samos et al. (1998). Mendelzon (n.d.) has published a rather comprehensive on-line bibliography on this subject. Further up-to-date information can be found on specialized websites, for instance Greenfield (n.d.) and Pendse & Creeth (n.d.).

It should be said that some of the models cited in this section cannot be classified as "conceptual" in the sense specified earlier. However, they are mentioned to provide a general overview of the state of the art in both the research community and commercial systems.

According to the classification proposed by Pedersen (2000), data warehousing models can be divided into three main categories: *cube models*, *multidimensional models*, and *statistical models*. In the first category are simple models that provide the notion of cube, but in which the concept of dimension is modeled to only a limited extent. Conversely, multidimensional models allow representation of dimensions in structured (although different) ways. With the statistical model we finally denote the large body of work in the area of statistical database modeling, which is strictly related to the multidimensional approach (Shoshani, 1997).

Cube Models

Simple cube models (Datta & Thomas, 1997; Gray et al., 1996; Gyssens & Lakshmanan, 1997; Kimball, 1996) treat data in the form of *n*-dimensional cubes. They all have a more or less explicit notion of fact, measure, and dimension. However, the hierarchy between the various levels of aggregation in a dimension is not explicitly captured by the schema, so the user cannot infer from the schema that, for instance, City rolls-up to State and not the opposite. The *star schema* approach (Kimball, 1996) and its variants—like the *snowflake* scheme, in which a central relational table represents the fact on which the analysis is focused, and a number of tables, usually de-normalized, represent the dimensions of analysis—should also be considered a cube model as they are semantically equivalent, although at a lower level of abstraction.

The majority of models adopted by commercial systems (Pendse & Creeth, n.d.; Oracle Corporation, 1998) should also be included in this category. Modeling aspects are covered by commercial systems in a pragmatic way. The representation used in ROLAP (*Relational* OLAP) systems is the star schema (Kimball, 1996), whose limit in representing the multidimensional aspects of OLAP applications at the right level of abstraction has already been discussed. In MOLAP (*Multidimensional* OLAP) systems (Colliat, 1996), information is represented directly in multidimensional form, but the structure of a dimension is usually hard coded in the physical index structures used to access data.

Multidimensional Models

Multidimensional models (Agarwal, Gupta, & Sarawagi, 1997; Cabibbo & Torlone, 1997, 1998a; Dyreson, 1996; Franconi & Sattler, 1999; Jagadish, Lakshmanan, & Srivastava, 1999; Lehner, 1998; Li & Wang, 1996; Mendelzon & Vaisman, 2000; Microsoft Corporation, 2000; Nguyen, Tjoa, & Wagner, 2000; Pedersen & Jensen, 1999; Vassiliadis, 1998) capture the hierarchies in the dimensions explicitly, providing a better understanding of the application and a support for easy data cube manipulation. This information may also be useful for query formulation and optimization.

Interestingly, while the basic features are more or less covered by these models, each of them represents the dimension structure very differently, e.g., by using grouping relations (Li & Wang, 1996), dimension merging functions (Agrawal, Gupta, & Sarawagi, 1997), measure graphs (Dyreson, 1996), roll-up functions (Cabibbo & Torlone, 1998a; Mendelzon & Vaisman, 2000), level lattices (Vassiliadis, 1998), hierarchy schemes and instances (Jagadish, Lakshmanan, & Srivastava, 1999), or an explicit tree-structured hierarchy as part of the cube (Lehner, 1998; Microsoft Corporation, 2000).

A number of data models have also been defined by extending traditional conceptual data models (Sapia et al., 1998). Others have used known paradigms, e.g., object-orientation (Abello, Samos, & Saltor, 2000) and nested structure models (Dekeyser et al., 1998), or specific metaphors, e.g., tapes (Gebhardt, Harke, & Jacobs, 1997). Finally, several data models have been proposed with the main goal of studying specific data warehousing application problems, such as incomplete information (Dyreson, 1998; Pedersen, Jensen, & Dyreson, 1999), efficiency issues (Harinarayan, Rajaraman, & Ullman, 1996; Jagadish, Lakshmanan, & Srivastava, 1999), heterogeneous dimensions (Hurtado & Mendelzon, 2001), dimension updates (Hurtado, Mendelzon, & Vaisman, 1999), and temporal OLAP queries (Mendelzon & Vaisman, 2000), and so are well suited for them.

Statistical Models

The last group is statistical database models (Bezenchek, Rafanelli, & Tininini, 1996; Rafanelli, 1995; Rafanelli & Ricci, 1993; Rafanelli & Shoshani, 1990; Shoshani, 1997; Tininini, Bezenchek, & Rafanelli, 1996).

A great deal of relevant work has already been done in this area. Shoshani (1997) made a very interesting comparison of work done in statistical and multidimensional databases. This revealed that after taking apart the terminology used, the two areas have a lot of overlap, even if each of them has emphasized different aspects. In particular, research in statistical databases has focused on the treatment of complex classification structures, management of certain special dimensions (e.g., spatial and geographic), and on the important issues (especially from the statistical point of view) of privacy and summarizability. On the other hand, OLAP literature has emphasized data warehouse design, query processing, and above all, efficiency issues. It is clear, however, that though the emphasis is on different aspects, work done in one area can greatly benefit the other (Shoshani, 1997).

A statistical data model is usually based on the notions of *summary table, summary attribute,* and *category attribute.* Actually, there is a close correspondence between these notions and the concepts used in multidimensional data models. Specifically, a summary table corresponds essentially to a data cube, a summary attribute to a measure, and a category attribute to a dimension. As in multidimensional models, a category attribute is always associated with a hierarchy of concepts. A number of operators are usually introduced in statistical models to manipulate, concatenate, and aggregate summary tables.

Notable examples of conceptual statistical models are *STORM* (Rafanelli & Shoshani, 1990) and *Mefisto* (Rafanelli & Ricci, 1993). In particular, Mefisto introduces the important notion of *statistical entity,* the conceptual counterpart of the notion of summary table.

In statistical models, a structured classification hierarchy is almost always coupled with an explicit aggregation function on a single measure to produce a sort of pre-defined object capable of answering a specific set of queries. This approach is sometimes less flexible than the approaches usually taken by multidimensional models, but unlike most of these, it can provide an effective way to avoid incorrect results from queries.

FUTURE TRENDS AND CONCLUSIONS

In this chapter, we have discussed the requirements that an ideal conceptual multidimensional model should fulfill. These are suggested by general information system modeling principles and by the specific characteristics of OLAP applications. Starting from these requirements, we have presented a simple conceptual multidimensional data model, called MD, which can be used to describe the basic aspects of a business application in a way that is easy to understand and independent of the criteria for actual data organization in the various systems. With this model, we have tried to capture both the conceptual means used in business applications to describe information and the core of the various multidimensional data models proposed in the scientific literature or adopted by commercial systems. The model relies on two principal, agreed-upon concepts: the *dimension* and the *data cube*. A dimension represents a business perspective under which data analysis is to be performed and is organized in a hierarchy of *levels*. The levels of a dimension correspond to different ways of grouping dimension members. A *data cube* represents the factual data on which the analysis is focused and associates *measures* with *coordinates*, defined over a set of dimension levels. Using these concepts as a reference, we have summarized the general features that a multidimensional conceptual model should support and mentioned the various multidimensional models which have been proposed.

Clearly, much work remains to be done in this area. First of all, the use of conceptual data models still has difficulties to overcome in the applicative area, and the research community should clearly demonstrate the benefits to be gained by adopting them. Moreover, with such a proliferation of data models, a commonly accepted formalism is strongly advisable. This is fundamental for support of interoperability and standardization. Another problem that still needs to be solved is the definition of an effective and general methodology for the development of OLAP applications, an important aspect which has received little attention (Cabibbo & Torlone, 1998a; Golfarelli, Maio, & Rizzi, 1998; Kimball, 1996). This would also lead to the development of CASE tools which, in contrast to the present situation, were not strictly related to a specific OLAP system. Devising a common standard declarative language is also of high importance, and the use of a conceptual multidimensional model (independent of the underlying physical model) could give useful results in the area of logical optimization and caching rules (in order to exploit the possibility of reusing existing data cubes for the computation of new ones). Finally, there are a number of specific problems, such as the characterization of summarizability, for which a definitive solution has not yet been given.

REFERENCES

Abello, A., Samos, J., and Saltor, F. (2000). Benefits of an object-oriented multidimensional data model. *Proceedings of the 14th European Conference on Object-Oriented Programming (ECOOP'00).* Lecture Notes in Computer Science, No. 1944. City: Springer-Verlag, 141-152.

Agarwal, S., Agrawal, R., Deshpande, P., Gupta, A., Naughton, J. F., Ramakrishnan, R. & Sarawagi, S. (1996). On the computation of multidimensional aggregates. *Proceedings of the 22nd International Conference on Very Large Data Bases,* Bombay, 506-521.

Agrawal, R., Gupta, A., & Sarawagi, S. (1997). Modeling multidimensional databases. *Proceedings of the 13th International Conference on Data Engineering,* 232-243.

Batini, C., Ceri, S., & Navathe, S. (1992). *Conceptual Database Design: An Entity-Relationship Approach.* City: Benjamin/Cummings.

Bezenchek, A., Rafanelli, M., & Tininini, L. (1996). A data structure for representing aggregate data. *Proceedings of the 8th International Conference on Scientific and Statistical Database Management (SSDBM'96).* City: IEEE Computer Society Press, 22-31.

Blaschka, M., Sapia, C., Höfling, G., & Dinter, B. (1998). Finding your way through multidimensional data models. *Proceedings of the 9th International Conference on Database and Expert Systems Applications (DEXA).* Lecture Notes in Computer Science, No. 1460. City: Springer-Verlag, 198-203.

Cabibbo, L., & Torlone, R. (1997). Querying multidimensional databases. *Proceedings of the 6th International Workshop on Database Programming Languages (DBPL '97).*

Cabibbo, L., & Torlone, R. (1998a). A logical approach to multidimensional databases. *Proceedings of the 6th International Conference on Extending Database Technology (EDBT'98).* City: Springer-Verlag, 183-197.

Cabibbo, L., & Torlone, R. (1998b). From a procedural to a visual query language for OLAP. *Proceedings of the 10th International Conference on Scien-*

tific and Statistical Database Management (SSDBM'98). City: IEEE Computer Society Press, 74-83.

Chatziantoniou, D., & Ross, K. (1996). Querying multiple features of groups in relational databases. *Proceedings of the 22nd International Conference on Very Large Data Bases,* Bombay, 295-306.

Chaudhuri, S., & Dayal, U. (1997). An overview of data warehousing and OLAP technology. *ACM SIGMOD Record, 26*(1), 65-74.

Codd, E.F., Codd, S.B., & Salley, C.T. (n.d.). *Providing OLAP (On-Line Analytical Processing) to User-Analysts: An IT Mandate.* Arbor Software White Paper. Available on-line at: http: //www.arborsoft.com.

Colliat, G. (1996). OLAP, relational, and multidimensional database systems. *ACM SIGMOD Record, 25*(3), 64-69.

Datta, A., & Thomas, H. (1997). A conceptual model and algebra for On-Line Analytical Processing in decision support databases. *Proceedings of the Seventh Annual International Workshop on Information Technologies and Systems (WITS'97),* 91-100.

Dekeyser, S., Kuijpers, B., Paredaens, J., & Wijsen, J. (1998). Nested data cubes for OLAP. *Proceedings of the International Workshop on Data Warehousing & Data Mining,* Singapore, 129-140.

Dyreson, C.E. (1996). Information retrieval from an incomplete data cube. *Proceedings of the 22nd International Conference on Very Large Data Bases,* Bombay, 532-543.

Franconi, E., & Sattler, U. (1999). A data warehouse conceptual data model for multidimensional aggregation. *Proceedings of the International Workshop on Design and Management of Data Warehouses (DMDW).*

Gebhardt, M., Jarke, M., & Jacobs, S. (1997). A toolkit for negotiation support interfaces to multi-dimensional data. *Proceedings of the ACM SIGMOD International Conference on Management of Data,* 348-356.

Golfarelli, M., Maio, D., & Rizzi, S. (1998). Conceptual design of data warehouses from E/R schemes. *Proceedings of the 31st Hawaii International Conference on System Sciences.*

Gray, J., Bosworth, A., Layman, A., & Pirahesh, H. (1996). Data cube: A relational aggregation operator generalizing group-by, cross-tab, and sub-totals. *Proceedings of the 12th IEEE International Conference on Data Engineering,* Vienna, 152-159.

Greenfield, L. (n.d.). Data Warehousing Information Center. Available on-line at: *http: //www.dwinfocenter.org/.*

Gyssens, M., & Lakshmanan, L.V.S. (1997). A foundation for multidimensional databases. *Proceedings of the 23rd International Conference on Very Large Data Bases,* 106-115.

Harinarayan, V., Rajaraman, A., & Ullman, J. (1996). Implementing data cubes efficiently. *Proceedings of the ACM SIGMOD International Conference on Management of Data,* 205-216.

Hull, R. (1997). Managing semantic heterogeneity in databases: A theoretical perspective. *Proceedings of the 16th ACM SIGACT SIGMOD SIGART Symposium on Principles of Database Systems*, 51-61.

Hurtado, C., & Mendelzon, A. (2001). Reasoning about summarizability in heterogeneous multidimensional schemas. *Proceedings of the 8th International Conference on Database Theory (ICDT)*. Lecture Notes in Computer Science, No. 1973. City: Springer-Verlag, 375-389.

Hurtado, C., Mendelzon, A., & Vaisman, A. (1999). Maintaining data cubes under dimension updates. *Proceedings of the 15th International Conference on Data Engineering*, 346-355.

Inmon, W.H. (1996). *Building the Data Warehouse* (2nd Ed.). New York: John Wiley & Sons.

Jagadish, H.V., Lakshmanan, L.V.S. & Srivastava, D. (1999). What can hierarchies do for data warehouses? *Proceedings of the 25th International Conference on Very Large Data Bases,* Bombay, 530-541.

Kimball, R. (1996). *The Data Warehouse Toolkit*. New York: John Wiley & Sons.

Lehner, W. (1998). Modeling large-scale OLAP scenarios. *Proceedings of the 6th International Conference on Extending Database Technology (EDBT)*. Lecture Notes in Computer Science, No. 1377. City: Springer-Verlag, 153-167.

Lehner, W., Albrecht, J., & Wedekind, H. (1998). Normal forms for multidimensional databases. *Proceedings of the 10th International Conference on Scientific and Statistical Database Management (SSDBM'98)*. City: IEEE Computer Society Press, 63-72.

Lenz, H.J., & Shoshani, A. (1997). Summarizability in OLAP and statistical databases. *Proceedings of the 9th International Conference on Scientific and Statistical Database Management (SSDBM)*, 132-143.

Li, C., & Wang, X.S. (1996). A data model for supporting On-Line Analytical Processing. *Proceedings of the Conference on Information and Knowledge Management*, 81-88.

Mendelzon, A.O. (n.d.). *Data Warehousing and OLAP: A Research-Oriented Bibliography*. Available on-line at: http: //www.cs.toronto.edu/~mendel/dwbib.html.

Mendelzon, A.O., & Vaisman, A.A. (2000). Temporal queries in OLAP. *Proceedings of the 26th International Conference on Very Large Data Bases,* Cairo, Egypt, 242-253.

Microsoft Corporation. (2000). *OLE DB for OLAP 2.0*. Microsoft Technical Document.

Nguyen, T.B., Tjoa, A.M., & Wagner, R. (2000). Conceptual multidimensional data model based on MetaCube. *Proceedings of the 1st International Conference on Advances in Information Systems (ADVIS),* 24-33.

Oracle Corporation. (1998). *Oracle OLAP Products: Adding Value to a Data Warehouse*. Oracle Technical Document.

Pedersen, T.B. (2000). *Aspects of Data Modeling and Query Processing for Complex Multidimensional Data.* PhD Thesis, Faculty of Engineering & Science, Aalborg University, country.

Pedersen, T.B., & Jensen, C.S. (1999). Multidimensional data modeling for complex data. *Proceedings of the 15th International Conference on Data Engineering (ICDE).* City: IEEE Computer Society Press, 336-345.

Pedersen, T.B., Jensen, C.S., & Dyreson, C.E. (1999). Supporting imprecision in multidimensional databases using granularities. *Proceedings of the 11th International Conference on Scientific and Statistical Database Management (SSDBM'99).* City: IEEE Computer Society Press, 90-101.

Pedersen, T.B., Jensen, C.S., & Dyreson, C.E. (2001). A foundation for capturing and querying complex multidimensional data. *Information Systems*, 26(5), 383-423.

Pendse, N., & Creeth, R. (n.d.). *The OLAP Report.* Available on-line at: http: // www.olapreport.com.

Rafanelli, M. (1995). Aggregate statistical data: Models for their representation. *Statistics and Computing*, 5(1), 3-24.

Rafanelli, M., & Ricci, F.L. (1993). Mefisto: A functional model for statistical entities. *IEEE Transactions on Knowledge and Data Engineering*, 5(4), 670-681.

Rafanelli, M., & Shoshani, A. (1990). STORM: A statistical object representation model. *Proceedings of the 5th International Conference on Scientific and Statistical Database Management (SSDBM).* Lecture Notes in Computer Science, No. 420. City: Springer-Verlag, 14-29.

Rao, S., Badia, A., & Van Gucht, D. (1996). Providing better support for a class of decision support queries. *Proceedings of the ACM SIGMOD International Conference on Management of Data*, 217-227.

Samos, J., Saltor, F., Sistac, J., & Bardés, A. (1998). Database architecture for data warehousing: An evolutionary approach. *Proceedings of the 9th International Conference on Database and Expert Systems Applications (DEXA).* Lecture Notes in Computer Science, No. 1460. City: Springer-Verlag, 746-756.

Sapia, C., Blaschka, M., Höfling, G., & Dinter, B. (1998). Extending the E/R model for the multidimensional paradigm. *Proceedings of the Advances in Database Technologies, ER'98 Workshops.* Lecture Notes in Computer Science, No. 1552. City: Springer-Verlag, 105-116.

Shoshani, A. (1997). OLAP and statistical databases: Similarities and differences. *Proceedings of the 16th ACM SIGACT SIGMOD SIGART Symposium on Principles of Database Systems*, 185-196.

Stanford Technology Group, Inc. (1995). *Designing the Data Warehouse on Relational Databases.* Unpublished Manuscript.

Tansel, A.U., Clifford, J., Gadia, S.K., Jajodia, S., Segev, A., & Snodgrass, R.T. (1993). *Temporal Databases: Theory, Design, and Implementation.* City: Benjamin/Cummings.

Tininini, L., Bezenchek, A., & Rafanelli, M. (1996). A system for the management of aggregate data. *Proceedings of the 7th International Conference on Database and Expert Systems Applications (DEXA).* Lecture Notes in Computer Science, No. 1134. City: Springer-Verlag, 531-543.

Vassiliadis, P. (1998). Modeling multidimensional databases, cubes and cube operations. *Proceedings of the 10th International Conference on Scientific and Statistical Database Management (SSDBM'98).* City: IEEE Computer Society Press, 53-62.

Vassiliadis, P., & Sellis, T.K. (1999). A survey of logical models for OLAP databases. *SIGMOD Record,* 28(4), 64-69.

Chapter IV

Hierarchies

Elaheh Pourabbas
Istituto di Analisi dei Sistemi ed Informatica – C. N. R., Italy

Maurizio Rafanelli
Istituto di Analisi dei Sistemi ed Informatica – C. N. R., Italy

ABSTRACT

In this chapter we will focus on the rules of aggregation hierarchies in analysis dimensions of a cube. We give an overview of the related works on the basic concepts of the different types of aggregation hierarchies. We then discuss the hierarchies from two different points of view: mapping between domain values and hierarchical structures. In relation to them, we introduce the characterization of some OLAP operators on hierarchies and give a set of operators that concern the change in the hierarchy structure. Finally, we propose an enlargement of the operator set concerning hierarchies.

INTRODUCTION

Hierarchies play a fundamental role in knowledge representation and reasoning. They have been considered as the structures created by *abstraction processes.* According to Smith & Smith (1977), an abstraction process is an instinctively known human activity, and abstraction processes and their properties are generally used for multi-level object representation in information systems. An *abstraction* of a system is a model of that system in which certain details are deliberately omitted with the objective of allowing users to notice details of the system which are relevant to the application, but ignore other details and consequently simplify the model.

Sometimes, a system may have too many relevant details for a single abstraction to be intellectually manageable. However, manageability can be provided by decomposing the model into a *hierarchy of abstraction.* One advantage of this hierarchy of abstraction is the capability of accessing the model at different levels of abstraction.

An abstraction of an object denotes its essential characteristics which distinguish it from all other objects. It can be defined as a mental ordering imposed on the environmental model depending on the goal or task of the user. An abstraction can be understood as a selection of a set of attributes, objects, or actions from a much larger set of attributes, objects, or actions according to certain criteria determined by the above-mentioned goal or task. Repeating this selection several times, that is, continuing to choose, from each subset of objects, another subset of objects with even more abstract properties, we create other levels of (semantic) details of objects. In Smith & Smith (1977), two kinds of abstraction are defined. The former, called *aggregation*, refers to an abstraction in which a relationship between objects is regarded as a higher level object. The latter, called *generalization*, refers to an abstraction in which a set of similar objects (i.e., a class of objects) is regarded as a generic object.

The complete structure created by the abstraction process is a *hierarchy,* and the type of hierarchy depends on the operation used for the abstraction process and the relations. As for the relations the best known in the literature are *classification*, *generalization*, *association* (or *grouping*), and *aggregation*. Their main characteristics are briefly listed in the following.

Classification is a simple form of data abstraction in which an object type is defined as a set of instances. It introduces *an instance-of* relationships between an object type in a schema and its instances in the database.

Generalization is a form of abstraction in which similar objects are related to a higher level generic object. It forms a new concept by leaving out the properties of an existing concept. With such an abstraction the similar constituent objects are specializations of the generic objects. At the level of the generic object, the similarities of the specializations are emphasized, while the differences are suppressed. This introduces an *is-a* relationship between objects. The *is-a* relation between types is a partial order and organizes types into an *is-a* hierarchy. This relation covers a wide range of categories which are used in other frameworks, such as inheritance, implication, and inclusion. It is the most frequent relation resulting from subdividing concepts, called *taxonomies* in lexical semantics. The inverse of the generalization relation, called *specialization*, forms a new concept by adding properties to an existing concept.

A particular type of generalization hierarchy, named *filter hierarchy*, is defined by the so-called filtering operation. This operation applies a *filter function* to a set of objects on one level and generates a subset of these objects on a higher level. The main difference from the generalization hierarchy is that the objects that do not pass the filter will be suppressed at the higher level.

Association or *grouping* is a form of abstraction in which a relationship between member objects is considered as a higher level set of objects. With this relationship the details of member objects are suppressed and properties of the set object are emphasized. This introduces the *member-of* relationship between a member object and a set of objects.

Aggregation is a form of abstraction in which a relationship between objects is considered as a higher level aggregate object. Each instance of an aggregate object can be decomposed into instances of the component objects. This introduces a *part-of* relationship between objects. The type of hierarchy constructed by this abstraction is called an *aggregation hierarchy*.

Like data warehousing and OLAP, the above-mentioned aggregation hierarchies are widely used to support data aggregation. Thus, an aggregation hierarchy is an effective form of knowledge representation for encoding prior domain knowledge relevant to a data cube. In a simple form, such a hierarchy shows the relationships between domains of values. Each operation on a hierarchy can be viewed as a mapping from one domain to a smaller domain. In the OLAP environment, hierarchies are used to conceptualize the process of generalizing data as a transformation of values from one domain to values of another smaller/bigger domain by means of drill-down/roll-up operators.

In this chapter, attention will be focused on the rules of aggregation hierarchies in analysis dimensions of a data cube. In this context, some different types of hierarchies, as well as their characterization, will be discussed. These hierarchies divide a single dimension into different levels of aggregation. In relation to them, we discuss the characterization of some OLAP operators which refer to hierarchies in order to maintain data cube consistency. Moreover, we propose a set of operators for changing the hierarchy structure. The issues discussed provide modeling flexibility during the schema design phase and correct data analysis.

The chapter is structured as follows: the next section gives an overview of the related works on the basic concepts of the different types of aggregation hierarchies. We then discuss the hierarchies from two different points of view: mapping between domain values and hierarchical structures. The next section introduces the characterization of some OLAP operators on hierarchies and gives a set of operators that concern the change in the hierarchy structure. We then proposes an enlargement of the operator set concerning hierarchies, and conclude the chapter.

Related Works

The core of the aggregation hierarchy revolves around the partial order, a simple and powerful mathematical concept to which a lot of attention has been devoted (see Birkhoff, 1979; Davey & Priestley, 1993). The partial ordering can be represented as a tree, with the vertices denoting the elements of the domains and the edges representing the ordering function between elements. The notion of levels has been introduced through the idea that vertices at the same depth in the tree belong to the same level of the hierarchy. Thus, the member of levels in the hierarchy corresponds to the depth of the tree. The highest level is the most abstract of the hierarchy and the lowest level is the most detailed.

A partially ordered set is a pair $(V, \leq)$ where V is a set and $\leq$ is a reflexive, anti-symmetric, and transitive binary relation defined on . Accordingly, the concept of hierarchy can be sketched by the following definition V:

Definition: A *hierarchy* is an ordered structure, where the order is established between individuals or between classes of individuals; the ordering function may be any function defining a partial order:

$$v_1 \leq v_2 \quad iff \quad \exists f : v_2 = f\{v_i\}; \quad v_1 \in \{v_i\}$$

A vertex v_1 in a graph is on a lower level than v_2 if and only if there exists a function f, such that v_2 can be calculated by applying f to a set of v_i of which v_1 is a member. The function f must be reflexive, anti-symmetric, and transitive.

Now, suppose that T is a subset of V. Then, $t \in T$ is a *maximal element* of T if $t \leq x \in T$ implies that $t = x$, and t is the *greatest* element of T if $t \geq x$ for each $x \in T$. Minimal and least elements can be defined dually. If V has one minimal (maximal) element, then it will be called the bottom (top) and it is usually indicated by $\perp$(T).

As in data warehousing and OLAP, the notion of partial ordering is widely used to organize the hierarchy of different levels of data aggregation along a dimension. Sometimes, hierarchies have been perceived structurally as trees, i.e., no generic object is the immediate descendant of two or more generic objects, and where the immediate descendants of any node (supposing any hierarchy is represented by a graph) have classes which are mutually exclusive. A class with a mutually exclusive group of generic objects sharing a common parent is called a *cluster*. Generally speaking, many real cases cannot be modeled by the hierarchies having the previous characteristics. For instance, in Figure 1 two illustrative examples of this situation are shown.

For this reason, usually, a dimension hierarchy is represented as a direct acyclic graph. Sometimes, it can be defined with a unique bottom level and a unique top level, denoted by ALL (see Gray et al., 1997).

Figure 1: Examples of non-hierarchical structures

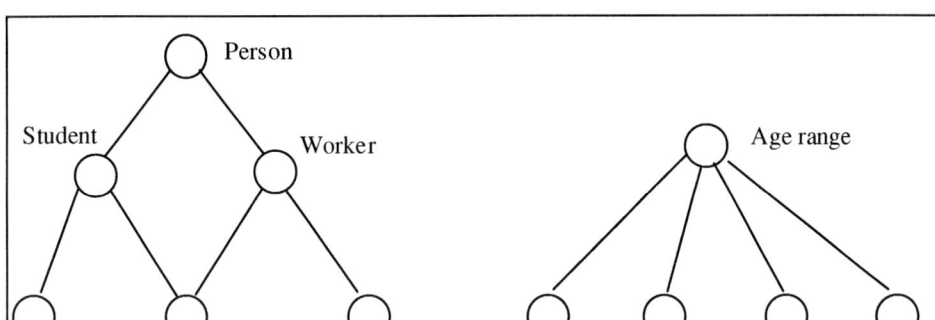

One of the most important issues related to the aggregation hierarchy is the *correct aggregation* of data, and this has been considered since 1980 (see Chen & Shoshani, 1981; Rafanelli & Shoshani, 1990; Lenz & Shoshani, 1997). It is known as *summarizability*, which intuitively means that individual aggregate results can be combined directly to produce new aggregate results.

Rafanelli & Shoshani (1990), speaking on the Statistical Object Representation Model (STORM), give some concepts and definitions regarding the summarizability of a category attribute (for a given statistical object, the term by which they denote a multidimensional aggregate data, MAD), the fullness and completeness of a mapping between category attributes which belong to the same hierarchy (represented in the graphical model by nodes labeled C, so that this mapping is called C-mapping), and finally, the well-formedness of a MAD.

Some observations on null values are made and the "unknown" and "non-existing" values are studied. In particular, in the former (unknown) case, the total with respect to the category attribute, for which there are unknown values, is not the total of the attribute itself. In the latter (non-existing) case, although it is true that, for example, the total number of cancer of the prostate cases is the total reported in the marginal (summarizing with regard to "sex"), the information on these cases are the total number of males alone and not the entire population is lost. For this reason, for the summary data of the tables in output, the authors distinguish two null values: non-available value (symbol "NA") and structural zero (symbol "-").

Structural zeros are also important in the case in which the cross-product among the different category attributes which describe the summary data is incomplete (for example, in the case of a relation). Suppose we have the MAD Occupation in the Computer Science Department in 1990 at UCLA, and suppose that the structure of this department is the following: <Research, Technical, Administrative>. Moreover, suppose that the Professional/ Academic titles admitted for each area are respectively <PhD, Degree>, <PhD, Degree, Diploma (General Certificate of Education)>, and <Degree, Diploma (General Certificate of Education)>. This situation is shown in Figure 2 by a bipartite graph and by a G-relation (discussed in Chapter 2 and proposed in Su, 1983), where the last column is the summary data. Therefore, the authors, after having defined the classification instance as a C-mapping between one category attribute instance of the higher level and one or more category attribute instances at the lower level, give the following definitions:

> **Definition:** A classification instance is *full* if all the category attribute instances of the lower level exist (according to the semantics of the classification as determined by the database administrator).

Note that fullness is a semantic concept.

> **Corollary:** A C-mapping is full if each classification instance is full.

As an example, the authors consider the mapping between the two category attributes "city" and "state" of the hierarchy "Political classification." In order to summarize correctly over cities, they observe that it is necessary to know that there are no missing values in its definition domain. However, it is also reasonable to assume that some small towns or other villages were not included in the list of cities. In this case the summarization from city to state level will have incorrect results. To compensate for such situations, often the instance "other" is included in the definition domain.

The graphical conceptual STORM model uses nodes to represent a MAD structure. In particular, it uses one S node to represent the summary values of the MAD (its measure), one A node to represent the cross-product among the different category attributes (i.e., the descriptive space of the MAD) which describe the measure, and C nodes to represent the category attributes mentioned above. Non-oriented edges link two nodes (in the case of a hierarchy, the edges consist of double lines).

The authors define the S-mapping as the mapping between an A node and the relative S node (i.e., between a tuple of the descriptive space of the MAD and the corresponding measure instance); then, they give the definitions of completeness, quasi-completeness, and incompleteness of an S mapping as follows:

> **Definition:** An S-mapping is *complete* if for every tuple of values in A a unique, defined (i.e., not "non-available" or "structural zero") value in S exists.

> **Definition:** An S-mapping is *quasi-complete* if for some tuple in A a "structural zero" exists, that is a non-existent value (represented by "-" in the summary attribute) in S.

Figure 2: Example of the MAD occupation

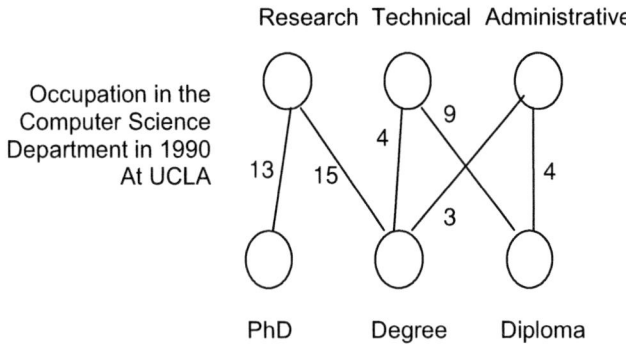

Research Technical Administrative

Occupation in the
Computer Science
Department in 1990
At UCLA

PhD   Degree   Diploma

	Structure	Title	Number of
Occupation in the Computer Science Department in 1990 At UCLA	research	PhD	13
	research	degree	15
	technical	degree	4
	technical	diploma	9
	administr.	degree	3
	administr.	diploma	4

For example, in a database on cancer rates, a value for prostate cancer for females (regardless of any other category attributes) does not exist.

Definition: An S-mapping is *incomplete* if for every tuple of values in A a unique, defined, or non-available (represented by "NA" in the summary attribute) value in S exists.

Note that an incomplete S-mapping can become complete (for example, it may be that not all the data referring to a given fact is available at present, but within a given time they will all be available).

They give the definition of a summarizable category attribute in the following way:

Definition: A category attribute of a given MAD is *summarizable* if:

a)   there is no overlapping among the values of its definition domain;

b)   there is no overlapping among the numeric values of the summary attribute;

c)   there is no overlapping along the time dimension among the population (people, cars, etc.) considered in the phenomenon studied by this statistical object;

d)   the edge starting from it, if it takes part of a classification hierarchy, defines a complete C-mapping toward the upper level C node;

e)   the unit of measure is the same for each instance of the definition domain and for each category attribute of a possible partition of which it takes part;

f)   no unknown (NA, not available) value appears in the summary attribute.

For a clear description of the above-mentioned cases, we give an example for each of them:

Case A: Suppose we have a statistical object in which the examined phenomenon is "people with a given disease," and that this phenomenon is classified by "state," "sex," and "disease." For a given state and a given sex, the total number of people that had at least one disease can be lower than the total obtained from the sum of the number of people that had a given disease (it is sufficient to think of a person that had two or more diseases).

Case B: Suppose we have a statistical object in which a category attribute is, for example, "age range." If the domain instances are, for example, < 0-10, 6-18, 19-29, 30-60, over 60 >, an overlapping among these values exists. This fact does not necessarily determine an overlapping among the numeric values of the summary attribute (we do not know if, with regard to this statistical object, somebody exists in the range "6-10"), but not knowing it determines the non-summarizability of the category attribute.

Case C: It is sufficient to consider, for this case, two different time series which describe, over the years, the different phenomena "Production of cars" and "Moving cars," by model. In the former there is no overlapping, but in the latter a large number of cars which moved in a given year, move also during the following year, and so on.

Case D: This is the case in which a sub-area of a category attribute is completely lacking (for example, all the states of "west" in the statistical object "Car production" by "year" and "state"); obviously the summarization of the category attribute "state" will not be possible with regard to the upper C node "country," even if, in general, it will be possible, by an estimation based on more or less complex criteria, to evaluate the lacking values. Another situation is the classification hierarchy "state"-"city," in the statistical object "Population of the USA" by state and year. If only the population of San Francisco and Los Angeles appears in the statistical object, we cannot summarize "city" because the population of California obtained in such a way would be an erroneous value.

Case E: This is the case of the category attribute "dairy" in which you can have two definition domains, each of them with different units of measure.

Case F: In this case, non-available (NA) values appear in the summary attribute of a statistical object as "holes" (for example, we can have car production in California for the years 1985, 1986, 1987, and 1990, but do not have it for the years 1988 and 1989. Therefore, the value of the production of cars in California for the years 1985-1990 is "NA"). This situation is like the previous Case D, and it is also generally solved by estimated values (for example, by an interpolation).

The authors give the definition of well-formedness MAD, based on the previous concept of summarizability:

Definition: A MAD is *well formed* if each of its category attributes is summarizable.

As subsequently discussed in Lenz & Shoshani (1997), summarizability of OLAP and multidimensional aggregate databases is an extremely important property because violating this condition can lead to erroneous conclusions and decisions. The summarization operation inherits the concept proposed in Rafanelli & Shoshani (1990) and discussed above. It can reduce the dimensions of the multidimensional space of a MAD when the dimension consists of a simple category attribute (without a hierarchy, for example, race or sex), or cannot reduce it when the summarization is carried out along the same dimension (i.e., when a hierarchy exists), thus decreasing the granularity of this dimension.

Summarizability conditions are the conditions under which the summarization operation produces the correct result. Suppose we have applied the aggregation process described in Chapter 2 to a relational (micro) database. In this way the characteristics of all the multidimensional aggregate data (MAD) of the multidimensional aggregate database are defined (i.e., name of each fact described by each MAD and belonging phenomenon, their summary type, their base relation (i.e., their descriptive space), their subject, etc.). The authors affirm that three necessary conditions of summarizability have to be satisfied:

1) Disjointness of the category attributes: The disjointness condition means that the category attributes must form disjoint subsets over the individuals/objects. Note that the testing of this disjointness condition is carried out at the instance level of the microdata. This condition is shown schematically in Figure 3(a), where the broken line violates the disjointness condition. It is necessary to examine this condition for two different cases: a category attribute that is part of the multidimensional space (the category attribute is a leaf node), and a category attribute that belongs to a classification hierarchy only (the category attribute is a non-leaf node). In the former, the disjointness test is relative to the individuals of the "microdata framework." In the latter, the non-leaf nodes form a category hierarchy, so that the disjointness test is made relative to the node below it. For example, in the hypothetical hierarchy "state-city," "state" is a non-leaf node, so that we have to test whether states form disjoint sets over cities. If it turns out that cities have globally unique names, this condition holds, but if cities do not have unique names (i.e., the same city name is sometimes used in two or more states), this summarizability condition is violated.

2) Completeness: Another summarizability condition to verify is whether the grouping of individuals/objects into clusters is "complete." The test for completeness in this context has two parts. The first is to verify that all the elements of the clusters exist (i.e., the union of all clusters constitutes the entire set of the individuals/objects). The second is to verify that each individual/object is assigned to a category. These two parts of the test are shown schematically in Figures 3(b) and (c). An important aspect of this condition is that it depends on the summary attribute. As was the case for Condition 1, the tests for the two types of category attributes (i.e., leaf and non-leaf nodes) are different.

3) Summary type of the individual/object measure: Summarization is performed quite differently with temporal categories as opposed to non-temporal ones. Summary attributes can be classified either as a *flow* (or *rate*), a *stock* (or *level*), or a *value-per-unit*. Flows refer to periods and are recorded at the end of these periods (they record the cumulative effect over a period), while stocks are

Figure 3: Example of Non-completeness

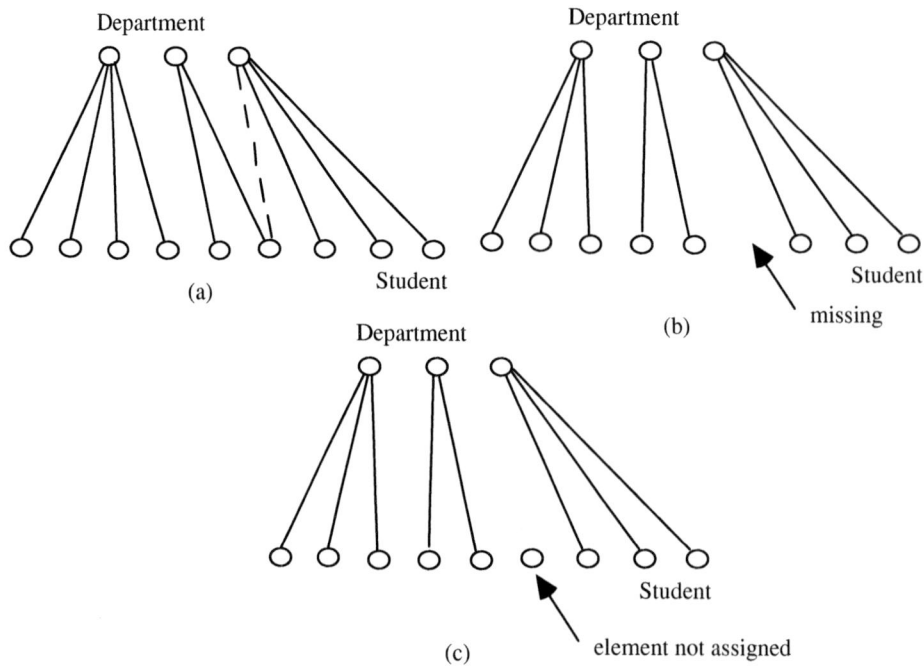

Figure 4. Temporal Summarization by Function and Summary Attribute Type

	stock	flow	value-per-unit
min	ok	ok	ok
max	ok	ok	ok
sum	not ok	ok	not ok
avg	ok	ok	ok
range	ok	ok	ok

measured and recorded at particular points in time. Value-per-units are determined by a fixed time, too, but are not similar to stocks in so far as their unit is different. While temporal summarizability of flows is semantically meaningful, temporal summarizability for stocks and value-per-unit depends on the aggregation function being applied.

The effects of the type of function and the type of attribute on temporal summarizability is shown in Figures 4 (by the type of function and summary attribute) and 5 (non-temporal summarizability dependent upon type of function and attribute).

In conclusion, the third condition of summarizability checks for the compatibility of the three parts of a summarization function: the category attribute, the summary attribute, and the aggregation function. It is necessary to determine the category attribute type (temporal, non-temporal), the summary attribute type (stock, flow, value-per-unit), and determine whether summarizability holds for the aggregation function associated with the MAD.

Figure 5. Non Temporal Summarization by Function and Summary Attribute Type

	stock	flow	value-per-unit
min	ok	ok	ok
max	ok	ok	ok
sum	ok	ok	not ok
avg	ok	ok	ok
range	ok	ok	ok

More recently, the problem of heterogeneity in aggregation hierarchy structures and its effect on data aggregation has attracted the attention of the OLAP database community. The term heterogeneity, as introduced by Kimball (1996), refers to the situation where several dimensions representing the same conceptual entity, but with different categories and attributes, are modeled as a single dimension. According to this description, which has also been called *multiple hierarchy* (see Agrawal, Gupta, & Sarawagi, 1997; Pourabbas & Rafanelli, 2000), dimension modeling may require every pair of elements of a given category to have parents in the same set of categories. In other words, the roll-up function between adjacent levels is a total function. The hierarchies with this property are known to be regular or *homogeneous*. For instance, in a homogeneous hierarchy, we cannot have some cities that roll-up to provinces and some to states, i.e., the roll-up function between City and State is a partial function.

In order to model these irregular cases, some authors introduced *heterogeneous* dimensions. For instance, the city Washington, DC, cannot roll-up to *State* because it rolls-up to *Country* without passing through *State*. In the context of irregular or heterogeneous hierarchies, there are several solutions which deal with summarizability.

Lehner, Albrecht, & Wedekind (1998) tackled heterogeneity by proposing different solutions. Their proposal consists of transforming heterogeneous dimensions into homogeneous dimensions in order to be in *Dimensional Normal Form* (DNF). This transformation is actually performed by considering categories, which cause the heterogeneity, as attributes for tables outside the hierarchy. On the flattened child/parent relation, summarizability is achieved for dimension instances.

Pederson & Jensen (1999) considered a particular class of heterogeneous hierarchies, for which they proposed their transformation into homogeneous hierarchies by adding null members to represent missing parents. In their opinion, summarizability occurs when the mappings in the dimension hierarchies are *onto* (all paths from the root to a leaf in the hierarchy have equal lengths), *covering* (only immediate parent and child values can be related), and *strict* (each child in a hierarchy has only one parent). Other conditions required for summarizability are that the relationships between facts and dimensions are many-to-one and that facts are always mapped to the lowest level in the dimensions. The proposed solutions consider a restricted class of heterogeneous dimensions, and null members may cause a waste of memory and increase the computational effort due to the sparsity of the cube views.

Hurtado & Mendelzon (2001) extended the notion of summarizability for homogeneous dimensions in order to tackle summarizability for heterogeneous dimensions. They classified four classes of dimension schemas, which are:

- *homogeneous*, if, along a dimension, the roll-up function from the member set of a level to the member set of its immediate higher level is a total function;
- *strictly homogeneous*, if it is homogeneous and it has a single bottom level;

- *heterogeneous*, if it is not homogeneous;
- *hierarchical*, which allows heterogeneity but keeps a notion of ordering between each pair of levels and their instances;
- *strictly hierarchical*, if the ordering relationship among different levels is transitive.

Then, in order to study summarizability in heterogeneous schemas, the authors introduced a class of constraints on dimension instances. These are statements about possible categories to which members of a given category may roll-up. However, their proposal is only suitable for inferring summarizability for a particular class of heterogeneous dimensions.

In a recent work, Hurtado & Mendelzon (2002) re-examined inferring summarizability in general heterogeneous dimensions. They introduced the notion of *frozen* dimensions, which are minimal homogeneous instances representing different structures that are implicitly combined in a heterogeneous dimension. This notion is used in an algorithm for efficiently testing implication of dimension constraints. The main idea behind their proposal is: a given category C of dimension Δ is summarizable from a set of categories $\{C_1, ..., C_n\}$ of dimension Δ if, for every fact table and every distributive aggregate function, the cube view for C can be computed from the cube views on the C_i's.

CHARACTERIZATION OF HIERARCHIES

In this section, we discuss the hierarchies from two different perspectives: the first is the mapping between domain values, i.e., total and partial classification hierarchies, and the second is the hierarchical structure. The latter case introduces multiple hierarchies and multiplicity of hierarchies. Note that the dimension analysis which will be discussed in the rest of this chapter concerns only regular hierarchies; in other words we will consider only hierarchies in which no overlapping exists among the domain instances of each category attribute or variable.

Classification Hierarchies

As we mentioned before, dimensions have often been associated with different hierarchically organized levels. These levels correspond to different granularities of viewing data. The name of each level is expressed by the corresponding *variable* name. Generally, the shift from a lower (more detailed) level to a higher (more aggregate) level is carried out by a mapping. A mapping between two variables can be complete or incomplete. In the first case the hierarchy is called a *total classification* hierarchy, and in the second case it is called a *partial classification* hierarchy. We give the following definitions:

Definition: A mapping between two variables of a hierarchy defines a *containment function* if each variable instance of a lower level corresponds to only one variable instance of a higher level and each variable instance of a higher level corresponds to at least one variable instance of a lower level. In such a case, this mapping is called *full mapping*.

Definition: A *total classification hierarchy* on a given dimension is a hierarchy in which there is a full mapping between each adjacent couple of variables.

The containment function respects the summarizability conditions (disjointness and completeness) of multidimensional databases described in Lenz & Shoshani (1997). As known in the literature, a hierarchy is intentionally represented by a partially ordered set. Therefore, a total classification hierarchy is any subset that defines a total order.

Example 1: Let us consider a nationwide drink company that owns chain stores located in all cities. Let as assume that all stores in the chain sell the same beverages. Sales data are collected yearly, i.e., at the end of each year, each member store reports the total number of sales of each drink to the regional headquarters. Figure 6 shows part of the data reported in 1997 and 1998.

The hierarchy along the location and beverage dimensions are represented in Figure 7 both in intentional level and extensional level.

Figure 6: Example of a Data Cube

Sales						
	Class	*City*	*Vendor*	*Year*	1997	1998
	Alcoholic	Los Angeles	Smith		10,000	12,000
		New York	Wong		20,000	16,000
		Washington, DC	McDonald		23,000	17,000
		Atlanta	Laurent		50,000	60,000
		...	...	...	...	...
		Dallas	Backer		21,000	32,000
		Detroit	Clifford		90,000	18,000
	Non-Alcoholic	Los Angeles	Smith		20,900	14,500
		New York	Wong		12,300	32,009
		Washington, DC	McDonald		87,000	23,890
		Atlanta	Laurent		23,100	49,000
		...	...	...	...	...
		Dallas	Backer		56,000	34,500
		Detroit	Clifford		21,000	30,000

Figure 7: The Hierarchy Along Dimensions: Beverages and Location (on the Left) and the Relative Domain Value (on the Right)

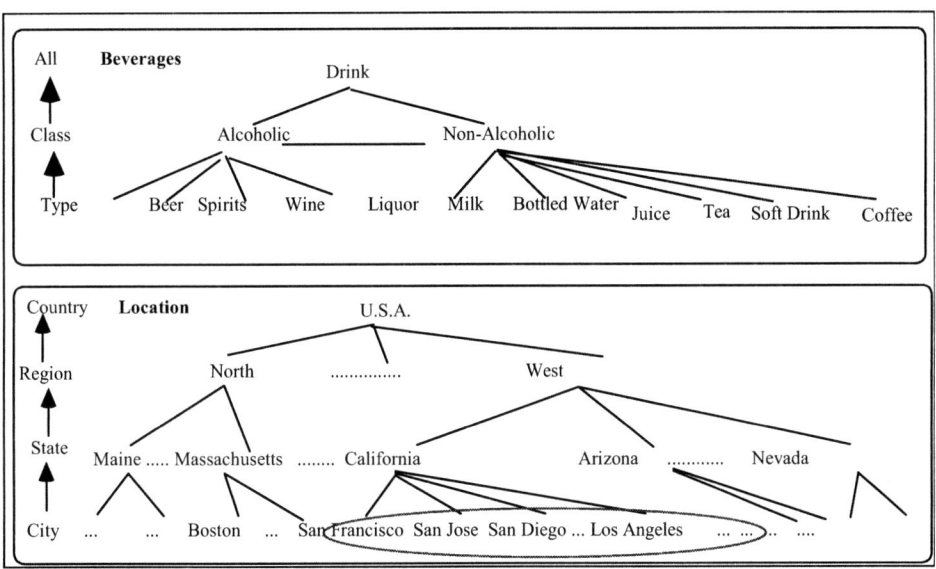

As shown in Figure 7, for a domain value of a level on a location dimension, all domain values of the lower level are defined, i.e., it is a classification hierarchy. This is completely in accordance with the hypothesis made in Example 1, where in all cities of the given country such a drink store is located.

Definition: A *partial classification hierarchy* on a given dimension is a hierarchy in which there is no full mapping between at least one adjacent couple of variables.

Example 2: Let us consider the chain store example we gave in Example 1. Suppose that the chain stores of the above-mentioned company in the state of California are located only in some of its cities (see Figure 8).

Therefore, the domain value of the *City* level along the *Location* dimension is restricted compared to that shown in Figure 4. Accordingly, these cities are not listed in the table of *Drink Sales*.

These two types of hierarchies will influence the result of queries for which the summarization operations will be needed. Details of this fact are discussed in a further section.

Figure 8: Domain Values of the Level City

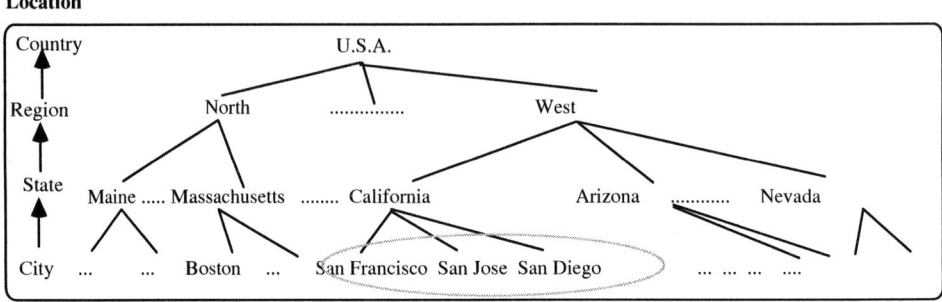

Multiplicity of a Hierarchy and Multiple Hierarchies

One of the most important problems regarding hierarchies is their definition. In this section we propose a set of definitions as a reference point for their study.

First of all, we distinguish between the multiplicity of a hierarchy and a multiple hierarchy.

Definition: Let H and H_1 be two hierarchies. H_1 is a *multiplicity* of H if its level domains are the same as the H level domains, and the variable name associated with each level of H_1 is a specialization of the variable name associated with the corresponding levels of H.

Example 3: Let us consider a location hierarchy defined as: City $\rightarrow$ Province $\rightarrow$ Region. A possible multiplicity of this hierarchy is City of residence $\rightarrow$ Province of residence $\rightarrow$ Region of residence.

Definition: Let H_1, H_2, ..., H_n be a set of hierarchies. This set forms a *multiple hierarchy* if each of them has at least one variable in common with another hierarchy of the same set.

Example 4: Let us suppose that we have four hierarchies, labeled (a), (b), (c), and (d), as illustrated in Figure 9. The hierarchy labeled (d) is a multiple hierarchy, where the level *Province* is the same for (a) and (b), and the level *Region* is the same for (a) and (c).

Definition: Let H be a hierarchy. The hierarchy obtained from deleting one or more non-terminal variables or levels of H is a *derived hierarchy*.

Figure 9: Example of a Multiple Hierarchy

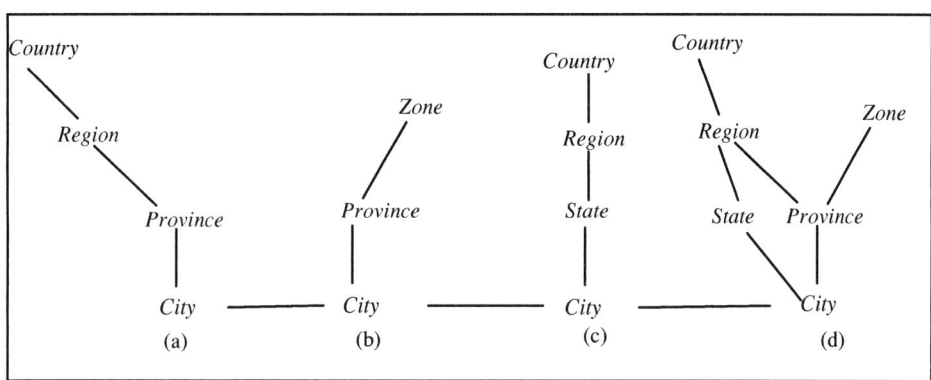

Example 5: From the multiple hierarchy (d) shown in Figure 8, we obtain the following derived hierarchies: *City → Province → Country, City → Region → Country, City → Country, City → Zone, City → Region → Country, City → Province → Country, and City → State → Country.*

More specifically, in the case of partial classification hierarchies, the variable instances of derived hierarchies are the instances of variables that are adjacent to the instances of deleted levels and between which a connected path can be defined. For example, in Figure 10(b) the variable instances of derived hierarchies obtained from variable instances of the partial classification hierarchy illustrated in Figure 10(a) that satisfy the above mentioned condition are reported.

CHARACTERIZATION OF OLAP OPERATORS ON HIERARCHIES

Recently different authors proposed a set of OLAP operators, which are defined on the data cube and which produce a new cube as output (see Codd, Codd, & Salley, 1993; Agrawal, Gupta, & Sarawagi, 1997; OLAP Council, 1997; Cabibbo & Torlone, 1998). In this section, we discuss the operators involved in manipulating dimensions with hierarchies in order to introduce some important modifications and specializations.

The Roll-up Operator

As mentioned above, this operator decreases the detail of the measure, aggregating it along the dimension hierarchy. A problem arises when a variable relative to a level of the hierarchy is not complete (i.e., in the case of a partial classification hierarchy).

Figure 10: Example of Path Generation Between Levels

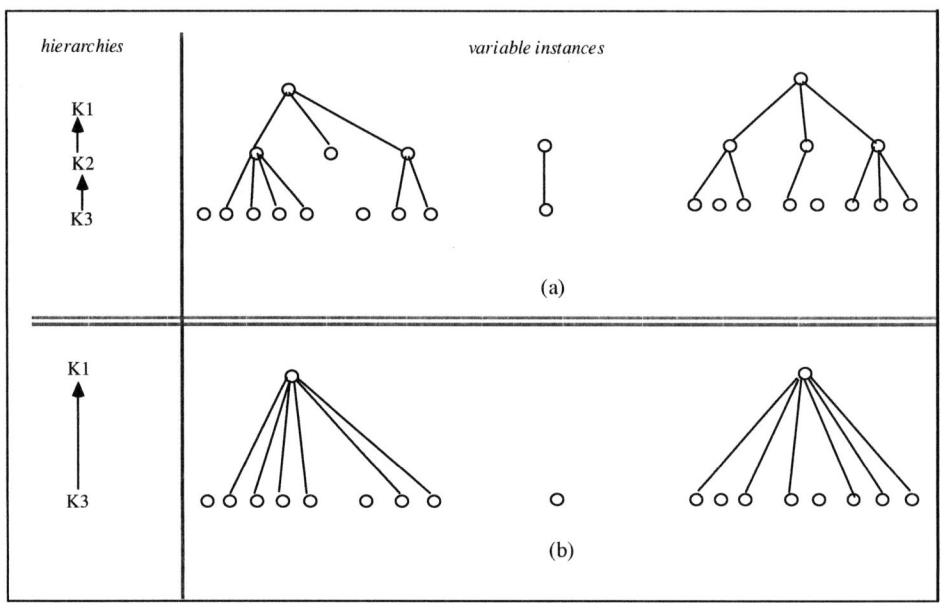

In the following we consider what happens when this operator is applied to a total classification hierarchy and then to a partial classification hierarchy.

Example 6: Let us consider the data cube represented in Example 2. In Figure 11, its "multidimensional" view is illustrated.

Let us suppose to formulate a query defined as below:
"Select Vendors for which the total Sales is >10000 units in each State in the West"

This query is solved in the following way:
Roll-up from City to Region, *Select* Region = West, *Drill-down* from Region to State, *Push* Vendors, *Pull* # of Drink sales, *Dice* # of Drink sales "≤10000."

Note that in some cells of the resulting cube (see Figure 12), null values can appear. This demonstrates that some instances of "Vendors" are not defined.
Let us suppose, now, that the classification hierarchy relative to "Region → State → City" has, as domains, the cities of California, but with only the instances "San Francisco, San Jose, San Diego." This is a subset of the primitive domain of City, in which all the cities of California (San Francisco, San Jose, San Diego, Fresco, Los Angeles, etc.) are stored.

Figure 11: A Multidimensional View of the Drink Sales Data Cube

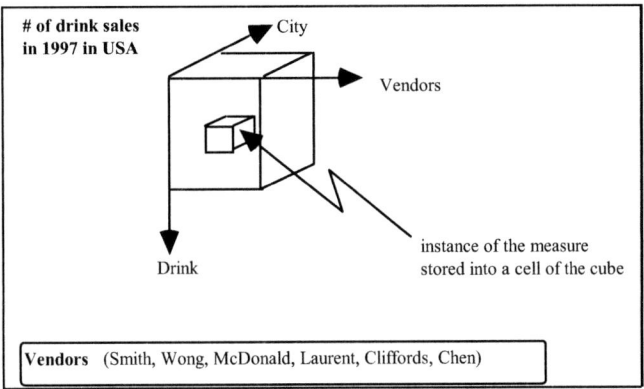

In particular, when the *roll-up* operator from City to State is applied, no information about the non-completeness of the City domain relative to California is stored. This means that for California, the number of vendors for which the total drinks sold in 1990 is >10,000 units refers only to the cities of San Francisco, San Jose, and San Diego, and not to all the cities in California. Therefore, since this information is not specified anywhere, the answer for this state is wrong.

A solution to this is to save the information about the domain values that cause the non-completeness of the hierarchy. This can be achieved in two different ways. The first is to add a *Note* (the clause *where <variable name> is-a subset of the primitive domain*) to the title of the cube. In the case of Figure 8, the title becomes "Vendors with total drink sales >10,000 units in each state in the West in 1997 in USA, *where city of California is-a subset of the primitive domain*." The second is to

Figure 12. The result of the query

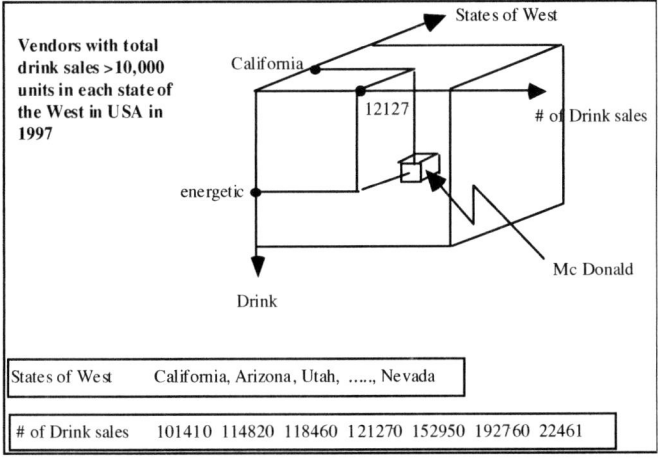

add the same *Note* to each variable of the hierarchy whose level is higher compared to that of the variable with the incomplete domain. In the same figure, we have to add the clause *where city of California is-a subset of the primitive domain* to the variables State, Region, and Country.

The Slice Operator

As mentioned above, the *slice* operator reduces the dimensions (or cardinality) of a cube by eliminating one dimension through its multidimensional space. This fact is not always true because, if we delete a dimension whose domain is a subset of the primitive domain, we lose information and the resulting cube of this operation contains incorrect data.

Before discussing this situation, we need to introduce the *implicit dimension* definition.

> **Definition:** We call any dimension of a data cube which has only one instance in its definition domain an *implicit dimension*. This instance can be one value or multi-valued.

> **Example 7:** Let us consider the cube in Figure 6, where the primitive domain of the dimension Year assumes the values <1990, 1991, 1992, 1993, 1994, 1995, 1996, 1997, 1998>. This means that they are all the possible values which this dimension can assume in the database. Instead, the value domain of Year in the considered cube is <97, 98>.

> Let us suppose that the following query is carried out:
> "Give me the drink sales in all Cities by Class and Vendor"
> It is solved in the following way:

Slice Year

In this case if the slice operator deletes the dimension Year, the result seems to refer to the whole primitive domain of Year. This means that we lose the exact information on the real period to which the result should refer. To overcome this mistake, a specialization of the *slice* operator, called Implicit Slice (or *I-Slice*) was introduced (see Chapter 2). We remember its definition below.

> **Definition** The *I-Slice* operator removes the dimension on which it is applied, transforming it into an implicit dimension. The only value of the implicit dimension domain is all the values which formed the domain of the removed dimension.

> Similarly to the solution proposed for the roll-up operator, the same *Note* is added to the title of the cube.

In accordance with this definition, the above query is now solved in the following way:

I-Slice Year

The result is a cube with the same dimensions as the primary one, where the title becomes "Drink sales by Class, Vendor, and Year *where Year is a subset of the primitive domain*." For the symmetric reasoning of terminology, we use the term Total Slice (or *T-Slice*) for the well-known *slice* operator.

If, instead, the Year domain in the cube under consideration coincided with its primitive domain, the previous query would be solved in the following way:

T-Slice Year

The cardinality of the resulting cube is now decreased by one dimension, since, by removing Year, no information is lost.

AN ENLARGEMENT OF OPERATORS FOR HIERARCHIES

In this section we propose a set of operators which are able to change the primary configuration of a hierarchy by extending or reducing its number of levels, adding a new multiplicity, and creating a multiple hierarchy. To formalize them, let us consider that and are two adjacent levels of a given hierarchy defined as . Let us discuss them.

Insert level

The *Insert level* operator makes it possible to add a new level to a hierarchy (giving the variable name, the domain instances, and the relationships between this level and, respectively, the higher and lower adjacent levels in the hierarchy). The insertion of a new level, denoted by l_i, between the above-mentioned levels, is represented through the symbol $Insertlevel_{l_1}^{l_2} l_i(I_{i,1}, ..., I_{i,n}; R_i)$, where $I_{i,1}, ..., I_{i,n}$ represents the inserted level domain instances and

$$R_i = (I_{l_2,1} (I_{l_i,1}(I_{l_1,1}, ..., I_{l_1 a_1}), ..., I_{l_i, p_1}(I_{l_1 a_1+1}, ..., I_{l_1 a_2})),$$
$$.., I_{l_i \cdot q_k = h} (I_{l_1 q_{Ph-1} +1}, ..., I_{l_1 a_n = n})))$$

(where the instances of levels l_2, l_i, l_1, are divided, respectively, into k, h, and n subsets and p_j, q_v) also represents a generic set of l_i and l_1 level instances with $j = 1$, ..., h and $v = 1, ..., n$.

Example 8: Let us consider the Location dimension shown in Figure 6. Let us suppose we insert the variable *County* between the variables *City* and *State*. We have to define its domain values, as well as the mapping relationship between *City* and *County*, and between *County* and *State* (see Figure 13). This is obtained by the following formula:

$$Insertlevel \ \substack{State \\ City} County \ (Green \ , Orange \ , ... ; R \ _{County})$$

where

$$R \ _{County} = (..., (California \ (Green \ (San \ Francisco \ , ..., Richmond \ ),$$
$$... , Orange \ (Los \ Angels \ , ..., Oxnard \ )), ...)$$

Delete Level

The *Delete level* operator redefines a hierarchy as a subset of the existing one by deleting a variable with its relative domain. This operation re-creates the mapping relationships between the two levels (higher and lower) adjacent to the deleted variable. The deletion of a level l_i is denoted by $Deletelevel \ _{l_1}^{l_2} l_i(I_{i,1}, ..., I_{i,n})$, where l_1 and l_2 are, respectively, the relative lower and higher level in the given hierarchy. This is the converse operation of the insert level operation. Note that in this case the relationships between levels are defined automatically by the system.

Add Multiplicity

The *Add multiplicity* operator duplicates a given hierarchy, by changing the name of the variables in order to specialize them, as already discussed in a previous section. Let H be a hierarchy. This operator is represented through *AddmultiplicityH (additionalname)* where *additionalname* is a string to be added to each variable name of the hierarchy in order to specialize the previous variable name.

Figure 13: Example of an Insert Level

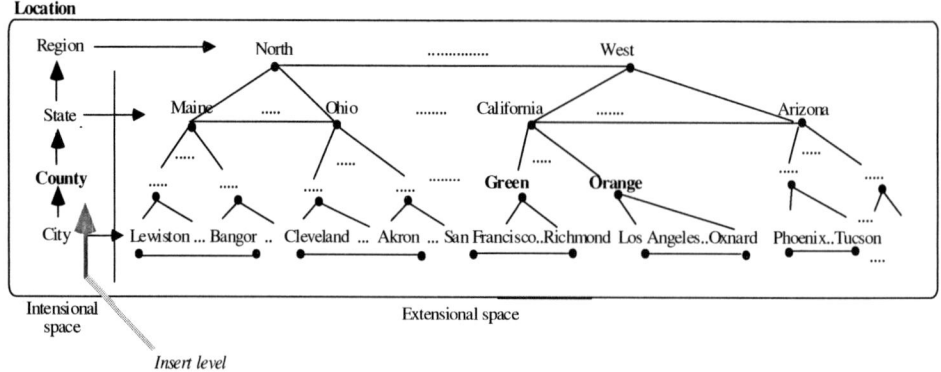

Example 9: Let us consider Example 3. The hierarchy City of residence $\rightarrow$ Province of residence $\rightarrow$ Region of residence is obtained by *Addmultiplicity Location (of residence)*

Add Level

The *Add level* operator creates a new hierarchy starting from a given hierarchy. It creates a new variable and its relative domain, and defines the mapping relationship between this and the variable of the starting hierarchy. These two hierarchies define, in this way, a multiple hierarchy. Let l_n be a level to be added to a hierarchy defined as $l_1 \rightarrow l_2 \rightarrow ... \rightarrow l_{n-1}$. The *Add level* operator is represented through:

$$Addlevel_{l_1 \rightarrow l_2 \rightarrow ... \rightarrow l_{n-1}} \ l_n(I_{i1}, ..., I_{in}, R_i)$$

and gives, as result, a hierarchy $l_1 \rightarrow l_2 \rightarrow ... \rightarrow l_{n-1} \rightarrow l_n$

Example 10: An example of a multiple hierarchy has been already shown in Figure 9(d).

CONCLUSIONS

In this chapter, the issue of the characterization of a hierarchy is proposed. We distinguished between hierarchies in which the mapping between different levels is full and hierarchies in which this condition is not verified. Then, we defined the concepts of multiplicity of a given hierarchy and of multiple hierarchies.

Based on the definitions and concepts proposed in this paper, we discussed the characterization of the OLAP operators involved in the hierarchy manipulation. In particular, depending on the type of hierarchy, we studied the different behavior of the roll-up and the drill-down operators in order to keep consistent data that are the result of the queries. We also characterize the slice operator, defining the implicit dimension concept and specifying two different types of operator: P-slice and T-slice. Finally, we proposed an enlargement of the operator set, specifically for hierarchy manipulation, which are the Insert level, the Delete level, the Add multiplicity, and the Add level operators. For each situation discussed, clear examples are given.

REFERENCES

Agrawal R., Gupta A., & Sarawagi, S. (1997). Modeling multidimensional databases. *Proceedings of the 13th International Conference on Data Engineering (ICDE'97),* Birmingham, UK, 232-243.

Birkhoff, G. (1979). *Lattice Theory* (3rd Ed.). Volume 25 of Colloquium Publications. Providence, RI: American Mathematical Society.

Cabibbo, L., & Torlone, R. (1998). From a procedural to a visual query language for OLAP. *Proceedings of the 10th International Conference on Scientific and Statistical Database Management (SSDBM'98),* Capri, Italy. Los Alamitos, CA: IEEE Computer Society Press, 74-83.

Chan, P., & Shoshani, A. (1981). SUBJECT: A directory-driven system for organizing and accessing large statistical databases. *Proceedings of the 7th International Conference on Very Large Data Bases (VLDB'81),* Cannes, France, 553-563.

Cruse, D. (1986). *Lexical Semantics.* Cambridge: CUP.

Codd, E.F., Codd, S.B., & Salley, C.T. (1993). *Providing OLAP (On-Line Analytical Processing) to User-Analysts: An IT Mandate.* Technical Report, E.F. Codd and Associates.

Dawey, B.A., & Priestley, H.A. (1993). *Introduction to Lattice and Order.* Cambridge, England: Cambridge University Press.

Gray, J., Bosworth, A., Layman, A., & Pirahesh, H. (1996). Data cube: A relational operator generalizing group-by, cross-tab, and roll-up. *Proceedings of the International Conference on Data Engineering (ICDE'96),* New Orleans, Louisiana. Los Alamitos, CA: IEEE Computer Society Press, 152-159.

Hurtado, C.A., & Mendelzon, A.O. (2001). Reasoning about summarizability in heterogeneous multidimensional schemas. *Proceedings of the 8th International Conference on Database Theory (ICDT'01).* Lecture Notes in Computer Science, No. 1973. Berlin Heidelberg, Germany: Springer-Verlag, 375-389.

Hurtado, C.A., & Mendelzon, A.O. (2002). OLAP dimension constraints. *Proceedings of the 21st ACM SIGACT-SIGMOD-SIGArT Symposium on Principles of Database Systems (PODS'02),* Madison, Wisconsin, 169-179.

Kimball R. (1996). *The Data Warehouse Toolkit.* New York: John Wiley & Sons.

Lehner, W., Albrecht, H., & Wedekind, H. (1998). Normal forms for multidimensional databases. *Proceedings of 8th IEEE International Conference on Scientific and Statistical Database Management (SSDBM'98),* Capri, Italy. Los Alamitos, CA: IEEE Computer Society Press, 63-72.

Lenz, H.-J., & Shoshani, A. (1997). Summarizability in OLAP and statistical databases. *Proceedings of 9th International Conference on Scientific and Statistical Data Management (SSDBM'97),* Olympia, Washington. IEEE Computer Society Press, 132-143.

OLAP Council. (1997). *The OLAP Glossary.* Available on-line at: http://www.olapcouncil.org.

Pourabbas, E., & Rafanelli, M. (2000). Hierarchies and relative operators in the OLAP environment. *Journal of the ACM SIGMOD Record,* 29(1), 32-37.

Pedersen, T.B., & Jensen, C.S. (1999). Multidimensional data modeling for complex data. *Proceedings of the 15th International Conference on Data Engineering (ICDE'99),* Sydney, Australia. Los Alamitos, CA: IEEE Computer Society Press, 336-345.

Rafanelli, M., & Shoshani, A. (1990). STORM: A statistical object representation model. *Proceedings of the 5th International Conference on Statistical and Scientific Database Management (5th SSDBM),* Charlotte, North Carolina. Lecture Notes in Computer Science, No. 420. Berlin Heidelberg, Germany: Springer-Verlag, 14-29.

Smith, J.M., & Smith, D.C.P. (1977). Database abstractions: Aggregation and generalization. *Journal of the ACM Transactions on Database Systems,* 2(2), 105-133.

Chapter V

Operators for Multidimensional Aggregate Data

Maurizio Rafanelli
Istituto di Analisi dei Sistemi ed Informatica – C. N. R., Italy

ABSTRACT

In this chapter the author proposes the different approaches for defining operators able to manipulate this multidimensional structure. In particular, he initially considers operators for multidimensional aggregate data which extend relational algebra and relational calculus (the so-called enlarged relational model). Then he discusses operators for multidimensional aggregate data defined in a tabular environment. In both the cases the author defines such data as statistical (aggregate) data. Subsequently he introduces the operators for OLAP applications, giving a terminology correspondence between the multidimensional aggregate (statistical) databases and OLAP areas. Then he defines the fundamental operators deduced from the previous ones, which form the basic algebra for the manipulation of multidimensional aggregate data, giving their formal definitions and some explanatory examples.

A data model consists of a data structure, a set of operators which define an algebra, and the semantics that such operators have for this data structure. In the case of multidimensional aggregate databases, the classic relational algebra operators do not support all the possible operations needed to manipulate these complex data. Since 1982 Klug understood this limitation and proposed, in Klug (1982), an innovative operator (aggregate formation). He extended the relational algebra and the relational calculus with aggregate functions and showed that this extended language was equivalent in expressive power. He defined the *aggregate formation* operator in the following way. First it partitions tuples of relation R so that tuples having the same X component are in the same partition. Then the function f is applied to component A of the tuples in each partition, and the X-value and the associated aggregate value are output for each partition. Formally:

Let $R \in D$, $X \subseteq Atr (R)$, $|X| = k$. Let f be an aggregate function, and A, $A \in Atr (R)$, be simple-valued. Then

$$R < X, f_A > = \{ t \mid X \mid o \ y \mid t \in R \ and \ y = f_A (\{ t' \mid t' \in R \ and \ t' [X] = t [X] \}) \}$$

where '**o**' denotes concatenation.

Starting from this idea, other authors began to propose different models for multidimensional aggregate data (called, at that time, statistical data) and, consequently, many operators (and algebras) able to solve the problems which arose when a user tried to manipulate this kind of data. Two different approaches were followed. The first approach consisted essentially of modifying the relational algebra, and extending its operators. Among the different proposals made, those proposed in Ozsoyoglu & Ozsoyoglu (1983), in Ozsoyoglu, Ozsoyoglu, & Mata (1985), in Su (1983), in Ghosh (1986), and in Ozsoyoglu, Ozsoyoglu, & Matos (1987) are the most significant.

At the same time, other authors suggested a different approach. It consisted of formulating new operators which had an array, i.e., a multidimensional table, rather than a relation, as a reference structure. They also had the characteristics of not distinguishing between rows and columns (as a relation did) and of considering two different types of attributes with particular peculiarities: the descriptive attributes and the summary attributes. Among the proposals which refer to this second approach, we remember those presented in Rafanelli & Ricci (1984, 1985), in Fortunato, Rafanelli, Ricci, & Sebastio (1986), in Rafanelli & Shoshani (1990), and in Rafanelli & Ricci (1993). As already mentioned, these data structures can be represented in different ways: relations, istograms, graphics, cakes, tables, cubes, etc. The most widely used representation is a multidimensional table, also because it is the most common way of seeing (on a paper or on a screen) this kind of data. This manner of representing multidimensional aggregate data can often create confusion from different points of view. For example, their multidimensionality is not so evident. Different representations of the same structure (for example, exchanging rows and columns) can make the described fact seem different, while it is only the visual representation that is different—sometimes the difference between an attribute and an instance of an attribute is not clear—as explained in Rafanelli & Shoshani (1990).

For many years different algebra, and, therefore, many ad-hoc operators, were proposed, based both on a relational structure, and on a tabular structure.

In these last years, with the coinage of the new acronym OLAP (Codd, Codd, & Salley, 1993), the interest from important academic and industrial sectors has increased the amount of research in the multidimensional database area, especially from the data analysis point of view. At the heart of OLAP applications is the ability to simultaneously aggregate across many sets of dimensions. Data analysis on-line is part of this need. The request for simultaneous multidimensional aggregation was the main reason for the extension of SQL, including the "cube" operator, as proposed in Gray, Bosworth., Layman, & Pirahesh (1996) and subsequently in Gyssens & Lakshmanan (1997) and in Nguyen, Tjoa, & Wagner (2000). We will discuss the different proposals which refer both to the enlarged relational algebra and to the operators for tables, and to OLAP operators.

OPERATORS FOR MULTIDIMENSIONAL AGGREGATE DATA WHICH EXTEND RELATIONAL ALGEBRA AND RELATIONAL CALCULUS

In the literature on aggregate multidimensional (called initially statistical) databases, there are two different approaches for defining operators able to manipulate this multidimensional structure, from now on called *multidimensional aggregate data* (MAD): the enlarged relational model and the multidimensional tabular model.

Historically, the first approach refers to the enlarged relational model, and it is described in different papers. For example, in Klug (1982) the author describes the equivalence which exists between relational algebra and relational calculus with regard to the query languages which have aggregate functions. He gives a precise definition of aggregate functions and extends the relational algebra and relational calculus in a general and natural fashion to include aggregate functions. He also shows that the languages extended in such a way have equivalent expressive power. Finally, the author proposes a new operator, called *aggregate formation*, described at the beginning of Chapter 1.

In Su (1983) the author proposes the G-relation structure (G stands for 'Generalized'), and a set of operators based on this structure which enlarge the classic relational operators. A G-relation is formally defined as follows: given a collection of concept types A_1, ..., A_n, B_1, ..., B_m, each of which contains a set of occurrences belonging to the same summary type, a G-relation type, denoted as

$R(A_1, ..., A_n \| B_1, ..., B_m)$ and defined on these n+m sets, is a set of ordered n+m tuples $<a_1, ..., a_n, b_1, ..., b_m>$ such that $a_i \in A_i$, for i = 1 to n, and $b_j \in B_j$, for j = 1 to m. Ai are called identifying domains and B_j are called summary domains. A number of restructuring operations and reviewed algebraic relational operations are defined and illustrated, such as *aggregation* (similar to summarization) or its inverse, *disaggregation*.

In Ghosh (1996) the author discusses an extension of relational algebra in order to execute a set of operations on a table (or two tables for joints), and to still have a table as a result. He also explains that it is possible to define both the classic relational operations of projection, selection, ordering, union, intersection, join, and outer join on domains of the relation R, and, simultaneously, also to define numerical operations of sum, subtraction, multiplication, division, vector dot products, restricted dot products, vector products, and restricted vector products on the rows of the relation.

In different papers, G. Ozsoyoglu and Z.M. Ozsoyoglu study an extension of relational algebra for summary tables, and a query language for manipulating these data. In particular, in Ozsoyoglu & Ozsoyoglu (1983) and Ozsoyoglu, Ozsoyoglu, & Mata (1985), they propose this extension, which is completed in Ozsoyoglu, Ozsoyoglu, & Matos (1987), where an extended relational algebra (ERA), containing aggregate functions and set-valued attributes, is proposed. A rigorous formalization of this algebra is presented and discussed, as well as some other operators, which are different from the traditional relational operators being proposed. In particular, starting from literals and relations, the operators discussed in the following are proposed: *projection*, *cross-product* (the Cartesian Product without the ordering constraint), *restriction*, *set union*, *set difference*, *aggregate formation* (imported from Klug, 1982), *pack* (which groups all the instances of the category attribute specified by the operator in one unique set-value), *unpack* (which is the inverse of the latter, i.e., splits the unique set-value instance of the category attribute specified by the operator in all the instances previously grouped), *construct* (which constructs a single-column set-valued relation using a range of values to form tuple components or tuples of the relation explicitly), *aggregation-by-template* (which groups tuples of the first relation (first operand) specifying both the attributes and the aggregation function). By the construct operation it generates a new relation where the summary values are obtained by applying the aggregation function to the instances, single values, or set-values, specified in the construct operator), *θ-join* and *selection*. They also discuss and formally define the calculus objects with aggregate functions, considering calculus expressions constructed by using variables, terms, formulas, range formulas, and by utilizing alpha expressions recursively. Finally, they demonstrate the equivalence between Algebra expressions and Calculus objects in both directions.

In Meo-Evoli, Ricci, & Shoshani (1992), operators that correspond to the relational algebra operators "select," "project," and "union," as well as "aggregate," were defined. They have been labeled "S-select," S-project," etc. (S stands for "statistical operator"). They have the following semantics in the context of an aggregate statistical object:

S-select: selects a subset of category values of a category attribute. This does not reduce the cardinality of the multidimensional space, except if a single value is selected.

S-project: summarizes over all values of a dimension. This reduces the cardinality of the multidimensional space by one.

S-aggregation: summarizes over the values of the classification hierarchy. One can specify a summarization over one or more levels. This also does not reduce the cardinality of the multidimensional space.

S-union: is used to combine multiple statistical objects which have overlapping (or partially overlapping) category values.

We will show that similar operators have been identified in the OLAP area.

OPERATORS FOR MULTIDIMENSIONAL AGGREGATE DATA DEFINED IN A TABULAR ENVIRONMENT

The second approach mentioned above was created when some authors did not consider the relational model and its algebra as the correct data structure and the correct operator set for this kind of complex data. In fact, the relational model did not consider differences among columns, but it distinguished the difference between a column and a row. The latter is called a "tuple" of the relation, while all the columns were considered "attributes" of the relation, equivalent to each other. Previously, other authors highlighted the distinction between parameter (descriptive variable or category attribute) and measured data (quantitative variable or summary attribute), which becomes quite important in multidimensional aggregate databases, as observed in Johnson (1981), Shoshani (1982), Rafanelli & Ricci (1983), and Su (1983). Recently, because of the new application type, i.e., the On-Line Analytical Processing of data, different authors have proposed models which provide symmetric treatment not only of all dimensions, but also of measures, by considering a measure as another dimension. This fact brought about the definition of new operators (for example, push or pull), but also some considerations on the possibility to use the same set of operators after the exchange between the measure and one dimension, as we will see in the section relative to the OLAP environment.

The operators described in this section are able to operate on MAD, working only on the descriptive part of this data structure, with the automatic recomputation, if needed, of the measure (summary attribute). The recomputation of the new summary values depends on the type of summary data, i.e., on the aggregate function applied in the aggregation process to obtain the MAD. Moreover, this recomputation happens in a manner transparent to the user (automatic management of the summary type).

These operators have the aim of generating, from the set of tables memorized in the multidimensional aggregate database, those on which to carry out statistical analysis or On-Line Analytical Processing of data, which however turns out to be a subsequent phase tied to the use of statistical packages or similar ones. They present the following advantages:

1)   flexibility and compactness in their use, in that they are independent from the single MAD and the single summary type (the control of the summary type is the responsibility of the user);

2)   logical independence, in that the user does not have to specify the calculation procedures of the summary values and can therefore work on the metadata which describe the MAD by means of (possibly visual) interfaces, which use direct manipulation techniques, as proposed in Rafanelli (1990);

3)   ease in verifying their properties (associative, commutative, etc.); in fact, it is not possible to set general conditions for the calculation procedures of the summary values and therefore verify the properties of the operators themselves.

It is important to underline the fact that no operator modifies the summary type of the measure, because to obtain this, it is necessary to use suitable statistical packages; in this case the user does not perform "data manipulation," but "data elaboration," even if, in the user activity, these two phases often intersect.

In Rafanelli & Ricci (1984, 1985), the authors proposed a statistical query language (STAQUEL) for the aggregate data definition and manipulation, and also the operators of *summarization* (which reduces the number of dimensions of a complex table by eliminating one or more category attributes) and of *reclassification* (which reduces the number of domain values related to category attributes by grouping them and, therefore, aggregating such groups). This reclassification is carried out along a predefined hierarchy, it is called (in OLAP applications) *roll up*.

In Fortunato et al. (1986), a distinction between operators independent of the summary type of a given table and operators dependent on this summary type is discussed. All these operators, for which that paper proposed a formal definition, are different from the traditional relational operators, because the data structure on which they work is different from a classical relation by being a more complex structure. In particular, as operators independent of summary type, they propose *macro-union* (on the same MAD schema, the union between the single pairs of equal attributes are performed), *macro-select* (single instances specified in a suitable table are selected), and *comparison* (two MAD or one MAD and one number are compared, according to a θ-operator (>, ≥, etc.), obtaining a new MAD with possible null values). As operators dependent on summary type, they propose *summarization* (one category attribute is deleted and the measure instances are recomputed), *restriction* (only the category attributes specified by a qualification condition are selected for the new MAD), *enlargement* (practically the inverse of restriction, obtained by a distribution law expressed by a table), and *reclassification* (it substitutes a set of category attributes with another set, according to a functional dependency expressed by a relation).

The Mefisto model, based on the previous aggregate data structure and on a reviewed set of operators, has been proposed in Rafanelli & Ricci (1988, 1990, 1991, 1993). In these papers a formalization of the above-mentioned elements of the model (data structure and operators) is illustrated. Part of these operators were further studied in later papers (see Rafanelli & Shoshani, 1990; Bezenchek, Rafanelli, & Tininini, 1996a, 1996b).

As mentioned in the introduction of this chapter, visualization and data analysis (which represent, together with query formulation and extracting aggregate data from a database, the four steps for data analysis applications, as discussed in Gray et al., 1997), use tools which carry out dimensionality reduction (aggregation or summarization) in order to represent the dataset as an n-dimensional space. Boolean operations or logical associations between data are not of prime importance to users; in addition, updating and deleting data is rare or forbidden (see Ghosh, 1986), because their statistical use means they are generally considered *static* data, that is, data which represent "events consolidated in time." In fact, the most common manipulation is related to the encoding of data summarization, or to the reclassification of the descriptive data. In order to schematically represent the operations formally described later in this chapter, we use the graph in Figure 1, where the orientated edges go from the input data structure to the output data structure, with regard to the operations associated with a single edge.

Figure 1

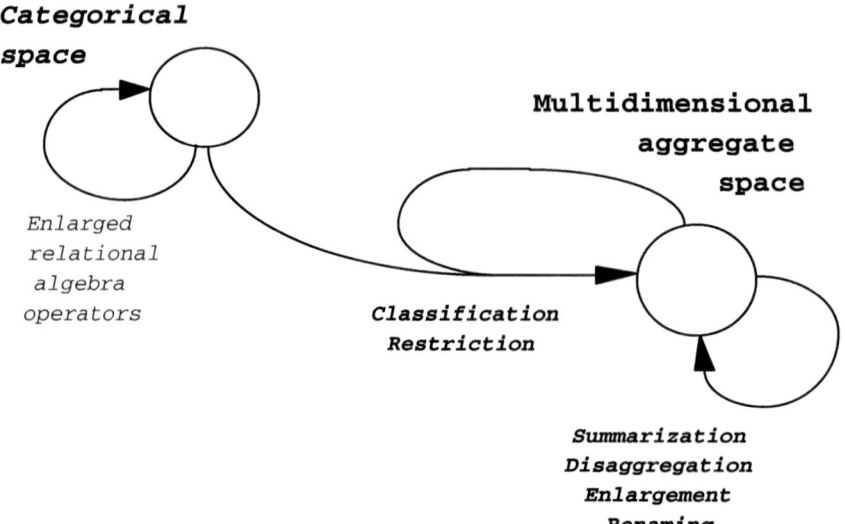

The algebra operations can have one or two multidimensional aggregate entities in input, or the couple <relation, multidimensional aggregate entity>; output is one multidimensional aggregate entity. The relational algebra operations (see, for example, Abiteboul, Hull, & Vianu, 1995) are necessary for the manipulation of relations resulting from the performance of the comparison operation; they are relations in non-first normal-form (that is, relations having a set like tuple components, relations such as those discussed in Ozsoyoglu, Ozsoyoglu, & Matos (1987).

The operators, described in the following, refer to the n-dimensional aggregate data structure, generally represented by a table. They are able to operate on this data structure, working only on its descriptive part. Furthermore the way of working on the summary values is included in each operation, that is, the computation of the summary values is automatic and therefore it happens in a manner transparent to the user.

OPERATORS FOR OLAP APPLICATIONS

Recently, in the literature, many authors have proposed operators and query languages which refer to OLAP applications. This acronym, coined and proposed in Codd, Codd, & Salley (1993) to characterize the requests for summarizing, consolidating, viewing, applying formulae to, and synthesizing data according to multiple dimensions, resumes the concept of enabling analysts, managers, and executives to gain insight into the performance of an enterprise through fast access to a wide variety of views of data organized to reflect the multidimensional nature of the enterprise data (Colliat, 1995). The previous approaches were the basis for the new operator proposals in the OLAP environment, changing the goals of the users, which focused greater attention on the dynamic analytical processing of data on-line, and therefore on the consequent problems linked to this application, such as access performance, or the number of materialized views to store. Depending on the initial approach, relational or multidimensional-tabular, the relative processes have been called ROLAP (Relational On-Line Analytical Process); see MicroStrategy (www.strategy.com), MOLAP (Multidimensional On-Line Analytical Process), and ArborSoft (arborsoft.com). As discussed in Shoshani (1997), MOLAP proponents claim that:

i)   Relational tables are unnatural for multidimensional data.
ii)   Multidimensional arrays provide efficiency in storage and operations.
iii)   There is a mismatch between multidimensional operations and SQL.
iv)   For ROLAP to achieve efficiency, it has to perform outside current relational systems, which is the same as what MOLAP does.

Instead, ROLAP proponents claim that:
i) ROLAP integrates naturally with existing technology and standards.

ii)   MOLAP does not support ad hoc queries effectively, because it is optimized for multidimensional operations.

iii)   Since data has to be downloaded into MOLAP systems, updating is difficult.

iv)   Efficiency of ROLAP can be achieved by using techniques such as encoding and compression.

v)   ROLAP can readily take advantage of parallel relational technology.

The claim that MOLAP performs better than ROLAP is intuitively believable, and in Zhao, Deshpande, & Naughton (1997), this was also substantiated by tests.

Now we examine the most relevant proposals, presenting, if necessary, the reference model (the data models have been discussed in Chapter 2).

In Gray et al. (1996), and subsequently in Gray et al. (1997), the authors propose the *data cube* operator as an extension to SQL which generalized the histogram, *group-by* (the classic SQL operator extended to support histograms and other function-valued aggregations), *cross-tabulation* (the 2-D data cube obtained crossing two category attributes of the n-dimensional cube structure), *roll-up* (going up the levels of a hierarchy), and *drill-down* (going down the levels of a hierarchy) constructs found in most reports. An example of these constructs is shown in Figure 2 (the original figure appears in Gray et al., 1997).

In Li & Wang (1996), the authors formalized a multidimensional data model for OLAP, and developed an algebra query language called *Grouping Algebra* which is the basis for a *multidimensional cube algebra*, proposed in order to facilitate the data derivation.

Figure 2

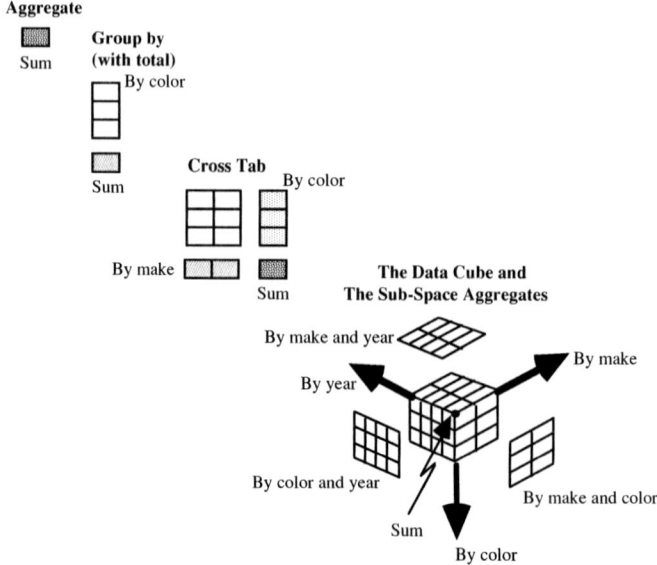

After the definition of the classic rename, roll, and aggregation operators, the authors introduce other operations, namely: *add dimension* (which generates a new cube which has a new dimension named D, whose relation scheme is the empty set, and the relation for the dimension only has one tuple, i.e., the empty tuple), *transfer* (which generates a new cube by transferring attribute A of dimension D_1 to a (new) attribute B on dimension D_2), *union, cube aggregation* (which generates a new cube by "compressing" the original cube into a smaller one), *rc-join* (which joins the relation *r* to dimension D_1), and *construct* (which generates a cube from relation r so that the new cube is one-dimensional).

In Gyssens & Lakshmanan (1997), the authors give a foundation for multidimensional databases. They define a slicing and dicing operation as "just a special case of relational *selection* extended to multidimensional data tables," and also propose the classic algebraic operators. In particular, they define three sets of classic algebraic operators. The first consists of three *unary* operators, *selection* σ_C, *projection* π_X, and *renaming* $\rho_{B \leftarrow A}$, where C is a valid selection condition, defined as usual, X is a subset of the attribute set of the table (i.e., the MAD), and A and B are attribute names. The authors define **op(τ)** = tab_S *(op(rep(τ)),* where they denote a table instance by τ, its tabular representation by tab_S (r) (i.e., a table instance with schema S), and the relational representation f(τ) of the table by *rep(τ)*. After having discussed the classic operators *union, intersection, difference,* and *Cartesian product*, they propose the following operators: *fold* (which reduces the table by one dimension, modifying the table schema), *unfold* (the inverse of the previous one), *classification* (defined on relations and on tables, it maps tuples of a relation to one or more groups), and *summarization* (where the structure of the output table is the same as the input table, as far as the dimensions and parameters are concerned; the only change is that sets of measure tuples are summarized according to the aggregate function and transformed into single values). Finally, they give a rigorous formal expression of *data cube* and *monotone roll-up*.

In Agrawal, Gupta, & Sarawagi (1997), the authors propose a data model which provides support for multiple hierarchies along each dimension and for ad hoc aggregates, as well as a few algebraic operators. In this paper, they deal more with multiple hierarchies, and introduce the multiplicity of a hierarchy as a semantic variant of the simple one. The operators which they propose are: *push* (it converts dimensions into elements that can be manipulated using function f_{elem}), *pull* (it is the converse of the push operator, and creates a new dimension for a specified member of the elements), *destroy dimension* (this operator removes a dimension *D* that had a single value in its domain), *restriction* (this operator operates on a dimension of a cube and removes the cube values of the dimension that do not satisfy the initial condition), *join* (it is used to relate information in two cubes; the result of joining an m-dimensional cube with an n-dimensional cube on k dimensions is cube C with n+m-k dimensions), the classic *Cartesian product, natural join, union, merge* (this is an aggregation operation which maps multiple product names to one or more categories, or similar mapping), and *associate* (depending on the associative function f_{elem}, it transforms the measure values of two cubes into another cube, with possible reductions of the domain instances of the dimensions, and the possible changing of the summary type of the cube).

In Cabibbo & Torlone (1997, 1998), the authors propose a logical model for OLAP systems that is usable in designing multidimensional databases, and independent of any specific implementation. the operators proposed in this paper are the classic Cartesian product, natural join, renaming, selection, (simple) projection, roll-up, aggregation, as well as level description, Scalar function application, and abstraction.

In Lehner (1998) the author discusses the design problem which arose when the OLAP scenarios became very large, and proposes a nested multidimensional data model useful during schema designing and multidimensional data analysis phases. The author points out that he will not propose another data model, but provide necessary extensions in different directions. After having defined the *nested multidimensional data cube*, he introduces operators on multi-dimensional objects (MOs). In particular, they have the *slicing* operator (it produces a new MO which inherits all components of the source MO except the context descriptor), *drill-down, roll-up, split, merge, explicit aggregation*, and *cell-oriented* operators (which are the classic unary operators -, abs, and sign, and binary operators *, /, +, -, min, max, and which work directly on cells).

In Nguyen, Tjoa, & Wagner (2000), the authors propose a conceptual multidimensional data model based on MetaCube, and three basic navigational metacube operators which are applied to navigate along a metacube C, corresponding to a dimension D, namely *jumping, rollingUp* and *drillingDown*.

In Pedersen & Jensen (1999), and subsequently in Pedersen, Jensen, & Dyreson (2001), after a survey of 14 multidimensional data models, the authors propose an extended multidimensional data model and algebraic query language. Then, they define the basic algebra of the model, where the operators are similar to the standard relational algebra operators. In particular, they propose *group* (which groups the facts characterized by the same dimension values), *union* (which performs a union of the categories and the partial orders, taking the set union of the facts and the fact-dimension relations), *selection* (which restricts the set of facts to those that are characterized by values where a given predicate *p* on the dimension types evaluates to true; dimensions and schema stay the same), *projection* (over k dimensions, without removing duplicate values), *rename* (used to alter the names of dimensions), *difference* (which obtains the set-difference of the facts, while the dimensions of the first argument are retained, and the fact-dimension relations are restricted to the new fact set), *identity-based join* (the new fact type is the type of pairs of the old fact types, while the new set of dimension types is the union of the old sets; the set of facts is the subset of the cross-product of the old fact sets, where the join predicate *p* holds), *aggregate formation* (see Klug, 1982), *value-based join* (a join between two MADs on common dimension values, by combining

Cartesian product, selection, and projection), *duplicate removal* (elimination of "duplicate values," i.e., facts characterized by the same combination of dimension values), *SQL-like aggregation* (it carries out the computation of an SQL aggregate function on a MAD grouped by a set of dimension categories, by first applying the aggregate formation operator with the given categories and the given function), *star-join*, *drill-down* (it gives more detail by descending the dimension hierarchies), and *roll-up* (it gives less detail by ascending, the dimension hierarchies). They also propose two operators for handling time, namely *valid-timeslice* and *transaction-timeslice*.

As noted in Shoshani (1997), there is a terminology correspondence between the multidimensional aggregate (statistical) databases and OLAP areas. The more common equivalent terms are:

Category Attribute	Dimension
Category Hierarchy	Dimension Hierarchy (table)
Category Value	Dimension Value
Summary Attribute	Measures (fact column)
Statistical Object	Data Cube (fact table)
(Multidimensional Aggregate Data)	
Cross-Product	Multidimensionality
Summary Table	Table/Data Cube

Before speaking of the operators—taken mainly from the set of all the operators previously presented, which will form the base algebra, together with other operators which, even if not strictly necessary, enlarge the base algebra in order to facilitate the definition of multidimensional aggregate query languages—we return to the symmetric treatment of dimension and measure of a MAD. This is for two reasons: i) in order to demonstrate that this symmetric treatment was also possible (even if not always necessary) in a statistical database, and ii) that the problems which arose in a statistical database repropose word for word in an OLAP environment. It is evident that a "described" data can become, in its turn, a "describing" data, i.e., to play either the role of parameter (or category attribute) or the role of measure (or summary attribute). Therefore, a measured data is not always also a "summary" data. Problems arise when, after an exchange between the measure (summary) and one dimension, we apply the same operators which normally apply to the original conditions.

For example, suppose we have the table shown in Figure 3.

In reality this table does not have aggregate data, i.e., it is a classic relation, as can be seen in Figure 4.

As can be easily seen, the only change is the different way of representing the data. In this situation an exchange between "Cost (per unit)" and another attribute (for example, "Product") is also easy to carry out. The more evident problem is the large quantity of null values which appear as "measure," as can be seen in Figure 5.

Figure 3

title Sales		Cost (1) per unit, x Sales	Year		1990			1991			...
Product	**City**		**Month**	**Jan**	**Feb**	...	**Dec**	**Jan**	**Feb**	... **Dec**	...
PC	Montreal			700,70	670,80	...	640,46				
	Toronto			528,48	631,51	...	665,43				
	...			...	...	...	...				
Laser writer	Montreal			...	...	...	...				
	Toronto			...	...	...	...				
	...			...	...	...	...				
...	...			...	...	...	...				

(1) - Cost in US $

Different problems arise when we try to apply the aggregate operators to this relation. For example, if we *group* the instances of the attribute "Month" in the table in Figure 3 by the value-range "Quarter," we do not know what the PC cost is in the

Figure 4

Product cost	Product	City	Year	Month	Cost (per unit)
	PC	Montreal	1980	January	700,00
				February	670,80
				....	
				December	640,46
		Toronto		January	528,48
				February	631,51
				....	
				December	665,43
			1981		
	....	....	....	......	......
	Laser writer	Montreal	1980	January	....
	......	....	.....	....	....

Figure 5

Product sold		Year	1990				1991				...
		Month	Jan	Feb	...	Dec	Jan	Feb	...	Dec	...
City	Cost (1) per unit										
Montreal	...										
	640,46		---	---	...	PC	...				
	670,80		---	PC	...	---	...				
	700,70		PC	---	...	---	...				
	...										
Toronto	528,48		PC	---	...	---	...				
	631,51		---	PC	...	---	...				
	665,43		---	---	...	PC	...				
	...										
Montreal	...		Laser writer	...	...	...	...				
Toronto	...		...								
...											

(1) - Cost in US $

different cities in the first quarter, or the second quarter, etc. At the same time, in the table in Figure 5, if we *group* the "Cost per unit" values into two value-ranges, "≤650" and ">650" respectively, and then we apply the *roll-up* operator to the hierarchy Year-Month, the PC instance of the attribute "Product" appears in both the rows "≤650" and ">650" relative to city = Montreal, as shown in Figure 6.

Analogous problems happen if we only use the operator *roll-up* applied, for instance, to the hierarchy "Year-Month." In fact, if we delete the attribute "Month" in Figure 3, what is the cost measured in 1990, in 1991, and so on? If instead we have a MAD, i.e., an array with a measure obtained by an aggregation process and, therefore, as a first result, *numbers* as measures in the cells of the MAD, we can still exchange the numeric measure with another dimension.

Applying the group-by operator on this new MAD, some inconsistencies can be produced. For example, if we group into two ranges ("≤15" and ">15"), the values relative to "Sold Qty," the measure "Vendor" appears in the cells as shown in Figure 9(a).

Figure 6.a

| Product sold | | | | 1990 | | 1991 | Year |
City	Cost range	January	February	....	December	.....	Month
Montreal	≤ 650	---	---		PC		. . . .
	> 650	PC	PC		---		. . . .
Toronto	≤ 650	PC	PC		---		. . . .
	> 650	---	---		PC		. . . .
. . . .		. . . .	. . . .		. . . .		

Figure 6.b

| Product sold | | | | | | |
City	Cost range	1990	1991	. . . .		Year
Montreal	≤ 650	PC	. . . .	. . . .		
	> 650	PC	. . . .	. . . .		
Toronto	≤ 650	. . . .	. . . .	. . . .		
	> 650	. . . .	. . . .	. . . .		
. . . .		. . . .	. . . .	. . . .		

If, then, we group the cities again by region, we can have two different situations: i) the vendors B and C appear both in the row "≤15" and in the row ">15."

In this case it is very difficult to define a general rule by which if B (or C) is defined by two (or more) different values of the same dimension, it is automatically defined only by the greater value (in this case, ">15"). Also, the vendor A is described only by one value of a dimension (in our case, "≤15").

In this case it is impossible to find a general rule by which if A is defined by only one value of a dimension, then applying the grouping operator to another dimension, it remains as described by the same previous value.

For example, if A has sold ≤15 IBM computers in all the cities of the region Lazio, grouping these cities in region=Lazio, it is impossible to understand if the total number of IBM computers sold by A in Lazio is still ≤15 or not (see Figure 9(b)). This is analogous for the vendor C and Macintosh computers.

Suppose we consider the MAD in Figure 7. Suppose we also exchange "Vendor" with "Sold quantity." The new MAD is shown in Figure 8.

Figure 7

Quantity sold in 1999	IBM						Macintosh				
	Lazio			Tuscany		.....	Lazio			Tuscany	
	Roma	Viterbo	Latina	Florence	Siena	.....	Roma	Viterbo	Latina	Florence	Siena
A	15	2	1	....	....	....	55	1	0	....	....
B	35	4	1	....	....	....	5	2	1	....	....
C	80	7	2	....	....	....	15	4	1	....	....

Vendor

Figure 8

Vendors in 1999	IBM						Macintosh				
	Lazio			Tuscany		.....	Lazio			Tuscany	
Quantity sold	Roma	Viterbo	Latina	Florence	Siena	.....	Roma	Viterbo	Latina	Florence	Siena
0	--	--	--	...	...	...	--	--	A	....	....
1	--	--	A, B	....	....	....	--	A	B, C	....	....
2	--	A	C	....	....	....	--	B	--	....	....
4	--	B	--	....	....	....	--	C	--	....	....
5	--	--	--	....	...	...	B	--	--	....	....
7	--	C	--	....	...	...	--	--	--	....	....
15	A	--	--	....	...	...	C	--	--	....	....
35	B	--	--	....	...	...	--	--	--	....	....
55		--	--	....	...	...	A	--	--	....	....
80	C	--	--	....	...	...	--	--	--	....	....

Figure 9.a

Vendors in 1999	IBM						Macintosh				
	Lazio			Tuscany		.....	Lazio			Tuscany	
Quantity sold	Roma	Viterbo	Latina	Florence	Siena	....	Roma	Viterbo	Latina	Florence	Siena
≤15	A	A, B, C	A, B, C	...	...	...	B, C	A, B, C	A, B, C	....	....
>15	B, C	--	--	....	....	....	A, C	--	--	....	....

Figure 9.b

Vendors in 1999	IBM			Macintosh	
	Lazio	Tuscany	.....	Lazio	Tuscany
Quantity sold					
≤ 15	A, B, C	...	...	A, B, C	....
>15	B, C	....	....	A, C	....

This means that it is always possible to exchange a numeric measure (obtained by an aggregation process from microdata) with a dimension, but on the new MAD it is very often impossible to operate again with the classic operators for aggregate data without risking having inconsistent or wrong data.

This observation is implicitly confirmed in Hurtado, Mendelzon, & Vaisman (1999), where the authors note that in a relational implementation of OLAP (namely ROLAP), facts (viewed as a mapping from a point in a space of dimensions into one or more spaces of measures) are stored in fact tables, while each dimension is described in a dimension table. The usual assumption is that fact tables (generally *summary data*) reflect the dynamic aspect of the database (or, in general, of a data warehouse), while data in dimension tables (the descriptive space) are relatively static.

THE BASIC ALGEBRA FOR MULTIDIMENSIONAL AGGREGATE DATA

In this section we define the fundamental operators deduced from the previous ones, which form the basic algebra for the manipulation of multidimensional aggregate data. Part of these operators are, to all appearances, similar to the standard relational algebra. However, we still have to consider the fundamental difference existing between the classic relation (where differences between attributes do not exist) and the multidimensional aggregate data (represented by tables, but that could be represented also by cubes, or any other visual representation conforming to the wishes of the user), where a fundamental difference exists between the descriptive attributes and the summary measure, as previously explained.

Obviously, when the semantics of the operator allow the exchange of roles between one descriptive attribute and the summary measure, for example, in order to answer a given query, this operation is legal. But if we wish to apply different operators to the result of the above-mentioned exchange, often this sequence of operations can produce inconsistent or wrong information. For this reason the important concept of *summarizability* was introduced by many authors. An excellent discussion is made in Lenz & Shoshani (1997).

Summarizability depends on different situations, like the summary type (a count or a sum is summarizable, while an average is summarizable only if we have the corresponding count and sum data) or the hierarchy type.

An important, simple definition used in the following refers to the *f-expression.*, as defined in Cabibbo & Torlone (1998). If F is a MAD (or a fact table, or f-table) having a schema schema $[A_1, A_2, ..., A_n] \rightarrow [M_1, M_2, ..., M_m]$, then F is an *f-expression* over the same schema as F, that is, an f-expression over the attributes $A_1, A_2, ..., A_n$ and the measures $M_1, M_2, ..., M_m$.

Now we will resume, in a unique set, all the main proposals of operators for multidimensional aggregate databases made in the literature, and give their formal definition.

SUMMARIZATION

One of the predominant operations on multidimensional aggregate data is that "to remove" a dimension from a multidimensional aggregate data (obtaining, for example, a "Population by year and age-groups" from a "Population by year, age-groups, and sex"). Such an operation is often called *summarization*. This operator works with only one operand, and it produces a recomputed measure (in the case of numerical values) or instances formed by sets of alphanumeric values (in the case of non-numerical values). The first formal proposal of summarizing an attribute, reducing the number of dimensions of a MAD (in that case, a table), was made in Rafanelli & Ricci (1984, 1985), and subsequently in Fortunato et al. (1986), Rafanelli & Shoshani (1990), and Rafanelli & Ricci (1993). In Gyssens & Lakshamanan (1997), the authors study this operator both on relations, and on tables.

Other different terms have been used for this operation. Among them, we remember *aggregation*, informally discussed in Shoshani & Wong (1985), where, among the different concepts discussed, there is that of the "collapsing" of multidimensional data structures in order *to remove* a certain dimension; *attribute removal by aggregation* in Ozsoyoglu, Ozsoyoglu, & Mata (1985), *slice* (term especially used for OLAP applications) in Gyssens & Lakshamanan (1997) and in Shoshani (1997), and *destroy dimension* in Agraval, Gupta, & Sarawagi (1997). Since the term "aggregation" has been widely used in this chapter to denote a different concept, in the following we will use the terms *summarization* (which often refers to the statistical databases) and *removing* with the same meaning. Often, when referring to the relational algebra, this operator is called *projection*, as in Ozsoyoglu, Ozsoyoglu, & Matos (1987), Pedersen & Jensen (1999), and Pedersen, Jensen, & Dyreson (2001), with very few differences.

As already mentioned, this operator deletes one category attribute of a MAD, with consequent recomputation of the summary attribute values. This recomputation is not always possible: for example, if the measure is not numeric, or if, in the case of numeric values, the summary type of the MAD is "average." In this latter case we need the relative "count" and "sum" aggregate summary values, or the raw data, to which to apply the aggregation process again. Since a multidimensional aggregate data structure represents a *functional link* between sets of raw data (rather than n-tuples of dimension instances) and measures, in our framework *summarization* is the operation that allows the user to (implicitly or explicitly) delete one attribute (which, in this case, represents one dimension of the MAD), or to transform it into an implicit one, and to recompute the measures accordingly.

In the following, in order to avoid ambiguity, we will distinguish the *total summarization* or **T-SUMMARIZATION** (which implies the removal of the dimension) from the *implicit summarization* or **I-SUMMARIZATION** (which transforms the dimension from explicit to implicit, and the set of definition domain instances in only one set-value, which resume all the values of the original domain, but not all the values of the primitive attribute definition domain). The descriptive space of the MAD reduces itself to one dimension without loss of information only in the first case.

In Bezenchek, Rafanelli, & Tininini (1996a) and, subsequently, in the ADAMO model (see Bezenchek, Rafanelli, & Tininini, 1996b) in Chapter 1, the above-mentioned distinction between *total summarization* and *implicit summarization* was made. Therefore, with the introduction of the "implicit attribute" concept, the summarization operator has been refined. The (total or implicit) summarizability of a category attribute, or its non-summarizability, depends on three interdependent factors, namely:

1) the partitioning characteristics of the category attribute A_i ;
2) the fact described by the MAD;
3) the aggregation function type applied to the raw data to obtain this MAD.

In particular, it has been shown that the partitioning characteristics of A_i depend on A_i itself and on the specified instances, but also on the particular fact described by the MAD. For example, the attribute *continents,* with its instances Africa, America, Asia, Europe, and Oceania, partitions the fact domain corresponding to "population" but does not partition (since it does not cover) the fact domain corresponding to "terrestrial surface." In principle, the system can automatically determine whether a given attribute $\mathcal{A}_i$ in a MAD a is (T- or I-) summarizable, provided that it has an adequate knowledge regarding the instances of the category attribute $\mathcal{A}_i$ combined with the specified fact and aggregation function type. However, this is not always true in practice, and the summarizability of a category attribute also depends on the particular survey which has produced the MAD, i.e., on a collection of metadata, which has to be produced together with the aggregate data themselves. Note that, when speaking of *dimension,* in general we intend one level (i.e., one category attribute) of one of the possible hierarchies in a dimension; for the sake of simplicity, when ambiguities are not possible, we will call it *dimension,* collapsing the more complex meaning of the term into the simpler term of a descriptive variable.

For example, let us consider the MAD "Number_of_cars_produced_in_Japan" in Figure 10, described by "model" and "years" (but also by "country," where this dimension is "implicit" because it has only one value, "Japan," which appears in the title of the MAD). Suppose we wish to have the total number of cars produced only per "years." In this case we have to apply the summarization operation to the category attribute "model." Because the instances of the category attribute model are <Corolla, Civic, Corona>, and because these are not the only car models produced in Japan in that period, the operator applied will be *I-SUMMARIZATION.*

Figure 10

#_OF_CARS_PRODUCED_IN_JAPAN (by Model and Years) (absol. values, by thousands)	MODEL		
	Corolla	Civic	Corona
1980	427	341	220
1981	458	373	249
1982	499	401	285
YEARS 1983	536	438	317
1984	601	496	369
1985	710	580	421
....	...	...	...

Figure 11

Number_OF_CARS_PRODUCED_IN_JAPA (by years) (absolute values, by thousands)		
		#
	1980	988
	1981	1,080
YEARS	1982	1,185
	1983	1, 291
	1984	1,466
	1985	1,711
	. . . .	. . . .

Note: The car number refers only to the models
"Corona, Civic, and Corolla."

In this way the category attribute "model" will be transformed into an implicit attribute and a note will be added to the MAD, as shown in Figure 11.

We remember that in the Chapter 1 we gave the formal definition of a simple MAD s_1. Therefore, given a phenomenon x and given the set of all the relations R_x (of the micro database) involved in the production of all the MAD which describe this phenomenon, we considered the subset of R_x formed only by the relations involved in the building of fact F_1. As illustrated in Chapter 1, we call this subset an *aggregation relation*, and denote it by R_1^x, where $R_1^x = \{R_{1,1}, ..., R_{1,s}\}x$. The *base relation* R_{B1} of s_1 (that is, the descriptive space which describes fact F_1) is the subset of the aggregation relation R_1 of this fact which has all and only the descriptive attributes A_{B1j} (with j = 1, ..., s) of the fact F_1.

Let us suppose the summarizability conditions, discussed in the previous Chapter 4, have been verified. Then s_1 is a MAD defined on the base relation R_{B1}

(subset of the aggregation relation R_1^x which refers to fact $\mathcal{F}_1$). Therefore, its descriptive space is the base relation $\mathcal{R}_{B1}$ mentioned above, has its components formed by the set $\{A_{1j}\}$, with $j = 1, ..., s$, where s is the number of category attributes (cardinality) of MAD s_1. Let $<P, N, f, \mathcal{A}, \mathcal{S}, s>$ be the six-tuple which defines s_1. Let $\mathcal{A}$ be formed by $e_1 = \{E_1, E_2, ..., E_M\}$, i.e., the set of its *explicit category attributes*, each of them with its corresponding ordered instance domain $\Delta (E_1) = \{i_{E_1,1}, i_{E_1,2}, ..., i_{E_1,P_1}\}, ..., \Delta (E_M) = \{i_{E_M,1}, i_{E_M,2}, ..., i_{E_M,P_M}\}$, with $M \leq s$, and by i_1, i.e., the set of its *implicit category attributes*.

The **summarization** of s_1 with respect to $A1x$ (with $x \in \{1, ..., M\}$) produces a new MAD

$$s'_1 = <P', N, f, \mathcal{A}', \mathcal{S}, s'>,$$

where $N, f, \mathcal{S}$ are the same and P' is the new name of the MAD; $\mathcal{A}' = \mathcal{A} - \mathcal{A}_{1x}$ and s' is a set of recomputed summary values. This recomputation depends on the type of aggregation function (e.g., count, sum, average) applied to the original microdata.

A_{1x} becomes an implicit category attribute of s'_1. It can completely disappear from the new descriptive space $\{A_{1j}\}$, with $j' = 1, ..., x-1, x+1, ..., s$, if its definition domain $\Delta (E_x)$ completely covers the definition domain of the unique top-level category attribute (denoted by ALL, see Gray et al., 1997) of the hierarchy to which it belongs. If, instead, A_{1x} does not belong to any hierarchy, it disappears only if its definition domain coincides with its primitive definition domain.

MACRO-UNION

The *macro-union* operator of MAD is proposed in Fortunato et al. (1986). It was subsequently reproposed with the name of *fusion*. It works on two operands and, if necessary, recomputes the numeric values of the measure. In other words, given two MADs with common schemas, this operator provides a new MAD in output by taking the set-union of the facts described by the two MADs in input, and of the fact-dimension relations. This means that the definition domains of each category attribute of the output MAD are the union of the definition domains of the corresponding category attributes in the input MAD. Partial overlappings can happen between two category attributes with the same primitive definition domain. Null values can appear, with the meaning of "not available measured data" ('NA'). Operating only on one dimension, it performs the union of the categories and the partial orders. An example of a macro-union operator application is shown in Figures 12 and 13.

Figure 12.a

Percentage of Employees in the United States in 1985 (by CA & OR)			CA	OR	State
Graduated	M		11.8	3.8	
	F		6.8	3.1	
Not Grad.	M		27.3	17.3	
	F		16.8	13.1	
Qualification	Sex				100

Figure 12.b

Percentage of Employees in the United States in 1985 (by FL and NY)			FL	NY	State
Graduated	M		16.9	12.2	
	F		28.6	18.9	
Not Grad.	M		4,9	2.8	
	F		10.2	5.5	
Qualification	Sex				

It is similar to the *union* operator, proposed in Gyssens & Lakshamanan (1997), Pedersen & Jensen (1999), and Pedersen, Jensen, & Dyreson (2001), where it is applied only on one dimension, to the *set-union* operator, proposed in Ozsoyoglu, Ozsoyoglu, & Matos (1987), and to *enlargement*, proposed in Rafanelli & Ricci (1993).

Obviously, in order to have a MAD in output which makes sense, each pair of category attributes combined together have to belong to the same dimension (generally the same hierarchy and often the same level of this hierarchy).

It is important to note, similarly to the Restriction operator discussed in the following, that also this operator depends on the summary type of its measure. In fact, if its summary type had been "percentage," as in Figure 13, the single values inside the cells of the MAD would have had to be recomputed in order to maintain consistency in the information. For this recomputation we need a predefined distribution function, represented by a table T_D, or, often, by an expression which shows the rate which defines either the "weight" by which to compute the data in the MAD in output, or the way to relate this weight to the summary data in the MAD in input.

For example, let T_1 be the MAD, shown in Figure 12 (a), which we wish to enlarge, and let T_2 be the MAD to unite with T_1. If TD is represented by the expression "$100\,(T1) = 36,7\,(T_{out})$," the resulting MAD in output is shown in Figure 13.

Let $s_1 = <P_1, N, f, \mathcal{A}_1, S, s_1>$ be a MAD which describes a given fact $\mathcal{F}_1$ of a given phenomenon and which is defined on its MAD schema $\mathbb{A}_1 = \{A_{1j}\}$, and let $s_2 = <P_2, N, f, \mathcal{A}_2, S, s_2>$ be another MAD which describes another fact F_2 of the same phenomenon and which is defined on its MAD schema $\mathbb{A}_2 = \{A_{2j}\}$.

The union of s_1 and s_2 produces a new MAD

$$s_3 = <P_3, N, f, \mathcal{A}_3, S, s_3>,$$

where N, f, and S are the same of and s_1 and s_2,

P_3 is the new name of the MAD;

$\mathcal{A}_3$ is the set formed by the two subsets e_3 (explicit category attributes) and i_3 (implicit category attributes), each of them obtained by the union of e_1 with e_2 and of i_1 with i_2, respectively;

Figure 13

Percentage of Employees in the United States in 1985		CA	OR	FL	NY	State
Graduated	M	4.3	1.4	3.7	5.1	
	F	2.5	1.1	2.2	3.3	
Not Grad.	M	10.0	6.4	12.3	19.8	
	F	6.2	4.8	6.1	10.8	
Qualification	Sex					100

s_3 is a union of the sets of the summary values of s_1 and s_2. Notice that these values, sometimes, have to be recomputed, for example, if the summary type is "percentage" and the grand total is normalized to 100. This recomputation is transparent to the user.

For this operator the term *concatenation* has also been used. This operator (with this name), was proposed in Ozsoyoglu, Ozsoyoglu, & Mata (1985), and carries out a column category attribute forest of ordered sets which is the concatenation of the two MAD (called *summary tables* in that paper) in input and an ordered cell attribute set (concatenation of the two cell attribute sets of the previous column category attribute forests) so that, if this cell is in the first MAD, then its cell attribute is in the first ordered cell attribute set; otherwise, in the second ordered cell attribute set, this applies to each cell in the MAD.

Finally, another term used for this operator is *juxtapoint*. It refers to two different situations. The former regards MAD with a different summary type of the measure. In this case the result of the operation consists of a new *composite* MAD, and the juxtaposition happens along one dimension common to the two MADs. The latter consists of a new MAD for which, if the summary attribute of the original MAD is not summable (e.g., *average*), the instances inside its cells have to be recomputed, according to the aggregation function applied to the microdata to obtain it.

Figure 14

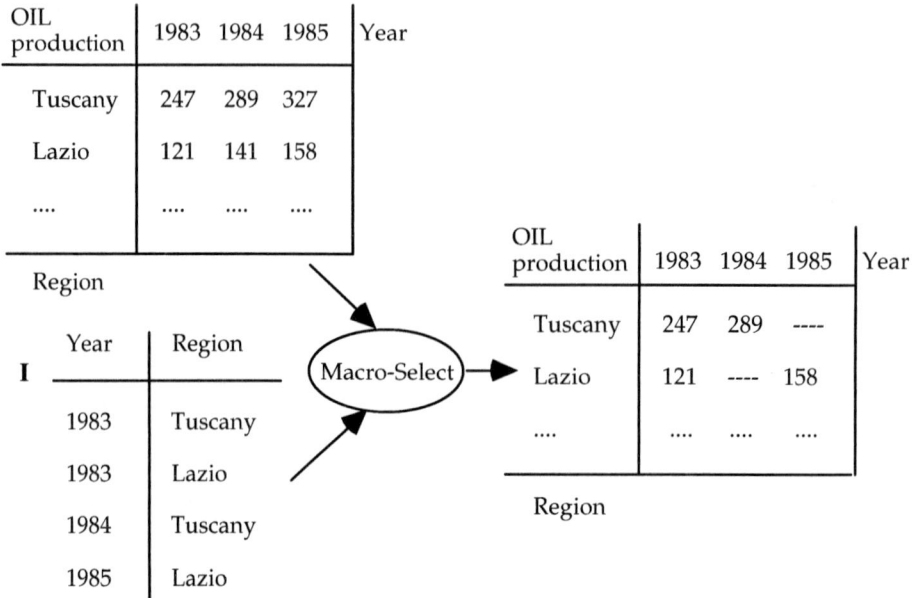

MACRO-SELECT

This is a unary operator, proposed in Fortunato et al. (1986), which selects one or more single values of the measure depending on the relation I given in input. The result is a new MAD defined on the same MAD schema and the same definition domains of the MAD category attributes, with its descriptive space equal to the active domain (see Maier, 1983) of the relation I. No recomputation of the measure instances is required.

Therefore, let $s_1 = <P, N, f, \mathcal{A}', \mathcal{S}, s>$ be the considered MAD and let I the relation formed by a subset of tuples of the base relation $\mathcal{R}_{B1}$ (which forms the descriptive space of s_1). The macro-select is carried out depending on if the relation I is the new MAD $s'_1 = <P, N, f, \mathcal{A}', \mathcal{S}, s'>$, where the only changing is the measure, in which the only summary attribute values, characterized by the tuples of the relation I, appear. The other summary values are substituted by the null value ' - .'

An example is shown in Figure 14.

RESTRICTION

The *restriction* operator was proposed in Fortunato et al. (1986) and in Rafanelli & Ricci (1993). In Ozsoyoglu, Ozsoyoglu, & Matos (1987), a similar operator, with the term *selection*, was proposed and subsequently, the same term was used in Pedersen & Jensen (1999) and in Pedersen, Jensen, & Dyreson (2001). In these papers the authors define the *selection* operator as a restriction of the set of facts to those that are characterized by valuing a given predicate *p* on the dimension types evaluated as true. The fact-dimension relations are restricted accordingly, while the dimensions and the schema remain the same (note that the term "dimension" stands for "category attribute," as previously explained).

Recently the term *dice* was coined and used in Agrawal (1997), Gyssens & Lakshamanan (1997), and Shoshani (1997), especially with regard to OLAP applications. It is an unary operator which selects only the category attribute instances specified by a qualification condition. These instances can be expressed by a complex Boolean expression, where predicates and logical operators (And, Or, Not) appear. The predicate form is <category attribute name> θ <variable | constant>, being θ a logical-mathematical operator of comparison or a set of comparison operators. The fact described by the MAD remains the same, and possible duplicate values are not removed. Possible recomputation of measure instances depends on the summary type of the measure. Therefore, this operation gives, in output, a MAD in which a category attribute of the MAD descriptive space is restricted to the elements of a set, defined as an operand of the operator, together with the name of the MAD. The result of its application is a new MAD defined, based on the same definition schema as the MAD in input.

Figure 15

OIL product-ion in Italy in 1995	Jan.	Febr.	...	Dec.	Month
Tuscany	2.3	2.3	...	2.4	15.1
Lazio	1.2	1.4	...	1.5	11.6
Pidemont	...	...	...	...	...
....	...	...	...	...	...
Region	11.1	11.3	...	12.0	100 %

OIL product-ion in Italy in 1995	Jan.	Febr.	...	Dec.	Month
Tuscany	2.3	2.3	...	2.4	15.1
Lazio	1.2	1.4	...	1.5	11.6
Region	3.5	3.7	...	4.1	26.7%

OIL product-ion in Italy in 1995	Jan.	Febr.	...	Dec.	Month
Tuscany	8.61	8.61	...	8.98	56.55
Lazio	4.49	5.24	...	5.61	43.45
Region	3.5	3.7	...	4.1	100 %

Restriction

N-Restriction

| Region |
| Tuscany |
| Lazio |

If the summary type of the measure is *count* or *sum*, no problem arises, but if it is, for example, *percentage*, we have to know if the values in each cell have to be recomputed (i.e., the *grand total* of the MAD remains 100) or if such values remain the same. In the first case we say that the restriction is normalized. We call this operator **N-RESTRICTION** and analogously for the summary data "average," where the grand total changes. An example is shown in Figure 15.

Therefore, given a (simple) MAD s_1, let us suppose the summarizability conditions have been verified. Let $<P_1, N, f, A_1, S, s_1>$ be the six-tuple which defines s_1, where A is formed by e_1 and i_1, respectively its explicit and implicit category attributes. Let A_{1x} be a category attribute chosen among the category attributes A_{B1j} of the *base relation* R_{B1} of s_1 (with $j = 1, ..., k$, being k the cardinality of s_1, i referring to the ith fact described by s_1 and with $x \in \{1, ..., k\}$).

The *restriction* of s_1 by A_{1x} is a new MAD s'_1 defined on the same descriptive space schema, but with the category attribute with the same name of A_{1x} having as definition domain a subset of the original one, i.e., formed by the instances of A_{1x}.

Then, given $s_1 = <P, N, f, \mathcal{A}, \mathcal{S}, s>$ and a_1 a subset of E_1, with $t \in k$ and $E_t \subset e_1 = \{ E_1, E_2, ..., E_M \}$, i.e., the set of its *explicit category attributes* of s_1, the restriction of new MAD s'_1 is defined by the six-tuple $<P, N, f, \mathcal{A}', \mathcal{S}, s>$, where $\mathcal{A}' = \{E_1, E_2, ..., E'_t = A_{1x}, ..., E_k\}$.

Notice that we can generalize the operator considering, instead a_1, all the set of category attributes A_{B1j} and, for each attribute, a subset of its definition domain. The operator defined above will be applied iteratively to the first, the second, ..., the k-th attribute.

The eventual recomputation of the summary value of the MAD depends on the summary type of the MAD as well as on the type of restriction (no recomputation if it is a regular restriction, eventual recomputation if it is a *normalized* restriction).

RECLASSIFICATION

The *reclassification* (or, simply, classification) operator, proposed in Rafanelli & Ricci (1984, 1985), and subsequently in Fortunato et al. (1986) and in Rafanelli & Ricci (1993), substitutes a set (or possibly all) of the category attributes in a MAD, with another set, according to the relation existing between two sets of category attributes. This fact means that there is a *redefinition* of the data in the MAD, with a computation of the summary values (for this reason it is often referred to by the term *redefine*). This operation classifies one category attribute of a MAD according to a given relation in which the new classification criteria are specified. For example, the category attribute "months" can be classified in "quarter" according to the relation "{January, February, March}" → "1st quarter," etc. This operator is also able to reclassify a set (or possibly all) of the category attributes of the MAD with another set of attributes, according to a given relation, by means of a substitution. The schema of this relation contains both the category attributes which must be substituted, and the category attributes which must substitute the above-mentioned attributes. The resulting MAD is a redefinition of the initial data, with (in general) an aggregation of data (recalculation of the summary values).

Let us consider, for example, the MAD "number_of_cars_produced_in_Japan" of Figure 16(a), characterized by the category attributes "model" and "years"; if we wish to have the same MAD described by the category attributes "piston displacement" and "years" (the link between "model and "piston displacement" is the relation "r" in Figure 16(b)), we perform the classification of "number_of_cars_produced_in_Japan" by the relation "r" along the category attributes "piston displacement" and "years," and we obtain the MAD of Figure 16(c).

This operator is more powerful than the *roll-up* operator (discussed in the following), because it allows aggregation only along a defined hierarchy, while the reclassification operator also allows another grouping of the lower level of the

Figure 16.a

| #_OF_CARS_PRODUCED_IN_JAPAN (by Model and Years) (absol. values, by thousands) | MODEL | | |
	Corolla	Civic	Corona
1980	427	341	220
1981	458	373	249
1982	499	401	285
YEARS 1983	536	438	317
1984	601	496	369
1985	710	580	421
. . . .	. . .	. .	. . .

hierarchy to be defined ex-novo, giving a grouping relation not defined a priori, but proposed by the user. Generally, it changes the cardinality of the MADs, decreasing them according to a functional dependency defined by the relation mentioned above.

Analogous to the previous operators, let R_x be the set of all the relations (of the micro database) involved in the production of all the MADs which describe a given phenomenon; we considered the subset of R_x formed only by the relations involved in the building of fact $\mathcal{F}_1$. As already explained, we call this subset an *aggregation relation*, and denote it by the expression $R_1x = \{\mathcal{R}_{1,1}, \ldots, \mathcal{R}_{1,s}\}^x$. Let $\{A_{1j}\}$ be the set of category attributes which define the descriptive space of the MAD s_1. Let $<P, N, f, \mathcal{A}, \mathcal{S}, s>$ be the six-tuple which defines s_1. Let $\{Ah\}$ be a set of attributes, different from the set of category attributes which form A, such that a functional dependency between a subset $\mathcal{A}^x \subset \mathcal{A}$ and $\{Ah\}$ exists. We denote this functional dependency by $\mathcal{A}^x \Rightarrow \{A_h\}$. Then, the *reclassification* of s_1 respective to the functional dependency $\mathcal{A}^x \Rightarrow \{A_h\}$ produces a new MAD

$$s'_1 = <P, N, f, \mathcal{A}', \mathcal{S}, s'>,$$

where $P, N, f, \mathcal{S}$ are the same and

$$\mathcal{A}' = \mathcal{A} - \mathcal{A}^x + \{Ah\} \text{ and}$$

Figure 16.b

Figure 16.c

r	MODEL	CUBIC CAPACITY
	Corolla	1,200
	Civic	1,200
	Corona	1,800

	#_OF_CARS_PRODUCED (absol. values, by thousands)	CUBIC CAPACITY	
		1,200	1,800
YEARS	1980	768	220
	1981	831	249
	1982	900	285
	1983	974	317
	1984	1,097	369
	1985	1,290	421
	. . . .	. . .	. . .

s' is a set of recomputed summary values. This recomputation depends on the type of aggregation function (e.g., count, sum, average) applied to the original microdata.

Obviously, the new MAD s'_1 is a more aggregate representation of the fact $\mathcal{F}_1$ described by the MAD s_1. Moreover, in general the number of dimensions of s'_1 is less than s_1, i.e., its cardinality is smaller. It is important to note that part of the dimension changes, so that also the descriptive space changes. In the case of Figure 16, the dimension number remains the same, but the descriptive space changes from <Model, Year> to <Model, Cubic capacity>.

COMPARISON

This operator was proposed for the first time (and in a simpler form) in Fortunato et al. (1986), and consists of a family of operators.

The original proposal was subsequently enlarged, in order to distinguish between the **T-COMPARISON** (where T means *table*, see Figure 17(a) and 17(b)) and **N-COMPARISON** (where N means *single numeric constant*, see Figure 17(c)).

The former compares the MAD in input with another MAD (table), which can be the same, opportunely shifted along a dimension.

The latter compares each measure instance of the MAD in input with a single numeric value. Both these comparisons are performed by a set $\theta = \{>, \geq, <, \leq, =, \neq\}$ of operators.

The comparison between two MADs can be performed in two ways: 1) comparing between the values which are inside the two corresponding cells of the two MADs; 2) or comparing the value inside the first cell of the first MAD with all the measure values of the second MAD, the second cell of the first MAD with all the measure values of the second MAD, and so on, until the last cell of the first MAD (compared, like the previous ones, with all the measure values of the second MAD). The former is denoted by $T_{1:1}$-Comparison, the latter by $T_{1:n}$-Comparison.

By this operator we are able to answer queries of the type:

i)   In which region has the production of cereals been greater than the previous year?

ii)   In which region has the production of cereals been greater than "100 tons?"

The result of this operation can be of two types: logical (i.e., in every cell of the resulting MAD, one logic value—*true* or *false*—appears), or numerical (if the Ø-condition is verified, value=true, the original value remains inside the cell, otherwise the symbol — appears.)

The result of this operation between s_1 and s_2 consists of a new MAD

$s'_1 = <P_1, N, f, A', S, s_1>$, where N, f, A', S are the same of s_1 and P_1 is the new name of the MAD;

s_1 has its instance equal to the corresponding instance of s if the comparison verifies the θ-condition (where θ is one of the operator $<>, \geq, <, \leq, =, \neq$), while it has another instance equal to " - " (null value) if the θ-condition is not verified.

Obviously, depending on the type of *comparison* considered, the term *to the corresponding instance of* is substituted by the term *to the instance of* or with the term *to the number*.

Figure 17

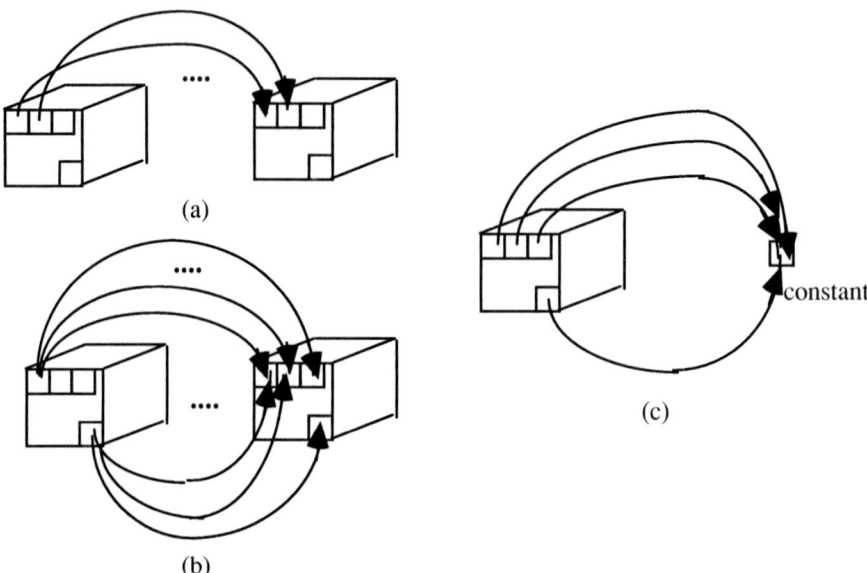

(a)

(b)

(c)

constant

ROLL-UP

This operator was proposed in Gray et al. (1996), and subsequently in Cabibbo & Torlone (1997), in Gray et al. (1997), in Lehner (1998), in Nguyen, Tjoa, & Wagner (2000), and in Pedersen, Jensen, & Dyreson (2001). A *roll-up* on a dimension hierarchy of a MAD means giving "less detail" by ascending this dimension hierarchy, aggregating with an implicit aggregation function. This corresponds to performing aggregate formation on "higher" category types with the given aggregate function. Sometimes, we also need a reference to the original MAD in this case. This is caused by the possible *non-summarizability* in the MAD, which means that we cannot necessarily combine the aggregate results from intermediate levels into higher-level results, but we need to compute the result directly from the lowest-level data (base data).

Reports commonly aggregate data at a coarse level, and then at successively finer levels. The car sales report in Figure 18 shows the idea. Data is aggregated by model, then by year, then by city. If the measure is summable and the MAD is *well formed*, as defined in Chapter 4, and if, obviously, another level at a higher level exists along the hierarchy, then the roll-up operator is applicable and the end result is a new MAD with more aggregate measure instances.

The *roll-up* carried out on the category attribute $A_{1X} \in \mathcal{A}$ of a given MAD s_1 produces a new MAD s'_1 defined on the same descriptive space schema, but with the category attribute E_x, on which the roll-up was applied, substituted by the category attribute E'_x of the higher level of the hierarchy defined in the relative dimension, i.e.:

$s'_1 = <P,' N, f, \mathcal{A}', \mathcal{S}, s'>$, where $N, f, \mathcal{S}$ are the same and P' is the new name of the MAD;

Figure 18

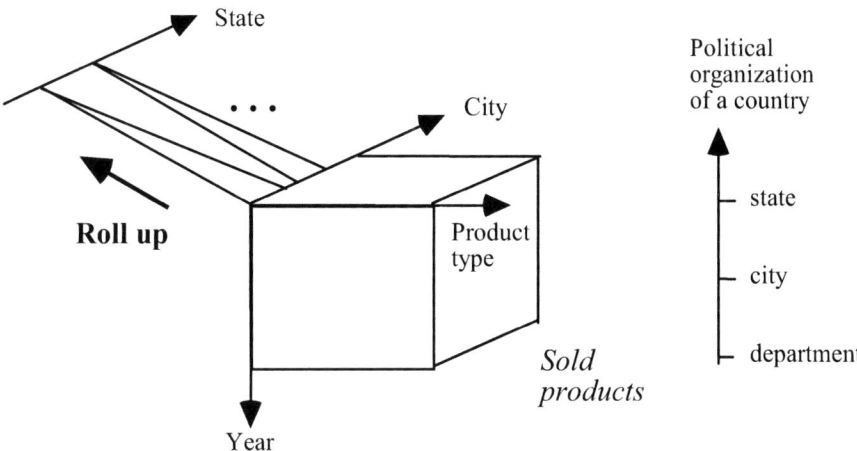

s' is a set of recomputed summary values. This recomputation depends on the type of aggregation function (e.g., count, sum, average) applied to the original microdata.

$\mathcal{A}' = e\cup_1 = \{E_1, E_2, \dots, E'_x, \dots, E_M\} \cup i_1$, where the hierarchy $\mathcal{H}$ of the dimension which the roll-up was applied to is ... $E_X \rightarrow E'_X$, ...

In Gray et al. (1996), and subsequently in Gyssens & Lakshamanan (1997), the authors propose a refinement of the above-mentioned operator, namely the *monotone roll-up* operator.

In many applications, in fact, only certain fragments of the MAD are of interest. In general, if X is the set of descriptive variables of the MAD and Y is a subset of X, then only aggregates regarding Y and all its subsets may be of interest.

GROUP-BY

The *group-by* operator is an unusual relational operator: it partitions the relation into-disjoint tuple sets and then aggregates over each set. It groups the facts characterized by the same dimension values together. The standard SQL-group-by operator does not allow a direct construction of aggregation over computed categories. For this reason its generalization (especially referring to the OLAP applications) was proposed in Gray et al. (1996, 1997), and subsequently in Lehner (1998), Hurtado, Mendelzon, & Vaisman (1999), and Nguyen, Tjoa, & Wagner (2000).

DRILL-DOWN

A *drill-down* on a MAD means giving "more detail" by descending the dimension hierarchies. It is proposed in Gray et al. (1996), and subsequently in Lehner (1998), Nguyen, Tjoa, & Wagner (2000), and Pedersen, Jensen, & Dyreson (2001).

An implicit aggregation function, e.g., *count* or *sum*, is assumed. Thus, a drill-down corresponds to performing aggregate formation on "lower" category types with the given aggregate function. To get to the lower category types, a reference to the *original MAD* is needed. In fact, this operator is, in reality, a facility for retrieving the original MAD after having applied the roll-up operator to it. It cannot be applied to any MAD without the previous operation, because we are not able to generate more disaggregate data than the original MAD carried out by the aggregation process. In other words, only if we know a disaggregative law which estimates the distribution of the summary data (probabilistic data) are we able to compute (to estimate) such data.

Figure 19

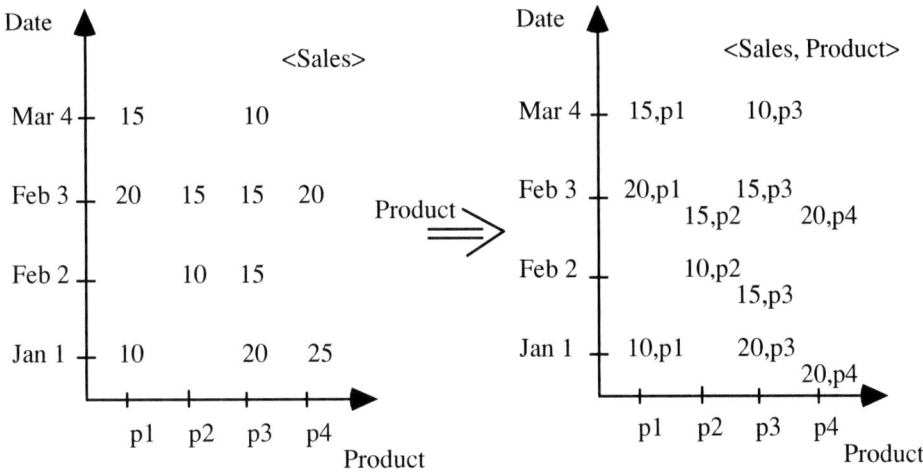

PUSH

This operator, proposed in Agrawal, Gupta, & Sarawagi (1997), is used to convert dimensions into elements that can then be manipulated using the combining function f_{elem}. This it a typical operator useful in OLAP application in order to allow dimensions and measures to be treated uniformly. An example (taken from Agrawal, Gupta, & Sarawagi, 1997) is shown in Figure 19.

PULL

This operation, proposed, like the previous one, in Agrawal, Gupta, & Sarawagi

Figure 20

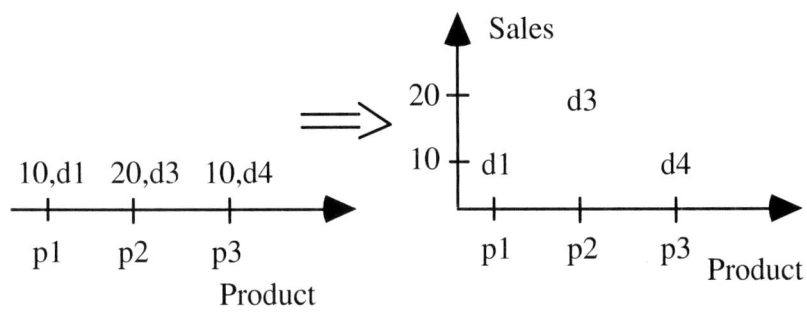

(1997), is the converse of the *push* operator. It creates a new dimension for a specified member of the elements. It is useful for converting an element into a dimension and, like the previous one, it allows dimensions and measures to be treated uniformly.

An example (taken from Agrawal, Gupta, & Sarawagi, 1997) is shown in Figure 20. In it the operator pulls the first member of each element as the new dimension "sales."

DATA CUBE

This operator was proposed in Gray et al. (1997), and, subsequently in Gyssens & Lakshamanan (1997) and in Nguyen, Tjoa, & Wagner (2000).

The traditional *group-by* can generate the core of the n--dimensional data cube. The *n-1* lower-dimensional aggregates appear as points (0-D), lines (1-D), planes (2-D), cubes (3-D), or hyper-cubes hanging off the data cube core.

The *data cube* operator builds a table containing all these aggregate values. The total aggregate using function f( ) is represented as the tuple: ALL, ALL, ALL, ..., ALL, f(*). Points in higher dimensional planes or cubes have fewer ALL values. Then, the cube operator is the n-dimensional generalization of simple aggregate functions. These concepts were represented in the Figure 2.

As discussed in Gray et al. (1997), creating a data cube requires generating the power set of the aggregation columns. Since the *cube* is an aggregation operation, it makes sense to externalize it by overloading the SQL group-by operator. Figure 21 (taken from Gray et al., 1997) includes an example of the cube syntax.

AGGREGATE FORMATION

This operator was defined in Klug (1982) and subsequently in Ozsoyoglu, Ozsoyoglu, & Matos (1987).

The *aggregate formation* operator first *partitions* tuples of relation R such that tuples having the same X component are in the same partition. Then the function f is applied to component A of tuples in each partition, and the X-value and the associated aggregate value are output for each partition.

Let $R \in \mathcal{R}$ be a relation included in the set of relations in the database, with $X \subseteq Atr(R)$, $|X| = k$. Let f be an aggregate function and A be simple-valued, with $A \in Atr(R)$. Then, $R <X, f_A>$ is a relation with degree k+1 and is defined as

$$R <X, f\ A> = \{t\,|X|\, \mathbf{o}\, y \,|\, t \in R\ and\ y = f_A(\{t' \,|\, t' \in R\ and\ t'\,[X] = t\,[X]\,\})\}$$

where '**o**' denotes *concatenation*.

AGGREGATION-BY-TEMPLATE

Proposed in Ozsoyoglu, Ozsoyoglu, & Mata (1985, 1987), the *aggregation-by-template* operator $R_1 <X, Y, f_A> R_2$ groups tuples of R_1 as follows. Let t be a tuple over attributes $(X \cup Z)$ such that $t[X] = t_1[X]$ for some t_1 in R_1 and $t[Z] = t_2[Z]$ for some tuple t_2 in R_2.

Each such tuple t defines a group G_t of tuples of R_1 such that a tuple v, with v $\in R1$, is in G_t if $v[X] = t[X]$, $v[Y_a] \in t[Z_a]$ and $v[Y_n] \subseteq t[Z_n]$. Then f is applied to attribute A of tuples in G_t.

Figure 21

SALES			
Model	Year	Color	Sales
Chevy	1990	red	5
Chevy	1990	white	37
Chevy	1990	blue	62
Chevy	1991	red	54
Chevy	1991	white	95
Chevy	1991	blue	49
Chevy	1992	red	31
Chevy	1992	white	54
Chevy	1992	blue	71
Ford	1990	red	64
Ford	1990	white	62
Ford	1990	blue	63
Ford	1991	red	52
Ford	1991	white	9
Ford	1991	blue	55
Ford	1992	red	27
Ford	1992	white	62
Ford	1992	blue	39

CUBE

DATA CUBE			
Model	Year	Color	Sales
Chevy	1990	red	5
Chevy	1990	white	95
Chevy	1990	blue	62
Chevy	1990	ALL	154
Chevy	1991	red	54
Chevy	1991	white	95
Chevy	1991	blue	49
Chevy	1991	ALL	198
Chevy	1992	red	31
Chevy	1992	white	54
Chevy	1992	blue	71
Chevy	ALL	blue	182
Chevy	ALL	red	90
Chevy	ALL	white	236
Chevy	ALL	ALL	508
Ford	1990	red	64
Ford	1990	white	62
Ford	1990	blue	63
Ford	1990	ALL	189
Ford	1991	red	52
Ford	1991	white	9
Ford	1991	blue	55
Ford	1991	ALL	116
Ford	1992	red	27
Ford	1992	white	62
Ford	1992	blue	39
Ford	1992	ALL	128
ALL	1990	blue	125
ALL	1990	red	69
ALL	1990	white	149
ALL	1990	ALL	343
ALL	1991	blued	106
ALL	1991	red	104
ALL	1991	white	110
ALL	1991	ALL	314
ALL	1992	blue	110
ALL	1992	red	58
ALL	1992	white	116
ALL	1992	ALL	284
ALL	ALL	blue	339
ALL	ALL	red	233
ALL	ALL	white	369
ALL	ALL	ALL	941

Figure 22.a

R_2	COUNTRY	GAME	*AGE	Count
	USA	Alpine-skiing	{11, ..., 40}	18
	USA	Alpine-skiing	{41, ..., 70}	3
	CAN	Nordic-skiing	{11, ..., 40}	20
	CAN	Nordic-skiing	{41, ..., 70}	5

Figure 22.b

R_3 *COUNTRY	SEX	*AGE
{USA, CAN}	M	{11, ..., 40}
{USA, CAN}	M	{41, ..., 70}
{USA, CAN}	F	{11, ..., 40}
{USA, CAN}	F	{41, ..., 70}

Figure 22.c

R_4 *COUNTRY	*SEX	*AGE
{USA, CAN}	{M, F}	{11, ..., 40}
{USA, CAN}	{M, F}	{41, ..., 70}

The value returned by f applied over an empty group is null. Formally:
Let R_1, $R_2 \in A$, $Y \subseteq Atr(R_1)$, $Z = Atr(R_2)$, where $|Y| = |Z| \geq 1$, and is defined as:

$R_1 <X, Y, f_A> R_2 = \{ t \mathbf{o} y \mid (\exists t_1)(\exists t_2)(t_1 \in R_1 \ and \ t_2 \in R_2 \ and \ t[X] = t_1[X] \ and \ t[Z] = t_2[Z] \ and \ y = f_A(\{ t' \mid t' \in R_1 \ and \ t'[X] = t[X] \ and \ t'[Y_a] \in t[Z_a] \ and \ t'[Y_a] \subseteq t[Z_a] \}))\}$.

Aggregation-by-template is more convenient than aggregate formation when there are prespecified groupings of attributes for aggregation (common in statistical database applications). Also, aggregation-by-template is based on grouping tuples (i.e., a tuple may belong to more than one group) while aggregate formation is based on partitioning tuples. However, each aggregation operator is expressible by an algebra expression utilizing the other aggregation operator.

PACK

This operator, as well as the following one, *unpack,* maps (packs) sets of tuples in R, whose (n-1) components for C_k are the same, into a single tuple.

Let $R \in \mathcal{R}$ be a relation included in the set of relations in the database, with $| \text{Atr}(R) | = n$, $A \in \text{Atr}(R)$ and $C_A = \text{Atr}(R) - \{A\}$. For each (n-1)-tuple $g \in \Pi_{C_A}(R)$, we define:

$$W_g[C_A] = g$$
$$W_g[A] = \{t[A] \mid t \in R \ and \ t[C_A] = g \} \quad \text{if A is simple-valued}$$
$$W_g[A] = \{x \mid \exists t)\ (t \in R \ and \ t[C_A] = g \ and \ x \in t[A]) \} \quad \text{otherwise}$$

then $P_A(R) = \{ W_g \mid g \in \Pi_{C_A}(R) \}$

For example, consider the relation R_1 in Figure 22(a). The relations R_3 and R_4, where $R_3 = P_1(\Pi_{1,2,3}(R_1))$ and $P_4(R_3)$, are shown in Figures 22(b) and (c).

UNPACK

This operator is part of the relational algebra of set-valued relations proposed in Ozsoyoglu, Ozsoyoglu, & Mata (1985). (The other algebraic operators are Cartesian product, project, select, natural join, set union, set difference, set intersection, set formation, aggregation-by-template, and construct.)

If A is simple-valued, then Unpack$k$ (R) maps each tuple t in R into a set of tuples such that each element in t[A] becomes the A-value of one resulting tuple.

Let $R \in D$ be a relation included in the set of relations in the database, with $A \in \text{Atr}(R)$, $CA = \text{Atr}(R) - \{A\}$. For each tuple $t \in R$, we define a set of tuples:

$$UN_A(\{t\}) = \{t\} \quad\quad \text{if A is simple-valued}$$
$$UN_A(\{t\}) = \{t' \mid t'[A] \in t[A] \text{ and } t'[C_A] = t[C_A]\} \quad \text{otherwise}$$

then $UN_A(R) = \cup_{t \in R}(UN_A(\{t\}))$.

Rename

This operator is defined in different papers, such as Fortunato et al. (1986), Rafanelli & Ricci (1993), Cabibbo & Torlone (1998), Pedersen & Jensen (1999), etc.

Given a simple MAD $s_1 = <P, N, f, \mathcal{A}, \mathcal{S}, s>$ having scheme $[A_1, A_2, ..., A_n] \to [M]$, A_i is an attribute of s_1 and A is a new attribute name, then $\rho_{Ai} \to_A$ is a new MAD over the scheme $[A_1, A_2, ..., A_{i-1}, A_{i+1}, ..., A_n, A] \to [M]$, with a new fact name s_1'. In theory, it works on one MAD and provides as output a new MAD in which the name of one category attribute is changed (i.e., it has a new MAD schema), as well as its fact name, but in which the subject, the measure, and the structure of the old schema remain the same.

For example, suppose you wish to change the name of the category attribute *Model* of the MAD in Figure 16(a) and call it by the new name *Modello* (the Italian word of "Model"). We have:

$\mathcal{S}$ (Model , Modello) (Number of cars produced in Japan by Modello and Year) s_1

DIFFERENCE

This operator, proposed in Gyssens & Lakshamanan (1997), Pedersen & Jensen (1999), and Pedersen, Jensen, & Dyreson (2001), makes it possible, given two MADs with common schemas, to take the set difference of the facts (the dimensions of the first argument MAD are retained) and to restrict the fact-dimension relations to the new fact set. Note that we do not take the set difference of the dimensions because this does not make sense. This operator is very similar to a previous operator, called *set difference*, proposed in Ozsoyoglu, Ozsoyoglu, & Matos (1997).

DUPLICATE REMOVAL

With this operator, proposed in Pedersen, Jensen, & Dyreson (2001), we can remove "duplicate values," i.e., several facts characterized by the same combination of dimension values, by performing a *set-count* aggre-gate formation on the $\perp$ categories, followed by projecting out the result dimension, where $\perp$ is the bottom ordering category of the hierarchy.

SQL-LIKE AGGREGATION

This operator, proposed in Pedersen, Jensen, & Dyreson (2001), computes the SQL aggregate function on a MAD, which is grouped by a set of dimension categories, by first applying the aggregate formation operator to the MAD with the given categories and the given function. Note that the categories not in the *group-by* clause are in the $\top$ categories of their dimensions. The dimensions not in the *group-by* clause are then projected out.

JOIN

For this operator many definitions have been proposed, with more or less little differences. We briefly discuss natural join, identity-based join, value-based join, S-join, S (f)-join and star-join:

Natural join: This operator was proposed for multidimensional aggregate data in Cabibbo & Torlone (1998). If E_1 and E_2 are MADs (or f-expressions) over the schema $[A_1, A_2, ..., A_k, A_{k+1}, ..., A_n] \rightarrow [M_1, M_2, ..., M_m]$ and $[A_1, A_2, ..., A_k, A'_{k+1}, ..., A'_n] \rightarrow [M'_1, M'_2, ..., M'_m]$, respectively, that is, having $A_1, A_2, ..., A_k$ as common attributes (defined over the same level) and no common measure, then $E_1 \bowtie E_2$ is a MAD with schema over the attributes $A_1, A_2, ..., A_k, A_{k+1}, ..., A_n, A'_{k+1}, ..., A'_n$ and measures $M_1, M_2, ..., M_m, M'_1, M'_2, ..., M'_m$. The result has an entry for each pair of entries in the two MADs with the same values in the common attributes. The corresponding measures are the juxtaposition of the measures in the original entries.

Identity-based join: This operator was proposed in Pedersen & Jensen (1999) and Pedersen, Jensen, & Dyreson (2001), and is used to combine information from several MADs. It produces a new fact type which consists of pairs from the old fact type, and a new set of dimension types which is the union of the old sets. The set of facts is the subset of the cross- product of the old sets of facts where the join predicate p holds. For p equal to $f_1 = f_2$, $f_1 \neq f_2$ and true, the operation is an *equi-join, non-equi-join,* and *Cartesian product,* respectively. For this instance, the set of dimensions is the set-union of the old sets of dimensions, and the fact-dimension relations relate a pair to a value if one member of the pair was related to that value before.

Then, given two MADs, s_1 and s_2, and a predicate $p(f_1,f_2) \in (f_1 = f_2, f_1 \neq f_2$, true$\}$, we define the *identity-based join* $\bowtie$ as:

$$s_1 \bowtie [p] \, s_2 = (S, \text{'} \, F, \text{'} \, D, \text{'} \, R \text{'}), \text{ where } (S\text{'} = (\mathcal{F} ,\text{'} \, \mathcal{A}\text{'}), \, \mathcal{F} \, \text{'} = \mathcal{F}_1 \times$$

$$\mathcal{F}_2, \, \mathcal{A}\text{'} = \mathcal{A}_1 \cup \mathcal{A}_2, \, F\text{'} = \{(f_1, f_2) \mid f_1 \in F_1 \wedge f_2 \in F_2 \wedge p(f_1, f_2)\},$$

$$D\text{'} = D_1 \cup D_2, \, R\text{'} = \{R\text{'}_i, \, i = 1, ..., n_1+n_2\} \text{and } R\text{'}_i = \{(f\text{'}_1, e) \mid$$

$$f\text{'}=(f_1, f_2) \wedge f\text{'} \in F\text{'} \wedge ((i \geq n_1 \wedge (f_1, e) \in R_{1i}) \vee (i > n_1 \wedge (f_2, e)$$

$$\in R_{2i-n1}))\}$$

Value-based join: This operator was proposed in Pedersen, Jensen, & Dyreson (2001). It is a join of two MADs on common dimension values, by combining Cartesian product (a special case of the identity-based join), selection, and projection. Natural join is another special case of this operator, where the selection predicate requires that values from the "matching" dimensions should be equal, followed by projecting "out" the duplicate dimensions. In practice, performing this operator on the common dimensions of two MADs means carrying out the drill-down operation from one MAD to another MAD.

S-join: This operator carries out a juxtaposition of two MADs, T_1 and T_2, which have the same *descriptive space* (S1=S2 where S1: <X1, Y1>, etc.), but in which the corresponding definition domains can be different ($dom(X1) = <x_1, x_2, x_3>$ and $dom(X2) = <x_1, x_3, x_4, x_5>$, etc.), and have as equal *summary type* (which has to be "summable") and the same *fact.* For example, let T1 and T2 be two given MADs shown in Figure 23.

The conditions mentioned above require the same descriptive space (in this case, "nations" and "year"), the same summary type (in this case, "count"), and the same described fact (in this case, "production of fruit.") Applying the S-join, we obtain MAD T3 in Figure 24.

The alternative could be that to obtain MAD T4 of Figure 25. In this second case, MAD T4 could be a complex MAD (see the ADAMO model in Chapter 1), whose summary values could have different summary types.

The conditions mentioned above require the same descriptive space (in this case, "country" and "year"), the same summary type (in this case "count"), and the same described fact (in this case "citrus fruit production").

Figure 23

T1	Lemon Production	Country	Year	# (count)
		x1	y1	v1
		x2	y1	v2
		x3	y1	v3
		x1	y2	v4
		x2	y2	v5
		x3	y2	v6
		x1	y3	v7
		x2	y3	v8
		x3	y3	v9

T2	Orange Production	Country	Year	# (count)
		x1	y1	w1
		x3	y1	w2
		x4	y1	w3
		x5	y1	w4
		x1	y4	w5
		x3	y4	w6
		x4	y4	w7
		x5	y4	w8

Figure 24

T3	Citrus fruit production	County	Year	#Oranges (count)	#Lemons (count)
		x1	y1	v1	w1
		x3	y1	v3	w2

Applying the S-join operator we obtain MAD T3 in Figure 24.

Note that T_3 is a composite table, where the two component MADs have the same dimensions but different subjects. Because these subjects are "part-of" of the higher level in the primitive hierarchy "Citrus fruit ← <Oranges, Lemons, …>," we can apply the roll-up operator to this hierarchy (treating the measure as another dimension), to obtain the MAD in Figure 25.

S (f)-join: A variant of the previous operator is *S (f)-join*. In this case the only condition (verifiable only by the user) which has to be satisfied is that the described fact must be the same. This means that both the descriptive spaces (S_1 and S_2) and their definition domains can be different (dom X_1 and dom X_2), and (like the second case of the S-join) their summary types can also be different. For example, let T_5 and T_6 be two MADs shown in Figure 26.

In this case the two descriptive spaces are different (even if they have a common subset, "nations" and "year"), and the cardinality of each MAD is also

Figure 25

T4	Citrus fruit production	County	Year	# (count)
		x1	y1	v1 + w1
		x3	y1	v3 + w2

Figure 26

T5 Car production	Country	Year	Model	Color	# (count)
	x1	y1	k1	h1	v1
	x2	y1	k1	h1	v2
	x3	y1	k1	h1	v3
	x1	y2	k1	h1	v4
	x2	y2	k1	h1	v5
	x3	y2	k1	h1	v6
	x1	y3	k1	h1	v7
	x2	y3	k1	h1	v8
	x3	y3	k1	h1	v9
	x1	y1	k2	h1	v10
	x2	y1	k2	h1	v11
	x3	y1	k2	h1	v12
	x1	y2	k2	h1	v13
	x2	y2	k2	h1	v14
	x3	y2	k2	h1	v15
	x1	y3	k2	h1	v16
	x2	y3	k2	h1	v17
	x3	y3	k2	h1	v18
	x1	y1	k1	h2	v19
	x2	y1	k1	h2	v20
	x3	y1	k1	h2	v21
	x1	y2	k1	h2	v22
	x2	y2	k1	h2	v23
	x3	y2	k1	h2	v24
	x1	y3	k1	h2	v25
	x2	y3	k1	h2	v26
	x3	y3	k1	h2	v27
	x1	y1	k2	h2	v28
	x2	y1	k2	h2	v29
	x3	y1	k2	h2	v30
	x1	y2	k2	h2	v31
	x2	y2	k2	h2	v32
	x3	y2	k2	h2	v33
	x1	y3	k2	h2	v34
	x2	y3	k2	h2	v35
	x3	y3	k2	h2	v36

T6 Car auto	Country	Year	Cubic capacity	# (count)
	x1	y1	t1	w1
	x3	y1	t1	w2
	x4	y1	t1	w3
	x5	y1	t1	w4
	x1	y4	t1	w5
	x3	y4	t1	w6
	x4	y4	t1	w7
	x5	y4	t1	w8
	x1	y1	t2	w9
	x3	y1	t2	w10
	x4	y1	t2	w11
	x5	y1	t2	w12
	x1	y4	t2	w13
	x3	y4	t2	w14
	x4	y4	t2	w15
	x5	y4	t2	w16

different (4 and 3, respectively). In order to apply the operator S-join, a "normalization" of the two descriptive spaces has to be made, obtaining only one schema. This means performing a *slice* operation (in this case, a P-slice), and possibly a *roll-up* operation (in case of the presence of hierarchies along the same dimension).

In the case in Figure 26 we have:

Dice T3 (Model, Color)

Dice T4 (Cubit capacity)

Therefore, we proceed as in the previous case, because no hierarchies are present in all the remaining dimensions.

Notice that every time a normalization is performed, a *note* has to be produced (which explains which dimension has become "implicit").

In the case in Figure 26, this *note* will contain information regarding what model (k1 and k2) and what colors (h1 and h2) refer to the summary data. If we delete model and color in T5, the summary values which refer to the pair x1-y1, that is, v_1, v_{10}, v_{19}, and v_{28}, are summed (obviously, we suppose that they are summable), as well as the summary values which refer to the pairs x_1-y_2, x_1-y_3, and so on and analogously for MAD T_6.

If, instead, the summary values are not summable (average, max, min, etc.), the *S (f)-join* has to be applied, where f represents the function which has been applied to microdata during the aggregation process which produced the aggregate data.

Obviously, it prefigures that we have all the microdata of the initial database at our disposal or, at least, the "basic" microdata (count and sum) necessary to the application of the formula (relative to the aggregation function) leads to aggregate data. The final result of the *P-slice* operation on the MAD in Figure 26 is shown.

In the case in which the summary values are summable, in Figure 27, the result of the S-join operation is shown in Figure 28.

Figure 27

T7	Car production	Country	Year	# (count)
		x1	y1	v1+v10+v19+v28
		x2	y1	v2+v11+v20+v29
		x3	y1	v3+v12+v21+v30
		x1	y2	v4+v13+v22+v31
		x2	y2	v5+v14+v23+v32
		x3	y2	v6+v15+v24+v33
		x1	y3	v7+v16+v25+v34
		x2	y3	v8+v17+v26+v35
		x3	y3	v9+v18+v27+v36

T8	Car production	Country	Year	# (count)
		x1	y1	w1+w9
		x3	y1	w2+w10
		x4	y1	w3+w11
		x5	y1	w4+w12
		x1	y4	w5+w13
		x3	y4	w6+w14
		x4	y4	w7+w15
		x5	y4	w8 +w16

Figure 28

T7	Car production	Country	Year	# (count)
		x1	y1	v1+v10+v19+v28+w1+w9
		x3	y1	v3+v12+v21+v30+w2+w10
		x1	y2	v4+v13+v22+v31+w5+w13
		x3	y2	v6+v15+v24+v33+w6+w14

Star-join: A star join, as described in Klug (1982), is merely a selection on the dimensions, usually combined with an aggregate formation with a given aggregate function on a set of category types.

FOLD

This operator was proposed in Gyssens & Lakshamanan (1997). They first define an n-dimensional aggregate data (table) schema S as a triple $<D, R, par>$, where $D = \{d_1, d_2, ..., d_n\}$ is a set of dimension names, $\mathcal{R} = \{A_1, A_2, ..., A_m\}$ is a set of attributes and $par: D \rightarrow 2^{\{A1, A2, ..., Am\}}$, such that: i) for all i, j = 1, ..., n, with i≠j, $par(d_i) \cap par(d_j) = 0$, and ii) $\cup_{d \in D} par(d) \subseteq \mathcal{R}$. The authors denote par (d_i) by X_i. The definition of the *fold* operator is:

Let *s* be a MAD with schema S and let *d* be one of the dimensions of D.

We define *fold*^{d} *(s)* as a MAD with schema S' = $<D \setminus \{d\}, R, par'>$, where, for all *di* in $D \setminus \{d\}$, $par'(d_i) = par(d_i)$, and with an instance tabS'(rep(s)), where rep(s) is the denotation of the relational representation f(s) of the table instance s, i.e., a relation with schema R.

UNFOLD

This operator was also proposed in Gyssens & Lakshamanan (1997). Analogous to the previous operator, let *s* be a MAD with schema S, let *d* be a new name of the set of names *N* appearing nowhere else in *s*, and let $X \subseteq M$ be a set of measure attributes (it is important to note that, defining M as $R - \cup_{1 \leq i \leq n} X_i$, the authors highlight the uniform or symmetric treatment of category attributes and measures).

We define *unfold*^d_X *(s)* as a table with schema S' $<D \cup \{d\}, \mathcal{R}, par'>$, where, for all di in D, $par'(d_i) = par(d_i)$ and $par'(d) = X$ and with instance tab$_s'$ (rep (s)).

Relate Levels

This operator, like the following three operators, has been proposed in Hurtado, Mendelzon, & Vaisman (1999).

The *relate levels* operator defines a roll-up function between two independent levels belonging to a dimension. A necessary condition for this is the existence of a function *f* between the instance sets of the levels being related, such that the dimension instance remains consistent. Otherwise, the *relate levels* cannot be applied. Moreover, when conditions for relating two levels l_a and l_b are met, we must delete all the redundant roll-up functions that may appear, which are not admitted in the model (we only include direct roll-ups in it). For instance, we must delete the roll-up functions between levels l and l_b such that $l \preceq^* l_a$ and $l \preceq l_b$.

Figure 29

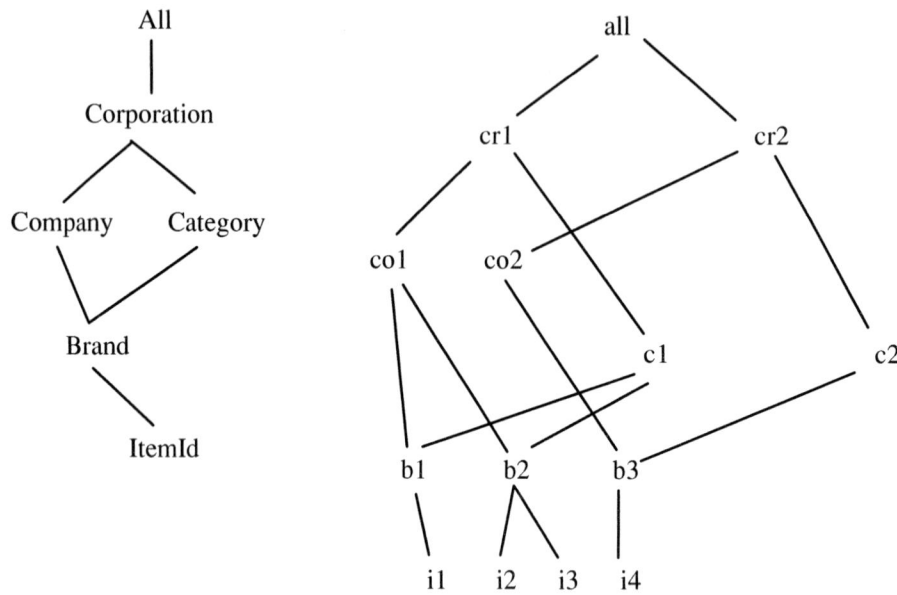

An example of application of relate levels is shown in Figure 29.

Unrelate Levels

The ***unrelate levels*** operator deletes a relation $\preceq$ between two levels l_a and lb, such that $l_a \preceq l_b$. The operator must guarantee that levels which are lower than la in the hierarchy will still be able to reach the same levels they reached before the unrelate operation. For instance, if $l_a \preceq l_b$ and $l_b \preceq l_c$, we must preserve $l_a \preceq l_c$ by making it explicit, in case we delete $l_a \preceq l_b$, because, again, this relation was only implicit in the model.

Delete Level

The *delete level* operator deletes a level and its roll-up functions. The level to be deleted cannot be the lowest one in the dimension (l_{inf}), unless it rolls-up to only one higher level. As was the case with the *relate levels* operator, taking into account that we only define the direct roll-ups, when deleting a level we must add the functions between levels above and below it.

Add Instance

The *add instance* operator inserts a new element, say x, into a level l_a (i.e., an element not belonging to the instance set of l_a). We must provide the operator with the pairs (l_i, x_i), such that every li is a level to which la directly rolls-up $(l_a \preceq l_i)$, and $RUP^{l_i}_{l_a}(x) = x_i)$, where RUP is a set of partial functions such that:

a) for each pair of levels l_1, l_2 such that $l_1 \preceq l_2$, there exists a roll-up function (partial function) $RUP^{l_2}_{l_1}$: dom $(l_1) \rightarrow$ dom (l_2);

b) for each pair of paths, in the graph with nodes in L and edges in $\preceq$, $\tau_1 = <l_1$, $l_2, ..., l_{n-1}, l_n>$ and $\tau_2 = <l_1, l'_2, ..., l'_{m-1}, l_m>, l_n = l_m$, we have $RUP^{l_2}_{l_1} \mathbf{o} ... \mathbf{o} RUP^{l_n}_{l_{n-1}}$ $= RUP^{l'_2}_{l'_1} \mathbf{o} ... \mathbf{o} RUP^{l'_m}_{l'_{m-1}}$; and

c) for each triple of levels $l_1, l_2, l_3 \in L$ (with L a finite set of levels) such that $l_1 \preceq l_2$ and $l_2 \preceq l_3$, ran $(RUP^{l_2}_{l_1}) \subseteq$ dom $(RUP^{l_3}_{l_2})$.

Delete Instance

The *delete instance* operator deletes an element belonging to the instance set of a level la. It is only defined when no element of any level li, such that $l_1 \preceq l_2$ rolls-up to the element being deleted.

An example of *add instance* and *delete instance* application is shown in Figure 30.

Figure 30
(a) Add instance   *(b) Delete instance*

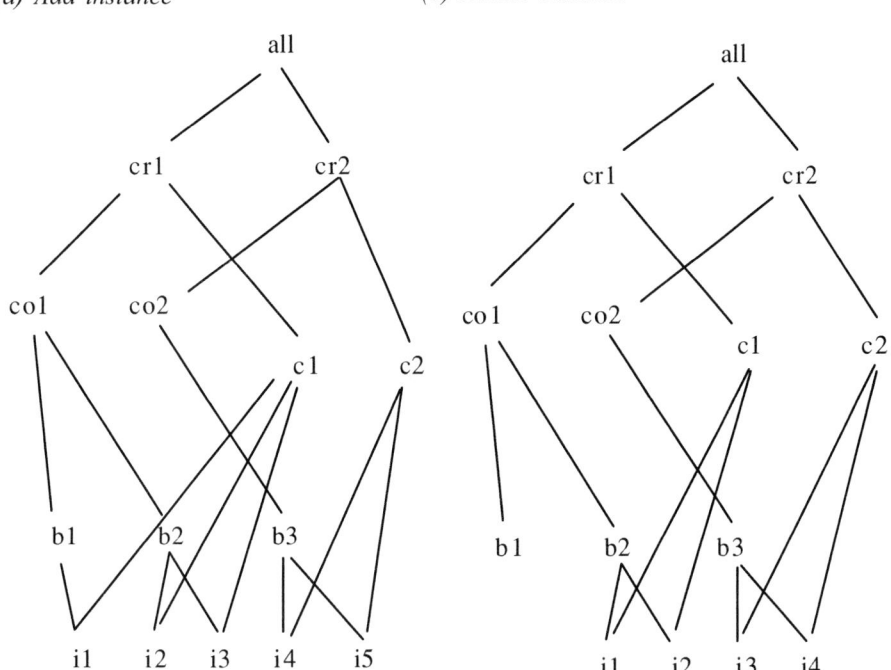

Other operators, which explicitly refer to "levels" (insert, add, and delete level) and *add multiplicity,* proposed in Pourabbas & Rafanelli (2000), have been described in Chapter 4.

REFERENCES

Abiteboul, S., Hull, R., & Vianu, V. (1995). *Foundations of Databases.* Addison-Wesley.

Agrawal R., Gupta A., & Sarawagi, S. (1997). Modeling multidimensional databases. *Proceedings of the 13th International Conference on Data Engineering (ICDE'97),* Birmingham, UK: IEEE Computer Society Press, 232-243.

ArborSoft. (n.d.). *Relational OLAP: Expectation & Reality, An Arbor Software White Paper.* Available on-line at: http://www.arborsoft.com/papers/rolapTOC.html.

Bezenchek, A., Rafanelli, M., & Tininini, L. (1996a). *ADAMO: A Conceptual Model for the Graphical Definition and Manipulation of Aggregate Data.* Technical Report IASI, No. 440, September.

Bezenchek, A., Rafanelli, M., & Tininini L. (1996b). A data structure for representing aggregate data. *Proceedings of the 8th International Conference on Scientific and Statistical Database Management (8th SSDBM),* Stockholm, Sweden. IEEE Computer Society Press, 22-31.

Cabibbo, L., & Torlone, R. (1997). Querying multidimensional databases. *Proceedings of the 6th International Workshop on Databases and Programming Languages (DBPL'97),* Estes Park, Colorado. Lecture Notes in Computer Science, No. 1369. Springer-Verlag, 319-335.

Cabibbo, L., & Torlone, R. (1998). From a procedural to a visual query language for OLAP. *Proceedings of the 10th International Conference on Scientific and Statistical Database Management (SSDBM'98),* Capri, Italy: IEEE Computer Society Press, 74-83.

Colliat, G. (1995). *OLAP, Relational and Multidimentional Database Systems.* Technical Report, Arbor Software Corporation, Sunnyvale, California.

Codd, E.F., Codd, S.B., & Salley, C.T. (1993). *Providing OLAP (On-Line Analytical Processing) to User-Analysts: An IT Mandate.* Technical Report, E.F. Codd and Associates.

Fortunato, E., Rafanelli, M., Ricci, F.L., & Sebastio, A. (1986). An algebra for statistical data. *Proceedings of the 3rd International Workshop on Statistical and Scientific Database Management (3rd SSDBM),* Luxembourg, 122-134.

Ghosh S.P. (1986). Statistical relational tables for statistical database management. *Journal of IEEE Transactions on Software Engineering,* 12(12), 1106-1116.

Gray, J., Bosworth, A., Layman, A., & Pirahesh, H. (1996). Data cube: A relational aggregation operator generalizing group-by, cross-tabs and subtotals. *Proceedings of the 12th IEEE International Conference on Data Engineering (ICDE'96),* New Orleans, Louisiana, 152-159.

Gray, J, Chaudhuri, S., Bosworth, A., Layman, A., Reichart, D., Venkatrao, M., Pellow, F., & Pirahesh, H. (1997). Data cube: A relational aggregation operator generalizing group-by, cross-tab and sub-totals. *Journal of Data Mining and Knowledge Discovery*, 1(1), 29-54.

Gyssens, M., & Lakshmanan, L.V.S. (1997). A foundation for multidimensional databases. *Proceedings of the 23rd International Conference on Very Large Data Bases (VLDB'97),* Athens, Greece, 106-115.

Hurtado, C.A., Mendelzon, A.O., & Vaisman, A.A. (1999). Maintaining data cubes under dimension updates. *Proceedings of the 15th International Conference on Data Engineering (ICDE'99),* Sydney, Australia: IEEE Computer Society Press, 346-355.

Johnson, R.R. (1981) Modeling summary data. *Proceedings of the ACM-SIGMOD'81 International Conference on Management of Data*, Ann Arbor, Michigan, 93-97.

Klug, A. (1982). Equivalence of relational algebra and relational calculus query languages having aggregate functions. *Journal of Association for Computing Machinery*, 29(3), 699-717.

Lehner, W. (1998). Modeling large-scale OLAP scenarios. *Proceedings of the 6th International Conference on Extending Database Technology (EDBT'98),* Valencia, Spain. Lecture Notes in Computer Science, No. 1377. Springer-Verlag, 153-167.

Lenz, H.-J., & Shoshani, A. (1997). Summarizability in OLAP and statistical databases. *Proceedings of 9th International Conference on Scientific and Statistical Data Management (SSDBM'97),* Olympia, WA: IEEE Computer Society Press, 132-143.

Li, C., & Wang, X.S. (1996). A data model for supporting On-Line Analytical Processing. *Proceedings of the International ACM Conference on Information and Knowledge Management (CIKM'96),* Baltimore, Maryland, 81-88.

Maier, D. (1983). *The Theory of Relational Databases*. Rockville, MD: Computer Science Press.

MeoEvoli, L., Ricci, F.L., & Shoshani, A. (1992). On the semantic completeness of macrodata operators for statistical aggregation. *Proceedings of the 5th International Conference on Statistical and Scientific Database Management (SSDBM'92),* Ascona, Switzerland, 239-258.

MicroStrategy. (n.d.). *The Case for Relational OLAP, A White Paper.* Prepared by MicroStrategy, Incorporated. Available on-line at: http://www.strategy.com/dwf/wp_b_al.htm.

Nguyen, T.B., Tjoa, A.M., & Wagner, R. (2000). Conceptual multidimensional data model based on metacube. *Proceedings of the 1st International Conference on Advances in Information Systems (ADVIS'00)*, Izmir, Turkey. Lecture Notes in Computer Science, No. 1909. Springer-Verlag, 24-33.

Ozsoyoglu, G., & Ozsoyoglu, Z.M. (1983). An extension of relational algebra for summary tables. *Proceedings of the 2nd International Workshop on Statistical Database Management*, Los Gatos, California, 202-211.

Ozsoyoglu, G., Ozsoyoglu, Z.M., & Mata, F. (1985). A language and a physical organization technique for summary tables. *Proceedings of the ACM-SIGMOD '85 International Conference on Management of Data*, Austin, Texas, 3-16.

Ozsoyoglu G., Ozsoyoglu Z.M., & Matos, Initial. (1987). Extending relational algebra and relational calculus with set-valued attributes and aggregate functions. *Journal of ACM Transaction on Database Systems*, 12(4), 566-592.

Pedersen, T.B., & Jensen, C.S. (1999). Multidimensional data modeling for complex data. *Proceedings of the 15th International Conference on Data Engineering (ICDE'99)*, Sydney, Australia: IEEE Computer Society Press, 336-345.

Pedersen, T.B., Jensen, C.S., & Dyreson, C.E. (2001). A foundation for capturing and querying complex multidimensional data. *Journal of Information Systems*, 26(5), 383-423.

Pourabbas, E., & Rafanelli, M. (2000). Hierarchies and relative operators in the OLAP environment. *Journal of the ACM Sigmod Record*, 29(1), 32-37.

Rafanelli, M. (1990). Statistical and scientific database management systems. In Kent, A., & Williams, J.G. (Eds.), *Encyclopedia of Computer Science and Technology*, 23(Suppl.8), 369-409. Pittsburgh, PA: Dekker Inc. Publishing.

Rafanelli, M., & Ricci, F.L. (1983). Proposal of a logical model for statistical databases. *Proceedings of the 2nd International Workshop on Statistical Databases*, Los Altos, California, 264-272.

Rafanelli, M., & Ricci, F.L. (1984). Statistical database: An interactive language for logical schema definition by means of a model based on graphs. *Proceedings of the 6th International Symposium on Computational Statistics (Compstat'84)*, Prague, Czechoslovakia: Physica Verlag, 279-284.

Rafanelli, M., & Ricci, F.L. (1985). STAQUEL: A query language for statistical macro-database management systems. *Proceedings of the International Conference on Convention Informatique Latine (CIL'85)*, Barcelona, Spain, 625-637.

Rafanelli, M., & Ricci, F.L. (1988). A statistical functional model for statistical tables. *Proceedings of the International Symposium on Modeling, Identification and Control*, Grindelwald, Switzerland: Acta Press, 136-139.

Rafanelli, M., & Ricci, F.L. (1990). A functional model for statistical entities. *Proceedings of the 1st International Conference on Database and Expert Systems Applications (DEXA'90),* Vienna, Austria, 513-520.

Rafanelli, M., & Ricci, F.L. (1991). A functional model for macro-databases. *Journal of ACM-Sigmod Record,* 20(1), 3-8.

Rafanelli, M., & Ricci, F.L. (1993). Mefisto: A functional model for statistical entities. *Journal of IEEE Transactions on Knowledge and Data Engineering,* 5(4), 670-681.

Rafanelli M., & Shoshani A. (1990). STORM: A statistical object representation model. *Proceedings of the 5th International Conference on Statistical and Scientific Database Management (5th SSDBM),* Charlotte, North Carolina. Lecture Notes in Computer Science, No. 420. Springer-Verlag, 14-29.
and
Bulletin of the IEEE Computer Society Technical Committee Data Engineering, 13(3), 1990, 12-18.

Shoshani, A. (1982). Statistical databases: Characteristics, problems and solutions. *Proceedings of the 8th International Conference on Very Large Data Bases (VLDB'82),* Mexico City, Mexico, 208-222.

Shoshani, A. (1997). OLAP and statistical databases: Similarities and differences. *Proceedings of the 16th ACM Symposium on Principles of Database Systems (PODS'97),* Tucson, Arizona, 185-196, 208-222.

Shoshani, A., & Wong, H.K.T. (1985). Statistical and scientific database issues. *Journal of IEEE Transactions on Software Engineering,* 11(10), 1040-1047.

Su, S.Y.W. (1983). SAM*: A semantic association model for corporate and scientific/ statistical databases. *Journal of Information Sciences,* 29(2&3), 151-199.

Zhao, Y., Deshpande, P., & Naughton, J.F. (1997). An array-based algorithm for simultaneous multidimensional aggregates. *Proceedings of the ACM-SIGMOD'97 International Conference on Management of Data,* Tucson, Arizona, 159-170.

Chapter VI

Time in Multidimensional Databases

Alberto O. Mendelzon
University of Toronto, Canada

Alejandro A. Vaisman
University of Buenos Aires, Argentina

ABSTRACT

In spite of the obvious importance of time in data warehousing and OLAP, current commercial systems do not support tracking the history of a data warehouse, either at the schema or instance level. In this chapter we address this issue, introducing the Temporal Multidimensional Model and a query language, denoted TOLAP, allowing expressing temporal OLAP queries at a high level of abstraction. Further, we show that previous work in temporal databases needs to be extended in order to handle evolution and versioning in OLAP. Finally, we present an implementation, along with preliminary experimental results.

INTRODUCTION

In previous chapters it was shown that in a relational implementation of OLAP *(ROLAP),* facts are usually stored in *fact tables,* while each dimension is described in a *dimension table.* Although time is obviously central to most data warehousing and OLAP applications, currently available commercial systems do not account for the data warehouse's history. In this chapter we concentrate on data warehouse evolution and versioning, introducing a *temporal multidimensional data model* and a temporal query language supporting it, which we called *TOLAP (Temporal*

OLAP), combining some of the temporal features of query languages like TSQL2 (Snodgrass, 1995) or SQL/TP (Toman, 1997) with some of the high-order features of languages like HiLog (Chen, Kifer, & Warren, 1989) or SchemaLog (Lakshmanan, Sadri, & Subramanian, 1993).

A data warehouse could be regarded as a materialized view of data located in multiple sources (Widom, 1995). Thus, it is not difficult to imagine a scenario in which the structure of these sources changes, a new source is added, or an old one dropped. Any of these changes may require updates to the structure of some dimensions. Moreover, as multidimensional views are designed according to requirements from end users, a redefinition of the initial requirements may cause a dimension update. For instance, let us consider Figure 1, which depicts a *geography* dimension where regions are defined within the same country. A business decision may relax this constraint, allowing regions to be spread across different countries. This may be represented in the lattice of Figure 1 by deleting the edge joining the *region* and *country* levels, and adding a new edge from *region* to the distinguished level *All.* Figure 2 shows the resulting dimension. Additionally, new salespersons may be hired or fired; new kinds of coverage introduced or discontinued; regions may be reorganized, merged, or split; etc. However, a user may be interested in querying the multidimensional database as of the instant depicted in Figure 1. Moreover, since the schemas of the fact tables are composed of attributes from associated dimensions, certain updates may trigger schema evolution over such fact tables. Suppose we wish to collect data at a granularity level finer than *city,* for instance *neighborhood.* Any fact table associated with the *geography* dimension would require its schema to be updated. We argue in this chapter that in an evolving scenario like this, OLAP systems need temporal features that allow tracking of the different states of a data warehouse throughout its lifespan.

Figure 1: A Geography Dimension

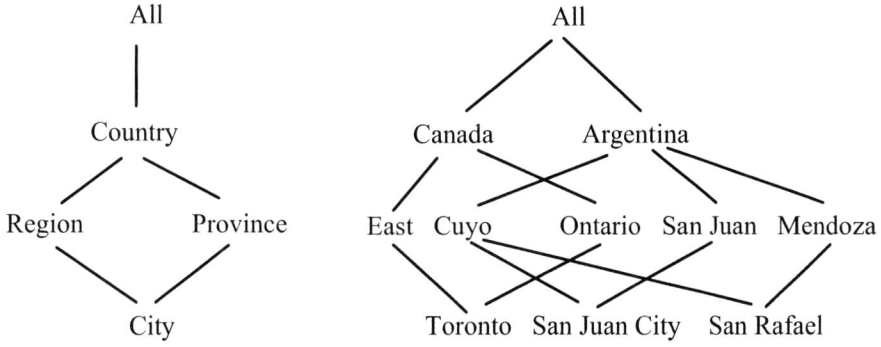

Figure 2: Updated Geography Dimension

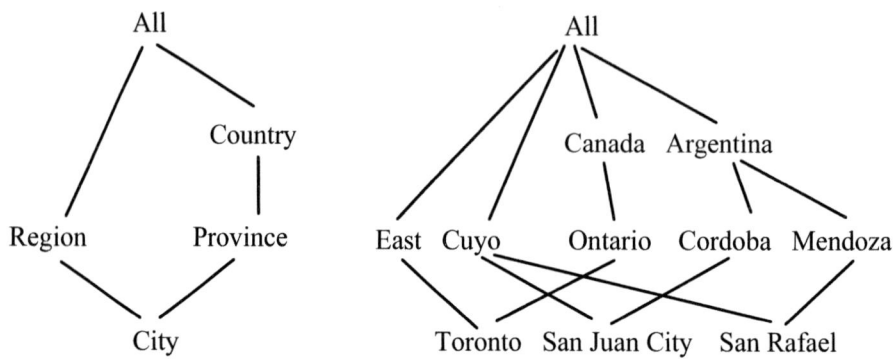

EXAMPLE

We will introduce the problem by means of a motivating example. For the remainder of this section, we will consider dimensions as relations representing data as of the current time. Let us suppose a retail data warehouse with the following dimensions: *time, salesperson, customer,* and *product.* Moreover, as dimensions are organized in hierarchies, let us also assume the hierarchy {*itemId → priceStatus, itemId → brand*}, and the following functions (denoted *roll-up functions*) from *itemId* to *priceStatus*: {(Panasonic DVD A120, reduced), (Sony STR DE675, regular), (Sharp MD-SR50, regular)} (let us ignore the dimension level *brand* at this time). The following table represents sales facts.

timeId	salesPersonId	customerId	itemId	salesAmount
10 - 10 - 01	S_1	Becker, J.	Panasonic DVD A120	250
10 - 11 - 01	S_2	Rodriguez, P.	Panasonic DVD A120	250
10 - 12 - 01	S_1	Ashfield, K.	Sony STR DE675	300
10 - 13 - 01	S_2	Finn, N.	Sharp MD - SR50	100

A query asking for the *total sales per salesperson and price status* would return the following table:

spId	itemType	sales Amount
S_1	reduced	250
S_2	reduced	250
S_1	regular	300
S_2	regular	100

Suppose now that on October 14, 2001, we decide to assign the Panasonic DVD a *regular* price status. A nontemporal star or snowflake schema will store < *Panasonic DVD* A120, regular > in the table representing dimension *Product*, replacing the tuple < *Panasonic DVD* A120, reduced >, i.e., there will be no memory of the price status of an item. If the user now poses the same query, as all the sales occurred before the revision, she would expect to get the same result. However, she gets the following:

spId	priceStatus	sales Amount
s_1	*regular*	550
s_2	*regular*	350

What happened is that there are no longer items with reduced prices.

Notice that in order to issue a query, the user needs to know the schema of the data warehouse, that is, which are the attributes in the fact and dimension tables. However, this schema may change over time. For instance, *itemId* may not always have been an attribute of the fact and/or dimension tables if in the early days of this data warehouse, data with granularity *itemId* was not available at the sources. In this case, the query will only consider total sales made since the time at which *itemId* was added to the fact table, although information is available to obtain the total sales over the whole lifespan of the data warehouse, at least at a coarser level.

As another example of inaccurate results a user could get when querying this data warehouse, suppose that a many-to-one relationship from customers to salespersons exists, such that each salesperson is assigned a set of customers to serve. At a certain time, a customer, say c_1, initially assigned to salesperson s_1, is reassigned to s_2. Suppose a sales manager wants to use the data warehouse to set future sales goals for each salesperson, basing the forecast on past sales data. Given the relationship between customers and salespersons, clearly this projection should be based on past volume of purchases made to customers *currently* assigned to each salesperson, no matter who was formerly assigned to whom, as a salesperson cannot expect anything from a customer no longer assigned to her.

The discussion above suggests that new models and query languages are needed in order to address temporal issues in OLAP. This will avoid building ad-hoc applications, as in current commercial OLAP systems, which have no built-in temporal capabilities.

BACKGROUND

The problem of handling "slowly changing dimensions" was mentioned by Kimball (1996), who suggested some partial solutions, like timestamping dimension tuples with their validity intervals. This proposal, however, does not consider schema versioning.

A work carried out by Bliujute et al. (1998) at the *Time Center* at the University of Arizona analyzes the performance of several SQL queries over three different approaches to the star schema: (a) "time series" fact tables; (b) "event" fact tables; (c) dimensions timestamped in the way proposed by Kimball (1996). This work was, to our knowledge, the first approach to the problem of temporal OLAP. Our work goes further, as we propose a model and a query language to address temporal issues at a higher abstraction level.

More recently, a multidimensional model for handling complex data was introduced by Pedersen & Jensen (1999) where the temporal aspect is considered a modeling issue, and is addressed in conjunction with other data modeling problems. At the time this is being written, we are not aware of any other work proposing a data warehouse evolution framework or a temporal query language for OLAP like the ones addressed in this chapter.

Recent work on maintenance of temporal views by Yang & Widom (2000) presents a view definition language operating over non-temporal data sources, along with techniques for maintaining temporal views. This work focuses on the data sources and on how a set of temporal views are obtained and maintained. We will not deal with these issues in this chapter.

Mendelzon & Vaisman (2000) discussed the need for a temporal query language for OLAP, introducing TOLAP, along with a temporal multidimensional data model.

THE TEMPORAL MULTIDIMENSIONAL MODEL

In the multidimensional model developed by Hurtado and the authors (Hurtado, Mendelzon, & Vaisman, 1999), dimensions are defined as non-temporal structures like the *geography* dimension of Figure 1. A dimension in this model has a schema, and instances associated to this schema which may evolve through time, although dimension versioning is not supported. In the *Temporal Multidimensional Model* we present in this section, we timestamp dimension elements at the schema and instance level in order to keep track of the updates that occur during the dimension's lifespan.

We will consider time as discrete; that is, a point in the timeline, called a *time point,* will correspond to an integer. We will also assume, except when noted, instant $t0$ to be the dimension's creation instant.

Temporal Dimensions. The following sets are defined: a set of level names **L**, where each level $l \in$ **L** is associated with a set of values $dom(l)$; a set of attribute names **A**, such that each attribute $a \in A$ is associated with a set of values $dom(a)$; a set of temporal dimension names **TD**; and a set of fact table names **F**.

Definition 1 (Temporal Dimension Schema): *A temporal dimension schema is a tuple (dname, L, λ, $\preceq$, A, $\gg$, μ) where:*

- *dname $\in$ **TD** is the name of the temporal dimension.*
- *μ is a level in the* Time *dimension. Intuitively, μ defines the granularity of the dimension* dname. *It represents* transaction time.
- *$L \subseteq$ **L** is a finite set of levels, which contains a distinguished level name* All, *s.t. dom(All) = {all}. This distinguished level is considered valid during the complete lifespan of the dimension.*
- *"λ" is a function with signature $dom(\mu) \to$ **L**, defining the instants when each level was part of the dimension.*
- *"$\preceq$" is a function with signature $dom(\mu) \to 2^{\mathbf{L} \times \mathbf{L}}$, such that for each $t \in dom(\mu)$, $\preceq_t$ is a relation such that $\preceq^*_t$, the transitive and reflexive closure of $\preceq_t$, is a partial order, with a unique bottom level, $l_{inf} \in \lambda$ (t), and a unique top level,* All, *where, for every level $l \in \lambda$ (t), $l_{inf} \preceq^*_t l$ and $l \preceq^*_t$ All hold.*
- *A is a finite set of attributes.*
- *"$\gg$" is a function with signature $dom(\mu)$ x $A \to$ **L**. For every level $l \in \lambda(t)$, a $\gg_t l$ means that if the function is applied to an attribute* a, *it returns the level l such that attribute* a *belongs(ed) to level l at time t.*

The structure introduced by Definition 1 suggests that, given an appropriate language, it would be possible to query the schema of a dimension (using the functions $\preceq$ and $\gg$). Note that l_{inf} is unique at any time instant t, although it may not be the case at different times. Also observe that taking a snapshot of the schema at a given time t, we get a non-temporal dimension schema.

Notation: In the figures of this chapter, a label t_i associated to an edge in a graph, will mean that the edge is valid for all $t \geq t_i$, and a label t^*_i, that the edge was valid for all $t < t_i$. If an edge has no label, it is considered valid for all of the dimension's lifespan. An interval $[t_i, t_j)$ means that the edge is valid from $t \geq t_i$ to $t < t_j$.

Definition 2 (Temporal Dimension Instance): *A temporal dimension instance is a tuple (D, TRUP, TDESC), where D is a temporal dimension schema, and:*

- *TRUP (temporal roll-up) is a set of functions, satisfying the following conditions:*
  - *For every instant $t \in dom(\mu)$, and for each pair of levels $l_1, l_2 \in \lambda(t)$ such that $l_1 \preceq_t l_2$, there exists in TRUP a roll-up function: $dom(l_1) \to dom(l_2)$. Thus, a function is defined for every snapshot taken at any instant $t \in dom(\mu)$.*
  - *For every instant t in the dimension's lifespan, and for every pair of paths τ_1 and τ_2 in the graph with nodes in $\lambda(t)$ and edges in $\preceq_t$, such that $\tau_1 = <l_1, l_2, ..., l_k, l_n>$, and $\tau_2 = <l_1, l'_2, ..., l'_k, l_n>$, we have $\rho[t]_{l_1}^{l_2} \circ ... \circ \rho[t]_{l_k}^{l_n} = \rho[t]_{l_1}^{l'_2} \circ ... \circ \rho[t]_{l'_k}^{l_n}$.*
  - *At every instant t of the dimension's lifespan, and for each triple of levels $l_1, l_2, l_3 \in \lambda(t)$ such that $l_1 \preceq_t l_2$ and $l_2 \preceq_t l_3$, $ran\,(\rho[t]_{l_1}^{l_2}) \subseteq dom\,(\rho[t]_{l_2}^{l_3})$.*
- *TDESC (temporal description) is a set of functions such that for every instant $t \in dom(\mu)$, and for each level $l \in \lambda(t)$ and for each attribute a such that $a \gg_t l$, there exists in TDESC a function with signature $\xi[t]_l^a : dom(l) \to dom(a)$.*

We will call the second condition in Definition 2 *snapshot consistency*.

Figure 3: The Temporal Dimension "Store"

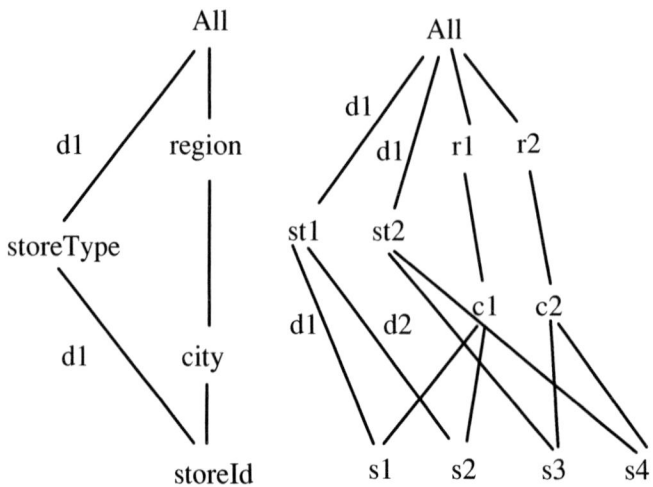

Example 1: *Figure 3 depicts a schema and an instance for a dimension* Store. *Here, dname = * Store, *L = {storeId, storeType, city, region, All}, μ = day. Initially "≼" contains the following pairs: storeId ≼ city, city ≼ region, region ≼ All. The addition of a new level* storeType *above level* storeId *at time* d_1 *will modify "≼" as follows: ≼ = {storeId ≼ city, city ≼ region, storeId ≼$_{d_1}$ storeType, region ≼ All, storeType ≼$_{d_1}$ All}.*

Figure 3 also shows a temporal dimension instance for dimension Store. *We see that the roll-up functions with no label are valid for the whole lifespan of* Store, *while for example* $\rho^{storeType}_{storeId}$ $(s_1) = st_1$ *is valid from* d_1 *on.*

Definition 3 (Active Instance Set): *Given a temporal dimension* d, *a level* $l \in L$, *and an instant t, the set of elements belonging to dom(l) at time t is called the* Active Instance Set *of l. We denote it* Ainstset (l,t) .

Example 2: *In Figure 3,* Ainstset (stereId,d)= $\{s_1, s_2, s_3, s_4\}$, *for all d. However, deleting s_1 at time d_3 will yield* Ainstset (storeId,d_3)= $\{s_2, s_3, s_4\}$.

The previous definitions set the basis for a temporal data warehouse. We will introduce the idea through the following example.

Example 3: *Figure 4 shows a sequence of updates to a temporal dimension* Product. *Initially,* Product *consisted only of level* itemId, *and the distinguished level* All. *After that, the* brand *was added to the dimension, although the initial state is not lost (Figure 4(b)). Later, the type of the item is inserted, with level name* itemType. *The* company *to which an item belongs is also added above level* brand *at time d_5, and level* brand *is deleted at time d_9 (Figure 4(d)). Note that an edge is added from level* itemId *to level* brand, *valid from time d_9 on.*

A slice or snapshot (in temporal database terminology) of a dimension, taken at a given instant *t*, defines the state of the dimension at that time. For instance, Figure 5 shows a snapshot of dimension *Product* at d_4.

Definition 4 (Temporal Fact Table): *A temporal fact table schema is a tuple s = (fname, f, m, μ), where m is a level name, called the* measure *of the fact table, μ is a level in the* Time *dimension, and f is a function with signature dom(μ) →* 2^L.

Given a temporal fact table schema (fname, f, m, μ), a set of levels L in the range of f, and a level μ in the Time *dimension, a mapping from each level l_i ∈ (L ∪ μ) to dom(l_i) is called a* point.

Given a temporal fact table schema s = (fname, f, m, μ), a temporal fact table instance *over it is a partial function named f name which maps points of s to elements in dom(m).*

Figure 4: A Series of Updates to Dimension "Product"

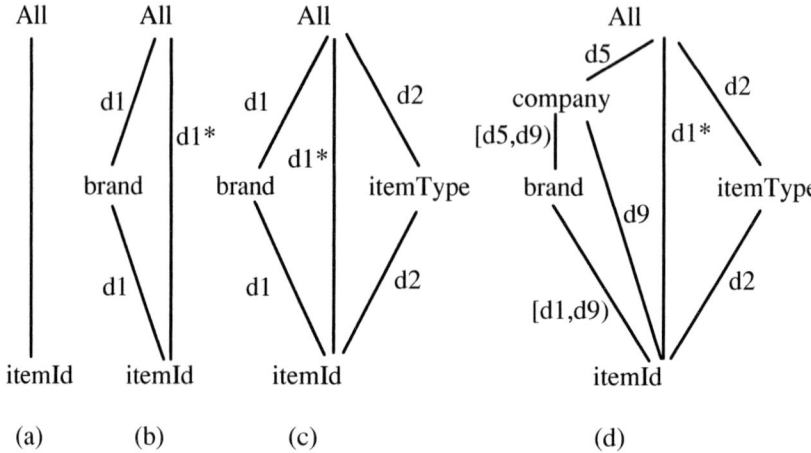

(a)   (b)   (c)   (d)

Definition 5 (Temporal Base Fact Table): *Given a set* **D** *of temporal dimensions, a temporal base fact table is a temporal fact table with schema (fname, f_D, m, μ), where f is a function with signature dom(μ) → 2^L, such that for each t ∈ dom(μ), every level in $f_D(t)$ is a bottom level of the dimension it belongs to. Thus, a temporal base fact table is a temporal fact table such that its attributes are the bottom levels of each one of the dimensions in* **D**.

Example 4: *Given D = {Store, Product} where dimensions* Store *and* Product *are those of Figures 3 and 4 respectively, the Temporal Base Fact Table associated to D would have $f_D(t)$ = {storeId, itemId} for all t. If updates occur such that, for instance, at time d12, brand becomes the bottom level of dimension* Product, $f_D(d12)$ = {storeId, brand}.

Figure 5: A Snapshot at "d4"

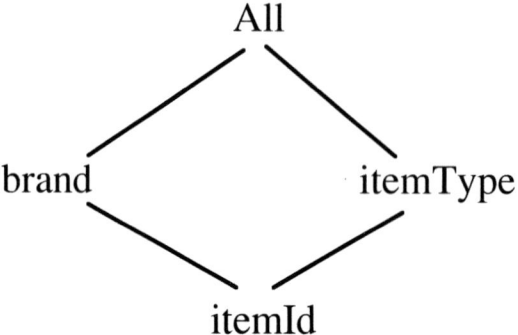

Definition 6 (Temporal Multidimensional Database):

- *A temporal multidimensional database schema, denoted* $\mathbf{B}_s$, *is a pair* $(\mathbf{D}_s,$ $\mathbf{F}_s)$, *where* $\mathbf{D}_s$ *is a set of temporal dimension schemas, and* $\mathbf{F}_s$ *is a set of temporal fact table schemas.*

- *A* multidimensional database instance $I(\mathbf{B})$, *is a tuple* $(\mathbf{F}_I, \mathbf{D}_I)$, *where* $\mathbf{D}_I$ *and* $\mathbf{F}_I$ *are dimension and fact table instances defined as above.*

Valid and Transaction Times in the Temporal Multidimensional Model

In temporal databases, in general, instances are represented using *valid* time, this is, the time when the fact being recorded became valid. On the contrary, a database schema holds within a time interval that can be different from the one within which this schema was valid in reality. Changes in the real world are not reflected in the database until the database schema is updated. Schema versioning, thus, is often related with *transaction* time. In our temporal multi-dimensional model, things are different than above: an update to a dimension schema modifies the structure as well as the instance of the dimension. Figure 3 shows a dimension where schema and instance updates occurred. When *storeId* becomes generalized to *storeType,* the associated roll-ups must correspond to the instants in which they actually hold. Thus, we consider that temporal dimensions are represented by *valid* time. It is straight-forward to extend the model for supporting temporal dimensions with valid and transaction times.

Fact table instances can be represented in our model using valid and transaction times. In order to accomplish this, two time dimensions must be defined. Fact table schema versioning is represented using transaction time. When a bottom level of a dimension is deleted or a level is added below it, the schemas of the associated fact tables change and new versions are defined for them.

TOLAP: A Temporal OLAP Query Language

In the previous section we introduced a model accounting for temporal dimensions, i.e., dimensions that evolve across time, allowing us to keep track of the dimension's history. Let us call this *dimension versioning.* We will show that typical OLAP queries involving temporal dimensions admit different interpretations, which should be considered in order to give the user the correct answers to queries. Further, we will introduce the *TOLAP* query language, which supports the model presented in the previous section.

Do We Need a Temporal OLAP Query Language?

One might argue, at first sight, that a generic temporal query language like TSQL2 (Snodgrass, 1995) could be used instead of defining a special-purpose one like *TOLAP.* There are two main reasons supporting the idea of introducing a new language. First, a language designed specifically for the multidimensional model makes typical OLAP queries much more concise and elegant. In a generic language, queries would have to be laboriously encoded using detailed knowledge of the low-level relational structures used to encode the dimensional data. Second, the best-known temporal query languages such as TSQL2, support only a minimal level of schema versioning (Snodgrass, 1995, p.29).

Another alternative would have been adding temporal features to other languages with schema management features, such as HiLog (Chen, Kifer, & Warren, 1989) or SchemaLog (Lakshmanan, Sadri, & Subramanian, 1993). Again, using a language specifically designed for OLAP yields much simpler syntax and semantics, and just the high-order features that are needed to support schema evolution.

The following example will show that a temporal OLAP query language is needed in order to capture the particular characteristics of OLAP queries expressed over a set of temporal dimensions.

Let us consider again a retail data warehouse, with a set of temporal dimensions $D = \{Product, Store\}$, and a base fact table with schema $(Sales, f, sales, day)$. Dimensions $Store$ and $Product$ are the ones in Figures 3 and 6, respectively. The dimension levels in the fact table are $\{itemId, storeId\}$, the bottom levels of the dimensions in D. Notice that in $Product$, item i_1 has type ty_1 until day d_4, and type ty_2 since then. For the $Time$ dimension we assume: $\rho_{day}^{week} = \{d_1 \rightarrow w_1, d_2 \rightarrow w_1, d_3 \rightarrow w_2, d_4 \rightarrow w_2, d_5 \rightarrow w_2\}$ (and the roll-ups from week to All). Finally, we have the following instance for the $Sales$ fact table $(day$ is displayed for the sake of clarity, but could have been omitted, like in TSQL2):

Let us now suppose the query: "List the weekly total sum of sales, by city and item type." Two interpretations could be given to this query. The usual one would compute the sum of sales considering the type an item had when it was sold. In this case, for instance, item i_1 would contribute to the aggregation in the following way: the first three tuples, with a total of 800, will add to the group $\{ty_1, c_1, w_1\}$, while the last one will contribute to $\{ty_2, c_1, w_2\}$.

itemId	storeId	day	sales
i_1	s_1	d_1	600
i_2	s_2	d_1	100
i_2	s_1	d_2	100
i_3	s_2	d_2	100
i_3	s_3	d_3	100
i_3	s_4	d_4	100
i_1	s_1	d_5	100

The second interpretation, the **only** one currently supported by non-temporal systems, would compute the sum of the sales considering that each sold item has the current type, regardless of the time the sale occurred. The result a user would get under this interpretation is given by the table below, and was computed in the following way: the roll-up function for every occurrence of item i_1 is set to:

Figure 6: Dimension Product for the Running Example

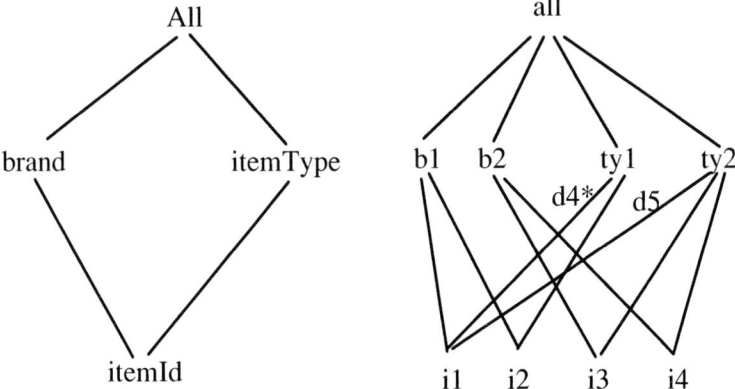

The results will be given by the following table.

itemType	city	week	sales
ty_1	c_1	w_1	800
ty_2	c_1	w_1	100
ty_2	c_2	w_2	200
ty_2	c_2	w_2	100

$\rho_{itemId}^{itemType}$ $(i_1) = ty_2$. Thus, all the i_1 tuples will contribute to type ty_2. For instance, the first tuple will now contribute to the group $\{ty_2, c_1, w_1\}$. The following table shows the result the user would obtain under the second interpretation.

TOLAP by Example
We will first present *TOLAP* by means of examples, and then define its syntax and semantics. It is interesting to notice that the queries we will present operate at a high level of abstraction, without requiring low-level knowledge about the database design that underlies the dimensional model.
We will work with a set of dimensions **D** = {*Store, Product*}, where *Store* is the dimension in Figure 3 and *Product is* the dimension depicted in Figure 4; the base fact table *Sales* is the one introduced at the beginning of this section.

itemType	city	week	sales
ty_1	c_1	w_1	200
ty_2	c_1	w_1	700
ty_2	c_2	w_2	200
ty_2	c_2	w_2	100

Simple Queries. We begin with queries not involving aggregates.

Example 5: *A query returning the sales for stores in Buenos Aires, on a daily basis, will be expressed in* TOLAP *as*

$$BASales(p,s,m,t) \longleftarrow Sales(i,s,m,t), \overset{t}{\rightarrow} city:'Buenos\ Aires'$$

In TOLAP, *the query above returns the tuples in* Sales *such that* s *rolled-up to 'Buenos Aires' at the time of the sale, where* s *represents an element in the instance set of the lowest level of the dimension* Store. *The query is interpreted as follows: a tuple in* Sales *will be in the result if an item* i *was sold in store* s *on a day* t *if* s *was settled in Buenos Aires on that day (recall that the granularity of the fact table is* day).

We assume a fixed order of the attributes in the base fact tables. For instance, in the base fact table *Sales,* the first position from the left will always correspond to dimension *Product.*

Queries with Aggregates. In order to address queries involving aggregation, we adapted a non-recursive data log with aggregate functions (Consens & Mendelzon, 1990) which, in turn, was based on the approach of Klug's relational calculus with aggregates (Klug, 1982).

Example 6: *Consider the query: "list the total sales per item and region," where we want aggregates to be computed using temporally consistent values (i.e., a sale in a given store is credited to the region that corresponded to that store at the time of the sale).*

$$IR(it,re,SUM(m)) \longleftarrow Sales(i,s,m,t), \overset{t}{\rightarrow} region:re.$$

This query is interpreted as follows: let us suppose a pair of tuples $t_1 : < i_1$, *TaosBranch*, 100, 10/10/01 >, and $t_2 :< i_1$, *DallasBranch1*, 100, 10/10/01 >, in fact table Sales; moreover, assume that the store branches identified by *TaosBranch* and *DallasBranch1* were assigned to the southern region on October 10, 2001. As the expression s $\overset{t}{\rightarrow}$ region:re will evaluate to *true* when variables s, t, and re become instantiated with values *TaosBranch* (or *DallasBranch1*), 10/10/01, and *Southern*, respectively, the tuple $< i_1$, *Southern*, 200 $>$ will be in the result.

It is worth noting that in case the time granularities of the fact table *Sales* and the *Store* dimension differ (for instance, if the dimension's granularity was *month*), the user would not have to make the granularity conversion explicit, allowing a limited form of "schema independence." Later examples show the use of variables that range over level names, pushing this independence farther.

Example 7: *We now introduce descriptive attributes of dimension levels. Suppose we want the total sales by store and brand, for stores with more than 90 employees. Assume that the level* storied *is described by an attribute* nbrEmp *(number of employees).*

$$SB(br,st,SUM(m)) \longleftarrow Sales(i,s,m,t), i \xrightarrow{t} brand:br,$$
$$s \xrightarrow{t} storedId:st, s.nbrEmp \geq 90.$$

Metaqueries.TOLAP also allows querying the system's metadata, supporting queries with no fact table in the body of the rules. We call these queries *metaqueries*.

- "*Give me the time instants at which store* s_1 *belonged to the Southern region,*" expressed as:

$$StoreTime(t) \longleftarrow Store: storeId: 's'_1 \xrightarrow{t} region: 'Southern'.$$

* Note that we must specify the name of the dimension in the atom Store: storeId: 's_1,' because there is no fact table in the body of the rule to which the dimension's bottom level could be bound.

- "*List the months when 'Southern' was not a valid region.*"

$$noSouth(t) \longleftarrow !Store: region:'Southern' \xrightarrow{t} X:x.$$

In the example above, variable X ranges over level names in the *Store* dimension. Variable x is bound by the values in the levels of the dimension. The domain of t is the lifespan of the dimension referenced in the metaquery. For every instantiation of the variables X,x, and t, which makes the body rule of the rule true, the time instant t is listed.

- "*Were products categorized by brands two years ago?*"

$$ProdBrand() \longleftarrow Product:X:x \xrightarrow{1/1/98} brand:y.$$

Here again X ranges over level names. The expression above means that if any element, in any level in the *Product* dimension, is rolled-up to an element in level *brand* at the required date, the answer to the query will be "*yes.*"

- *"How were customers classified three years ago?"*

$$CustCat(cust,X,x) \longleftarrow Customer:customerId:cust \overset{1/1/97}{\rightarrow} X:x.$$

Data Warehouse Versioning in TOLAP

Earlier, we showed that our model supports different versions of a dimension schema over time (in temporal database terminology this is called *schema versioning*). For instance, consider an alternative history for dimension *Store* in our running example, which is depicted in Figure 7. We can see that the bottom level of *Store* was initially *city,* and at time d_5 *storeId* was inserted below it. Also assume that the fact table depicted below was in effect *before* d_5.

itemId	city	sales	day
i_1	c_1	100	d_1
i_2	c_2	100	d_2
i_1	c_3	100	d_3
i_2	c_1	100	d_1

After the update, the sales in the following table occurred (notice the new structure of the fact table).

itemId	storeId	sales	day
i_1	s_4	100	d_6
i_2	s_2	100	d_6
i_1	s_3	100	d_7
i_2	s_1	100	d_7

In *TOLAP,* an element not defined at a given instant will not contribute to the result. For instance, given the query *"list the total sum of sales by brand and storeId,"* we have:

$$SB(br,st,SUM(m)) \longleftarrow Sales(i,s,m,t), \ i \overset{t}{\rightarrow} brand:br,$$
$$s \overset{t}{\rightarrow} storeId:st.$$

The expression $s \overset{t}{\rightarrow} storeId:st$ means that if an element in any level, which was once a component of a base fact table, rolled-up to level *storeId* at time t, it contributes to the aggregation. Thus, the sales made before d_6 will not contribute to the aggregation in the head   (condition $s \overset{t}{\rightarrow} storeId:st$ will not be satisfied). Analogously, a query like *"total sales by store and itemId"* would return exactly the instance of the second fact table above. This query is expressed:

$$StIt(it,st,SUM(m)) \longleftarrow Sales(i,s,m,t), i \xrightarrow{t} itemId:it, s \xrightarrow{t} storeId:st.$$

For the dimension *Product,* Figure 4 shows that at time d_5, level *company* was added above level *brand.* Given the query, *"total sales by company and region,"* the sales occurring before d_5 will not contribute to the aggregation, as *company* was not a level of the dimension at that time. The query reads in *TOLAP*:

$$CR(c,reg,SUM(m)) \longleftarrow Sales(i,s,m,t), i \xrightarrow{t} company:c,$$
$$s \xrightarrow{t} region:reg.$$

Finally, at time d_9, level *brand* was deleted from *Product.* The query *"total sales by brand and region"* is expressed in *TOLAP* as:

$$BR(br,reg,SUM(m)) \longleftarrow Sales(i,s,m,t), i \xrightarrow{t} brand:br,$$
$$s \xrightarrow{t} region:reg.$$

Any sale that occurred after d_9 will not be considered, as *brand* is not a level of the dimension anymore.

TOLAP Programs

A *TOLAP* program allows computing the result of a rule and using it as a predicate in the body of another rule. The predicate's name is the name of the predicate in the head of the rule which computes it.

For instance, let us suppose the following query: *"List the total sales by brand, for those brands that sold more than $100,000 in Buenos Aires."* We want to answer this query precomputing a view holding the total sales in Buenos Aires, by brand.

$$BASales(c,SUM(m)) \longleftarrow Sales(i,s,m,t), i \xrightarrow{t} brand:c,$$
$$s \xrightarrow{t} city:x, x.name="Buenos Aires."$$

Figure 7: Data Warehouse Versioning

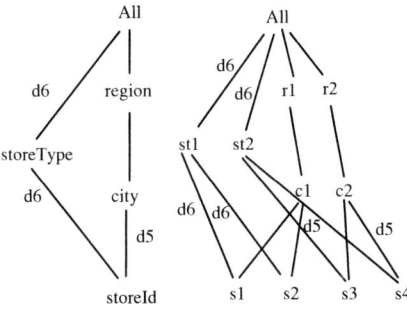

Then, we compute the query, using predicate *BASales*

$$Q(c,SUM(m)) \longleftarrow Sales(i,s,m,t), \; BASales(c,q), \; i \xrightarrow{t} brand:c, q \geq 100,000.$$

For each brand in dimension *Product* matching a brand in *BASales,* in the second rule, if the amount sold was greater than 100,000, the sales contributes to the aggregation.

Syntax

We will formally define the syntax of a *TOLAP* rule. We will first give some definitions which will be used below, and then formalize the concepts previously introduced in the section, *TOLAP by example.*

Preliminary Definitions. Given a set *T*, and a discrete linear order <, with no endpoints, we define a *point-based temporal domain* as the structure $T_p = (T, <)$. Analogously, given $T_p = (T, <)$, we define the set $I(T) = \{(a, b) \mid a \leq b, \; a, b \in T \cup \{-\infty, +\infty\}\}$. Let us denote by θ the set of the usual interval comparison operators. Then, $T_I = (I(T), \theta)$ is an *interval-based temporal domain* corresponding to T_p. The roll-up functions in TOLAP will be defined over T_p.

Atoms, Terms, Rules, and Programs. Assume $\mathbf{B_s} \; (\mathbf{D_s}, \mathbf{F_s})$ and $I(\mathbf{B})$ represent a multidimensional database schema and instance, respectively, as already defined. Let $\mathbf{V_L}$ and $\mathbf{V_D}$ be a set of level and data variables, respectively. We have also the sets $\mathbf{C_L}$ and $\mathbf{C_D}$ of level and data constants, respectively. Let $\mathbf{P}$ be a set of predicate names and $\mathbf{F_{Ag}}$ a set of aggregate function names.

Definition 7 (Terms):

- A data term *is either a variable in* $\mathbf{V_D}$ *or a constant in* $\mathbf{C_D}$;
- a roll-up term *is an expression of the form* d:X:x, X:x, *or* x, *where X is a level name variable in* $\mathbf{V_L}$ *or constant in* $\mathbf{C_L}$, x *is a data term, and* d *is a constant in* $\mathbf{C_L}$;
- a descriptive term *is an expression of the form* x.a *where* x *and* a *are data terms;*
- an aggregate term *is an expression of the form* f(d) *such that* f *is a function name in* $\mathbf{F_{Ag}}$ *and* d *is a data term.*

A term *is a* data, roll-up, descriptive, *or* aggregate *term.*

Definition 8 (Atoms):

- A fact atom *is an expression of the form* $F(X_1, ..., X_n, M, t)$, *where F is a fact table name in* $\mathbf{F_s}$ *and* $X_1, ..., X_n$, M *and* t *are data terms;*

- *a* roll-up atom *is an expression of the form* $X \xrightarrow{t} Y$, *or* $X \rightarrow Y$, *where* X *and* Y *are roll-up terms, and t is a data term;*
- *a* descriptive atom *is an expression of the form* $x \stackrel{t}{=} y$, *where* x *is a descriptive term, and* y *and t are data terms;*
- *an* aggregate atom *is of the form* $Q(R,...,Z)$ *s.t.* $Q \in \mathbf{P}$, *and* R,...,Z *are data terms s.t. at least one is an aggregate term;*
- *a* constraint atom *is an expression* $t_1 \theta t_2$, *where* t_1 *and* t_2 *are data terms, and* θ *is one of* $\{<, =\}$;
- *if* $g : N \times ... \times N \rightarrow N$ *is a scalar function*, $g(n_1, ..., n_m)$, *where* n_i *are data terms, is a* scalar atom;
- *an* intentional (extensional) predicate atom *is an expression of the form* $p (X, ..., Z)$ *where* X,Y,Z *are data terms, and* p *is an intentional (extensional) predicate name in* **P**. *In what follows, we will refer to these kinds of atoms simply as* predicate *atoms when possible.*

An atom *is a fact, roll-up, descriptive, aggregate, constraint, scalar, or predicate atom. An expression* $!t_1$, *where* t_1 *is an atom, is a* negated atom.

Definition 9 (TOLAP Rules): *A* TOLAP rule *is a formula of the form* A $\longleftarrow$ $A_1, A_2, ..., A_n$, *where* A *is an intentional (possibly aggregate) positive atom, and* A_i, *i = 1...n are non-aggregate atoms. A TOLAP rule* Γ *satisfies the following conditions:*

- *if a variable appears in the head of the rule, it must also appear in its body;*
- *for every level/data variable v, all the roll-up terms in which v appears are associated to the same dimension; moreover, all the variables in a roll-up atom belong to the same dimension;*
- *if there is an aggregate atom* $Q(a_1, ..., a_n)$ *in the head of the rule, for all atoms in the body, of the form* $d:X:x \xrightarrow{t} y:a_i$, $x \xrightarrow{t} y:a_i$, *y is a constant data term (thus, aggregation is performed over constant level names);*
- *every variable* x *in a roll-up atom of the form* $x \xrightarrow{t} y:a$ *must appear in a fact atom in the body of the rule;*
- *if* x.a *is in the body of* Γ, *at least one roll-up term in the body is of the form* d:X:x, X:x, *or* x;
- *every variable appearing in a predicate atom must appear in a fact or roll-up atom in the body of the rule;*
- *in a roll-up term of the form* $d:X:x \xrightarrow{t} y:a$, d *is a constant data term;*
- *if there is no fact atom in the body of* Γ, *all the roll-up atoms must be of the form* $d:X: x \xrightarrow{t} y:a$;
- *In a negated* fact *atom, at least one of its terms must be a constant in* $\mathbf{C_D}$;
- *if* $t_1 \theta t_2$ *is a constraint atom,* t_1 *and* t_2 *are both variables in a fact, roll-up, or descriptive atom in the body, or one of them is a constant in* $\mathbf{C_D}$ *and the other is a variable in a fact, roll-up, or descriptive atom;*

- *if the head of a rule is of the form* $Q(a_1, ..., a_n)$, Q *cannot appear in the body of the same rule. From the rules above, it follows that there exists a function, call it* dim, *that maps each data variable* x *to a unique dimension* dim(x), *and each level variable* X *to a unique dimension* dim(X). *Furthermore, there exists a function level, that maps each instant* t *to a unique level* level(x,t) *of the dimension* dim(x), *and to a unique level* level(X,t) *of the dimension* dim(X);

- *the granularity of a fact table F must be finer than the finest granularity of any of its component dimensions involved in a roll-up atom in the body of the rule. This rule prevents a fact from being counted more than once. For example, consider the query:*

$$Q(rs, SUM(m)) \longleftarrow F(i, s, m, t), s \xrightarrow{t} d{:}rs.$$

Assume that the fact table granularity is "year," and the granularity of dim(s) *is "day." A fact in an instance of* F *could be of the form* $<i_1, s_1, 100, 1998>$ *The dimension may have a pair of tuples* $<s_1, rs_1, 1\text{-}1\text{-}1998, 5\text{-}5\text{-}1998>$ *and* $<s_1, rs_2, 5\text{-}6\text{-}1998, 12\text{-}31\text{-}1998>$. *Thus, the fact would be counted twice.*

Definition 10 (Mutual Recursion): *Given a set R of TOLAP rules, a precedence graph G is built as follows: for each aggregate, predicate, or fact atom P, there is a node named P in G. If P and Q are nodes in G, add an edge from P to Q if there is a rule* Γ *in R such that P and Q occur in the body and the head of* Γ, *respectively. Following Abiteboul, Hull, & Vianu (1995), we say that two aggregate, fact, or predicate atoms R and S are mutually recursive if R and S participate in the same cycle in G.*

Definition 11 (TOLAP Programs): *A finite set of* TOLAP rules *which does not contain mutually recursive atoms is called a* TOLAP *program.*

Semantics

We will use point-based semantics (Toman, 1995) for the roll-up functions. This means, for instance, that in a dimension such as *Store* of our running example, a value for a roll-up, say storeId:s:'i1' $\xrightarrow{d}$ city:c:'c$_1$,' exists for each month in its validity interval.

Let us assume that for each dimension instance, we have a pair of relations, call them R_D and D_D, representing the sets *TRUP* and *TDESC* of Definition 2, respectively. The multidimensional database is defined over three different domains: D, N, T_p, where variables ranging over D belong to an uninterpreted sort, the ones ranging over N belong to an interpreted sort (), and the temporal variables range over T_p, as previously defined. For each dimension d, we will consider that T_p is limited to the lifespan of d.

A *valuation* θ, for a *TOLAP* rule Γ, is a tuple (θ_S, θ_I), where θ_S is called a *schema valuation,* and θ_I is an *instance valuation.* Valuation θ_S maps the level and attribute variables in Γ to level and attribute names in $\mathbf{B}_S(\mathbf{D}_S, \mathbf{F}_S)$, while θ_I maps domain variables to values in $I(\mathbf{B})$.

Definition 12 (Valuations): *A* schema valuation *for a rule Γ, denoted $\theta_S(\Gamma)$ maps level and attribute variables in the atoms of Γ as follows:*

- *given a roll-up atom of the form* $\mathtt{d:X:x} \overset{t}{\rightarrow} \mathtt{Y:y}$, *$\theta_S$ maps d to a dimension name in* $\mathbf{D}_s$, t *to a value* $w \in T_p$, *and X and Y to a pair of values (v, u) s.t.* $v \preceq^*_\omega u$ *holds in* $\mathtt{d}$;

- *if the roll-up atom is of the form* $\mathtt{d:X:x} \overset{t}{\rightarrow} \mathtt{Y:y}$, *$\theta_S$ maps t to a value $\omega \in T_p$, and X and Y to a pair of values v, u s.t. $v \le u$ holds in* $\mathtt{dim(X)} \in \mathbf{D}_s$;

- *if the roll-up atom is of the form* $\mathtt{x} \overset{t}{\rightarrow} \mathtt{Y:y}$, *$\theta_S$ maps t to a value $\omega \in T_p$, and Y to a value u s.t.* $\mathtt{level(x,w)} \preceq^*_\omega u$ *holds in* $\mathtt{dim(X)} \in \mathbf{D}_s$;

- *for a roll-up atom of the form* $\mathtt{x} \rightarrow \mathtt{Y:y}$, *$\theta_S$ maps Y to a dimension level u in* $\mathtt{dim(Y)}$, *s.t.$l_{inf} \le^* u$ holds in* $\mathtt{dim(X)}$;

- *given a descriptive atom of the form* $\mathtt{x.A} \overset{t}{=} \mathtt{y}$, *$\theta_S$ maps t to a value $\omega \in T_p$, and A to an attribute name $u \in \mathbf{A}$, s.t. $u \gg_t \mathtt{level(x,w)}$ in* $\mathtt{dim(X)} \in \mathbf{D}_s$.

Given a rule schema valuation $\theta_S(\Gamma)$ for a rule Γ, an instance valuation *is a function θ_I s.t.:*

- *θ_I maps the domain variables x and y in the roll-up atoms defined above to values in R_D over levels defined by θ_S;*

- *θ_I maps variable x in the descriptive atom $\mathtt{x.A} \overset{t}{=} \mathtt{y}$, to values in D_D, over levels defined by θ_S, and y to a value in D or N;*

- *θ_I maps a fact atom $F(x_1, ..., x_n, M, t)$ as follows: the rightmost term t in F is mapped to a value $\omega \in T_p$; each domain variable x_i in F to a value in dom $(\mathtt{level(x_i, w)})$; and the data term M in F to a value in N (M is the measure of the fact table);*

- *a constraint atom $x \{<, =\} y$ evaluates to* true *whenever $\theta_I(x) \{<, =\} \theta_I(y)$ is true. Recall that every variable in a constraint atom must appear in a roll-up or fact atom in the body. Thus, either θ_I maps x and y in the way described above, or they are mapped to a value in D, N or T_p;*

- *a negated roll-up atom is evaluated using the Close World Assumption. Thus, $!(x \overset{t}{\rightarrow} Y:y)$ is true if, given a valuation θ s.t. $\theta(x) = u$, $\theta(t) = \omega$, $\theta(Y) = l$, and $\theta(y) = v$, then there does not exist a roll-up in $\mathtt{dim(x)}$ s.t. $\rho^t_{level(x,w)}[\omega](u) = v$; recall that every variable must be in a positive atom in the body of the rule;*

- *a* negated descriptive atom $!\mathtt{x.A} \overset{t}{=} \mathtt{y}$ *is treated as explained above for negated roll-up atoms; thus, !x.A ⊥ y is true if, given a valuation θ s.t. θ(x)* *= u, θ(t) = ω, θ(A) = a, and θ(y) = v, a description does not exist in* dim(x) *such that* $\xi[t]_{level(x,w)}^{a}(u) = v;$

- *a negated constraint atom is evaluated in the standard way (i.e., θ(!x = b)* *= θ(x <> b));.*
- *a negated fact atom !F(x, y, . . ., "a," m, t) is true if for a valuation* $θ_1(x) =$ *u, $θ_1(y) = v$, ..., the tuple < u, v....... "a," ... > is not in F;*
- predicate atoms *are valuated as in standard data log (Abiteboul, Hull, &* *Vianu, 1995).*

Let *AGG* be the set of aggregate functions, with extension *AGG* = {*MIN, MAX, COUNT, SUM, AVG*}, and *r* a relation. The *aggregate operation* (Consens & Mendelzon, 1990) $\gamma f_{A(X)}(r)$ is the relation

$$\gamma f_{A(X)}(r) = \{t : t \text{ is an XA-tuple, } t[X] \in \pi_x(r), t[A] = f_A(\sigma_{X=t[X]}(r))\},$$

over *XA*, s.t. *XA* ∈ *schema* (*r*), *f* ∈ *AGG*, and $f_A(r)$ denotes the aggregation of the values in *t*[A], *t* ∈ *r*, using *t*.

We can now define the semantics of a *TOLAP* rule Γ of the form

$$Q(a_1, a_2, ..., a_n, AGG(m)) \longleftarrow A_1, ..., A_m$$

as follows:

For each level or data variable v_i in the body of Γ, and for a valuation θ of the variables in the rule's body, we have:

$$r_\Gamma = \{< θ(v_1), . . ., θ(v_n) > | θ \text{ is a valuation of } Γ\}.$$

Then $Q = \gamma_{AGGm(a_1, ..., a_n)}(r_\Gamma)$

Safety

Queries expressed as *TOLAP* rules studied in the previous subsections always lead to finite answers. This follows from the syntactic restrictions and from the semantics already defined. The existence of functions dim(x) and level(x, t) makes every variable in a rule *bounded* by the active domain of some level in a dimension. This is analogous to the concept of *range restricted variable* defined by Abiteboul, Hull, & Vianu (1995). Thus, for instance, if a negated roll-up atom appears in the body, safety is always granted. The same occurs with negated fact atoms. Also, limiting the temporal domain to the lifespan of the dimensions avoids infinite query answers. Thus, we conclude that *TOLAP* rules are *safe*.

What Can Be Expressed in TOLAP?

Intuitively, it is not hard to see that *TOLAP* has at least the power of first-order query languages with aggregation. However, in a sense it goes beyond this class. Note that in the temporal multidimensional data model, only the direct roll-ups are stored. Thus, to evaluate a roll-up atom like $d:X:x \overset{t}{\to} Y:y$, we need to compute the transitive closure of the roll-up functions in dimension d. It is a well-known fact that this cannot be done in first order, even after adding aggregate functions (Libkin & Long, 1997). However, as long as the dimension schema is fixed, this computation can be done in first order, because for a fixed schema, the number of joins needed to transitively close the roll-up functions is known in advance.

Observe that not only the structure of a dimension is subject to updates. There are common real-life situations in which an instance of a dimension may be modified in a non-trivial fashion. For instance, suppose we add the *Geography* dimension of Figure 1 to our running example data warehouse, and pose the following query: *"total sales per item and region, using only the currently existing regions."* In *TOLAP* we would write:

```
SB(p,r,SUM(m) ⟵ Sales(i,s,m,t), i ⟶ itemId:
               t
p, st  ⟶  region:r.
   Now
```

The second roll-up atom filters out regions not valid today.

Suppose now the following constraint: if a region where a sale occurred does not exist today, we want in the result a descendant of such region, if it exists. This query cannot be expressed in *TOLAP*. To show this, suppose the *Cuyo* region is split into *CuyoEast* and *CuyoWest*. Later, *CuyoEast is* merged with another region, say *Pampa*. In the meantime, maybe some region could have been deleted. With the tools defined so far, we could not find the "descendants" of *Cuyo*, because this is a transitive closure problem even though the schema remains fixed. *TOLAP* can be extended in order to be able to express the class of queries exemplified above. This problem, however, is beyond the scope of this chapter.

TOLAP Implementation

We now present a preliminary *TOLAP* implementation. We describe the data structure supporting the system, and how *TOLAP* atoms, rules, and programs are translated to SQL statements. We also discuss different implementation alternatives.

A first *TOLAP* implementation was developed at the University of Buenos Aires (Vaisman, 2001). Vaisman and Mendelzon (2001) give details of the implementation, and discuss preliminary results using a real-life case study, which we will briefly present in the next section.

Relational Representation

We will analyze two different data structures for representing temporal dimensions: a "fixed schema" versus a "non-fixed schema" approach. In both of them, a relation with schema *(dimensionId, loLevel, toLevel, From, To)* represents the structure of each dimension across time. Each tuple in this relation means that in the dimension identified by *dimensionId,* level *loLevel* rolled up to level *toLevel* between instants *From* and *To.* For example, the schema of the retail data warehouse will be represented by a relation with tuples like $<D1, itemId, itemType, d_1, Now>$, $<D1, itemId, brand, d_2, Now>$, and so on.

We briefly discuss how instances of the dimensions are stored under both kinds of representations.

Fixed Schema. In a fixed-schema relational representation, a dimension instance is a relation with tuples of the form *(loLevel, upLevel, toVal, upVal, From, To).* Each tuple represents a roll-up $\rho_{loLevel}^{upLevel} [(From, To)] (loVal) = upVal$. As an example, the instances of dimension *Product* in our retail data warehouse would be represented by a relation with tuples of the form $<itemId, itemType, i_1, ty_1, d_1, d_3>$, $<itemId, brand, i_1, b_1, d_1, Now>$, and so on. Thus, as new levels are added to a dimension, new tuples are inserted in the relation representing the dimension, but the relation's schema does not change. For instance, adding a new level *company* above *brand* just requires adding tuples like $<brand, company, b_1, c1, d_6, Now>$ (of course, in the relation storing the schema, a tuple $<D1, brand, company, d_6, Now>$ must be inserted).

This is probably the most natural representation, because it implements the relation R_D straightforwardly. However, translation to SQL would be awkward, and the resulting SQL queries will be far from optimal, because in order to compute aggregation over non-bottom dimension levels, the transitive closure of the roll-up functions must be computed, requiring self-joining temporal relations.

Non-fixed Schema. In this case there is also one relation for each dimension, but each dimension level is mapped to an attribute in this relation. A single tuple captures all the possible paths from an element in the bottom level, to the distinguished element "all." Two attributes, *From* and *To*, indicate, as usual in the temporal database field, the interval within which a tuple is valid. For example, a relation representing dimension *Product will* have schema: {*itemId, itemType, brand, From, To*}.

This structure requires more complicated algorithms for supporting dimension updates. For instance, adding a dimension level above a given one, or below the bottom level, induces a schema update of the relation representing the dimension instance: a new column will be added to this relation. If a level is deleted, no schema update occurs (i.e., the corresponding column is not dropped, in order to preserve the dimension's history. In the example above, a column *company* would be added. However, the translation process is simpler, and the subsequent query performance better, because computing the transitive closure of the roll-up functions reduces to a single relation scan.

Fixed vs. Non-fixed Schemas. To make the ideas above more clear, let us show how a *TOLAP* query is translated to SQL. Later, we will give full details of this process.

Consider the query *"total sales by company,"* posed to the earlier example of a data warehouse. This query reads in *TOLAP:*

$$COMP(c, SUM(qty)) \longleftarrow Sales(i, st, qty, t), i \overset{t}{\rightarrow} company:c.$$

Assuming that no schema update affected the fact table *Sales,* the SQL equivalent of the *TOLAP* query above in the fixed schema approach will look like [1]:

```
SELECT P1.upLevel, SUM(sales)
FROM Sales S, Product P, Product P1, Time T
WHERE
    S.itemId = P.loVal AND P.loLevel = 'itemId' AND
    P.upLevel = 'brand' AND P1.upLevel = 'company' AND
    P.upLevel = P1.loLevel AND
    S.day between P.From AND P.To AND
    S.day between P1.From AND P1.To
GROUP BY P1.upLevel
```

In the non-fixed schema representation, the SQL equivalent for the query is:

```
SELECT P.company, SUM(sales)
FROM Sales S, Product P
WHERE
    S.itemId = P.itemId AND
    S.day between P.From AND P.To
GROUP BY P.company
```

It is easy to see that the result is much more elegant and efficient in the second approach. The computation of the roll-up from *itemId* and *company* is straightforward, while in the first approach, the self-join of the table *Product* is required.

These arguments lead to us conclude that the non-fixed schema approach would be the better choice for a *TOLAP* implementation.

Translating TOLAP into SQL

The non-fixed schema approach discussed above was followed in the implementation of *TOLAP*. A parser performs a first pass over a rule, checking that the syntactic conditions of Definition 9 are met. In case an error is found, execution halts. In this first pass, the different atoms are identified and a symbol table built. The second pass translates each atom into an SQL statement and builds the equivalent SQL query.

We will show how a *TOLAP* rule of the form

$$Q(x,y,Ag(m)) \longleftarrow F(x_i, y_j, m, t), \; x_i \xrightarrow{t} 1_i:x, \; y_j \xrightarrow{Now} 1_j:$$
$$y, \; Dim:1:r \xrightarrow{t} p:z;$$

is translated to SQL. The translation of atoms, rules, and programs will be explained in that order.[2]

TOLAP Atoms. For each roll-up *atom* like $x_i \xrightarrow{t} 1_i:x$, bound to a variable in a fact table, a selection clause is built as follows:

$F.i = Dim_i.bottom$ AND $F.time$ BETWEEN $Dim_i.From$ AND $Dim_i.To$

The expression Dim_i is the table representing the dimension D_i. The first conjunct joins this table with the fact table, on the attribute representing the bottom level of the dimension. The actual name of the column $F.i$ is taken from the fact table's metadata. The second conjunct corresponds to the join between the fact table and the "Time" dimension.
- Each *time constant is* translated as a selection clause. The second roll-up atom in the rule above will be translated as:[3]

$F.j = Dim_j.bottom$ AND
$Dim_j.To = Now$

Dim_j is the dimension such that variable y_j is bound to.
- If a roll-up atom of the form $x \rightarrow Y:x$ corresponds to a user-defined time dimension, the atom is translated as:

$F.j = Dim_j.bottom$

- A roll-up atom $Dim:1:r \xrightarrow{t} p:z$, not bound to any fact table, is translated as an EXISTS clause.

EXISTS
    SELECT *
    FROM Dim

```
WHERE
    F.time between Dim.From AND Dim.To
```

The WHERE clause is omitted if the join with the time dimension is not required.

- The roll-up from the bottom levels of the dimensions to the levels 1_i and 1_j, corresponding to the variables in the head of the rule above (the aggregation levels), is computed as a projection and an aggregation in the way shown below. Thus, the SQL query generated by the *TOLAP* query of the beginning of this section, will look like:

```
SELECT  Dim_i.1_i,  Dim_j.1_j,Ag(measure)
FROM F_1, Dim_i, Dim_j
WHERE
    F-1.i = Dim_i.bottom AND
    F-1.j = Dim_j.bottom AND
    F-1.time between Dim_i. From AND Dim_i.To AND
    Dim_j.To = Now AND
    EXISTS
        (SELECT *
         FROM Dim
         WHERE
         F-1.Time between Dim.From AND Dim.To)
GROUP  BY  Dim_i.1_i,  Dim_j.1_j
```

The term measure is the measure in the fact table, bound to variable m. The fact table sub-index represents the version of the fact table. In this case there is only one version, as no schema update occurred. We will come back to this shortly.

- *A constraint atom* is translated as a selection condition in the WHERE clause. If the constraint atom is *negated,* this condition is treated in the usual way (a NOT predicate is added).
- *A negated roll-up atom* is translated as a NOT EXISTS clause. Suppose that in the query above we add a negated atom as follows[4]:

$!(y_j \xrightarrow{Now} 1_1\text{:'a'})$

where 'a' represents a constant. This atom is converted into an SQL expression of the form:

```
NOT EXISTS (
    SELECT *
    FROM Dim_j
    WHERE
    Dim_j.To=Now AND Dim_j.1_1 = 'a' AND F.j = Dim_j.bottom)
    where 1_1 is the attribute representing level 1_1.
```

- *A predicate atom is* translated as a table in the FROM clause, with the conditions which arise from the variables or constants in it.

TOLAP Rules. So far we tackled the problem of translating each atom in a *TOLAP* rule separately. The next step will be the study of the translation of the whole rule.

Above, we gave an example of a simple rule, assuming no schema update occurred in the fact table. However, we claimed that one of the main features of *TOLAP* is the ability to deal with different schema versions triggered by the insertion or deletion of a bottom level. Given a fact table F, versions F_1, F_2, and so on may be created, each one with a different schema.

Given a *TOLAP* query Γ involving a fact table F, with versions $F_1,..., F_n$, the SQL query Q equivalent to Γ will be such that

$$Q = Q_1 \cup Q_2 \cup .... \cup Q_n,$$

where Q_i are queries involving facts that occurred in the intervals I_i in which each F_i holds. If the query involves an aggregation function, call it f_{AGG}, one more aggregation must be performed, in order to consider duplicates in each subquery. Thus

$$Q_{AGG} = f'_{AGG} (Q)$$

TOLAP programs are treated analogously, as a series of *TOLAP* rules.

Join elimination. There are cases in which the join between dimensions and fact tables is not needed. This situation arises when a variable in the head of the rule belongs to a level which is the bottom level of a fact table in the body (or was the bottom level at least during some interval I_i). The current *TOLAP* implementation takes advantage of this fact, and does not generate the join.

Example 8: *Let us consider the situation stated earlier in which the fact table Sales of our running example is split into Sales$_1$, and Sales$_2$, each one holding before and after an instant d_s, as a result of the insertion of level storeId below level city. In* TOLAP, *the query "total sales by itemId and city" reads:*

```
IC(it,ci,SUM(m))⟵ Sales(i,s,m,t), i →ᵗ itemId:it,
                    s  →ᵗcity:ci.
```

Two SQL subqueries will be generated, one for each fact table.

```
SELECT itemId,city,SUM(sales)
FROM (
   SELECT itemId,city, SUM(sales)
   FROM Sales₁
   GROUP BY itemId,city


   UNION ALL


   SELECT itemId,city, SUM(sales)
   FROM Sales₂,Store
   WHERE
   Sales₂.Time between Store.From AND Store.To AND
   Sales₂.storeId=Store.storeId
GROUP BY itemId,city )
GROUP BY itemId,city
```

Notice that, as city *and* itemId *were the bottom levels of Sales*₁*, no join is needed in the first subquery.*

Subquery Pruning. If a *TOLAP* rule contains a constraint atom with a condition over time such that the lifespan of a version of the fact table does not intersect with the interval determined by the constraint, the subquery corresponding to that version of the fact table is not generated by the translator, as it will not yield any tuple in the output. For instance, in Example 8, adding the constraint $t < d_6$ will prevent the first subquery from being generated. We call this step *subquery pruning.*

A Case Study

In order to illustrate the need for temporal management in OLAP discussed in this chapter, let us present a real-life case study, a medical center in Argentina (Vaisman & Mendelzon, 2001), where each patient receives different services, including radiographies, electrocardiograms, medicine, disposable material, and so on. These services are denoted "Procedures." Data was taken from different tables in the clinic's operational database. Figure 8 shows the final state of the dimensions. We built a temporal data warehouse using six months of data taken from medical procedures performed on patients at the center.

Dimension *Procedure,* with bottom level *procedureId,* and levels *procedureType, subgroup,* and *group,* describes the different procedures available to patients. Dimension *Patient,* with bottom level *patientId,* and levels *yearOfBirth* and *gender,* represents information about the person under treatment. Age intervals are represented by a dimension level called *yearRange.* Patients are also grouped according to their health insurance institution. Moreover, these institutions are further grouped into types (level *institutionType),* such as private institutions, labor unions, etc. Dimension *Doctor* gives information about the

Figure 8: Dimensions in the Case Study

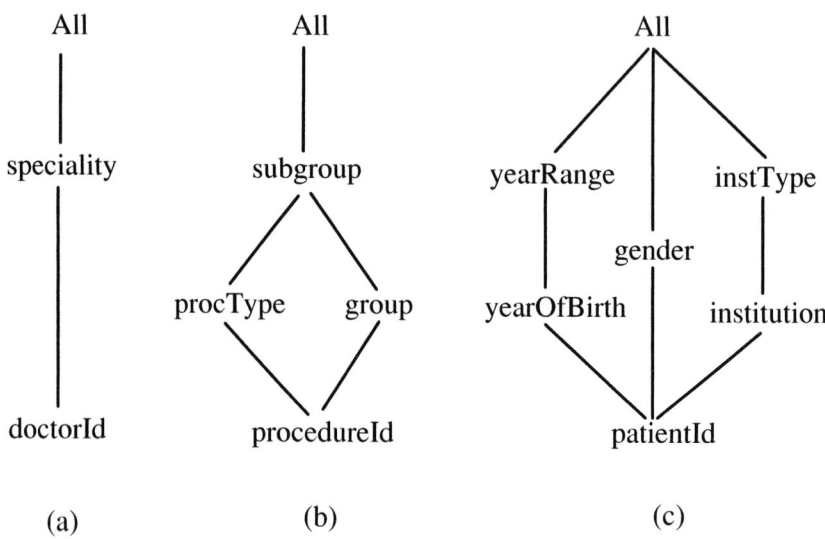

(a)   (b)   (c)

available doctors (identified by *doctorId*) and their *specialties* (a level above *doctorId*). There is also a user-defined dimension *Time*, with levels *day*, *week*, and *month*.

A fact table holds data about procedures delivered to patients by a certain doctor on a certain date. We call this fact table *Services*, and its schema contains the bottom levels of the dimensions above, plus the measure. The fact table schema is: *Services (doctorID, procedureId, patientID, day, quantity, t)*, where *t* represents a system-maintained time dimension, and *quantity* is the measure; attribute *day* represents a user-defined time dimension. For instance, a tuple <docl, 10.01.01, pat120, 10, 2, "10/ 10/2000"> means that doctor "doc1" prescribed two special radiographies to patient "pat120" on day "10" which corresponds to October 10, 2001. In this case study, the system-maintained and user-defined times are synchronized, meaning that there is a correspondence between the values of attributes "day" and "t" (e.g., day "11" corresponds to October 11, 2000). This synchronization, however, is not mandatory in the model. Thus, both times can take independent values.

The scenario was prepared as follows. Dimension *Procedure* was created with bottom level *procedureId*. Subsequent operations generalized this level to level *subgroup*. Level *procedureId* was then generalized to *procType*. Finally, *subgroup* was generalized to *group*, and *procType* related to *group*. In the *Patient* dimension, the initial bottom level *patientId* represents information about the person under treatment. This level was later generalized to *year0fBirth*, *gender*, and *institution*, in that order. Levels *yearOfBirth* and *institution* were further generalized into *yearRange* and *institutionType*, respectively. For dimension *Doctor*, in order to show how schema versioning is handled in *TOLAP,* we assumed that although facts were recorded at the *doctorId* level, this level was temporarily deleted, originating the second version of the fact table, called *Services_2.* Thus, during a short interval, the dimension's bottom level was *specialty,* later specialized into level *doctorId,* which triggered the creation of the third version of the fact table, denoted *Services_3.*

Finally, a user-defined *Time* dimension was created, with granularity *day,* allowing expressing aggregates over time. The dimension's hierarchy is the following: *day* rolls-up to *week* and *month*; *week* and *month* roll-up to *All.* The three resulting fact table versions have the following schemas: (a) *Services_1:{procId, patientId, doctorId, day, quantity, tmp},* where *tmp* represents the built-in time dimension, and *quantity* is the number of procedures delivered; (b) *Services_2:{procId, patientId, specialty, day, quantity, tmp};* (c) *Services_3: {procId, patientId, doctorId, day, quantity, tmp}.* The fact tables were populated off-line.

The following queries reflect *TOLAP*'s expressive power. Suppose a user wants to analyze the doctors' workloads, in order to estimate future needs. The following query measures how the arrival of a new doctor influences the number of patients served by a doctor named *Martinez*: "list the total number of services delivered weekly by Dr. Martinez while Dr. Feinsilver was not working for the clinic."

```
Q1MartFei(w,SUM(qty))  ⟵── Services(doc,proc,pat,
                           day,qty,t), day→week:w,
                           doc[t]→doctorId:d,
                           d.name='Martinez,'
                           !Doctor:doctorId:dd[t]→All:all,
                           dd.name='Feinsilver.'
```

Notice that the negated atom is not bound to the fact table. Also notice the use of the user-defined *Time* dimension.

The following query returns the number of services delivered by Dr. Martinez while both doctors were employed at the clinic.

```
Q2MartFei(w,SUM(qty))  ⟵── Services(doc,proc,pat,day,
                           qty,t),day→week:w, doc[t]→
                           doctorId:d,d.name='Martinez,
                           Doctor:doctorId:dd[t]→All:
                           all, dd.name='Feinsilver.'
```

The next query illustrates how to check patients who were served when they were affiliated to 'MEDICUS' and are currently affiliated to 'OSDE' (two private health services in Argentina).

```
QchangePlan(pat)     ←——  Services(doc,proc,pat,day,m,t),
                          pat[t]→institution:'MEDICUS,'
                          pat[Now]→institution:'OSDE'.
```

The reader must notice that the queries above would require several lines of SQL code. Moreover, without an underlying temporal multidimensional data model, these queries cannot be expressed, unless an ad-hoc design of the data warehouse is carried out.

Visualization

The *TOLAP* implementation includes a graphic environment developed taking advantage of the temporal multidimensional model. The graphic interface allows to:
- browse dimensions across time, and watch how they were hierarchically organized throughout their lifespan; further, dimension instances could also be browsed;
- perform dimension updates;
- import roll-up functions from text files;
- browse different versions of a fact table;
- send *TOLAP* programs to the system, and display their results without leaving the environment, including seeing the generated SQL query.

Figure 9: Browsing Dimension "Patient"

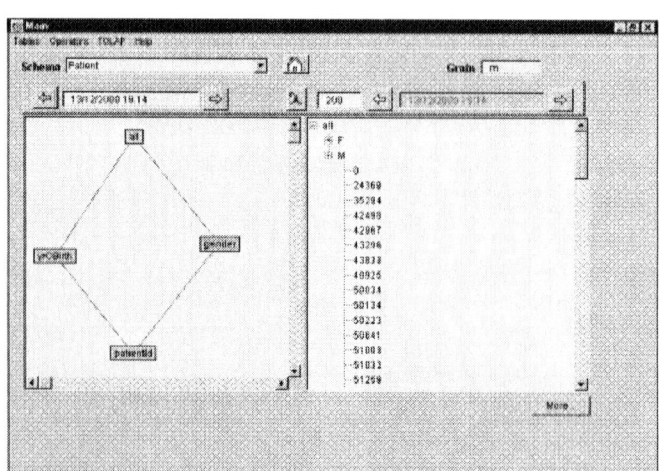

Figure 9 shows a typical system screen depicting a dimension *Patient,* taken from the case study presented above, as of December 13, 2000. The window on the left presents the dimension's structure. The arrows upon the window allow browsing forward and backward across time. This is not allowed if the system is in "initial" mode. The window on the right shows the dimension's instances. This window is synchronic with respect to the one on the left. Thus, the instances being displayed correspond to the structure on the left although they can vary as instance updates occur. However, while the screen of the left remains unchanged, several instance updates may occur, which can be displayed in the window on the right. The little box in the upper middle indicates the number of elements in the bottom level which are going to be displayed, allowing partial loading of the instance graph in main memory. This feature is crucial in making the tool usable in real applications, because loading the entire instance graph would be very expensive, even for not very large dimensions.

CONCLUSION

In this chapter we have argued that dimensions in a data warehouse are not static entities, but are subject to updates, either at the schema or instance level. We presented a *temporal multidimensional model,* and a query language supporting it which we denoted *TOLAP. TOLAP* accounts for schema versioning, allowing expressing complex queries at a high abstraction level. We also show a first implementation of the temporal model.

TOLAP can be extended in order to allow the definition of constraints, which could be easily introduced within our visualization tool. Also, there is space for studying query optimization in *TOLAP.* Another issue which deserves attention is adding update support to *TOLAP,* allowing bulk updates like *"delete all customers who had not completed any transaction since 1998."* Also, transactions and update expressions in *TOLAP* must be addressed. For example, the expression above could be followed by: *"classify all customers who did not perform any transaction since 1999 as 'low-priority' customers."*

A second version of *TOLAP* is currently under construction. It will provide a totally platform-independent system, with a three-tier architecture. A front-end web-enabled application will communicate with an application server, in which XML metadata will be stored. The back end will be any database management system where the data warehouse will be stored.

REFERENCES

Abiteboul, S., Hull, R., & Vianu, V. (1995). *Foundations of Databases.* Reading, MA: Addison Wesley.

Bliujute, R., Saltenis, S., Slivinskas, G., & Jensen, G. (1998). *Systematic Change Management in Dimensional Data Warehousing.* Technical Report TR-23, Time Center, Computer Science Department, University of Arizona.

Chen, W., Kifer, M., & Warren, D.S. (1989). HiLog as a platform for database language. *Proceedings of the 2nd International Workshop on Database Programming Languages,* Oregon Coast, Oregon, 315-329.

Consens, M., & Mendelzon, A.O. (1990). Low complexity aggregation in Graphlog and Datalog. *Proceedings of the 3rd International Conference on Database Theory.* Lecture Notes in Computer Science, No. 470, 379-394.

Hurtado, C., Mendelzon, A.O., & Vaisman, A. (1999). Maintaining data cubes under dimension updates. *Proceedings of 1999 International Conference on Data Engineering (IEEE/ICDE'99),* Sydney, Australia, 346-355.

Hurtado, C., Mendelzon, A.O., & Vaisman, A. (1999). Updating OLAP dimensions. *Proceedings of the 1999 Workshop on Data Warehousing and OLAP (DOLAP'99),* Kansas City, 60-66.

Kimball, R. (1996). *The Data Warehouse Toolkit.* New York: John Wiley & Sons, Inc.

Klug, A. (1982). Equivalence of relational algebra and relational calculus query languages having aggregate functions. *Journal of the ACM,* 29(3), 699-717.

Lakshmanan, L.V.S, Sadri, F., & Subramanian, I.N. (1993). On the logical foundations of schema integration and evolution in heterogeneous database systems. *Proceedings of the 3rd International Conference on Deductive and Object-Oriented Databases (DOODS'93).* Lecture Notes in Computer Science, No. 760, 81-100.

Lakshmanan, L.V.S, Sadri, F., & Subramanian, I.N. (1997). Logic and algebraic languages for interoperability in multidatabase systems. *Journal of Logic Programming,* 33(2), 101-149.

Libkin, L., & Wong, L. (1997). On the power of aggregation in relational query languages. *Proceedings of the 6th International Workshop on Database Programming Languages (DBPL'97),* East Park, Colorado, 270-280.

Mendelzon, A.O., & Vaisman, A. (2000). Temporal queries in OLAP. *Proceedings of the 26th International Conference on Very Large Databases (VLDB'00),* Cairo, Egypt, 242-253.

Pedersen, T.B, & Jensen, C. (1999). Multidimensional data modeling for complex data. *Proceedings of 1999 International Conference on Data Engineering (IEEE/ICDE'99),* Sydney, Australia, 336-345.

Snodgrass, R. (1995). *The TSQL2 Temporal Query Language.* Boston, MA: Kluwer Academic Publishers.

Toman, D. (1997). A point-based temporal extension to SQL. *Proceedings of the 7th International Conference on Deductive and Object-Oriented Databases (DOODS'97),* Montreaux, Switzerland, 103-121.

Vaisman, A. (2001). *Updates, View Maintenance and Time Management in*

Multidimensional Databases. PhD Thesis. Available on-line at: http://www.cs.toronto.edu/ avaisman/publications.

Vaisman, A., & Mendelzon, A.O. (2001). A temporal query language for OLAP: Implementation and a case study. *Proceedings of the 8th International Workshop on Database Programming Languages (DBPL'01),* Rome, Italy, 49-64.

Widom, J. (1995). Research problems in data warehousing. *Proceedings of the 4th International Conference on Information and Knowledge Management,* Baltimore, Maryland, 25-30.

Yang, J., & Widom, J. (1998). Maintaining temporal views over non-temporal information sources for data warehousing. *Proceedings of the Sixth International Conference on Extending Database Technology,* Valencia, Spain, 389-403.

Yang, J., & Widom, J. (2000). temporal view self-maintenance in a warehousing environment. *Proceedings of the 7th International Conference on Extending Database Technology,* Konstanz, Germany, 345-412.

ENDNOTES

1   For the sake of clarity we do not show here how time granularity is handled in the translation; usually, this will depend on the underlying DBMS.

2   **Notation:** The dimensions' sub-indices represent their position in the fact table to which they are bound. In the implementation, constructions of the form x 4 Y:x, are replaced by x[t] -+ Y:x , in order to simplify parsing.

3   In an actual implementation, Now can be replaced by Sysdate() or any function returning the current time.

4   Actually, the parenthesis is not required.

Chapter VII

Dynamic Multidimensional Data Cubes

Mirek Riedewald
University of California, Santa Barbara, USA

Divyakant Agrawal
University of California, Santa Barbara, USA

Amr El Abbadi
University of California, Santa Barbara, USA

ABSTRACT

Data cubes are ubiquitous tools in data warehousing, online analytical processing, and decision support applications. Based on a selection of pre-computed and materialized aggregate values, they can dramatically speed up aggregation and summarization over large data collections. Traditionally, the emphasis has been on lowering query costs with little regard to maintenance, i.e., update cost issues. We argue that current trends require data cubes to be not only query-efficient, but also dynamic at the same time, and we also show how this can be achieved. Several array-based techniques with different tradeoffs between query and update cost are discussed in detail. We also survey selected approaches for sparse data and the popular data cube operator, CUBE. Moreover, this work includes an overview of future trends and their impact on data cubes.

INTRODUCTION

For modern data warehouses it is common to manage Terabytes (Tb) of data. According to a survey by the Winter Corporation (2001), for instance, the decision support database of SBC, a telecom company, reaches a size of 10.5 Tb. Furthermore, 17% of the surveyed companies expect a tenfold increase of the size of their databases over the next three years. Already today retail companies like Kmart and Wal-Mart prepare their data warehouses to accommodate dozens of Terabytes of business data.

Human analysts cannot "digest" such large amounts of information at a detailed level. Instead they rely on the system to provide a summarized and task-specific view of some part of the data. Consequently, efficient summarization and aggregation of large amounts of data play a crucial role in On-Line Analytical Processing (OLAP) and Decision Support Systems (DSS). Such aggregated information is often maintained in data cubes which also provide an easy-to-understand conceptual model. Queries can be described as regions in the data cube. Pre-computed aggregate values that speed up query execution for group-bys, cross-tabs, and sub-totals can be easily included in the model (e.g., CUBE operator as proposed in Gray et al., 1997).

The primary goal of data warehouses is to support data analysis, i.e., queries. Update costs were often considered to be unimportant. Hence systems are typically oriented towards batch updates which are applied to the warehouse during times of low system load, e.g., overnight. The corresponding data cubes would be batch-updated as well, or even be re-computed from scratch. However, for very large data collections, incrementally maintainable *dynamic* data cubes are clearly preferable. There are several arguments why there should also be data cubes that support efficient single updates, not only in large batches.

Batch updates might effectively reduce the *average* cost per operation, but processing the whole batch makes the data cube inaccessible to analysts for a long period of time. Today businesses demand a 24/7 availability of their important data, hence finding a suitable time slot for a large batch *update window* becomes increasingly difficult. In most applications, especially decision support and stock trading, having the latest information incorporated as early as possible can give a decisive advantage over competitors that have to wait for an overnight batch update processing. Also, OLAP is inherently an *interactive* information exploration process that includes what-if scenarios. What-if analysis requires real-time and hypothetical data to be integrated with historical data for the purpose of instantaneous analysis and subsequent action.

One possible solution to address the above concerns is to use additional data structures that temporarily maintain updates until they are applied to the cube in a batch or lazy manner. In contrast, this chapter examines truly dynamic solutions that do not require batch processing and additional update structures. The presented techniques offer different tradeoffs between query, update, and storage costs, allowing an administrator to find the matching approach for each application. We mainly focus on array-like structures, but will also mention alternative approaches for sparse and high-dimensional data. It will be shown that efficient queries and updates can both be supported for multidimensional data cubes.

BACKGROUND

The term "data cube" is not used consistently in the data warehousing research literature. We will use the following rather general definition. A data set is conceptually modeled as a multidimensional hyper-rectangle, or *data cube* for short. The *d* functional attributes that describe a data item in the set are the *dimensions*. Some of them are hierarchical, e.g., year-quarter-month-day for the time dimension (sometimes multiple hierarchies for an attribute exist, e.g., week-day is another hierarchy for the time dimension). A *d*-tuple of dimension values defines a *cell* of the data cube, similar to a multidimensional array. Cells contain the values of additional attributes, the *measure* attributes. For simplicity and without loss of generality, we will further on assume that the cube has only a single measure attribute.

Figure 1: Data Cube Example

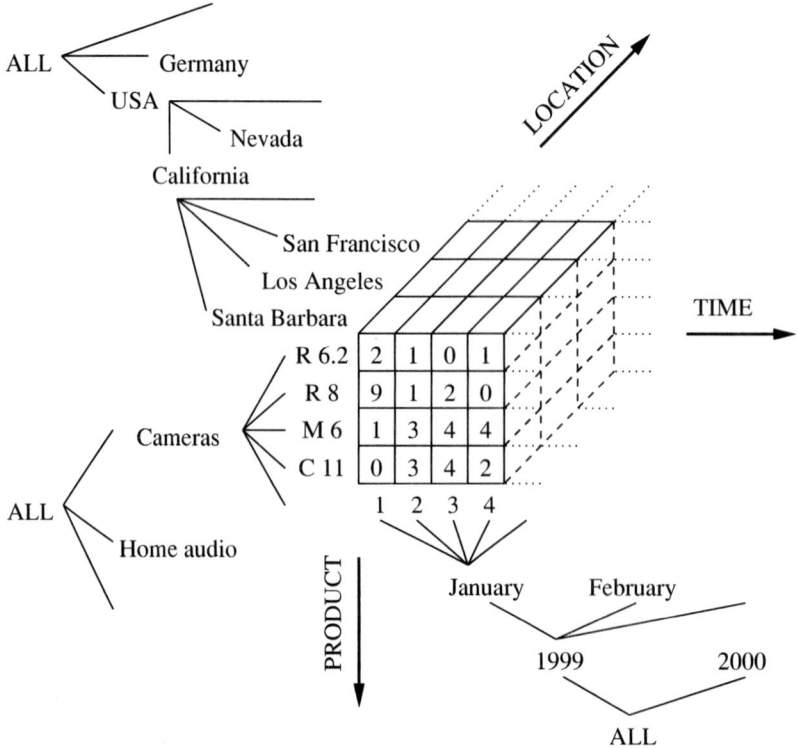

Note that we do not advocate any particular implementation for data cubes. Possible solutions were examined elsewhere, e.g., in Goil & Choudhary (1997), Roussopoulos, Kotidis, & Roussopoulos (1997), and Sarawagi & Stonebraker (1994). The model is also not restricted to a certain OLAP architecture like Multidimensional OLAP (MOLAP) or Relational OLAP (ROLAP) (for a discussion of ROLAP and MOLAP, see Pendse & Creeth, 2001). However, our discussion will mainly focus on MOLAP techniques.

A simple example is a sales data cube with dimensions product, time (date of sale), and location (place of sale), and the measure attribute amount (value of sales transaction). Hence for each product, time, and location, the corresponding amount of sales is stored. Figure 1 shows a possible instance with dimension hierarchies indicated (ALL represents the whole domain of an attribute). If there is no sales transaction for a certain combination of dimension values, the corresponding cell is *empty*. In the figure empty cells contain the value 0.

As summarized in Chaudhuri & Dayal (1997), dominant aggregate query types on data cubes are roll-up/drill-down (increases, respectively decreases, level of aggregation along one or more attribute hierarchies) and slice-and-dice (selects attributes and cells of interest).

These queries are typically *range queries* or sequences of related range queries (e.g., hierarchical range queries as discussed in Koudas, Muthukrishnan, & Srivastava, 2000). Hence, we are mainly concerned with support for aggregate range queries which select a hyper-rectangular region of the data cube and return an aggregate, e.g., SUM or COUNT, of the measure values of the selected cells. A formal definition is given below.

Let A denote a d-dimensional data cube such that dimension i ($1 \leq i \leq d$) has N_i cells. Without loss of generality, let the domain of dimension i be $\{0,1,\ldots,N_i-1\}$. A cell $c = [c_1,\ldots,c_d]$, where each c_i is an element of the domain of the corresponding dimension, contains the measure value $A[c]$. With $e{:}f$ we denote a hyper-rectangular *region* of the data cube, more precisely the set of all cells c that satisfy $e_i \leq c_i \leq f_i$ for all $1 \leq i \leq d$. Cell e is the *anchor* and cell f the *endpoint* of the region. The anchor and endpoint of the entire data cube hence are $[0,\ldots,0]$ and $[N_1 - 1,\ldots, N_d - 1]$, respectively. The term op($A[e] : A[f]$) denotes the result of applying the aggregate operator op to the measure values in region $e : f$, i.e., it defines an *aggregate range query*. For example, SUM($A[e] : A[f]$) is a *range sum*. The range sum SUM($A[0,\ldots, 0] : A[f]$) will be referred to as a *prefix sum*.

The term *original cube* denotes a data cube which is obtained as the straightforward projection of a data set to the d-dimensional space defined by the cube's dimensions. If any of the cube's cells contain values which are aggregates of multiple cells of an original cube (e.g., sub-totals), it will be referred to as a *pre-aggregated cube*.

DATA CUBES FOR INVERTIBLE
AGGREGATE OPERATORS

Our research is mainly concerned with invertible aggregate operators, i.e., where the inverse of a value is well defined and unique. More precisely, an operator is invertible if the domain of the measure attribute forms an abelian group under the operator. Popular examples are important SQL operators like SUM, COUNT, and AVG (when expressing it with SUM and COUNT). The existence of inverse operations enables the construction of elegant techniques for speeding up queries on MOLAP data cubes that do not require additional storage compared to materializing the original cube.

We will also explain the techniques for the operator SUM. Other invertible operators are handled in a similar way. Let A denote the original data cube. The unit of cost used is the number of cells accessed by an operation. Hence, updating a single cell in A results in a cost of 1. A range query Q that selects range r_i of size $|r_i|$ in dimension i has a cost of $\Pi_{i=1}^{d}|r_i|$, i.e., a worst-case cost of $\Pi_{i=1}^{d} N_i$. Using A for OLAP represents a perfectly dynamic solution, but incurs high query costs.

Achieving Constant Query Costs

In a query-dominated environment, simply storing A is not the best choice. In the worst case all the cells of A have to be accessed in order to answer a range query. On the other hand the query only outputs a single aggregate value, the sum over all selected cells.

Ho et al. (1997) addressed this issue with their Prefix Sum (PS) technique. The main idea is to use a prefix sum cube PS instead of A. Each cell $PS[c_1, \ldots, c_d]$ contains the corresponding prefix sum $\sum_{0 \leq x1 \leq c1} \sum_{0 \leq x2 \leq c2} \ldots \sum_{0 \leq xd \leq cd} A[x_1, x_2, \ldots, x_d]$. Based on the principle of inclusion-exclusion, each range sum query on A can be computed from up to 2^d corresponding prefix sums. Figure 2 shows an original data cube A and the corresponding PS cube. The cells accessed by query SUM($A[4, 2] : A[6, 5]$) are shaded. Consider the lower right corner PS[6, 5] in PS. The value represents the sum over all cells in A that are dominated by, i.e., to the upper left of, cell $A[6, 5]$. Hence, to answer the range query, the upper right (PS[3, 5]) and lower left (PS[6,1]) cells need to be subtracted. This causes the cells in $A[0, 0] : A[3,1]$ to be removed twice, therefore the upper left cell PS[3, 1] must be added. This approach can be easily generalized to higher dimensionality (see Ho et al., 1997). Consequently, the worst-case query cost using PS is reduced to 2^d which is independent of the size of the data cube and the size of the query region. At the same time the update cost increases dramatically. An update to a cell $A[c]$ affects all cells $PS[x]$ that dominate c, i.e., all cells x where $x_i \geq c_i$ for all dimensions i. In the worst case the whole PS cube could be affected.

To avoid PS's expensive cascading updates, Geffner et al. (in Geffner, Agrawal, El Abbadi, & Smith, 1999, and in Geffner, Riedewald, Agrawal, & El Abbadi, 1999), introduced the Relative Prefix Sum (RPS). In this chapter we present a later version of RPS which removed RPS's initial unnecessary storage overhead and appeared in Riedewald, Agrawal, El Abbadi, and Rajarola (2000). RPS retains the constant query time but considerably reduces the update cost by removing some of the dependencies of the pre-aggregated values. The technique conceptually partitions A into smaller hyper-rectangular chunks, called *boxes*. Essentially *inner* cells of a box store prefix sums local to the box, while the *border* cells in selected surfaces also aggregate outside values. Figure 3 shows examples for aggregation regions of box cells. Note that any *prefix* range query can be answered by accessing up to 2^d cells (2^d - 1 border cells, 1 inner cell) in the box the endpoint of the query-region falls into.

Figure 4 shows an example for an RPS cube and how a prefix sum can be partitioned such that the corresponding box values provide the result. Compared to the PS technique, more values are accessed to obtain a prefix sum, but it is still much less costly than adding the values of the original data cube on-the-fly. Arbitrary range sum queries are answered by combining up to 2^d prefix sums as for the PS technique. The cascading updates are restricted to the box that contains the updated cell. Outside the box only some of the border cells of other boxes are affected, as shown for an example in Figure 4. It can be shown analytically that the optimal tradeoff between query and update costs occurs when the side-length of a box is chosen to be $\sqrt{N_i}$ in dimension i. Then the worst-case costs are 4^d for queries and $\prod_{i=1}^{d} 2\sqrt{N_i}$ for updates. Note that PS is a special case of RPS when boxes of size 1 are selected.

Figure 2: Original Data Cube A and Corresponding Prefix Cube PS with Range Sum Query

A	0	1	2	3	4	5	6	7	8
0	3	5	1	2	2	4	6	3	3
1	7	3	2	6	8	7	1	2	4
2	2	4	2	3	3	3	4	5	7
3	3	2	1	5	3	5	2	8	2
4	4	2	1	3	3	4	7	1	3
5	2	3	3	6	1	8	5	1	1
6	4	5	2	7	1	9	3	3	4
7	2	4	2	2	3	1	9	1	3
8	5	4	3	1	3	2	1	9	6

PS	0	1	2	3	4	5	6	7	8
0	3	8	9	11	13	17	23	26	29
1	10	18	21	29	39	50	57	62	69
2	12	24	29	40	53	67	78	88	102
3	15	29	35	51	67	86	99	117	133
4	19	35	42	61	80	103	123	142	161
5	21	40	50	75	95	126	151	171	191
6	25	49	61	93	114	154	182	205	229
7	27	55	69	103	127	168	205	229	256
8	32	64	81	116	143	186	224	257	290

Query: SUM(A[4,2]:A[6,5])

Result: 1+3+3+4+3+6+1+8+2+7+1+9 = 48

SUM(A[4,2]:A[6,5]) = PS[6,5]-PS[3,5]-PS[6,1]+PS[3,1]

Result: 154-86-49+29 = 48

Figure 3: Aggregation Regions of Cells for RPS

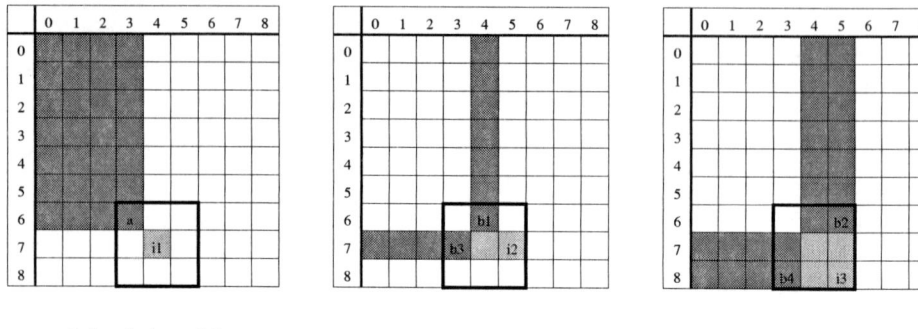

Anchor cell a, inner cell i1   Border cells b1 and b3, inner cell i2   Border cells b2 and b4, inner cell i3

Making the Cubes More Dynamic

For environments with high update volumes, Geffner, Agrawal and El Abbadi (2000) propose a technique that balances query and update costs such that they are both polylogarithmic in the data size. However, the original proposal resulted in considerable storage overhead. This overhead was removed by Riedewald, Agrawal, El Abbadi, and Pajarola (2000). We present an overview of the latter approach which will be referred to as the Dynamic Data Cube (DDC).

Like RPS, DDC conceptually partitions the original cube into boxes and uses the same aggregation regions for a box's border cells. However, the number of boxes is bounded by a constant (e.g., divide each dimension in half). Hence the side-length of a box is linear in the domain size. For inner box cells a recursive approach is taken. To be more specific, the same partitioning into boxes is recursively applied to the region of inner box cells. As defined so far, the cumulative nature of the border cells in a box would still lead to high update costs. Interestingly, a recursive encoding similar to the inner cells solves the problem.

Figure 4: Original Data Cube A and Corresponding RPS Cube with Range Sum Query (Middle) and Update (Right)

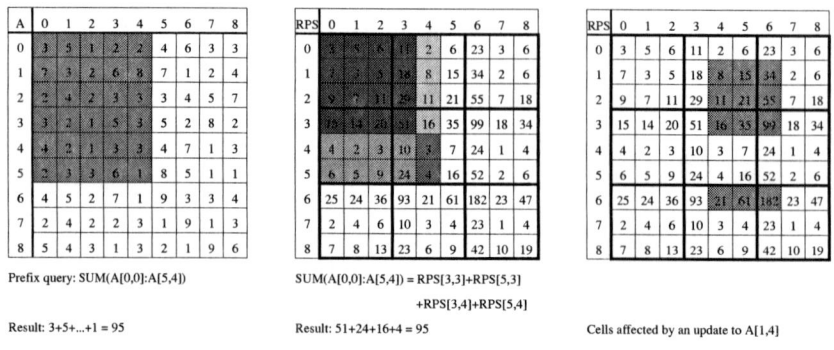

Prefix query: SUM(A[0,0]:A[5,4])   SUM(A[0,0]:A[5,4]) = RPS[3,3]+RPS[5,3]
                                                    +RPS[3,4]+RPS[5,4]

Result: 3+5+...+1 = 95   Result: 51+24+16+4 = 95   Cells affected by an update to A[1,4]

Figure 5: Original Data Cube A and Corresponding DDC Cube with Range Sum Query

A	0	1	2	3	4	5	6	7	8	9
0	3	5	1	2	2	4	6	3	3	1
1	7	3	2	6	8	7	1	2	4	2
2	2	4	2	3	3	3	4	5	7	4
3	3	2	1	5	3	5	2	8	2	1
4	4	2	1	3	3	4	7	1	3	2
5	2	3	3	6	1	8	5	1	1	2
6	4	5	2	7	1	9	3	3	4	1
7	2	4	2	2	3	1	9	1	3	3
8	5	4	3	1	3	2	1	9	6	5
9	6	1	2	4	2	1	3	1	5	2

Prefix query: SUM(A[0,0]:A[7,8])

Result: 3+5+...+3 = 256

DDC	0	1	2	3	4	5	6	7	8	9
0	3	5	1	8	2	17	6	3	12	1
1	7	3	2	11	8	33	1	2	7	2
2	2	4	2	3	3	17	4	5	7	4
3	12	9	5	28	14	69	7	15	35	7
4	7	2	1	6	3	17	7	1	11	2
5	18	19	10	54	20	126	25	20	65	12
6	4	5	2	14	1	28	3	3	10	1
7	2	4	2	8	3	14	9	1	13	3
8	11	13	7	30	7	60	13	9	39	9
9	6	1	2	7	2	16	3	1	9	2

SUM(A[0,0]:A[7,8]) = DDC[5,5]+DDC[6,5]

+DDC[7,5]+DDC[5,8]+DDC[6,8]+DDC[7,8]

Result: 126+28+14+65+10+13 = 256

Figure 6: Tree Structure of the DDC Cube (Query Endpoint Hatched, Accessed Cells Shaded)

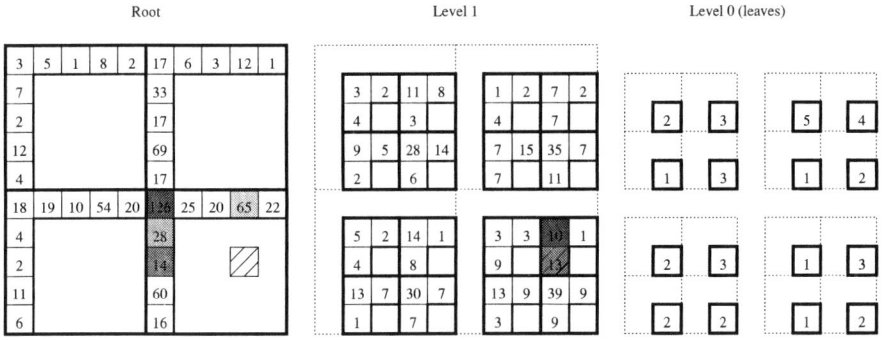

The recursive DDC approach defines a hierarchy. Queries and updates are processed by conceptually descending this tree. If the recursive box partitioning always divides boxes in half, then the technique guarantees worst-case costs of $\prod_{i=1}^{d} 2\log_2 N_i$ and $\prod_{i=1}^{d} \log_2 N_i$ for queries and updates, respectively.

The main disadvantage is that, due to its recursive nature, the technique is not easy to analyze (e.g., for average costs) and to implement. Figure 5 shows a DDC cube; in Figure 6 the tree structure is illustrated, along with how a query is processed.

Offering Multiple Query-Update Tradeoffs

Chan and Ioannidis (1999) are the first to explicitly explore the possibility of letting the user choose a query-update tradeoff by parameterizing the construction of the pre-aggregated cube. They apply ideas from bitmap index design and define an interesting class of so-called Hierarchical Cubes (HCs). Like for DDC the data cube is hierarchically partitioned into smaller boxes of equal size. According to the partitioning, the cells are assigned to classes. Two techniques, Hierarchical Rectangle Cubes (HRCs) and Hierarchical Band Cubes (HBCs) are proposed that define aggregation regions for the cells depending on their classes. Having different hierarchical partitionings and different ways of assigning aggregation regions, a variety of tradeoffs between query and update cost can be generated. Figure 7 shows an example for HBC when each dimension is recursively split in half. The shading in the middle cube indicates the level of each cell (darkest at root level). The right cube indicates how the query result is obtained as the sum of two pre-computed values that are the range sums for the corresponding regions. Explicit cost formulas enable an analyst to choose the right HBC or HRC parameter setting for an application. However, the formulas are quite complex (cf., Chan & Ioannidis, 1999). Finding good configurations therefore requires an experimental evaluation.

Combining Efficiency, Flexibility, and Simplicity

RPS, DDC, and HC are sophisticated multidimensional pre-aggregation techniques. Hence the corresponding algorithms are typically complex (cf., Chan & Ioannidis, 1999; Geffner, Agrawal, & El Abbadi, 2000), and it is difficult to prove their correctness and to analyze their performance. The different dimensions of a data cube are treated uniformly in the sense that even though box sizes might vary, the same general scheme is used for each dimension. However, in practice attributes have very diverse characteristics. For instance, the gender attribute of a census data set has only two possible values, while income covers a large range. For some attributes a priori semantic information makes one aggregation technique preferable

Figure 7: Original Data Cube A and a Possible HBC Cube

A	0	1	2	3	4	5	6	7
0	3	5	1	2	2	4	6	3
1	7	3	2	6	8	7	1	2
2	2	4	2	3	3	3	4	5
3	3	2	1	5	3	5	2	8
4	4	2	1	3	3	4	7	1
5	2	3	3	6	1	8	5	1
6	4	5	2	7	1	9	3	3
7	2	4	2	2	3	1	9	1

HBC	0	1	2	3	4	5	6	7
0	3	5	6	2	12	4	10	3
1	7	15	13	20	26	37	34	39
2	9	12	26	11	40	14	65	10
3	3	17	6	22	14	33	21	39
4	19	16	23	19	80	23	43	19
5	2	21	8	33	15	46	28	48
6	6	24	42	32	34	40	102	23
7	2	30	8	42	13	54	23	47

HBC	0	1	2	3	4	5	6	7
0	3	5	6	2	12	4	10	3
1	7	15	13	20	26	37	34	39
2	9	12	26	11	40	14	65	10
3	3	17	6	22	14	33	21	39
4	19	16	23	19	80	23	43	19
5	2	21	8	33	15	46	28	48
6	6	24	42	32	34	40	102	23
7	2	30	8	42	13	54	23	47

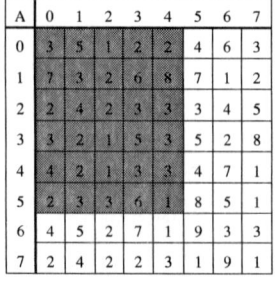

Prefix query: SUM(A[0,0]:A[5,4])

Result: 3+5+...+1 = 95

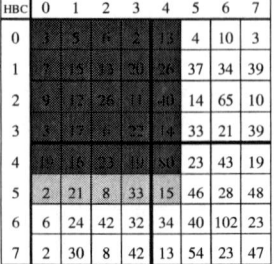

SUM(A[0,0]:A[5,4]) = HBC[4,4]+HBC[5,4]

Result: 80+15 = 95

Figure 8: Original Array A and Corresponding RPS (Block Size 3) and PS Arrays (Query Range and Updated Cell in A Are Framed, Accessed Cells Are Shaded)

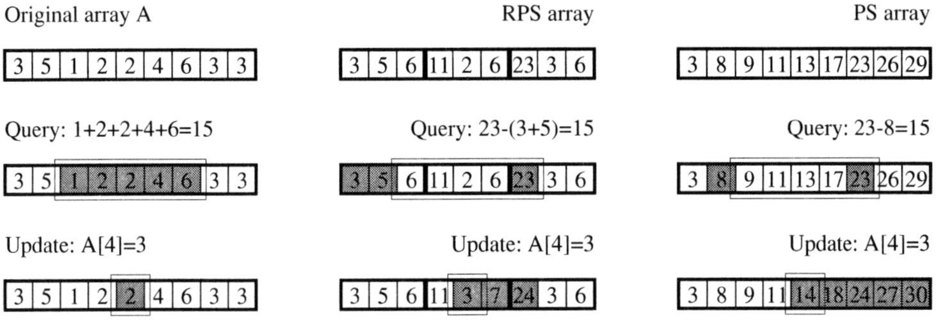

Original array A

| 3 | 5 | 1 | 2 | 2 | 4 | 6 | 3 | 3 |

RPS array

| 3 | 5 | 6 | 11 | 2 | 6 | 23 | 3 | 6 |

PS array

| 3 | 8 | 9 | 11 | 13 | 17 | 23 | 26 | 29 |

Query: 1+2+2+4+6=15

| 3 | 5 | 1 | 2 | 2 | 4 | 6 | 3 | 3 |

Query: 23-(3+5)=15

| 3 | 5 | 6 | 11 | 2 | 6 | 23 | 3 | 6 |

Query: 23-8=15

| 3 | 8 | 9 | 11 | 13 | 17 | 23 | 26 | 29 |

Update: A[4]=3

| 3 | 5 | 1 | 2 | 2 | 4 | 6 | 3 | 3 |

Update: A[4]=3

| 3 | 5 | 6 | 11 | 3 | 7 | 24 | 3 | 6 |

Update: A[4]=3

| 3 | 8 | 9 | 11 | 14 | 18 | 24 | 27 | 30 |

over another. For instance, when a natural hierarchy exists for an attribute, users typically query according to this hierarchy (e.g., it is more likely that a query aggregates monthly sales figures than sales figures for some 30-day period that starts on the third of a month). For such dimension attributes DDC with its hierarchical aggregation structure could be the technique of choice. For another attribute PS or HBC might be better.

This is the motivation behind the iterative data cubes (IDCs) proposed by Riedewald et al. (2001). The approach provides a modular framework for combining one-dimensional pre-aggregation techniques to create space-optimal multidimensional pre-aggregated data cubes. By selecting the appropriate one-dimensional techniques, the specific properties of the dimensions can be taken into account. Not only can the advantages of different pre-aggregation schemes be combined within a single cube, also the overall cost functions can easily be obtained based on the corresponding one-dimensional costs. Hence, analyzing and implementing a d-dimensional IDC is essentially reduced to the one-dimensional case. When a new dimension attribute with specific properties is to be included into the framework, all one has to do is develop a new one-dimensional pre-aggregation approach for it.

For the following discussion let Θ be a one-dimensional pre-aggregation technique and A be a one-dimensional array with N cells. Technique Θ generates a pre-aggregated array A_Θ of size N, such that each cell of A_Θ stores a *linear combination* of the cells of A:

$$\forall 0 \leq j \leq N - 1 : A_\Theta [j] = \sum_{k=0}^{N-1} \alpha_{j,k} A[k] .$$

The values of the variables $\alpha_{j,k}$ are determined by the pre-aggregation technique. Figure 8 shows an example. The array RPS is the result of applying the

RPS technique with block size 3 to the original array A. RPS pre-aggregates a one-dimensional array as follows. A is partitioned into blocks of equal size. The anchor of a block $a : e$ (its leftmost cell) contains the corresponding prefix sum of A, i.e., $RPS[a] = SUM(A[0] : A[a])$. Any other cell c of the block stores the "local prefix sum" $RPS[c] = SUM(A[a + 1] : A[c])$. Consequently, the coefficients $\alpha_{j,k}$ in the example are $\alpha_{0,0} = 1$, $\alpha_{1,1} = 1$, $\alpha_{2,1} = \alpha_{2,2} = 1$, $\alpha_{3,k} = 1$ for $0 \le k \le 3$, $\alpha_{4,4} = 1$, $\alpha_{5,4} = \alpha_{5,5} = 1$, $\alpha_{6,k} = 1$ for $0 \le k \le 6$, and $\alpha_{j,k} = 1$ for all other combinations of j and k.

To construct an IDC for a d-dimensional original data cube A, a pre-aggregation technique Θ_i is selected for each dimension i. First each one-dimensional vector $A[0, x_2, x_3,..., x_d] : A[N_1, x_2, x_3,..., x_d]$, $x_i \in \{0, 1,..., N_i\}$ for $2 \le i \le d$, is processed with technique Θ_1. Let the result be cube A_1. A_1 is then processed similarly, applying technique Θ_2 to all vectors along dimension 2, and so on. Interestingly, if PS is selected for each dimension, the final IDC cube is identical to the d-dimensional PS cube. The same holds for using RPS and DDC for all dimensions. However, IDC can also combine these and other techniques. For example for a three-dimensional data set, one could select PS for the first, DDC for the second, and HBC for the third dimension. Figure 9 shows how the RPS cube from Figure 4 is constructed using the IDC approach. First A's rows are processed using one-dimensional RPS, then the columns of the intermediate result are processed with RPS as well.

A nice property of IDC is that the query and update algorithms for a d-dimensional data cube also can be obtained as a simple combination of the one-dimensional algorithms. The designer of a one-dimensional pre-aggregation technique only has to ensure that the technique is *reversible*. Let A be a one-dimensional array, Θ be a one-dimensional pre-aggregation technique, and A_Θ be the resulting pre-

Figure 9: Original Data Cube A, Intermediate Result, and Final IDC Cube (Fat Lines Indicate Partitioning into Blocks by RPS)

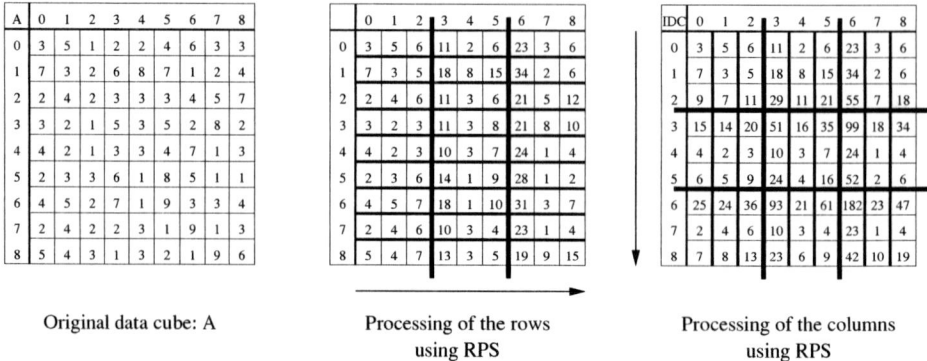

A	0	1	2	3	4	5	6	7	8
0	3	5	1	2	2	4	6	3	3
1	7	3	2	6	8	7	1	2	4
2	2	4	2	3	3	4	4	5	7
3	3	2	1	5	3	5	2	8	2
4	4	2	1	3	3	4	7	1	3
5	2	3	3	6	1	8	5	1	1
6	4	5	2	7	1	9	3	3	4
7	2	4	2	2	3	1	9	1	3
8	5	4	3	1	3	2	1	9	6

	0	1	2	3	4	5	6	7	8
0	3	5	6	11	2	6	23	3	6
1	7	3	5	18	8	15	34	2	6
2	2	4	6	11	3	6	21	5	12
3	3	2	3	11	3	8	21	8	10
4	4	2	3	10	3	7	24	1	4
5	2	3	6	14	1	9	28	1	2
6	4	5	7	18	1	10	31	3	7
7	2	4	6	10	3	4	23	1	4
8	5	4	7	13	3	5	19	9	15

IDC	0	1	2	3	4	5	6	7	8
0	3	5	6	11	2	6	23	3	6
1	7	3	5	18	8	15	34	2	6
2	9	7	11	29	11	21	55	7	18
3	15	14	20	51	16	35	99	18	34
4	4	2	3	10	3	7	24	1	4
5	6	5	9	24	4	16	52	2	6
6	25	24	36	93	21	61	182	23	47
7	2	4	6	10	3	4	23	1	4
8	7	8	13	23	6	9	42	10	19

Original data cube: A   Processing of the rows   Processing of the columns
                                    using RPS   using RPS

Figure 10: Processing Queries and Updates on an Iterative Data Cube (RPS Technique Used for Both Dimensions)

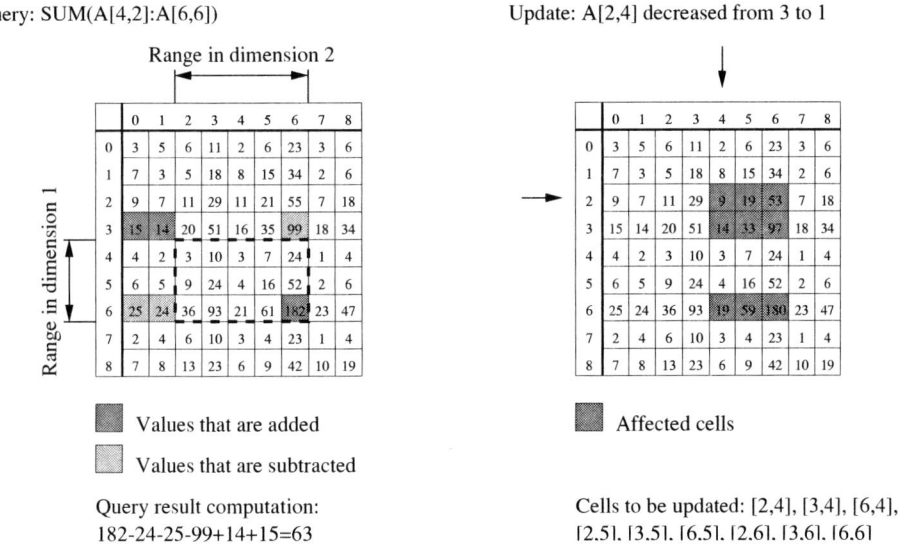

Query: SUM(A[4,2]:A[6,6])   Update: A[2,4] decreased from 3 to 1

Range in dimension 2

Range in dimension 1

Values that are added

Values that are subtracted

Query result computation:
182-24-25-99+14+15=63

Affected cells

Cells to be updated: [2,4], [3,4], [6,4],
[2.5], [3.5], [6.5], [2.6], [3.6], [6.6]

aggregated array. Θ is reversible if and only if for each range $r = A[i] : A[j], 0 \leq i \leq j \leq N$, there are variables $\beta_{r,l}$ such that SUM($A[i] : A[j]$) = $\sum_{l=0}^{N-1} \beta_{r,l} A_\Theta[l]$. In other words, the pre-aggregation technique has to guarantee that each range sum on A can still be computed on A_Θ. A range query on a d-dimensional original data cube A is then translated to a query on the IDC cube by determining the relevant cells (those with non-zero betas) for each dimension. Note, that there are no inter-dimensional dependencies, i.e., the betas in a dimension only depend on the technique used and the range selected in this dimension. The update process is similar.

Figure 10 shows an example for a query that computes SUM($A[4, 2] : A[6,6]$) on the IDC cube from Figure 9. For range 4 : 6 in dimension 1 and range 2 : 6 in dimension 2, the indices with non-zero beta values are obtained together with the betas. For instance for range 2 : 6 in dimension 2, the relevant indices are 0, 1, and 6 with beta values of -1, -1, and 1, respectively (see also RPS query translation in Figure 8). Similarly, for dimension 1 the indices 3 and 6 with beta values of -1 and 1 are obtained. Note that selecting a different pre-aggregation technique or query range in dimension 2 would not affect the indices or beta values in dimension 1, and vice versa. Combining the one-dimensional results is straightforward. The relevant indices in the d-dimensional cube are the combinations of the one-dimensional result sets. The beta values are multiplied (for details see Riedewald, Agrawal, & El Abbadi, 2001).

The query and update *costs* for a d-dimensional IDC cube are the product of the numbers of non-zero betas in each dimension. Hence, for example, the worst-case query cost of an IDC cube is the product of the worst-case query costs of the one-dimensional pre-aggregation techniques used for its dimensions. Table 1 shows a selection of techniques with their respective worst-case costs for vectors with N_i elements.

The possibility of combining arbitrary, one-dimensional pre-aggregation techniques allows users to generate a great variety of IDC instances. Their query and update characteristics are given by simple formulas. For instance, if PS and RPS are used for dimensions 1 and 2 of a two-dimensional data cube, the worst-case query and update costs are 8 and $2N_1\sqrt{N_2}$, respectively. If DDC and PS are used, the costs are $4\log_2 N_1$ and $\log_2 N_1$, respectively.

Note that this simplicity and flexibility comes at the cost of a restriction of the solution space compared to arbitrary multidimensional pre-aggregation techniques. However, it can be shown that IDC either matches or improves the query-update tradeoffs of previously proposed techniques. PS, RPS, and DDC are special cases of IDC.

DATA CUBES FOR OTHER AGGREGATE OPERATORS

Popular aggregate operators like MIN and MAX cannot be handled by the techniques discussed in the previous section due to the lack of an inverse operation. This section summarizes approaches that work for these two operators as well.

Table 1: Query-Update Cost Tradeoffs for Selected One-Dimensional Techniques

One-dimensional technique	Query cost (worst case)	Update cost (worst case)
Original Array	N_i	1
Prefix Sum (PS)	2	N_i
Relative Prefix Sum (RPS)	4	$2\sqrt{N_i}$
Dynamic Data Cube (DDC)	$2\log_2 N_i$	$\log_2 N_i$

Some of them take advantage of specific monotonicity properties, while others apply whenever the domain of the measure attribute forms a commutative semi-group under the aggregate operator.

In Ho et al. (1997), a technique for the computation of range max queries is proposed. The data cube is recursively partitioned into smaller hyper-rectangular chunks (similar to a multidimensional quad tree). For each chunk the position of the maximum value in the chunk is stored. The corresponding tree is traversed top-down, using pruning techniques to reduce the number of accesses. Lee, Ling, & Li (2000) use a similar data structure but a different pruning strategy. They obtain constant average query cost and logarithmic update cost in the size of the data cube (for fixed dimensionality).

Poon (2001) proposes a general approach for commutative semi-groups. Similar to iterative data cubes, Poon's approach combines one-dimensional schemes to obtain a multidimensional pre-aggregation technique. Query, update, and storage costs are $O(\prod_{i=1}^{d} 4L_i)$, $O(\prod_{i=1}^{d} 12 L_i^2 N_i^{\frac{1}{L_i}} \gamma(N_i))$, and $O(\prod_{i=1}^{d} 6N_i\gamma(N_i))$, respectively. Here $L_i \in \{1,..., \log N_i\}$ denotes a user-controlled parameter and $\gamma(N_i)$ a slow-growing function. Applied to invertible operators the query, update, and storage costs are $O(\prod_{i=1}^{d} 2L_i)$, $O(\prod_{i=1}^{d} 2L_i \ N_i^{\frac{1}{L_i}})$, and $O(\prod_{i=1}^{d} 2N_i)$, respectively.

Note, however, that techniques like IDC—which are optimized to take advantage of the inverse operator—achieve better cost tradeoffs for invertible operators. For instance, for $L_i=1$, an IDC that uses PS in each dimension achieves the same query but lower update and storage costs (cf., Table 1). Similarly IDC using RPS (DDC) in each dimension offers a better solution than the above technique for $L_i = 2$ $(L_i = \log N_i)$.

SPARSE DATA SETS

The techniques presented to this point are array-based and would not be efficient for very sparse data sets. There has been extensive research regarding range queries on sparse data in the computational geometry community (e.g., see de Berg et al., 2000; Chazelle, 1988, 1990; Fredman, 1981; Willard, 1986; Willard & Lueker, 1985). Both range reporting (return all points in the selected range) and range aggregation queries were examined. However, most of the techniques cause space overhead that is super-linear in the number of data points, n (e.g., $O(n \log^{d-1} n)$) which is not desirable for large data warehouses. Also, the data structures are typically quite involved, and hence these approaches are rarely used in practice.

Recently the authors (Riedewald, Agrawal, & El Abbadi, 2000) and Lazaridis & Mehrotra (2001) proposed frameworks to support efficient on-line aggregation on virtually any hierarchical index structure. The main idea is to add pre-computed

Figure 11: pCube with R-Tree Based Structure for the Aggregate Function COUNT

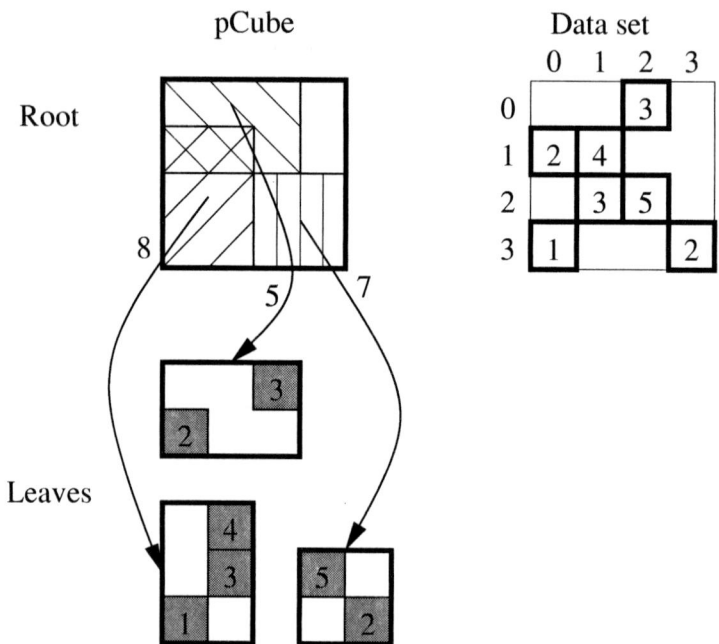

aggregates to the nodes of an index structure. These aggregates are used to reduce the number of node accesses and to return progressively refined approximate results with absolute error bounds to the user. Both approaches are very similar, mainly differing in the order of the node accesses. Here we describe the Progressive Data Cube (pCube) (see Riedewald, Agrawal, & El Abbadi, 2000).

The pCube can be designed to incorporate any hierarchical multidimensional index structure, e.g., R-tree (see Guttman, 1984), quad tree (see Gaede & Günther, 1998), etc. Figure 11 shows an R-tree-based example of pCube that supports the aggregate function COUNT. Each node entry of the R-tree, consisting of a bounding box and a pointer to a child node, contains the number of points in the corresponding sub-tree. Since there are overlapping bounding boxes of sibling nodes, it has to be ensured that no point is counted more than once.

A query processes the pCube tree top-down, starting at the root and then descending to all children that intersect the query. Based on the information in the nodes visited so far, an approximate answer is returned. The aggregates in the nodes are selected such that the approximate answer is accompanied with *absolute* error bounds. The further the query descends the tree, the more accurate the result. Figure 12 shows an example for a simple quad-tree-like pCube whose node entries store the sum of the values in the sub-tree. The query is shaded, the output is given in the figure.

Note that the approximate result is obtained by assuming uniform distribution of the values. To compute the error bounds, the extreme cases are evaluated (all/no non-empty cells fall into the query). A user or application can select an individual tradeoff between query cost and accuracy. When the query reaches the leaf nodes, the exact result is obtained smoothly from the same structure. The continuous feedback is especially helpful for interactive analysis. However, in contrast to the techniques previously discussed in this chapter, non-trivial worst-case costs are not guaranteed. The query cost generally is proportional to the number of nodes intersected by the query. Savings can be expected for practical applications when the query completely contains a node. Then this node's information already provides the exact result without further descending the corresponding sub-tree (e.g., upper left of the internal nodes in Figure 12).

The update cost of a pCube is typically identical to updating the corresponding index structure without pre-computed aggregates. More precisely, this holds true if the sets of aggregate values in pCube's nodes are *self-maintainable* with respect to insertions and deletions. A set of aggregate functions is self-maintainable if the new values of the functions can be computed solely from the old values and from the changes caused by the update operation (e.g., see Mumick, Quass, & Mumick 1997). Consequently COUNT and SUM are self-maintainable. AVG is not self-maintainable, but can be made so by computing it from SUM and COUNT. MAX, MIN, and median are not self-maintainable (except for the trivial case that all base values are stored). However, a pCube can efficiently handle MAX and MIN due to its hierarchical structure. Note that pCube is an example for compression on different levels which frees the system or developer from choosing the "best" granularity of the approximation.

Figure 12: Processing a Query on a Simple pCube for Aggregate Operator SUM

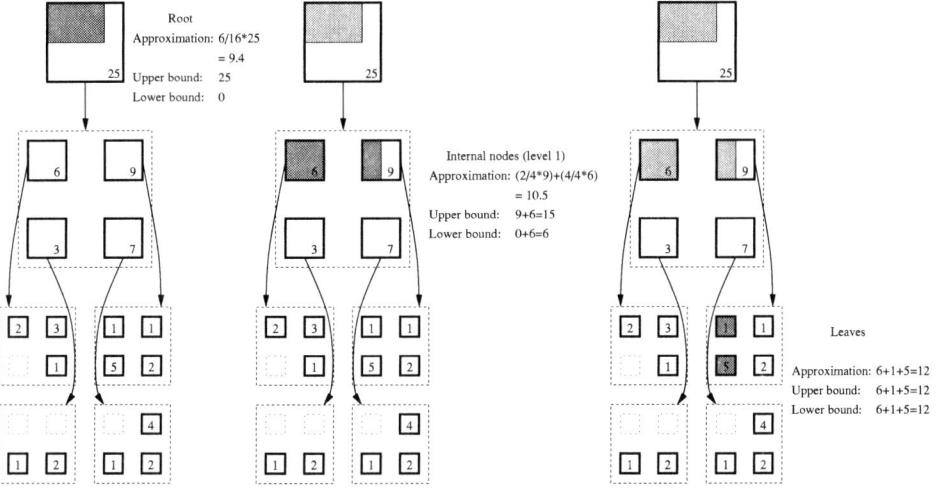

THE DATA CUBE OPERATOR CUBE

CUBE as proposed in Gray et al. (1997) generalizes SQL's GROUP BY operator. It computes the results of grouping a data set by all subsets of the specified d dimensions. Hence 2^d related groups are generated. We refer to each of these groups as a *cuboid* and to the overall result of CUBE as *Cube*. Figure 13 shows an example for a Cube for aggregate operator SUM that is computed for the dimensions time (T), product (P), and location (L), and measure attribute amount (A). The tables list all non-empty cells of the cuboids. In relational database terminology the cuboids are *views* generated by the corresponding GROUP BY queries as indicated in the figure. For simplicity we will use the dimension letters to refer to the cuboids; e.g., <TP> refers to the cuboid which is the result of grouping by time and product. Cuboid <TPL>, which has the most detailed information, is referred to as the *base cuboid*.

Cube's main purpose is to support aggregate queries. A query that only selects on a few attributes can be efficiently answered by processing the appropriate compact cuboid. For instance, finding the total sales volume in 1999 only costs a lookup on <T>, while computing the same result on the base cuboid would require finding and aggregating multiple tuples. Specialized index structures for Cube, e.g.,

Figure 13: Cuboids of a Three-Dimensional Example Cube

GROUP BY T, P, L

T	P	L	A
99	p1	SB	2
99	p1	SF	3
99	p2	LA	1
00	p1	SB	2
00	p1	SF	3
00	p2	LA	4
01	p1	SF	3
01	p2	LA	1

GROUP BY T, P

T	P	A
99	p1	5
99	p2	1
00	p1	5
00	p2	4
01	p1	3
01	p2	1

GROUP BY T, L

T	L	A
99	SB	2
99	SF	3
99	LA	1
00	SB	2
00	SF	3
00	LA	4
01	SF	3
01	LA	1

GROUP BY P, L

P	L	A
p1	SB	4
p1	SF	9
p2	LA	6

GROUP BY T

T	A
99	6
00	9
01	4

GROUP BY P

P	A
p1	13
p2	6

GROUP BY L

L	A
SB	4
SF	9
LA	6

GROUP BY { }: A = 19

Figure 14: Cube as a Multidimensional Data Cube; Cuboids of Figure 13 Indicated

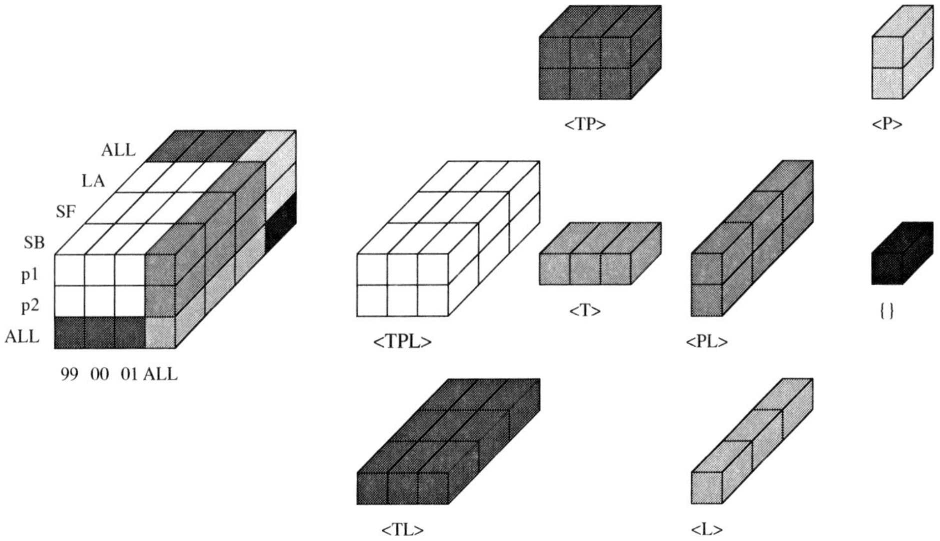

Johnson & Shasha (1997, 1999) and Roussopoulos, Kotidis, & Roussopoulos (1997), further reduce query time.

Note however, that Cube pays for this query support with a possibly large amount of redundancy. Each aggregate query that could be answered by a certain cuboid *C* could also be answered by the base cuboid (or any other cuboid whose set of group-by attributes contains *C*'s group-by attributes). For sparse high-dimensional data sets, materializing all views will lead to *database explosion* (cf., Pendse & Creeth, 2001). This is caused by the large number of cuboids (2^d) and the possibility that a cuboid does not perform a considerable amount of aggregation. We can already observe this effect in our (fairly dense) example in Figure 13. Cuboids <TPL> and <TL> contain the same number of tuples, i.e., the aggregation of <TPL> along the product dimension does not result in any aggregation. Higher dimensionality and greater sparseness tend to amplify the space explosion effect. At the same time the redundancy between the cuboids also increases update costs.

After Cube was proposed, most research focused on its efficient computation and space reduction. Bayer and Ramakrishnan (1999) propose a technique that restricts the generation of cuboids or partitions of cuboids to those that achieve "enough" aggregation. Other researchers (e.g., Gupta et al., 1997; Harinarayan, Rajaraman, & Ullman, 1996; Shukla, Deshpande, & Naughton, 2000; Smith et al., 1998) formulated the computation of Cube as an optimization problem. Their goal is to select and materialize only a subset of the cuboids (and indexes in case of Gupta et al., 1997) in order to minimize the query cost subject to storage constraints. Later work stressed the importance of also addressing Cube *maintenance* issues. Consequently the approaches in Baralis, Paraboschi, & Teniente (1997), Gupta (1997), and Gupta & Mumick (1999) extend the earlier work by explicitly taking update costs into account for the optimization problem.

Interestingly, the multidimensional data cube model provides a more compact description of the Cube by *conceptually* combining all 2^d cuboids into a single hyper-rectangle. Each dimension is augmented by the additional value "ALL," which represents the aggregation over the complete dimension. The additional surface layers for ALL coordinates are "padded" to the base cuboid and store the corresponding cuboids. Figure 14 illustrates this for the example from Figure 13.

CONCLUSIONS AND FUTURE TRENDS

We have discussed several approaches for supporting dynamic data cubes. The first part of the chapter mainly dealt with MOLAP data cubes and techniques that are particularly suitable for dense and low-dimensional data sets. Then techniques that explicitly address the sparseness issue were presented. The commonality between all approaches is that they try to find an appropriate balance between query, update, and storage cost. While earlier proposals mostly focused on query and storage aspects, large data sets with frequent updates created a need for more dynamic solutions.

In the future, support for *sparse* and *high-dimensional* data will become increasingly important. For example, business processes often span multiple organizational units of a company and involve multiple resources. Describing and analyzing a business process on a detailed level hence requires dealing with large numbers of attributes, i.e., high-dimensional data cubes. High dimensionality inevitably leads to sparseness since the number of data cube cells increases exponentially with the dimensionality.

Nevertheless, techniques for dense and low-dimensional data will remain important. For example, data distributions in practice are typically skewed and contain clusters with a higher density of non-empty cells. When the data cube is partitioned appropriately, some of the partitions might be dense enough to be handled like dense (sub)cubes. Also, analysts often specify selection conditions only for a subset of the dimensions. One could therefore maintain cubes for such lower-dimensional and denser projections of the data set. The Cube is another example of a structure containing dense data cubes, namely the cuboids for small numbers of grouping attributes.

Already today the size of data warehouses makes recomputation of aggregate information from scratch very costly. With increasing computing power and falling prices per storage unit, the current trend of *rapidly growing* data warehouses will continue and probably gain even more momentum in the near future. At the same time modern sensor technology and the ever-presence of computers in businesses result in the availability of a larger amount of information from virtually all units of a company. This poses new challenges for managing vast amounts of incoming data and fast analysis. Developing dynamic approaches for maintaining data cubes *incrementally* will become even more important.

ACKNOWLEDGMENTS

This work was partially supported by NSF grants EIA-98-18320, IIS-98-17432, and IIS-99-70700. We also would like to thank Steve Geffner and Renato Pajarola for their contributions to the OLAP research at UC Santa Barbara.

REFERENCES

Baralis, E., Paraboschi, S., & Teniente, E. (1997). Materialized view selection in a multi-dimensional database. *Proceedings of the International Conference on Very Large Databases (VLDB'97)*, 156-165.

Beyer, K., & Ramakrishnan, R. (1999). Bottom-up computation of sparse and iceberg CUBES. *Proceedings of the ACM International Conference on Management of Data (SIGMOD'99)*, 359-370.

Chaudhuri, S., & Dayal, U. (1997). An overview of data warehousing and OLAP technology. *SIGMOD Record*, 26(1), 65-74.

Chazelle, B. (1988). A functional approach to data structures and its use in multidimensional searching. *SIAM Journal on Computing*, 17(3), 427-462.

Chazelle, B. (1990). Lower bounds for orthogonal range searching: II. The arithmetic model. *Journal of the ACM*, 37(3), 439-463.

Chan, C.-Y., & Ioannidis, Y.E. (1999). Hierarchical cubes for range-sum queries. *Proceedings of the International Conference on Very Large Databases (VLDB'99)*, 675-686. Extended version published as Technical Report, University of Wisconsin.

de Berg, M., van Kreveld, M., Overmars, M., & Schwarzkopf, O. (2000). *Computational Geometry* (2nd Ed.). Springer Verlag.

Fredman, M.L. (1981). A lower bound on the complexity of orthogonal range queries. *Journal of the ACM*, 28(4), 696-705.

Gaede, V., & Günther, O. (1998). Multidimensional access methods. *ACM Computing Surveys*, 30(2), 170-231.

Geffner, S., Agrawal, D., & El Abbadi, A. (2000). The dynamic data cube. *Proceedings of the International Conference on Extending Database Technology (EDBT'00)*, 237-253.

Geffner, S., Agrawal, D., El Abbadi, A., & Smith, T. (1999). Relative prefix sums: An efficient approach for querying dynamic OLAP data cubes. *Proceedings of the International Conference on Data Engineering (ICDE'99)*, 328-335.

Geffner, S., Riedewald, M., Agrawal, D., & El Abbadi, A. (1999). Data cubes in dynamic environments. *Data Engineering Bulletin, 22*(4), 31-40.

Goil, S., & Choudhary, A. (1997). *BESS: Sparse Data Storage of Multi-Dimensional Data for OLAP and Data Mining.* Technical Report, Northwestern University.

Gray, J., Chaudhuri, S., Bosworth, A., Layman, A., Reichart, D., Venkatrao, M., Pellow, F., & Pirahesh, H. (1997). Data cube: A relational aggregation operator generalizing group-by, cross-tab and sub-totals. *Journal of Data Mining and Knowledge Discovery, 1*(1), 29-54.

Gupta, H., Harinarayan, V., Rajaraman, A., & Ullman, J.D. (1997). Index selection for OLAP. *Proceedings of the International Conference on Data Engineering (ICDE'97), 208-219.*

Gupta, H., & Mumick, I.S. (1999). Selection of views to materialize under a maintenance cost constraint. *Proceedings of the International Conference on Database Theory (ICDT'99), 453-470.*

Gupta, H. (1997). Selection of views to materialize in a data warehouse. *Proceedings of the International Conference on Database Theory (ICDT'97), 98-112.*

Guttman, A. (1984). R-trees: A dynamic index for spatial searching. *Proceedings of the ACM International Conference on Management of Data (SIGMOD'84), 47-57.*

Harinarayan, V., Rajaraman, A., & Ullman, J.D. (1996). Implementing data cubes efficiently. *Proceedings of the ACM International Conference on Management of Data (SIGMOD'96), 205-216.*

Ho, C., Agrawal, R., Megiddo, N., & Srikant, R. (1997). Range queries in OLAP data cubes. *Proceedings of the ACM International Conference on Management of Data (SIGMOD'97), 73-88.*

Johnson, T., & Shasha, D. (1997). Some index design for cube forests. *IEEE Data Engineering Bulletin, 20*(1), 27-35.

Johnson, T., & Shasha, D. (1999). Some index design for cube forests. *IEEE Data Engineering Bulletin, 22*(4), 31-40.

Koudas, N., Muthukrishnan, S., & Srivastava, D. (2000). Optimal histograms for hierarchical range queries. *Proceedings of the Symposium on Principles of Database Systems (PODS'00), 196-204.*

Lazaridis, I., & Mehrotra, S. (2001). Progressive approximate aggregate queries with a multi-resolution tree structure. *Proceedings of the ACM International Conference on Data (SIGMOD'01), 401-412.*

Lee, S.Y., Ling, T.W., & Li, H.G. (2000). Hierarchical compact cube for range-max queries. *Proceedings of the International Conference on Very Large Databases (VLDB'00), 232-241.*

Mumick, I.S., Quass, D., & Mumick, B.S. (1997). Maintenance of data cubes and summary tables in a warehouse. *Proceedings of the ACM International*

Conference on Data (SIGMOD '97), 100-111.

Pendse, N., & Creeth, R. *The OLAP Report.* http://www.olapreport.com/ Analyses.htm.

Parts available on-line in the current edition.

Poon, C.K. (2001). Orthogonal range queries in OLAP. *Proceedings of the International Conference on Database Theory (ICDT'01),* 361-374.

Riedewald, M., Agrawal, D., & El Abbadi, A. (2000). pCube: Update-efficient online aggregation with progressive feedback and error bounds. *Proceedings of the 12th International Conference on Scientific and Statistical Database Management (SSDBM'00),* 95-108.

Riedewald, M., Agrawal, D., & El Abbadi, A. (2001). Flexible data cubes for online aggregation. *Proceedings of the International Conference on Database Theory (ICDT'01),* 159-173.

Riedewald, M., Agrawal, D., El Abbadi, A., & Pajarola, R. (2000). Space-efficient data cubes for dynamic environments. *Proceedings of the International Conference on Data Warehousing and Knowledge Discovery DaWaK,* 24-33.

Roussopoulos, N., Kotidis, Y., & Roussopoulos, M. (1997). Cubetree: Organization of and bulk updates on the data cube. *Proceedings of the ACM International Conference on Management of Data (SIGMOD '97),* 89-99.

Sarawagi, S., & Stonebraker, M. (1994). Efficient organization of large multidimensional arrays. *Proceedings of the International Conference on Data Engineering (ICDE'94),* 328-336.

Shukla, A., Deshpande, P., & Naughton, J.F. (2000). Materialized view selection for multi-cube data models. *Proceedings of the International Conference on Extending Database Technology (EDBT'00),* 269-84.

Smith, J.R., Castelli, V., Jhingran, A., & Li, C.-S. (1998). Dynamic assembly of views in data cubes. *Symposium on Principles of Database Systems, PODS'98,* 274-283.

Willard, D.E. (1986). Lower bounds for dynamic range query problems that permit subtraction. *Proceedings of the International Colloquium on Automata, Languages and Programming.* Lecture Notes in Computer Science, No. 226. Springer-Verlag, 444-453.

Willard, D.E., & Lueker, G.S. (1985). Adding range restriction capability to dynamic data structures. *Journal of the ACM, 32*(3), 597-617.

Winter Corporation. (2001). *Database Scalability Program, 2001.* Results available on-line at http://www.wintercorp.com.

Chapter VIII

Materialized Views in Multidimensional Databases

Stefano Paraboschi
Università di Bergamo, Italy

Giuseppe Sindoni
National Institute of Statistics, Italy

Elena Baralis
Politecnico di Torino, Italy

Ernest Teniente
Universitat Politècnica de Catalunya, Spain

ABSTRACT

This chapter presents materialized views in the context of multidimensional databases (MDDBs). A materialized view is a view whose content is explicitly stored in the database. The advantage of materializing views is that it is not necessary to recompute the query every time the view is accessed. The shortcoming is that it has to be kept consistent with the updates on the base tables. However, efficient incremental maintenance techniques have been proposed. MDDBs are an ideal environment for materialized views because frequency of updates is low, MDDB data models permit easy adoption of incremental maintenance, and queries can be modeled in such a way to allow an easy definition of the view selection problem, i.e., the problem of selecting which query to materialize in an MDDB. Hence, we present the problems of choosing and maintaining materialised views with the corresponding solutions.

VIEWS

Views are database objects whose content is derived from data stored in the database. A view V can be considered as the result of the computation of a function f on the database state DB: $V = f(DB)$. In relational databases a view has the same structure of a relation, as it is labeled with a name (unique in the database schema among all tables and views), it is characterized by a set of attributes, and its instance consists of a set of tuples; the only difference is that its content is obtained executing a query on the database, and it is not explicitly created by database users.

Example 1: *View* TOTALSALES *has two attributes* Product *and* TotalAmount *and each tuple presents the sum of the amounts of all the sales for a particular product.*

```
create view TotalSales(Product,TotalAmount) as
select Product, sum(Amount)
from Sales
group by Product
```

Views have an important role in databases, as they are at the core of many significant database services. Using views, a database schema can be adapted to the needs of the different users of the database. Data independence, one of the main advantages of databases (Atzeni et al., 1999), is implemented at the logical level by means of views; when the database schema evolves, views are used to emulate previous schema versions, making the application code independent of the actual relational schema.

MATERIALIZED VIEWS

Views in databases typically are *virtual* objects, i.e., their content is computed only when the database receives a request to access them. A typical strategy consists in rewriting queries that refer to views, in order to express them in terms of the concrete relations available on the system.

Example 2: *A query like*

```
select Product, TotalSales, Brand
from TotalSales, Products
where TotalSales.Product = Products.Code
```

might be transformed by the query optimizer into query:

```
select Product, sum(Amount) as TotalSales, Brand
from Sales, Products
where Sales.Product = Products.Code
group by Product, Brand
```

An alternative strategy consists of explicitly storing the content of a view in the database, creating a *materialized view*. The advantage of materialized views is that access to the view will only require reading the blocks where the materialization is stored, without the need to compute the query from the base data. Since query computation is almost always more expensive than reading a materialization, this strategy is able to reduce the cost of query computation and to improve database throughput. For example, if view TOTALSALES is materialized, the query will be directly executed on the view and it will in general be more efficient, as it will require access to a smaller amount of data and will not need to compute the aggregation.

The main shortcoming of materialized views is that, to guarantee query correctness, their content must be equivalent to what can be obtained by executing the query associated with the view on the current database state. Then, every time a table used in the view computation is updated, materialized views have to be kept consistent, applying the effect of the base updates on the result of the query defining the view. In the example, the materialization of view TOTALSALES will have to be updated when tables SALES or PRODUCT are modified.

We formalize the benefit of materialized views with the following simple model. Let us suppose that for a view V, the cost of executing its query is Q_{DB} and the cost of accessing its materialization is Q_M. In general, it will be $Q_{DB} > Q_M$. We further represent the cost of storing the view as C_M and suppose that the query is executed with frequency f_Q. Updates on data that force a view recomputation occur with frequency f_U. To compute the benefit of view materialization, it will be necessary to evaluate the costs incurred executing the query directly on base data and compare them with the costs incurred using the materialization. When the materialization is not used, the cost of computing the query q_i with f_Q frequency will be equal to $f_Q \cdot Q_{DB}$. When the materialization is present, we have to sum up the cost of storing the materialization, the cost of computing it, and the cost of updating it f_U times; we obtain the formula $C_M + f_U \cdot Q_{DB} + f_Q \cdot Q_M$. The benefit B is the difference between the two expressions:

$$B = f_Q \cdot Q_{DB} - (C_M + f_U \cdot Q_{DB} + f_Q \cdot Q_M)$$

With an immediate algebraic transformation on the formula, we derive that the view materialization will offer a positive benefit only when

$$f_Q \cdot (Q_{DB} - Q_M) > C_M + f_U \cdot Q_{DB}$$

The formula confirms the intuition that the benefit increases with query frequency and with the advantage that the view materialization offers to view computation; the benefit decreases with the update frequency and with the cost of view materialization. If we disregard the materialization cost C_M, we obtain the simpler formula that identifies the situations when a view materialization is convenient:

$$ f_Q / f_U > Q_{DB} / Q_{DB} - Q_U $$

The advantages of materialized views can be significantly increased by the use of incremental maintenance techniques. These techniques have been the subject of a cosiderable amount of research (Gupta & Mumick, 1999). Many solutions have been identified that permit the efficient computation of the update that has to be applied on a materialized view to make it consistent with the updates that occurred on the base data. In this way, when updates are applied on the base tables, it is not necessary to recompute all the materialized views that have been modified from scratch.

For instance, for view TOTALSALES a complete recomputation would probably be extremely expensive. A more efficient solution may propagate updates on base tables to the view, e.g., by requiring that every insertion into SALES would generate an increment of the sale amount of the tuple in TOTALSALES describing the product sales. In typical database configurations, the size of updates is considerably smaller than the size of tables and views, and the use of incremental maintenance techniques offers a considerable potential. Extending on the previous formalization, if C_U is the cost of updating the view, the cost when using techniques for incremental maintenance of materialized views may be formalized as:

$$ f_Q \cdot Q_M + f_U \cdot C_U + C_M $$

with C_U being typically a small fraction of the cost Q_{DB} of computing the view from scratch.

In databases, there is a large experience in using cost models to optimize queries. The formula above constitutes a starting point for the implementation of view materialization solutions in a wide range of situations. What the formula does not show are the additional obstacles that make it difficult to exploit the potential of these solutions.

One of the obstacles to a wide adoption of these techniques is the great number of views that can help query execution. This aspect forces the system to do a preliminary and difficult choice of the views that may benefit the computation of queries. Indeed, the use of materialized views for query computation has a strong relationship with multi-query optimization techniques, where the goal is to build an efficient query execution plan for multiple queries reusing the same intermediate results in more than one query.

For instance, consider the two queries:

```
query 1:
select Product, sum(Amount)
from Sales, Products
where Sales.Product = Products.Code and
   Brand = 'SuperMagic'
group by Product
```

```
query 2:
select Brand, sum(Amount)
from Sales, Products
where Sales.Product = Products.Code
group by Brand
```

The computation of both queries would benefit from the identification and precomputation of TotalSales. The availability of the materialization of TOTALSALES permits us to rewrite the above queries into:

```
query1-rewritten:
select Product, TotalAmount
from TotalSales, Products
where Sales.Product = Products.Code and
   Brand = 'SuperMagic'
```

```
query2-rewritten:
select Brand, sum(TotalAmount)
from TotalSales, Products
where Sales.Product = Products.Code
group by Brand
```

In typical database applications queries arrive in a continuous stream to the database; the adoption of multi-query optimization would require us to analyze if there are common subparts in the arriving queries that may be computed only once. There are many obstacles that limit the applicability of multi-query optimization, which has met until now very limited success in commercial systems: the analysis has to be done in a relatively short time, the search space is extremely wide, and there is no opportunity to store the intermediate results for longer periods, as updates on the underlying data make them obsolete. When some of these restrictions are removed, multi-query optimization may become an interesting technique.

For instance, Mistry et al. (2001) demonstrate heuristics that can be used for multi-query optimization, in a situation where queries are known up front and updates arrive with a known frequency. Similar restrictions also apply to multidimensional databases, which have been until now the most interesting environment for the adoption of specific multi-query optimization and materialized views. The goal of this chapter is to explore the adoption of materialized views in this specific environment. The next section recalls some basic models and definitions. We then present the problem of incrementally maintaining a set of materialised views on a MDDB, and describe the problem of chosing an optimal set of materializations and possible solutions. Following that is an overview of current commercial implementations of materialized views, and we end with some conclusions.

MULTIDIMENSIONAL DATABASES

At the abstract level, considering its typical requirements, a multidimensional database (MDDB) is a data repository that provides an integrated environment for decision support queries that require complex aggregations on huge amounts of historical data (e.g., to identify trends in revenues). At the technical level, we assume for this chapter that an MDDB is a relational data warehouse, in which the information is organized following the star-model. Its basic structure may be represented with the simple entity-relationship diagram depicted in Figure 1, in which all the D_i entities represent the dimensions of the MDDB, while the connecting relationship F is the fact table.

Each *dimension table D_i* contains all the information that is specific only to the dimension itself (for example, product information), while the *fact table F* correlates all dimensions and contains information on the attributes of interest for the intersection of all the dimensions (for example, sales by product by day).

Multidimensional queries usually require the computation of aggregate functions on data grouped on an appropriate set of attributes. For multidimensional data processing, data is usually represented as a single (huge) fact table, on which all the interesting aggregates may be computed.

Figure 1: Entity-Relationship Representation of an MDDB

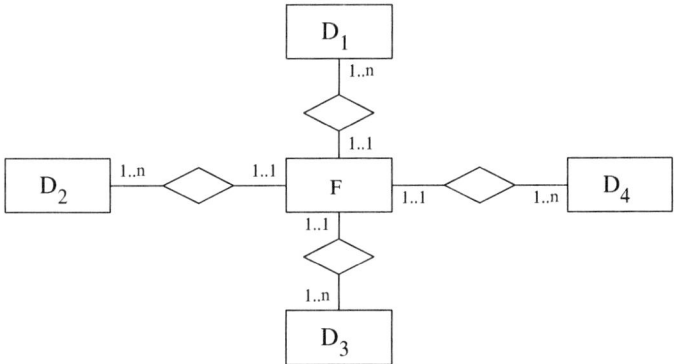

As shown in the practical example below, in addition to the fact table, MDDBs may have several dimensions, each of which is characterized by a considerable number of attributes, most of which may also be relevant for grouping computation.

A Practical Example

We consider as a practical example the MDDB for a large grocery store chain, taken from Kimball (1996). The company manages a large number of stores, each of which is a supermarket selling a wide variety of different products (e.g., grocery, frozen foods, bakery, etc.).

We can identify the following dimensions:

- *Product,* which can be characterized by more than 50 different attributes describing various features of the products, such as brand, category, diet-type (diet vs. non-diet), package type, weight, case size. Products belong to a merchandise hierarchy, which includes sub-category and category. This means that all the products that belong to the same sub-category also have the same value for attribute category.
- *Store,* which characterizes each point of sale with attributes such as complete store address and telephone, manager, description of store services, sizes of different departments. Stores are organized in a geographic hierarchy (zip code, county, state). The store dimension may have more than 20 attributes.
- *Time,* which provides the appropriate detail to allow accurate analysis of the MDDB data, that is indeed a time series. Time is represented with the granularity of day; days may be described as day in the month, in the quarter, and in the year. Furthermore, the indication of which days are holidays, weekend days, days in which special events took place (and many more) can be given. The time dimension may have more than 15 attributes.
- *Promotion,* which describes the characteristics of product promotions, such as the promotion type (ad, coupon, etc.) and several pieces of information on each promotion type, the promotion cost, and start/end date. Overall the promotion dimension is characterized by at least 10 attributes.

The fact table provides the sales information on which the actual financial analysis is performed. It includes the identifiers of all the dimensions and several attributes describing sales revenues (e.g., income, number of units sold, number of customer transactions performed, etc.). In this chapter we consider a subset of the attributes of each dimension as relevant attributes for grouping computations: we assume 15 attributes for dimensions *Product* and *Store*, nine attributes for *Time*, and 11 attributes for *Promotion*.

Materialized Views in MDDBs

MDDBs have three features that make them an ideal environment for the adoption of materialized views.

- MDDBs are implemented upon data warehouse architectures, where updates occur with a rigid low frequency (daily, weekly, or even monthly), in periods where the system load is low; thus, ratio f_Q/f_U is particularly high.
- MDDBs have a data model that permits the easy adoption of incremental view maintenance techniques; often, materialized views can be maintained considering only the updates and the current state of the materialization, with no need to access base tables.
- Queries in MDDBs can be characterized by a model that permits a precise and manageable definition of the view selection problem.

We now describe the multidimensional data model and the query model. We will then present the most important techniques for the maintenance of materialized views and the MD-mat problem, that is the problem of chosing an optimal set of views to materialize.

Multidimensional Database Model

We repeat for convenience some definitions which were previously introduced.

Definition 1: *A* **Multidimensional Database** *is a collection of relations D_1, ..., D_n, F, where 0*
- *Each D_i is a **dimension table**, i.e., a relation characterized by an identifier d_i that uniquely identifies each tuple (d_i is the primary key of D_i).*
- *F is a **fact table**, i.e., a relation connecting all dimension tables D_1, ..., D_n; the identifier of F is given by the foreign keys d_1, ..., d_n of all the dimension tables it connects; the schema of F contains a set of additional attributes V (representing the values on which the aggregate functions are applied).*

The dimension tables may contain hierarchies on the data.

Definition 2: *Let D be a dimension table with identifier d. An **attribute hierarchy** on D is a set of functional dependencies $FD_D = \{fd_0, fd_1, ..., fd_n\}$, where each fd_i is characterized by two sets of attributes $A^l_i \subset$ Attr(D) and $A^r_i \subseteq$ Attr(D) (respectively called left side and right side of the dependency); the dependency is represented as fd_i: $A^l_i \rightarrow A^r_i$.*

Each functional dependency fd_i is a constraint on the content of the dimension table D: for any pair of tuples $t_1, t_2 \in D$, $t_1[A^l_i] = t_2[A^l_i] \Rightarrow t_1[A^r_i] = t_2[A^r_i]$. A dependency fd_0 with $A^l_0 = \{d\}$ and $A^r_0 = \{$Attr(D) - d$\}$ will always be present in FD_D. A given set of attributes A^l_i may appear only once as the left side of a dependency. Functional dependencies must be acyclic, i.e., the graph obtained by drawing an arc from a_x to a_y, if $\exists fd_i \in FD_D \mid a_x \in A^l_i \wedge a_y \in A^r_i$, must be acyclic.

Definition 3: *An attribute hierarchy is* **tree-like** *(FD^T) if all the attributes appear at most once in the right side of a functional dependency and the left side always contains a single attribute.*

In this chapter we consider only tree-like attribute hierarchies.

Definition 4: *An* **MDDB attribute hierarchy** *FD$_{DB}$ is the union of the attribute hierarchies FD$_{Dj}$ of all the dimensions D$_j$ appearing in the MDDB.*

Business analysis queries usually require the computation of aggregate functions on data grouped on an appropriate set of attributes.

Example 1: *Consider again the multidimensional database of our example, MDDB = {Product, Store, Time, Promotion, F}, where each dimension table has its corresponding attributes and with fact table F = {p, s, d, r, f}, having p, s, d (time dimension is represented with the granularity of day), and r as foreign keys of the dimension tables, and f representing the amount of sales. The following queries can be requested on the MDDB:*

- q_1 = total sales per product
- q_2 = total sales per product and store
- q_3 = total sales per product and day
- q_4 = total sales per product, store and day

The queries we consider are select-join-group-by queries, with some restrictions on the allowed selection and join predicates. In particular, selection predicates are simple comparison predicates between a dimension attribute and a constant value (i.e., are of the type *attr* <operator> *const-value*). These predicates select a "slice" of data on which aggregates are actually computed. We envision a framework where the user provides a set of queries representative of the queries that the system must answer. For select queries we assume that only the attribute used is important and not the constants used in the comparison (which will generally change from query to query). To simplify the framework, we derive from a selection condition on an attribute a request for aggregating on that attribute. Consider a query q that returns the total sales of a particular store grouped by products. Query q can be answered by accessing the result of a query that returns the sales grouped by product and store, and selecting from them only the tuples relative to the specified store. In this way we can focus on queries computing aggregates. These considerations lead to the restriction that all the attributes appearing in a selection predicate must also appear as grouping attributes.

Join operations may be performed only between the fact table and any of its dimensions. Allowed predicates are equality predicates between a dimension identifier and the corresponding fact table foreign key, while join predicates involving non-key attributes are disallowed.

We consider the standard SQL notion of group-by and aggregate function (considered functions are `count, sum, avg, min, max`). Grouping attributes may be drawn both from dimension and fact table attributes.

> **Definition 5:** *Given a query q, we define the query as* **characterized** *by the set of its group-by attributes A. We may represent the query as q^A.*

Data-cube

A fundamental characteristic of MDDB queries is that it is often convenient to reuse the results of queries to answer other queries. In Example 1, we can use the result of query q_2 to answer query q_1, adding the sales across all stores to get the result. In a similar way, q_1 can also be computed in terms of the answers to queries q_3 and q_4, while q_2 and q_3 can be computed in terms of query q_4.

The reuse of queries is strictly related to the data-cube (Gray et al., 1996) operator. This operator, receiving as input a table T, a set of aggregating attributes A, and an aggregate function f, computes the union of the results of the queries evaluating function f, having as grouping attributes all possible combinations of attributes in A.

Consider the situation where only the fact table F is present. Applying the data cube operator on the foreign keys of F, we obtain a data cube lattice.

> **Definition 6:** *Given an MDDB = $\{D_1, ..., D_n, F\}$, the lattice* **Cube-lattice** *of MDDB is the lattice of the set of all possible grouping queries that can be defined on the foreign keys of F. This lattice is characterized by the following elements:*

- *an ordering relation defined as the comparison between the sets of grouping attributes (i.e., $q^{A1} \preceq " q^{A2} \leftrightarrow A_1 \subseteq A_2$);*
- *meet operator as union of the grouping attributes;*
- *join operator as intersection of the grouping attributes;*
- *the query grouping on all the foreign keys as top element;*
- *the query computing the aggregate function on all the tuples of F as bottom element (empty set of grouping attributes).*

Requested queries are associated with a subset of the views of the data-cube lattice of MDDB. In Agrawal, Chaudhuri, & Narasayya (2000), the subset of views is considered which appears in the execution plans of representative queries.

> **Example 2:** *Consider again the MDDB of Example 1. Figure 2 represents the data-cube lattice derived from the fact table F and shows the elements that have a query associated.*

The Multidimensional Lattice

The presence of dimensions significantly increases the size of the problem. The first aspect is the growth in the number of potential grouping attributes, which exponentially increases the number of elements of the lattice. The second aspect is the presence of hierarchies, which permits removal of some elements from the lattice. In fact, consider a query grouping on a dimension key d_i and also on an attribute a_j of the same dimension D_i. Since there exists a functional dependency from d_i to a_j, a query grouping on $\{d_i, a_j\}$ must produce the same result of the query grouping on $\{d_i\}$. This observation is the basis for the following generalization.

Definition 7: *Let q_i^{Ai} and q_j^{Aj} be two queries and FD_{DB} the MDDB attribute hierarchy. The operator* **ancestor** *(represented by the symbol $\oplus$) is defined by the following algorithm:*

Algorithm 1: Ancestor of two queries.

operator $\oplus$: $q_x^{Ax} \oplus q_y^{Ay} \Rightarrow q_z^{Az}$;
begin
    $A_z := A_x \cup A_y$;
    for each fd$_i \in FD_{DB}$
        for each $a_j \in A_i^r$
            if $(\{a_i^l\} \in a_j) \subseteq A_z$
                $A_z := A_z - a_j$;
return q^{Az};
end;

Figure 2: Data-Cube Lattice with Associated Queries

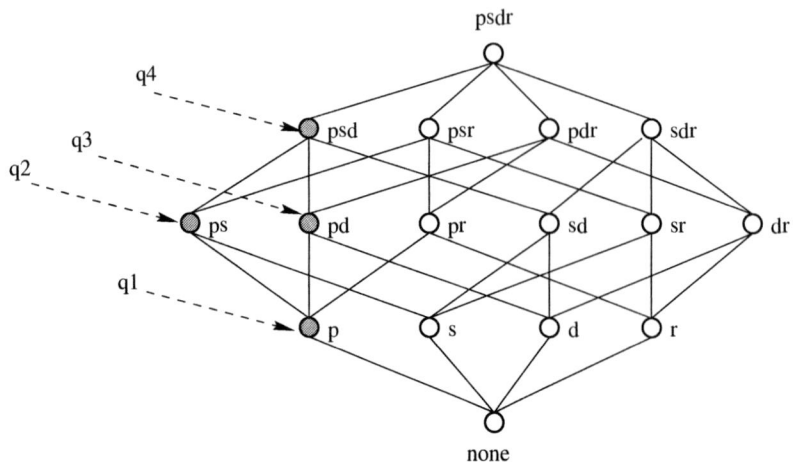

Algorithm 1 operates by building the union of the attributes characterizing the queries and eliminating all the elements for which there exists a functional dependency in FD_{DB}.

The result of applying the operator $\oplus$ to queries q_x and q_y is the smallest view that contains all the information necessary for answering q_x as well as q_y. If applied in a reflexive way (i.e., $q_x^{Ax} \oplus q_x^{Ax}$), it eliminates all redundant attributes.

Example 3: *Consider the* Store *dimension of the MDDB described earlier, with key s and restricted to a set of attributes {z, c, st, n} representing respectively zip code, county, state, and number of sale clerks. The dimension has the following tree-like attribute hierarchy: { fd_0: s →{z, c, st, n}, fd_1: z → c, fd_2: c → st}.*

The queries $q^{\{n\}}$, $q^{\{c\}}$, and $q^{\{st\}}$ can for example be computed from $q^{\{c,n\}}$ = $q^{\{n\}} \oplus q^{\{c\}} \oplus q^{\{st\}}$.

Definition 8: *The operator* **descendent** *(represented by symbol $\ominus$) is defined by the following algorithm:*

Algorithm 2: Descendent of two queries.

operator $\ominus$: $q_x^{Ax} \ominus q_y^{Ay} \Rightarrow q_z^{Az}$;
begin
    for each $fd_i \in FD_{DB}$
        if $\{a_i^l\} \subseteq A_x$
          $A_x := A_x \cup A_i^r$;
        if $\{a_i^l\} \subseteq A_y$
          $A_y := A_y \cup A_i^r$;
      $A_z := A_x \cap A_y$;
      return $q_z \oplus q_z$;
end;

Algorithm 2 first extends the parameters A_x and A_y with all the attributes that can possibly be derived from them. It then considers their intersection A_z and finally removes from it the right sides of dependencies whose left side is contained in A_z.

The descendent operator computes the "greatest" among the set of attributes characterizing the queries that can be computed by both q_x and q_y.

Definition 9: *Let $\{D_1, ..., D_n, F\}$ be a multidimensional database and FD_{DB} the MDDB attribute hierarchy. Consider the set of queries characterized by all the combinations among the attributes of $\{D_1, ..., D_n, F\}$, except the combinations that contain elements on both sides of a functional dependency $fd_i \in FD_{DB}$. This set of queries identifies a lattice where:*

- $\oplus$ is the meet operation;
- $\ominus$ is the join operation;
- the ordering relation is given by the following definition: $q^{A1} \preceq q^{A2} \leftrightarrow (A_1 \ominus A_2) = A_1 \leftrightarrow (A_1 \oplus A_2) = A_2$;
- the query grouping on all the foreign keys of F, $\{d_1, ..., d_n\}$, is the top element;
- the query computing the aggregate function on all the tuples of F is the bottom element (empty set of grouping attributes).

We call this lattice the **MD-lattice** *(multidimensional lattice).*

Comparing Definitions 6 and 9, it is easy to observe that without hierarchies the MD-lattice is equivalent to the Cube-lattice built on all the schema attributes.

Example 4: *Consider the* Store *dimension of Example 3. The MD-lattice for an MDDB where* Store *is the only dimension is represented in Figure 3.*

An MD-lattice defines all possible ways of computing queries that can be defined on an MDDB in terms of other queries defined also on MDDB. In fact, if a query $q_i \preceq q_j$ then q_i can be answered using q_j. This property can be generalized.

Definition 10: *Let* q_i *and* q_j *be two queries on an MDDB. Then, the* **least upper bound** *(l.u.b.) of* q_i *and* q_j *in the MD-lattice is the query obtained as a result of evaluating* $q_i \oplus q_j$, *and it represents a query (the most specialized) which can be used to answer both* q_i *and* q_j.

We now distinguish between queries and views. In the following we will use the term query to refer to the representative queries provided by the user, while we will

Figure 3: MD-Lattice of the Store Dimension

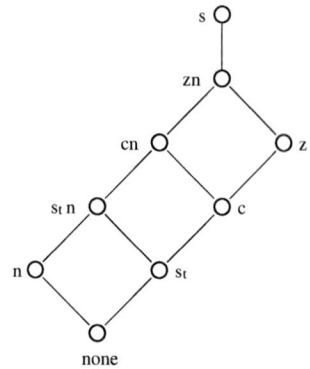

use the term view to refer to the elements of the lattices. When a query q computes the content of a view v of a lattice, we say that the query q is associated with view v. We also extend to views the notation for the queries and say that a view is characterized by a set of attributes A if A is the set of grouping attributes of v, represented as v^A.

> **Definition 11:** *A query q_i is said to* **depend** *on view v_j if the query can be answered using as only inputs the content of v_j together with any appropriate dimension D_i.*

There is a strict relationship between dependence of a query on a view and the ordering relation on an MD-lattice. In fact, if we consider a query q_i depending on view v_j and the view v_i associated with the query, it must happen that $v_i \preceq v_j$.

An important relationship exists between the ordering of the elements in the lattice given by relation $\preceq$ and the cardinalities (i.e., number of tuples) of the views. In fact, for each pair of views $v_i, v_j \in$ *MD-lattice*, if $v_j \prec v_i$, v_j can be computed by v_i with a further grouping. Since grouping can only reduce the number of tuples, it follows that $|v_j| \leq |v_i|$, where $|v|$ represents the cardinality of v.

Determining the Number of Nodes of an MD-lattice

The number of views of an MD-lattice depends on the number of attributes of the dimensions of the MDDB and on the number and structure of the attribute hierarchies.

In the absence of hierarchies on the dimensional tables, i.e., when each dimension has a hierarchy that contains only the functional dependency between the key of the dimension and all its attributes, the number of views of the MD-lattice is given by the formula:

$$n_{total} = \prod_i (2^{n_i} + 1)$$

where i is the number of dimensions of MDDB and n_i is the number of non-key attributes of each dimension D_i. In fact, the number of views of the MD-lattice is obtained as the product of the number of views of the lattices that can be built on every dimension. Each dimension D_i generates a lattice which has 2^{n_i} elements representing all the possible combinations of n_i attributes (exactly like a Cube-lattice), plus one view which represents the view characterized by the key d_i.

In the presence of hierarchies, the number of views in the local lattices is reduced and the above formula for n_{total} constitutes an upper bound. A formal treatment of the cardinality of the MD-lattice in the presence of hierarchies is beyond the scope of this chapter. We only provide as an example the determination of the number of elements for the MD-lattice of the MDDB previously described.

Example 5: *Consider the MDDB described earlier in this chapter. The number of views of the lattices that can be built on every dimension are:* $n_{Product}$ *= 12,289;* n_{Store} *= 8,193;* n_{Time} *= 129;* $n_{Promotion}$ *= 1025. The total number of views of the resulting MD-lattice is then the product of all these values, obtaining* $|MD\text{-}lattice| = 1.3313 \cdot 10^{13}$.

MAINTENANCE OF MATERIALIZED VIEWS

Maintaining a materialized view means to update it according to the changes which occurred to the base tables it refers to. Many techniques have been created for the efficient maintenance of materialized views. This has recently been an extremely active research area, which is well summarized in Gupta & Mumick (1999), which collects the most significant contributions to the area. The problem of efficient view maintenance can be separated in the two problems of computing the updates that have to be applied to the materialized view, and the problem of applying the update to the materialization. In general, maintenance can be immediate (updates are propagated from the base tables to the materializations as soon as they have occurred) or may be deferred (following one of many policies).

In this context we pay attention to the deferred choice, as we refer to the data warehouse environment. Among the many existing proposals, we succinctly describe the most important ones. The book (Gupta & Mumick, 1999) is fully dedicated to the topic of view maintenance and presents all of these solutions in detail.

One of the first proposals appears in Blakeley, Larson, & Tompa (1986), where a technique for the incremental maintenance of select-project-join views is presented. The paper first identifies criteria that identify irrelevant updates, i.e., updates on the base tables that cannot possibly have an impact on the view materialization. Then, it presents the use of counters to determine the number of computations from the base tables that justify the presence of a derived tuple in a view. The technique then demonstrates that the variations to the counters is derivable by substituting the sets of inserted/deleted tuples into the view definition formula. Many solutions have been identified for the incremental maintenance of recursive views (see, for example, Dong & Su, 2000). Other situations that received the interest of many researchers are views with duplicates (Griffin & Libkin, 1995), mediated data (Lu et al., 1995), chronicles (Jagadish, Mumick, & Silberschatz, 1995), and hypertext views (Sindoni, 1998; Labrinidis & Roussopoulos, 1999).

Focusing on our environment, the important contributions are the ones offered by Gupta, Mumick, & Subrahmanian (1993) and by Mumick, Quass, & Mumick (1997). In the first contribution, the authors present two techniques: a counting technique applicable to generic views, and a technique for recursive views. The counting technique extends the proposal of Blakeley, Larson, & Tompa (1986) to views defined using selection, projection, negation, union, and aggregations. The idea of the counting technique is to associate with every tuple in the materialized view a counter which represents the number of derivations for that tuple. Decomposability of operations guarantees that the determination of the update on the view can be obtained by comparing the updates to a base table with the content of all the other tables.

Mumick, Quass, & Mumick (1997) focus on the problem of the efficient maintenance of materialized views in a context similar to the one we depict in this chapter, where views are called *summary tables*. The solutions presented are all based on the observation that the regular structure of MDDB queries can be exploited to design efficient view maintenance techniques. The paper first identifies the two main issues that characterize the propagation of updates in the multidimensional data model we have introduced. The first problem is the propagation to a single summary view of the updates on the base table. The second problem is how to manage the presence of many summary tables.

The maintenance of a node of the MD-lattice can be based on the observation that only updates to the fact table are significant. We may assume that referential integrity constraints forbid the occurrence of tuples in the fact table that present a value for the dimension key which is not present in the corresponding dimension. This means tuples removed from a dimension cannot correspond with tuples in the fact table, and insertions into the fact table have to correspond to tuples in the dimension. Given the restrictions on the views appearing in the MD-lattice, operations on the dimension will not have an impact on the view, as long as updates on the fact table are always correctly managed. Consider dimension *Store* in the MDDB example and a generic view q_A. If referential integrity holds, insertions into *Store* will not contribute to q_A, as they have no corresponding element in the fact table; dually, deletions will only occur when no corresponding tuple is present in the fact table, thus they will not have an impact on q_A. We thus focus on updates to the fact table. To keep the model simple, we only consider insertions and deletions; an update is represented as a deletion followed by an update.

We now reconsider the classification of aggregate functions introduced in Chapter 2. Distributive and algebraic functions are immediately usable and permit an efficient recomputation of the view. Holistic functions limit the family of updates that can be propagated. An important property that distributive and algebraic functions offer is *self-maintainability*, i.e., the view can be updated based only on the analysis of its current state and of the updates to the fact table. In the example the sum function was used to compute the total sales amount. Since the sum function is distributive, it is possible to self-maintain a node of the lattice. Consider the node q_A that aggregates on a subset of the keys of the dimensions: the view can be updated considering the inserted/deleted tuples one by one and adding/subtracting the sales value represented in the generic inserted tuple t to the total stored in the view tuple t_v characterized by the same values of t for all the dimension keys. For example, an insertion of a tuple *Sale(StoreId: 15, Product: 58, Time: 1253, Promotion: 35, Amount: 120)* can be applied on MD-lattice node $q_{Store,Promotion}$ as an increase of 120 of the content *currentTotal* of the view tuple *(Store: 15, Product: 58, Total: currentTotal)*. It may happen that a tuple with the searched values is not present in the view. In this situation, a new tuple will be inserted into the view. In the example, a tuple *(Store: 15, Product: 58, Total: 120)* will be inserted into $q_{Store, Promotion}$.

For holistic functions, the situation is more variegated. There are aggregate functions that do not permit self-maintainance of views and always require consideration of the content of the *Sales* table (like the median and count(distinct)); the max and min functions instead exhibit a peculiar behavior, as they allow self-maintenance for insertions, whereas they may require access to the base table for deletions. This behavior derives directly from the analysis of the properties of min and max: if the materialization keeps only the min (or max) value combined with the counter, when a new value is inserted, it is sufficient to increment the counter and to determine if the new value is smaller (respectively, greater) than the value currently stored; in this case, the view is updated. For deletions, when the deleted value is different from the one stored in the view, the update does not require propagation; when the value to remove is exactly the one stored in the view, the value has to be recomputed considering all the possible values on the base tables. In some contexts, deletions may be considered as non-applicable and min or max views become self-maintainable.

A technique improving update propagation, proposed in Mumick, Quass, & Mumick (1997), requires the computation of a synthesis of updates from all the updates that are propagated up in the MD-lattice. We illustrate the idea with a simple example. Let us suppose that in our *Sales* MD-lattice, a set of sales is inserted.

Suppose we materialized the queries of Example 2. The propagation of this update to each materialized view is performed by the execution of two functions: a *propagate* function and a *refresh* function. The propagate function creates a view-delta table from the above set of changes: it represents the net changes to the materialized view due to the changes in the fact table. The refresh function applies the net changes to the materialized view. This also has the advantage of leaving the warehouse available for querying by clients during the propagate phase.

The maintenance of multiple materialised views is optimized by this technique by using a view-delta to compute view-deltas for other materialised views. In particular, once the view-delta for the view which is in the higher position in the lattice is computed, the remaining deltas are computed using their ancestor delta. So, in our example, the view delta for q_4 is computed, then the view-deltas for q_3 and q_2 are computed in terms of the one for q_4, and finally the view-delta for q_1 can be computed either in terms of that for q_3 or that for q_2.

Product	Store	Time	Promotion	Amount
Britney Spears GH	Milan	2002/10/10	Buy Music	14
Mozart K441	Milan	2002/10/20	Classic4all	9
Britney Spears GH	Rome	2002/10/10	Buy Music	14

Update Cost

The cost of performing the insertion of a tuple into a base table may be split into two parts: a) the cost of writing the inserted tuple in the affected table, and b) the cost of propagating the insertion to all the materialized views that are affected by the change. When the insertion is performed on the fact table, part "b" requires either the insertion of a new tuple, or the update of the aggregate value(s) for one tuple in each group-by. We assume that when a tuple is inserted on the fact table, it is immediately joined with all the dimensions. All the materializations are then accessed to apply the effect of the new fact on the materializations.

If we follow the model of Harinarayan, Rajaraman, & Ullman (1996), we can use the number of tuples read or written also as a metric for the update cost. We may assume that an access to an index costs as a tuple read. Then, the update cost $c_u(m_j)$ for a materialization m_j can be modeled considering an index read to identify the position of the tuple (requiring in general $log_x(|m_j|)$ reads, where x is a numerical value depending on the implementation) and a write of the updated tuple. When instead a tuple is inserted into a dimension table, the referential integrity constraint guarantees that no tuple is present in the fact table with the same identifier value. Hence, this insertion does not need to be propagated on materialized views.

Update and delete operations on both dimension and fact table are very rare. Hence, we do not include them in our cost model. When any such operation is performed, recomputation of several materialized views may be necessary.

Since the updates on the dimensions are not propagated and insertions into the fact table generate the update of a tuple in all the materializations, we may simplify our cost model assuming a unique update frequency f_u valid for all the materializations, where f_u represents the frequency of insertions into the fact table. The update cost function is:

$$C_M = f_u \times \sum_{m \in M} c_u(m_j)$$

Definition 12: *An update cost function c_u is* **monotonic** *if for all m_j, $m_k \in M$, $|m_j| < |m_k| \rightarrow c_u(m_j) \leq c_u(m_k)$, where $|m_j|$ represents the number of tuples of m_j. The update cost function we have described is monotonic.*

FINDING THE OPTIMAL SET OF MATERIALIZATIONS

When using materialized views in a MDDB, the most important problem to solve is to find the set of materializations that maximizes the performances of the MDDB in answering a given set of representative queries. The trade-off consists of choosing a set of materializations able to speed up query response time without requiring too much work to keep the materializations current with respect to the modifications on the tables of the MDDB.

> **Definition 13:** *Given an MDDB DB, a set of queries Q, and a set of frequency values F of queries in Q and updates on the tables of DB, the* **MDmat-problem** *(multidimensional database materialization problem) is represented by* $\Theta(DB, Q, F)$. *A solution to the problem* $\Theta(DB, Q, F)$ *is a set of views of the MD-lattice M (which can contain views that are not associated with queries in Q). A trivial solution, M = Ø, is always possible, which represents the situation where no additional materialization is available and all the queries must be answered directly by the fact table (the root of the hierarchy).*

To identify the optimal *M*, we must define a cost function. The optimal solution is the one characterized by the minimum cost. The cost function is composed of two parts: the *query cost* and the *update cost*.

> **Definition 14:** *Consider a set of queries Q specified by the user. Let F be a set of frequencies* f_{q_i}, *each associated with a query* $q_i \in Q$, *representing the frequency with which query* q_i *is asked. Let* $c_{q_i}(M)$ *be the cost to compute* q_i *from the set of materializations M (discussed in the second section).*

> *Then, the* **total query cost** $C_Q(Q, M, F)$ *is given by:*

$$C_Q(Q, M, F) = \sum_{q_i \in Q} f_{q_i} \cdot c_{q_i}(M)$$

> **Definition 15:** *Consider the same set of materialized views M. Let* f_{m_i} *be the frequency with which the materialized view* $m_i \in M$ *is modified and* $c_u(m_i)$ *its update cost. Then, the* **total update cost** $C_M(M, F)$ *is given by:*

$$C_M(M, F) = \sum_{m_i \in M} f_{m_i} \cdot c_u(m_i)$$

> **Definition 16:** *Given an MDmat-problem described by* $\Theta(DB, Q, F)$, *the* **cost of a solution** *M is the sum of query and update costs:*

$$C(Q, M, F) = C_Q(Q, M, F) + C_M(M, F)$$

Query Cost

We do not present here a complete cost model. The techniques we describe are applicable to a wide choice of cost models, from simple to complex ones. Very few restrictions have to be imposed on the cost formulas to permit the adoption of our results.

The function $cq_i(M)$ returns the cost of computing query q_i given a set of materializations M. We make two hypothesis about the query cost function: that each query cost depends on a unique element in M, and that the cost is monotonic with the size of the materialization on which the *query depends*.

> **Definition 17:** *A query cost function $cq_i(M)$ is restrictible if it is always equal to the least among the values obtained by considering $cq_i(m_j)$, for all the $m_j \in M$ on which query q_i depends.*

Every query of the type introduced in the fourth section of this chapter is restrictible (since the fact table is always present in the set of materializations).

> **Definition 18:** *Given a query q_i, a restrictible query cost function cq_i, and a set of materializations M, the materialization m_j such that $cq_i(M) = cq_i(m_j)$ is the* **least expensive materialization** *(for query q_i among the elements of M).*

> **Definition 19:** *A query cost function cq_i (MD-lattice) is* **monotonic** *if for all $v_j, v_k \in$ MD-lattice on which q_i depends, $|v_j| < |v_k| \rightarrow cq_i(v_j) \le cq_i(v_k)$, where $|v|$ represents the cardinality of v and the arrow denotes logical implication. From the observation on the cardinalities of the views in the lattice, a monotonic function also guarantees that $v_j \preceq v_k \rightarrow cq_i(v_j) \le c_{qi}(v_k)$.*

The simple cost model introduced in Harinarayan, Rajaraman, & Ullman (1996) is, for example, both restrictible and monotonic: the cost of answering a query q is set equal to the number of tuples read to return the answer. This cost model, although very simple, well characterizes the problem. An extension of this model permits consideration of the availability of indexes to accelerate the execution of queries, as described in Gupta et al. (1997). We do not treat indexes here, but a technique similar to Gupta et al.'s (1997) can be applied in this context as well, as it is also demonstrated in Agrawal, Chaudhuri, & Narasayya (2000).

The extension to an MDDB and join operations may require modification of this simple metric, which is still adequate when it is possible to keep in memory all the dimensions involved in the join. In fact, in this case the join requires reading only once all the tuples, and since the cardinality of the fact table is greater than that of any dimension, often by several orders of magnitude (Kimball, 1996), the cost of performing a join operation between any dimension and the fact table F can be modeled as the cost of reading F. When the join is more complex, a different cost function may be required. Sophisticated cost functions can be designed which adequately model this aspect of the system and still respect the requisites of restrictibility and monotonicity.

The cost of performing a group-by operation can be modeled as the cost of reading the table or materialized view on which the grouping is computed. This hypothesis does not consider the sort order which is used to store each materialization (see Agarwal et al., 1996), which has a relevant impact on the complexity of computing the aggregations; this aspect can also be modeled by a restrictible and monotonic cost function.

Identification of Candidate Views

Compared to the number of nodes that form the MD-lattice, the number of representative queries is extremely small. This considerable sparsity of queries among the views of an MD-lattice suggests that only some of these views are relevant when deciding which views to materialize in order to minimize the total cost. The idea of our reduction technique is to consider only those views of an MD-lattice that, when materialized, can provide some contribution to reduce the total cost. We call them candidate views, and define them as follows.

> **Definition 20:** *A view v_i belonging to an MD-lattice is a* **candidate view** *if one of the following two conditions holds:*
> * View v_i is associated with some query q_i.
> * There exist two candidate views v_j and v_k, such that v_i is the least upper bound (l.u.b.) of v_j and v_k.

Note that, from the previous definition, it follows that non-candidate views do not have an associated query.

> **Definition 21:** *Given a non-candidate view v_i with at least one candidate view depending on it, we call* **most directly dependent candidate view** *the unique candidate view v_j such that all the remaining candidate views that depend on v_i also depend on v_j.*

We can determine the most directly dependent candidate view of v_j by taking the set V' of all the candidate views that depend on v_j and computing the l.u.b. of all the views in V'. This view is unique (because an l.u.b. is always determined), is a candidate view (because it is an l.u.b. of candidate views), and belongs to V' (because $v_a \preceq v_i \wedge v_b \preceq v_i \rightarrow (v_a \oplus v_b) \preceq v_i$).

Intuitively, it is not difficult to see that candidate views may provide some benefit if they are chosen to be materialized. If the view is associated with a query, materializing it will decrease the total cost if the increment of the update cost is compensated by the reduction on the query cost. When the view is the l.u.b. of other candidate views v_j and v_k, which may be associated with queries q_j and q_k, the total cost will be minor if the cost needed to answer q_j and q_k using a materialization of

v_i is compensated by the reduction on the update cost of the materializations of v_i and v_j. We can analyze more formally these properties.

Let v_i be a candidate view of an MD-lattice. Then, choosing v_i for materialization may provide some benefit when looking for the solution that reduces the total cost. There are in fact two cases:

1. v_i has an associated query q_i.

It is trivial to show that the materialization of a view associated with a query can help the computation of the query. Starting from the definition of the cost of a solution, we obtain the following formula, which identifies the query frequency value fq_i that makes convenient the materialization of view v_i when a set of views M is already materialized:

$$f_{qi} > f_u \cdot c_u(v_i) / (cq_i(v_t) - cq_i(v_i))$$

$v_t \in M$ represents the least expensive materialization that can be used to answer q_i.

2.   There exist at least two candidate views, v_j and v_k, such that v_i is the l.u.b. of v_j and v_k.

It is enough to show that there exists at least one case in which materializing v_i provides some benefit. Assume that there exist two queries q_j and q_k associated with views v_j and v_k, respectively. The contribution of queries q_j and q_k and views v_j and v_k to the cost $C(Q, M, F)$, when v_j and v_k are materialized and v_i is not, is:

$$C_1 = f_u \cdot c_u(v_j) + fq_j \cdot cq_j(v_j) + f_u \cdot c_u(v_k) + fq_k \cdot cq_k(v_k)$$

The contribution to the cost $C(Q, M, F)$ if v_i is materialized and v_j and v_k are not, with v_i being the least expensive materialization for both q_j and q_k, is:

$$C_2 = f_u \cdot c_u(v_i) + fq_j \cdot cq_j(v_i) + f_u \cdot c_u(v_k) + fq_k \cdot cq_k(v_i)$$

Choosing v_i for materialization will decrease the total cost if $C_1 > C_2$, i.e., when:

$$f_u > (fq_j \cdot (cq_j(v_i) - cq_j(v_j)) + fq_k \cdot (cq_k(v_i) - cq_k(v_k))) / (c_u(v_j) + c_u(v_k) - c_u(v_i))$$

with the hypothesis that $c_u(v_j) + c_u(v_k) - c_u(v_i) > 0$. The intuition is that if the cost of updating views v_j and v_k is greater than the cost of updating v_i, for a high enough update frequency f_u, it may be convenient to materialize view v_i instead of v_j and v_k.

Moreover, if we want to ensure that candidate views are the only relevant views for the process of deciding which views to materialize, we must prove also that materializing a non-candidate view may never decrease the total cost. This is done in Theorem 1.

Theorem 1: *Let v_i be a non-candidate view of an MD-lattice. Then, the choice of v_i for materialization is always dominated by the choice of a candidate view.*

Proof: We will assume that there exists a non-candidate view v_i that belongs to the set of materializations M of the optimal solution, and we will get a contradiction. We distinguish two cases.

1. There is no candidate view depending on v_i.

    The contribution of view v_i to the cost $C_1 = C(Q, M, F)$ of the solution M for $\Theta(DB, Q, F)$ ($v_i \in M$) is represented only by the contribution to the update cost $f_u \cdot c_u(v_i)$, since no query can use v_i (otherwise there would be candidate views depending on v_i). The cost of the solution $M - v_i$ is equal to $C_2 = C(Q, M - v_i, F)$, where $C_1 = C_2 + f_u \cdot c_u(v_i)$. Since materializing v_i must provide some benefit, it must happen that $C_1 < C_2$, i.e., $f_u \cdot c_u(v_i) < 0$. Then, we get a contradiction because both f_u and $c_u(v_i)$ must be positive.

2. There exists at least one candidate view depending on v_i.

    We first identify view v_j, the most directly dependent candidate view of v_i. We then consider separately two sub-cases, represented in Figures 4 and 5.

    Case 1: Both v_j and v_i are materialized (represented in Figure 4). Let $M' = M - v_i - v_j$. The cost of this solution is:

    $$C_3 = \sum_{q_i \in Q} f_{qi} \cdot c_{qi}(M' \cup v_j \cup v_i) + f_u \cdot \sum_{v_k \in M' \cup v_j \cup v_i} c_u(v_k)$$

    We observe that view v_i is not used by any query $q_i \in Q$, because all the queries that depend on v_i also depend on v_j, and since $v_i \precsim v_j$, it follows that $c_{qi}(v_i) > c_{qi}(v_j)$. We can then remove it from the first term of the formula for C_3 obtaining:

    $$C_3 = \sum_{q_i \in Q} f_{qi} \cdot c_{qi}(M' \cup v_j) + f_u \cdot \sum_{v_k \in M' \cup v_j \cup v_i} c_u(v_k)$$

    The cost of the solution with materialization $M' \cup v_j$ is:

Figure 4: A Configuration of Materializations

$$C_4 = \sum_{q_i \in Q} f_{qi} \cdot c_{qi}(M' \cup v_j) + f_u \cdot \sum_{v_k \in M' \cup v_j} c_u(v_k)$$

The difference between C_3 and C_4 is represented by the single term $f_u \cdot c_u(v_i)$. In order for C_3 to be the optimum (i.e., $C_3 < C_4$), it must happen that $f_u \cdot c_u(v_i) < 0$. We get a contradiction since both f_u and $c_u(v_i)$ must be positive.

Case 2: v_i is materialized and v_j is not (represented in Figure 5). Let $M' = M - v_i$. The cost of this solution is:

$$C_5 = \sum_{q_i \in Q} fq_i \cdot cq_i(M' \cup v_i) + f_u \cdot \sum_{v_k \in M' \cup v_i} c_u(v_k)$$

In particular, this solution must be better than the solution where v_j is materialized but v_i is not, which has the cost C_4 defined in the above formula. But each term $cq_i(M' \cup v_j)$ in C_4 must be less than $cq_i(M' \cup v_i)$ in C_5, because all the queries that depend on v_i must also depend on v_j; v_i is smaller than v_j and the query cost function is monotonic. Term $c_u(v_j)$ must also be smaller than $c_u(v_i)$, for the monotonicity of update cost with view size. It is then impossible for C_5 to be smaller than C_4.

Once we have proven that candidate views are the only views that are relevant to the process of deciding which materializations minimize the total cost, we can define the sub-lattice obtained by considering only candidate views.

Figure 5: A Configuration of Materializations

Definition 22: *Given an MD-lattice and a set of queries Q, the set of its candidate views identifies also a lattice where the join and meet operations are identical to the ones defined for the MD-lattice. We call this sub-lattice the* **MDred-lattice**.

Given a set Q of queries and the MDDB attribute hierarchy, we can identify all the elements of the MDred-lattice. This is performed by means of the following algorithm.

Algorithm 3: *The MDred-Lattice Construction Algorithm*

```
Function MDred-lattice(Q): <set of elements>;
/* input: a finite set Q of queries */
/* output: the MDred-lattice obtained by Q */
begin
    L := Q;
    lastViews := L;
    newViews := Ø;
    while lastViews ≠ Ø
        for each v_i ∈ lastViews do
            for each v_j ∈ L, v_j ≠ vi do
                if v_i ⊕ v_j ∈ L
                    newViews := newViews ∪ (v_i ⊕ v_j);
                end for;
            end for;
        L := L ∪ newViews;
        lastViews := newViews;
        newViews := Ø;
    end while;
    return L;
end;
```

Algorithm 3 iteratively extends the set L. L is initially equal to the set of views in Q. The l.u.b. of all the pairs of elements in L are added to L, and the process iterates by considering the l.u.b. of all the pairs of views obtained by combining the new elements of L with all the elements of L, until a fixpoint is reached where no new l.u.b. can be added to L.

It is now convenient to describe another technique based on statistical estimates of the size of the materializations. Given its statistical foundation, the next technique is not as exact as the one just presented.

A Heuristic Reduction

When building the MDred-lattice, a simple heuristic technique can be used to further reduce the size of the lattice, removing views which are not expected to contribute to the optimal solution. This heuristics is based on the estimate of the size of the materialization. These estimates can be done following traditional query estimation techniques for aggregates. If the skewness of data is of concern, it is also possible to use the technique, described in Shukla et al. (1996), which obtains an estimate of the number of tuples that is quite precise for a wide range of skewness in the distribution.

According to the estimates on the size of views, it is possible to determine when the level of aggregation used is too detailed and the materialization offers very limited help in answering the query with respect to a materialization of a higher level view.

Example 6: *Consider a dimension A with 1,000 tuples. Let a view contain an aggregation for the pair of attributes $\{A_1, A_2\}$, where each attribute has 100 distinct values. There are 10,000 possible pairs of values of the attributes, but since there are only 1,000 tuples, at most 1,000 tuples can be present in the view. If the data is uniformly distributed, the estimate of the size of the view is quite close to 1,000. Instead of materializing this view, it could be convenient to use the view which has the key of dimension A as aggregating attribute. The great advantage is that it will be easier to reuse this view in the computation of other aggregates and the number of elements of the lattice will be, in general, reduced.*

We can easily modify the MDred-lattice construction algorithm to estimate at each step the dimension of a view, and substituting to it a higher level view if the reduction criteria are not met. In the next section we illustrate a few experimental results. We now describe more formally this technique.

Definition 23: *A **size estimating function** size(A) is a function that, applied to a set of attributes A of an MDDB, returns an estimate of the number of tuples of the result of an MDDB query q characterized by A (i.e., having A as grouping attributes).*

Definition 24: *A size estimating function is **correct** if for all the functional dependencies $fd_i \in FD_{DB}$, $size(A^r_i) \neq size(A^l_i)$.*

Intuitively, since there exists a functional dependency $A^l_i \rightarrow A^r_i$, for each combination of values of A^l_i, more than one combination of values for the attributes in A^r_i cannot exist. Thus, a query grouping on A^l_i cannot have more tuples than a query grouping on A^r_i.

Algorithm 4: *Heuristic Reduction Algorithm*

function heuristicRed(A,p): *<set of attributes>*;
begin
   repeat
     Stop := True;
     for each $fd_i \in FD_{DB}$
       if $size(A \cap A^r_i) < \rho \cdot size(A^l_i)$
         $A := A - A^r_i$;
         $A := A \cup A^l_i$;
         Stop := False;
     end for;
   until Stop;
   return A;
end;

Algorithm 4 considers all the functional dependencies to identify when the attributes in A that appear on the right side (or a subset of them) of a functional dependency produce a size estimate which does not differ more than a ratio ρ from the estimate of the size of the attributes on the left side. When this happens, the attributes on the right side are replaced by the attributes on the left side, effectively moving from a view v_i characterized by the set of attributes A to a view $v_j \preceq v_i$.

The parameter ρ represents the threshold on the amount of reduction in size above which the left side of a dependency should be used in place of the right side. The user should determine this value, depending on a trade-off between accuracy of the solution and reduction in the size of the solution space. Even high levels (e.g., $r = 0,95$) can be quite effective in simplifying the solution space.

COMMERCIAL SOLUTIONS

The approaches presented in this chapter have rapidly found success in DBMS systems, and today have an important role in the physical optimization of databases. The environment where materialized views offer the greatest advantages are data analysis solutions, where a commercial DBMS is used to support an OLAP application.

We briefly illustrate here the main features of materialized views for the main actors of the DBMS market (Microsoft, Oracle, and Red Brick). We base this analysis on the papers written by researchers working for DBMS vendors and on the analysis of the technical documentation accompanying these systems. As we did previously in the chapter, we are not considering special non-relational solutions, which are based on a completely different physical model, requiring a different set of techniques.

The main DBMS offering by Microsoft is represented by the SQL Server product. This DBMS offers, since version 2000, an Index Tuning Wizard, which is technically illustrated in Agrawal, Chaudhuri, & Narasayya (2000). The tool is based on the consideration of a number of representative queries, extracted from a workload trace registered on a live system or generated by hand. The tool evaluates which combination of materialized views and indexes offer the most support in the computation of the representative queries. The tool is not specific for use in a data warehouse or multidimensional database context, and it arose in the framework of a Microsoft initiative for the construction of tools for the automatic management of database systems, which produced a preliminary version of the tool where only index selection was realized. It uses an approach similar to the one presented in this chapter, except that views are considered together with indexes on the views themselves and base tables as components that can benefit the computation of the cubes, using a richer set of components that can contribute to improve system efficiency. Heuristic techniques are then used, with a configurable degree of precision, to identify the configuration to adopt. The heuristics are based on preliminary exhaustive search on all the configurations with a number k of elements, which is then extended incrementally by a greedy heuristic to identify the best configuration of n terms.

The support that Oracle Server offers for materialized views is presented in Bello et al. (1998). Materialized views may be incrementally updatable (if the view is a join-only view or it computes an aggregate function); if the view is complex and the system is not able to identify an efficient incremental update strategy, the view must be rematerialized from scratch. Materialized views are useful for many uses of Oracle Server, but are able to offer the greatest advantages on a data warehouse solution.

The Vista component of Red Brick Warehouse Server (Bunker et al., 2001) (now owned by IBM, which acquired Informix) is a specialized solution for aggregate computation and management. Red Brick Server is a specialized DBMS for the construction of data warehouse systems, customized for the management of multidimensional database schemas. The Vista component exploits the specialization of the model and is well integrated with many of the services that characterize the server (data versioning, star indexes, etc.).

CONCLUSIONS

The reasons for the success of view materialization in multidimensional databases are twofold: first, the extremely high ratio between queries and updates, typical of all data analysis systems, makes considerably easier the task to maintain all the materialized views; also, the restricted variety of queries typically supported by multidimensional databases permits the design of sophisticated models that are able to represent well the contribution that a view materialization can offer to view computation.

This chapter presented the main problems that arise in the adoption of materialized views for data analysis systems using conventional DBMS technology. The importance of these techniques is demonstrated by their rapid adoption in industrial DBMS systems.

REFERENCES

Agarwal, S., Agrawal, R., Deshpande, P.M., Gupta, A., Naughton, J.F., Ramakrishnan, R., & Sarawagi, S. (1996). On the computation of multidimensional aggregates. *Proceedings of VLDB'99.* 506-521.

Agrawal, S., Chaudhuri, S., & Narasayya, V.R. (2000). Automated selection of materialized views and indexes in SQL databases. *Proceedings of VLDB'00,* 496-505.

Atzeni, P., Ceri, S., Paraboschi, S., & Torlone, R. (1999). *Database Systems.* City: McGraw-Hill.

Baralis, E., Paraboschi, S., & Teniente, E. (1997). Materialized views selection in multidimensional database. *Proceedings of VLDB'97,* 156-165.

Bello, R.G., Dias, K., Downing, A., Feenan Jr., J., Norcott, W.D., Sun, H., Witkowski, A,. & Ziauddin, M. (1998). Materialized views in Oracle. *Proceedings of VLDB'98,* 659-664.

Blakeley, J.A., Larson, P., & Tompa, F.W. (1986). Efficiently updating materialized views. *Proceedings of SIGMOD'86,* 61-71.

Bunker, C.J., Colby, L.S., Cole, R.L., McKenna, W.J., Mulagund, G., & Wilhite, D. (2001). Aggregate maintenance for data warehousing in Informix Red Brick Vista. *Proceedings of VLDB'01,* 659-662.

Dong, G., & Su, J. (2000). Incremental maintenance of recursive views using relational calculus/SQL. *Proceedings of SIGMOD Record, 29*(1), 44-51.

Gray, J., Bosworth, A., Layman, A., & Pirahesh, H. (1996). Data cube: A relational aggregation operator generalizing group-by, cross-tab, and sub-totals. *Proceedings of 12th International Conference on Data Engineering,* New Orleans, Louisiana, 152-159.

Griffin, T., & Libkin, L. (1995). Incremental maintenance of views with duplicates. *Proceedings of SIGMOD'95,* 328-339.

Gupta, A., & Mumick, I.S. (1999). *Materialized Views: Techniques, Implementations and Applications.* City: MIT Press.

Gupta, A., Mumick, I.S., & Subrahmanian, V.S. (1993). Maintaining views incrementally. *Proceedings of SIGMOD'93,* 157-166.

Gupta, H., Harinarayan, V., Rajaraman, A., & Ullman, J.D. (1997). Index selection for OLAP. *Proceedings of 13th International Conference on Data Engineering,* April, Manchester, UK, 208-219.

Harinarayan, V., Rajaraman, A., & Ullman, J.D. (1996). Implementing data cubes efficiently. *Proceedings of ACM SIGMOD International Conference on Management of Data,* June, Montreal, Canada, 205-216.

Jagadish, H.V., Mumick, I.S., & Silberschatz, A. (1995). View maintenance issues for the chronicle data model. *Proceedings of PODS'95,* 113-124.

Kimball, R. (1996). *The Data Warehouse Toolkit.* New York: John Wiley & Sons.

Labrinidis, A., & Roussopoulos, N. (1999). On the materialization of WebViews. *WebDB (Informal Proceedings),* 79-84.

Lu, J.J., Moerkotte, G., Schü, J., & Subrahmanian, V.S. (1995). Efficient maintenance of materialized mediated views. *Proceedings of SIGMOD'95,* 340-351.

Mistry, H., Roy, P., Sudarshan, S., & Ramamritham, K. (2001). Materialized view selection and maintenance using multi-query optimization. *Proceedings of SIGMOD'01,* 307-318.

Mumick, I.S., Quass, D., & Mumick, B.S. (1997). Maintenance of data cubes and summary tables in a warehouse. *Proceedings of SIGMOD'97,* 100-111.

Shukla, A., Deshpande, P.M., Naughton, J.F., & Ramasamy, K. (1996). Storage estimation for multidimensional aggregates in the presence of hierarchies. In *Proceedings of VLDB'96,* 522-531.

Sindoni, G. (1998). Incremental maintenance of hypertext views. *WebDB,* 98-117.

Chapter IX

Querying Multidimensional Data

Leonardo Tininini
Italian National Institute of Statistics, Italy

ABSTRACT

A powerful and easy-to-use querying environment is certainly one of the most important components in a multidimensional database, and its effectiveness is influenced by many other aspects, both logical (data model, integration, policy of view materialization, etc.) and physical (multidimensional or relational storage, indexes, etc.). As is evident, multidimensional querying is often based on the metaphor of the data cube and on the concepts of facts, measures, and dimensions. In contrast to conventional transactional environments, multidimensional querying is often an exploratory process, performed by navigating along the dimensions and measures, increasing/decreasing the level of detail and focusing on specific subparts of the cube that appear to be "promising" for the required information.

In this chapter we focus on the main languages proposed in the literature to express multidimensional queries, particularly those based on: (i) an algebraic approach, (ii) a declarative paradigm (calculus), and (iii) visual constructs and syntax. We analyze the problem of evaluation, i.e., the issues related to the efficient data retrieval and calculation, possibly (often necessarily) using some pre-computed data, a problem known in the literature as the problem of rewriting a query using views. We also illustrate the use of particular index structures to speed up the query evaluation process.

INTRODUCTION

As shown in the previous chapters, multidimensional data modeling is based on the metaphor of the *data cube* and on the concepts of *facts, measures,* and *dimensions.* Analogously, the techniques to retrieve such data, which have been proposed in the literature and/or implemented in commercial systems, are based on the idea of determining the cube of interest and then navigating along the dimensions, by increasing or decreasing the level of detail (through the well-known operations of *roll-up* and *drill-down*) or selecting specific subparts of the cube (through the operations of *slice and dice*).

The query languages for multidimensional data support both these standard and additional operations for performance of more sophisticated computations. As is common in the literature, we distinguish among:

(i)   languages based on an algebra (usually an extension of the relational one), where queries are expressed by using operators that apply to the tables representing the facts, measures, and dimensions;
(ii)   languages based on a calculus (again usually an extension of the relational one), where queries are expressed in a more declarative way;
(iii) visual languages, usually relying on an underlying algebra, and based on a more interactive and iconic querying paradigm; this is the approach of most commercial OLAP products.

We analyze the characteristics of the main query languages proposed in the literature, along with the specific advantages and drawbacks, also emphasizing the common features. Particularly, we consider:

(i)   query languages based on a relational representation of multidimensional data, hence based on extensions of the relational algebra and calculus;
(ii)   query languages based on specifically designed multidimensional models, usually based on an abstraction of cubes or fact tables, on which the operators of the algebra are applied.

In the case of query languages based on specific models, we show that, while typical OLAP operations are expressed in a very straightforward manner, the expression of typical relational operation can become cumbersome, as well as the capability to *symmetrically deal with dimensions and measures.* As a consequence, several studies have been focused on minimizing the "impedance mismatch" between relational and multidimensional models.

We also focus on the problem of query evaluation, i.e., on how the query expressed in the chosen language can be translated into an efficient evaluation plan, which retrieves the necessary information and computes the required results. As already stressed in the previous chapter, the choice of a collection of pre-computed results (materialized views) has dramatic consequences on the overall performances of the query evaluation process.

From an abstract point of view, the determination of the evaluation plan can be seen as the process of transforming the source query into an equivalent target one, referring (possibly only) to the materialized views, and it is known in the literature as the *query rewriting problem*. We present some results on the equivalence and rewriting of aggregate queries that can be used to optimize the evaluation of queries on multidimensional data. We also introduce a formalism, based on the concept of *numerical dependencies*, allowing the user to formally describe the way measures are obtained, along with the interrelationships among different measures and among measures at different levels of detail.

Finally, we briefly outline how specific techniques of indexing can significantly improve the query evaluation process in a multidimensional context. Unlike in the traditional transactional context, it is shown that the various types of indexes can perform very differently, depending on the type of queries to be computed. In other words there is no *best* index for OLAP applications, but rather a set of different indexing strategies, that perform well in some cases and are inadequate in others.

The chapter is organized as follows. We begin with a review of the main definitions related to multidimensional querying and introduce a running example which is used throughout the whole chapter. "Key Features of Multidimensional Querying" illustrates the main features of multidimensional queries, particularly the role of aggregation functions and dimensions, as well as the differences between querying on conventional transactional systems and in multidimensional querying environments. The use of the relational model and (possibly extensions of) relational query languages to extract multidimensional data is the focus of "Expressing Multidimensional Queries by Extended Relational Query Languages." The next section is devoted to query languages based on models specifically designed for multidimensional data. Several issues related to the efficient evaluation of multidimensional queries are investigated in "Multidimensional Query Evaluation," particularly the problem of rewriting a query using some materialized views, the efficient computation of cubes and materialized views, and the use of indexing techniques.

BACKGROUND

In this section we review the main terms and concepts related to multidimensional data. We first give some fundamental definitions, then introduce an example, which is used throughout the chapter to illustrate several issues related to multidimensional querying.

Basic Definitions

Raw data (microdata). Multidimensional data are obtained by applying aggregations and statistical analysis functions over elementary data, usually called raw data or, with statistical terminology, microdata. The data representing each individual of a census survey, each customer call of a phone company, each medical treatment in a hospital, are possible examples of raw data. In most cases raw data are the product of complex activities of source integration, which are covered in detail in Chapter 12. Particularly, data have to be cleaned and reduced to a common integrated schema, solving possible conflicts in terms of both schema and contents.

Classification. An important issue in the production of raw data is that of classifying the properties of data. In this way, possibly continuous values are associated with a finite number of values, corresponding to the maximum degree of detail, by which each property of data is going to be analyzed. For example, in census data the birth date is transformed to an age in years, in hospitalization data a specific disease is classified and transformed into a(n) (alpha-)numerical code according to an international classification table.

Aggregation/grouping. As mentioned above, multidimensional data are obtained by applying aggregations and statistical functions on raw data. For example the datum "Number of inhabitants in Italy in 2001" is obtained by considering the set of microdata collected by the Italian 2001 Census and then applying a Count function on this set. Hence, an aggregation is basically a function mapping sets (or—in some cases—multisets, i.e., sets with duplicated elements) to numerical values. In multidimensional databases, aggregations are rarely applied on the whole set of microdata, but rather on several groups of data, each containing a subset of the data, homogeneous with respect to a given set of attributes.

For example, the data "Average duration of calls in 2001 by region and call plan" is obtained from the raw data corresponding to the phone calls in that year. Several groups are defined, each consisting of calls made in the same region and having the same call plan, and finally applying the average aggregation function on the duration attribute of the data in each group. The pair of values (region, call plan) is used to identify each group and associated with the corresponding average duration value. In multidimensional databases the attributes used to group data define the *dimensions* (in statistical terminology the *category attributes*), whereas the aggregate values the *measures* (in statistical terminology the *summary attributes*) of data.

Fact table. In many contexts the fact table coincides with the table of raw data. However, many authors usually define the fact table as constituted by dimension codes and measures, thus assuming that some form of classification and pre-aggregation on the data has been applied. Generally speaking, the fact table is the table comprising the measures of interest and the dimensions at the finest (allowed for querying) level of granularity.

Dimensions and dimension hierarchies. The term *multidimensional data* is due to the well-known metaphor of the *data cube*: for each of the *n* attributes, which is used to identify a single measure, a dimension of an *n*-dimensional space is considered. The possible values of the identifying attributes are mapped to points on the axis of the dimension, and in this way each point of this *n*-dimensional space is mapped to a single combination of the identifying attribute values and hence to a single aggregate value. The collection of all these points, along with all possible projections in lower dimensional spaces, constitutes the so-called data cube. In most cases, dimensions are structured in hierarchies, representing several levels of granularity of the corresponding measures. Hence a time dimension can be organized in days, months, and years; a territorial dimension in towns, regions, and countries; a product dimension in brands, families, and types.

When querying multidimensional data, the user specifies the measures of interest and the detail of the required information, by indicating for each dimension the desired level in the hierarchy. Note also that in a multidimensional environment querying is often an exploratory process, where the user "moves" along the dimension hierarchies, by increasing or reducing the granularity of displayed data. The operation of *drill-down* corresponds to increase the detail of data, e.g., by requesting the number of calls by region and month, starting from data about the number of calls by region and year, or from data about the number of calls by region. Conversely, the operation of *roll-up* allows the user to view data at a coarser level of granularity.

Star and snowflake schemes. In relational OLAP systems (also called ROLAP systems), facts and dimensions are usually mapped to (relational) tables: a central fact table and *n* surrounding dimension tables. This typical structure for multidimensional data is called *star scheme* and depicted in Figure 1(a). In star schemes dimension hierarchies are flattened (de-normalized) to minimize the number of joins required to express the queries and retrieve the data. In contrast, snowflake schemes (Figure 1(b)) use a normalized version of dimensions where each level in a hierarchy is associated with a distinct table. Snowflake schemes offer a more precise way to describe dimension hierarchies, but make the queries more complex to express, since they often require long sequences of join conditions, even for conceptually simple requests.

ROLAP and MOLAP systems. As shown above, ROLAP systems use

Figure 1: Star and Snowflake Schemes

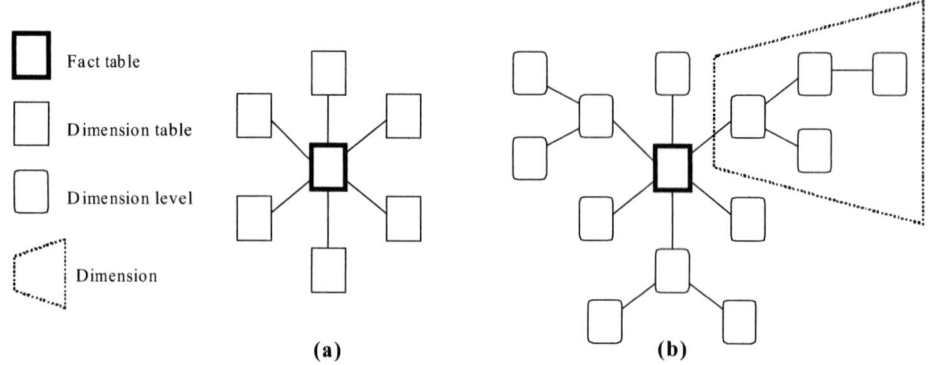

(a)   (b)

conventional relational tables to store fact tables. The schema of a fact table is constituted by n dimensional attributes (n being the number of dimensions) and m measure attributes, where m is the number of measures stored in the fact table. The collection of dimensional attributes constitutes the primary key of the fact table, hence each m-tuple of measures is identified by a combination of n dimensional values. In contrast, for the storage of the fact table, MOLAP systems use multidimensional arrays coupled with (often proprietary) techniques of clustering and compression. In this way the fact table is stored as a consecutive sequence of values, where the meaning of each value is defined by the position of the value in the sequence. This results in lower space occupation, and it has been exploited to achieve better performances in the retrieval process and query evaluation.

Running Example

We introduce an example, which is used throughout the chapter to illustrate the various issues related to multidimensional querying. A mobile phone company maintains a data warehouse containing the details on the traffic of its customers, in particular a table of raw data Calls on the schema:

(Id_call, Customer_no, Time, Ant_conn, Dur)

where:

- Id_call is an identifier of the phone call.
- Customer_no is the phone number of the calling customer.
- Time is the time (day/hour/minute/second), when the call started.
- Ant_conn is the antenna to which the calling customer was connected when the call started.
- Dur is the duration (in seconds) of the call.

The table Calls could be considered a fact table, where Dur is the measure and the other attributes (except id_call) are the dimensions. In practice, however, analyses are made at a level of detail that does not require a precision of seconds for the Time dimension and an additional measure is useful, representing the number of calls. We therefore pre-aggregate the Calls table into the fact table Fcalls on the schema:

(Customer, Time, Ant_conn, Dur_sum, No_of_calls)

where time has a maximum detail of days and Dur_sum is obtained by summing the duration of all calls for the same customer, day and antenna.

A tuple $[c,t,a,d,n]$ on Fcalls represents the fact that there were n calls of global duration d made by customer c connected to antenna a in the day t. The table Fcalls can be obtained from Calls by first building groups of tuples corresponding to the same customer, day, and antenna, then by applying the aggregation functions Sum (on the duration attribute) and Count on each group, and finally by associating the

Figure 2: The Dimension Hierarchies for Fcalls

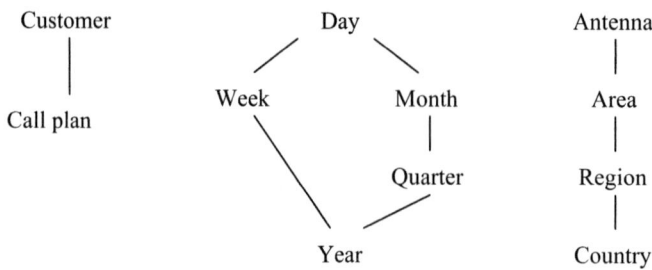

combination of values for customer, day, and antenna, which identifies the group, to the two computed aggregation values. The dimensions of Fcalls are Customer, Time, and Antenna, and we suppose to define the hierarchies depicted in Figure 2 on them.

In star schemes this corresponds to only three dimension tables, while in snowflake schema we would obtain 11 tables, one for each distinct dimension level. Typical queries on Fcalls may be: "Give me the global duration of calls by call plan, month, and region"; "Give me the number of calls for the first quarter of 2001 by call plan and area"; "Give me the average duration of calls in 2001 by region of the connecting antenna."

KEY FEATURES OF MULTIDIMENSIONAL QUERYING

In this section we briefly describe some of the distinctive features of OLAP querying with respect to traditional querying on transactional systems. We first make a comparison on some general aspects, e.g., the number of involved records, the proper indexing techniques, the typical user. Then, we consider some issues related to aggregate functions in query languages. Finally, the role of dimensions is illustrated, and it is shown that trade-offs are sometimes necessary to overcome the limitations of the relational model without excessively sacrificing the simplicity of the relational algebra and calculus.

OLTP vs. OLAP Queries

Number of records involved and the role of aggregation. One of the key differences between OLTP and multidimensional queries is the number of records required to compute the answer. OLTP queries typically involve a rather limited number of records, accessed through primary key or other specific indexes, which need to be processed for short, isolated transactions or to be issued on a user interface. In contrast multidimensional queries usually require the classification and aggregation of a huge amount of data.

Indexing techniques. Transaction processing is mainly based on the access of few records through primary key or other indexes on highly selective attribute combinations. The efficient access is easily achieved by well-known and established indexes, particularly B+-tree indexes. In contrast, multidimensional queries require a more articulated approach, since different techniques are required and each index performs well only on some categories of queries. An overview of indexing techniques for OLAP queries, particularly bitmap, projection, and bit-sliced indexes, will be addressed later in this chapter.

Current state vs. historical DBs. In OLTP operations the latest version of the database is required, the concurrent access/update of information is a critical issue, and the database usually represents only the current state of the system. In OLAP systems data do not need to be the latest available and should be time-stamped, thus enabling the user to perform historical analyses with trend forecasts. On the other hand, the presence of the temporal dimension poses relevant problems to the formulation and processing of queries, since schemes may evolve over time and conventional query languages are not adequate to cope with them. The problems related to queries over evolving schemes in OLAP systems are covered in detail in Chapter 6.

Target users. The typical users of OLTP systems are clerks, and the type of queries is rather limited and foreseeable. In contrast, multidimensional databases are usually the core of decision support systems, targeted for the enterprise executives. The types of queries are only partly predictable and often require highly expressive (and complex) query languages to be expressed. On the other hand, the user usually has little confidence even with "easy" query languages like basic SQL: the typical interaction paradigm is a spreadsheet-like environment based on iconic interfaces and on the graphical metaphor of the multidimensional cube.

Aggregate Functions

As already stressed above, the presence of aggregations is one of the most important distinctive features of OLAP systems with respect to conventional transactional systems. An important issue related to the definitions of aggregate functions in the relational model is that, while all standard relational operators take sets of tuples as input and produce sets of tuples as output, aggregate functions require a careful consideration of *duplicates*. Let us consider for example the sum of the durations of all calls in the Fcalls table. It is evident that the table may contain several tuples sharing the same value for Dur_sum. If we define the SUM aggregate function on the *set* of values of Dur_sum, we would incorrectly have a single contribution for each distinct value of Dur_sum. On the contrary, in order to correctly compute the global duration of calls, the contribution of each tuple should be considered, even if it is the same as that of another tuple. In other words the sum should not be applied on the set, but rather on the *multiset* (*bag*) of values of Dur_sum, i.e., on a collection of values where the same value can appear more than once (or, equivalently, where each value has an associated *multiplicity*).

The first important work on the extension of the relational model to include aggregate functions is by Klug (1982). In order to avoid multisets, the author considers a definition of aggregate function, which is based on a family of parameterized functions: for example the SUM aggregate function is defined by a family of SUM functions SUM_1, SUM_2,..., SUM_i,..., one for each column of the relation. Since the functions apply on the whole relation rather than on a single column, each value to be summed is part of a distinct tuple and there is no need to consider duplicates. Both an algebra and a calculus (with equivalent expressive power) are defined, having standard grouping capabilities.

The results were extended to relations containing complex objects in Özsoyoglu, Özsoyoglu, & Matos (1987), where a particular operator *aggregation-by-template* is also introduced, which can be considered as the first multidimensional operator: the aggregates are formed according to the groups defined in a particular table (representing the template), conceptually analogous to what we call now a dimension level.

More recently, the important results on query languages for bags (e.g., those in Albert, 1991; Libkin & Wong, 1993; Grumbach & Milo, 1996) have led to a more "natural" characterization of aggregate functions, closer to practical languages like SQL. In Gupta, Harinarayan, & Quass (1995), a new relational operator is proposed, *generalized projection*, which represents an extension of the classical relational projection (producing *sets* of tuples). Generalized projection captures both duplicate-eliminating projection (corresponding to classical projection and SQL DISTINCT projection) and duplicate-preserving projection (corresponding to SQL standard projection and useful to represent aggregations). In Libkin & Wong (1997), some query languages for bags and aggregate functions are analyzed, by investigating the relative complexity and expressive power. In Grumbach, Rafanelli, & Tininini (1999), a first-order language with real arithmetic and aggregation operators is proposed, offering a good trade-off between expressive power (enabling the user to express complex aggregations like median and average interest rates) and a tractable complexity.

Finally, in Cabibbo & Torlone (1999), a general framework for the description and investigation of aggregate functions in query languages is proposed. The abstract description of an aggregate function is obtained by considering a family of scalar functions g_0, g_1,..., g_i,..., corresponding to the computation of the aggregation function on a multiset of 0, 1,..., i ,... elements and such that the single g_is can be "easily" built (*uniform construction*). Depending on the complexity of this construction, different classes of aggregate functions and a hierarchy among them is defined.

The Role of Dimensions

The fact that the standard relational model and operators are inadequate to effectively represent and query multidimensional data was already shown in the early research on statistical databases (Chan & Shoshani, 1981; Su, 1983; Shoshani & Wong, 1985; Ghosh, 1986). This led to the distinction between *category attributes* (the dimensions) and *summary attributes* (the measures). One classical observation used to show the inadequacy of the relational model is that category attributes are "special" attributes which cannot be projected out as conventional attributes. For example the "number of calls by day" cannot be obtained by simply removing the category attribute Customer from the "number of calls by day and customer." On the contrary, this operation requires a *summarization* (roll-up) (Lenz & Shoshani, 1997; Hurtado & Mendelzon, 2001) of the category attribute Customer. Summarization may require a simple sum on the corresponding grouped values in some cases, or a complex series of operations in other cases, or may be even impossible to be performed without accessing the initial raw data, e.g., if the aggregation is a function like median.

The distinction between dimensions and measures is also at the basis of most models for OLAP systems. However, as noted by several authors, this distinction has some drawbacks, mainly because some operations that are easily expressible in the relational algebra become cumbersome in multidimensional models. Moreover, there are cases where a clear distinction between dimensions and measures prevents the user from expressing particular forms of queries. Consider for instance the query, "Give the number of calls classified by day and load level of the connecting antenna" where the load level of an antenna is classified as high, medium, or low depending on the number of calls that were routed through it in that day (let us suppose [0-1000] = low, [1001 - 5000] = medium, and over 5000 = high). It is obvious that the load level is a sort of dimension level for the dimension Antenna, but the actual instances of this dimension level depend on the measure "number of calls," so that in some sense the query *transforms* the measure into a dimension. Some authors have proposed a multidimensional model with a *symmetric treatment of measures and dimensions* to cope with this kind of problem.

A further issue related to the handling of dimension levels is that in some cases relations that were not in the initial design of the multidimensional schema, may be used to define additional dimension levels for querying. Consider for instance a relation Ant_type on the schema (antenna_id, type) which specifies the type of each antenna (e.g., the brand and model). This relation could be joined with the multidimensional data cube to express a query, "Give me the number of calls by day and type of antenna." This operation is straightforward in a relational environment (it is a simple join operation), but can be rather difficult to express in multidimensional models.

Finally, note that multidimensional querying is often an exploratory process, performed by navigating along the dimensions, increasing or decreasing the level of detail of the displayed data. So the user may decide to roll-up to less-detailed dimension levels to have a general view of the phenomenon under consideration (for example requesting the total number of calls by month) and then drill-down to investigate in more detail specific subsections of the cube, e.g., to determine which are the call plans that have produced a decrease in the total number of calls in June.

Expressing Multidimensional Queries by Extended Relational Query Languages

Several approaches to the problem of querying multidimensional data have been based on extensions of the relational algebra and calculus and/or of SQL, the most common relational query language. In Özsoyoglu, Özsoyoglu, & Matos (1987), an extension of the relation algebra is proposed to deal with set-valued attributes and aggregation functions. The authors propose an *aggregate formation* operator, which is a typical aggregate-group by operator and an *aggregation-by-template* operator, which anticipates some ideas of multidimensional operators and schemes. In practice a *template* is a table representing pre-specified groupings of attribute values. For instance a template may be a table defining groups of antennas of the same model, and the aggregation-by-template allows the user to perform aggregations on the groups of tuples specified by the template. The operator has evident analogies with aggregations on star schemas, with (specific columns of) dimension tables playing the role of templates.

In Gray et al. (1996), the CUBE operator is proposed as an extension of the conventional SQL GROUP BY clause. Since multidimensional applications often require the computation of several "points" of the n-dimensional data cube, the authors introduce a new operator which computes simultaneously all points of the n-dimensional cube and all possible roll-ups (note that this corresponds to 2^n distinct GROUP BYs!). In order to represent roll-up values, the polymorphic value ALL is associated with each dimension: ALL represents the union of all values in the dimension. As a consequence, in our fact table Fcalls, the tuple (ALL, Jan-4-2001, ALL, 20.5M, 945K) represents the information that in the indicated day 945K calls were globally made for a total duration of 20.5M seconds. The authors also propose an extended version of the GROUP BY operator, allowing the user to compute aggregations over computed categories (dimensions), e.g., to easily define the Day category of Fcalls from the time in seconds attribute of the original Calls table. Finally, for the (rather frequent) case, where the complete computation of the CUBE is prohibitive, the ROLLUP operator is introduced, which produces the result of n consecutive GROUP BYs, by progressively rolling-up each dimension of the data cube.

Simultaneous computation of the whole cube (roll-up) is advantagous, as some expensive operations required to calculate a GROUP BY with an aggregation can be "recycled" when computing the cube or the roll-up. Moreover, for many common aggregation functions, high-level aggregates can be computed by aggregating the lower level ones, thus avoiding the costly access to raw data, which are usually several orders of magnitudes larger than the aggregate data. Both the CUBE and the ROLLUP operators have been included in the SQL99 standard and are now implemented in the main commercial RDBMSs.

The CUBE operator is particularly useful when computing large collections of views to be materialized, but it is of little use when expressing single ad-hoc queries. In Chatziantoniou & Ross (1996), it is shown that some aggregate queries with groupings are very difficult to express with standard relational query languages, particularly SQL, even though conceptually simple and easily computable. What is particularly relevant is that, due to the complex syntax required to express the queries, the query optimizer is very unlikely to be able to compute them efficiently. A trivial example of this kind of query on the table Fcalls is "For each customer find the longest call and the day(s) when it was made." In standard SQL a nested query is required with an inner aggregate block which is very difficult to optimize. For this reason, the authors propose an additional operator for the relational algebra and an extension of SQL, namely the introduction of *grouping variables* ranging over the tuples of each group, and of a SUCHTHAT clause, defining the range of such grouping variables. It is shown that many aggregate queries requiring a complex syntax in standard SQL can be easily expressed in the extended language and that efficient query plans can be easily generated by the optimizer. The results are further elaborated in Ross, Srivastava, & Chatziantoniou (1998), where an extension of SQL is proposed combining both the characteristics of the CUBE operator and those of the grouping variables with SUCHTHAT clause.

Another interesting extension of SQL is the one adopted by Microsoft SQL Server and called MDX (for multidimensional expressions). The language is based on a symmetric treatment of dimensions and measures, and in fact the collection of measures constitutes a particular form of dimension, accessed through the reserved keyword **Measures**. Several keywords and clauses are introduced to access the single values (called members) of a dimension level, to refer to the hierarchical descendants/ancestors of a member, to compute new measures (e.g., a profit from a price and a cost measures) and sets of values, to manipulate time series, etc.

QUERY LANGUAGES BASED ON MULTIDIMENSIONAL MODELS

In this section we illustrate some of the most significant query languages for multidimensional data, based on specifically designed models, proposed in the literature. Since in most cases the models are based on an abstract data type, often called *cube*, we prefix the original term with a letter, in order to distinguish it from the general concept of cube used throughout this chapter and also to distinguish each "cube" from the "cubes" proposed by other authors. We therefore speak of G-cubes, S-cubes, etc. and also of f-dimensions, even if in the original literature they were simply referred to as cubes and dimensions. We do not consider here the standard operators of roll-up, drill-down, slice and dice, and pivot that have been covered in detail in an earlier chapter and which are obviously included in all models for multidimensional data. We focus instead on the peculiarities of each model and query language.

A Grouping Algebra for Cubes

The G-cube schemes proposed in Li & Wang (1996) have several analogies with relational star schemes. As usual in data models, the authors distinguish n-dimensional *G-cube schemes* and G-cube *instances* (*G-cubes* for brevity). Each dimension is represented by a name and a collection of attributes (as in dimension tables of star schemes). G-cubes are defined by a collection of dimensions and by a function that maps combinations of dimension instances to values (measures) in a domain of scalar values (integers, reals, etc.).

The key features of the model are the distinction between conventional and dimension attributes, along with the concepts of *grouping (relation) schemes* and *grouping relations*, by means of which the authors introduce an operator which extends the conventional SQL GROUP BY clause. The operator combines the classic grouping capabilities of relational languages with those related to cube dimensions. The proposed query language is basically an extension of the relational algebra and comprises: i) order-oriented operations to express queries like "Give me the 10 most used antennas in 2001"; ii) aggregate operations exploiting the extended capabilities of the grouping operator.

For example a grouping relation *g* on the scheme (Customer_no, Town_resid) could be used to express the town of residence of each customer. Then the expression

$$\Gamma_{Town\_resid}^{No\_of\_calls,g}(t)$$

represents for each town *t* the total number of calls made by customers resident in town *t*. The model assumes that each G-cube contains only one measure, thus we have supposed that a new cube No_of_calls has been defined from Fcalls by restricting only to that measure.

An Algebra for "Symmetric" Cubes

As already stressed, a desirable property for multidimensional query languages is the symmetric treatment of dimensions and measures, particularly the capability to use measures to classify data, i.e., to transform one or more measures into a dimension. Generally speaking, there are some typical OLAP operations (e.g., roll-up and drill-down) for which a clear distinction between dimensions and measures makes the query expression simpler and allows association of straightforward semantics. On the other hand, there are classes of queries on multidimensional data that are more clearly expressible if dimensions and measures are treated as simple attributes, just like in the relational model.

An interesting approach to this issue is proposed in Agrawal, Gupta, & Sarawagi (1997): the data model is based on the S-cube, which is constituted by a triple of components:

(i)   A collection of dimensions.

(ii) A mapping from the Cartesian product of the dimension domains to an element domain (which can itself be multidimensional). The values in the element domain generally correspond to measures, and the domain can itself be either Boolean or n-dimensional.

(iii) A tuple of names describing the elements of the S-cube.

The query language is based on an algebra. The operators take S-cubes as operands and produce S-cubes as output. The peculiarity of the model is in the element domain, since dimensions can be *pushed* into it and measures can be *pulled* from it, thus providing a powerful mechanism to symmetrically deal with dimensions and measures. The proposed algebra is shown to be at least as powerful as the relational algebra, and several common OLAP operations (e.g., roll-up, drill-down, slice and dice, star join) can be easily expressed by the proposed operators.

An MD Calculus for Fact Tables

The interaction between the user and a data warehouse is often based on a collection of operators that are applied on relations or cubes, and produce relations and cubes as output. In many cases, however, the use of a calculus for expressing queries is preferable, because it enables the user to express the manipulation in a more intuitive declarative way. The calculus MD-CAL proposed in Cabibbo & Torlone (1997) is based on a specific model for multidimensional data, called MD, which is closely related to snowflake schemes: fact tables and dimensions have a logical counterpart in the model, called *f-tables* and *f-dimensions*. Unlike most other multidimensional models, dimensions are described by normalized hierarchical structures, where the hierarchical relationships among dimension levels are modeled by *roll-up functions* R-UP.

MD-CAL queries define a mapping from instances over an input MD scheme to instances over an output MD scheme, where the input and the output schemes are defined over the same f-dimensions but distinct f-tables. Queries have the typical calculus form:

$$\{x_1, ..., x_n : x \mid \psi(x, x_1, ..., x_n) \}$$

where ψ is a first-order formula involving f-tables, roll-up functions, interpreted scalar functions (e.g., the standard arithmetic operators), and aggregate functions (e.g., min, max, sum, count, and avg). x is a special variable corresponding to the measure of the fact table and is called the *result variable*. The use of first-order expressions enables the symmetric treatment of measures and dimensions, i.e., the capability to transform measures into dimension components and vice versa, thus performing what the authors call an *abstraction* transformation.

Querying by n-Dimensional Tables

An important drawback of several models for multidimensional data is that, although providing powerful constructs to express typical OLAP operations, they often require a complex syntax to express conceptually simple queries that are instead easily expressed by relational query languages. In Gyssens & Lakshmanan (1997), a model is proposed whose aim is to provide a conceptual description of multidimensional data, remaining as close to the standard relational model as possible, therefore minimizing the above-mentioned difficulties in expressing standard relational manipulations.

The model is based on a conceptual view of multidimensional data, which is closely related to star schemes. Particularly, an *n-dimensional table schema* is analogous to the (relational) schema of the dimension tables, and the *instance* of an n-dimensional table schema is analogous to the instances of both the dimension and the fact tables in a star schema. The key idea of the model is to define a one-to-one mapping between table schemes and relational schemes, based on a correspondence between tables and "completed" relations. This mapping enables the users to go back and forth from the tabular to the relational formalism, expressing the queries in the most suitable environment. Some classical OLAP operators are defined, particularly to express classifications and summarizations (aggregations), and it is shown that many others can be derived in terms of them, e.g., cube and monotone roll-up.

In order to express "relational" operations on (n-dimensional) tables, the tables are first transformed into relations (by using the function of relational representation *rep*) and then the conventional relational operators are applied. Conversely, multidimensional operations on relations can be easily expressed by first transforming the relations into tables (by the function of tabular representation *tab*) and then using the specific multidimensional operators.

The symmetric treatment of measures and dimensions is obtained by a technique similar to the one of the S-cubes. Two operators are defined that affect the structure of tables: by the *unfold* operator a set X of measure attributes can be "pulled" and used to constitute a new dimension, while the *fold* operator can be used to "push" dimensions into the measure components.

Figure 3: Relational and Multidimensional Operations on Tables

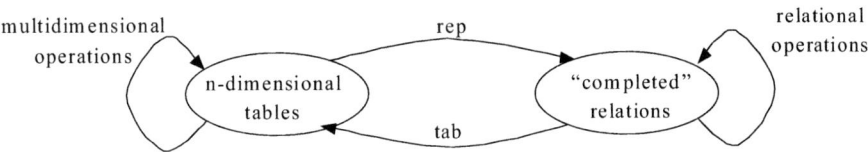

Embedding Hierarchies in the Relational Model

All multidimensional models considered so far are based on the definition of an abstract data type corresponding to the concept of data cube. In Jagadish, Lakshmanan, & Srivastava (1999), a different approach is pursued: the relational model is considered and the introduced abstract data type refers to dimension hierarchies. This enables the authors to overcome some modeling limitations related to star and snowflake schemes, particularly to describe:

- Unbalanced hierarchies, i.e., such that some of the paths from the root to the leaf instances traverse all nodes of the hierarchy, while other paths only traverse a subset of them. This is a typical problem when considering territorial hierarchies in different countries because, for example, territories are classified in regions, states, counties, and cities in the U.S., but simply in regions and cities in a smaller country.
- Heterogeneous hierarchies, i.e., such that the leaf nodes (dimension levels) differ according to the instance of a dimension in an upper level, e.g., antennas may be classified by type, (digital or analog) but digital antennas are further classified according to the number of bits of the transmitted signal, while analog are further classified according to the bandwidth.

The model is called SQL(*H*) and the query language is basically an extension of SQL with two added features: i) the DIMENSIONS clause, which extends the FROM clause to dimension tables, and is used to specify dimension levels to be included in the multidimensional query; ii) a collection of hierarchical predicates to represent conditions like "A is a descendant of B in the dimension hierarchy."

The problem of heterogeneous hierarchies in OLAP systems has also been analyzed in Lehner (1998), where *nested data cubes* are proposed for modeling and querying.

Visual Query Languages

Probably the first system enabling the user to query multidimensional (statistical) data by a graphical model was Subject (Chan & Shoshani, 1981). In Subject, multidimensional data are modeled by particular directed acyclic graphs (DAGs), which have some similarities with snowflake schemes. The nodes of the DAG have an inner label (X or C), representing the type of the node, and an external label representing its name. X nodes correspond to fact tables and some of the C nodes to dimension levels. Besides, C nodes are also used to group several X nodes just like "folders" of modern iconic operating systems. The query system is based on a hierarchical navigation on the DAGs, selecting the X node and the connected C nodes of interest (i.e., in the multidimensional terminology, the measures and the dimensions).

A similar approach is used in the ADAMO system, based on the homonymous graphical model for aggregate data (Rafanelli, Bezenchek, & Tininini, 1996): circle nodes and edges are used to represent dimension levels and hierarchies, while square nodes correspond to measures. Typical OLAP operations are directly performed on the graphical representations, e.g., adding a circle node to drill-down and removing it to roll-up or selecting specific values of a dimension level to slice and dice.

More recently, a visual query language was proposed (Cabibbo & Torlone, 1998a), based on some algebraic operators for the multidimensional model MD (Cabibbo & Torlone, 1998b). An elementary query for the running example used throughout this chapter is shown in Figure 4. It represents the query: "Give me the total number of calls in July, by day and call plan."

The chosen measures and dimension levels are visually represented by thick lines. Selection conditions are directly represented on the corresponding level, and the aggregation to be performed for the roll-up is specified below the corresponding

Figure 4: A Graphical Query of the MD Model

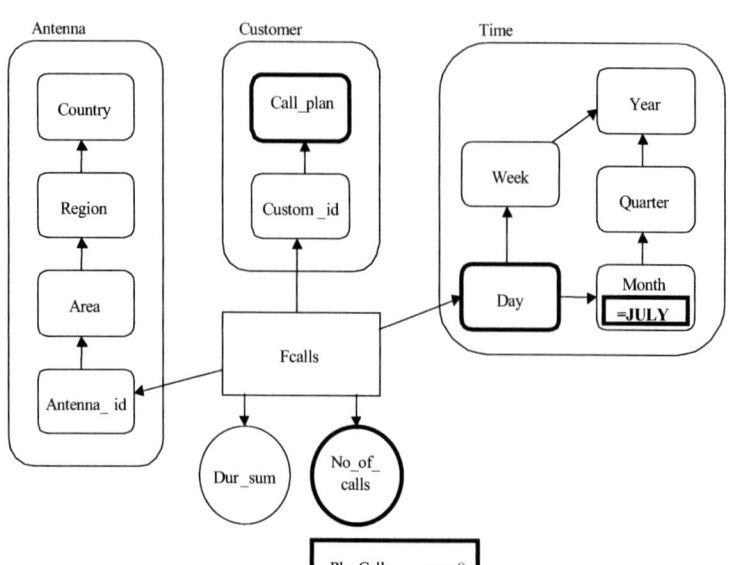

measure. Two or more queries of this form can be graphically composed (e.g., using scalar functions and aggregations) enabling the user to formulate arbitrarily complex queries.

Multidimensional Queries on Statistical Databases

Similarities and differences between OLAP and statistical databases are the main focus of Chapter 2. Here, we only outline some peculiarities that need to be taken into account when designing an interface for statistical database querying.

One of the crucial issues for statistical databases is the representation of complex territorial hierarchies which evolve over time. The problems related to the temporal evolution of dimensions and dimension hierarchies are covered in detail in Chapter 6. Some solutions adopted in an implemented statistical information system can be found in Tininini et al. (2002).

Another important issue is related to the fact that in statistical databases the number of "points" of the multidimensional space, corresponding to data allowable for publication, is very small compared with the global size of the multidimensional space itself. In other words the multidimensional array representing allowable data in the global multidimensional space is sparse, and there are two main reasons for that:

- In statistical databases it is crucial to preserve the privacy of the individuals, and excessive detail in aggregate data can produce disclosure of sensitive information about single individuals (see Chapter 11).
- Much of the data in statistical databases originate from sample surveys, and the corresponding aggregate data are not significant for the finest levels of detail. Typically, a collection of data is considered to be significant only if the cardinality of *each* group individuated by the dimensional product (GROUP BY) is greater than a given threshold value.

As a consequence, an interface enabling the user to freely navigate on the dimension hierarchies would often lead the user to express a query, corresponding to non-allowable data. Specific techniques are consequently required to establish a trade-off between the characteristic freedom of multidimensional navigation and the constraints of significance and disclosure prevention imposed by statistical databases.

MULTIDIMENSIONAL QUERY EVALUATION

We have seen in the previous section that the adoption of a particular query language can be useful not only to express the desired query in a more straightforward manner, but also to speed up the computation process. In this section we investigate the problem of efficiently computing the answer to a multidimensional query, possibly expressed in one of the languages introduced above.

A fundamental requirements of OLAP systems is the ability to perform multidimensional analyses in *on-line* response times. Since multidimensional queries usually involve huge amount of data to be aggregated, the only way to achieve this result is by pre-computing some queries, storing the answers permanently in the database, and reusing (almost exclusively) them to compute queries over the multidimensional database. These pre-computed queries are commonly referred to as *materialized views,* and several issues related to them, particularly which views to materialize and how to maintain them, have been analyzed in Chapter 8. In this section we show how materialized views can be used to efficiently process multidimensional queries. We also briefly outline some techniques for their efficient calculation, based on specifically designed indexing structures.

Query Answering vs. Query Rewriting

The problem of computing the answer of a query by using some pre-computed (materialized) views has been extensively studied in the literature and has been generically denoted as the problem of *answering queries using views* (see for instance Levy et al., 1995; Duschka & Genesereth, 1997; Halevy, 2000). Informally speaking the problem can be stated as follows: given a query Q and a collection of views V over the same schema s, is it possible to compute the answer of Q by using (only) the information provided by the views in V? A more rigorous distinction has also been made between *view-based query rewriting* and *query answering*, corresponding to two distinct approaches of the general problem (Calvanese et al., 2000c, 2000d; Halevy, 2000).

Query rewriting is based on the use of the view definitions to produce a new *rewritten query*, expressed in terms of the available view names and equivalent to the original one. The answer of the query can then be obtained by using the rewritten query and the view extensions (instances). Query answering is based instead on the exploitation of both the view definitions and the view extensions, trying to determine the best possible answer, possibly a subset of the exact one, which can be extracted from the view extensions. Several authors have studied the query answering problem for various classes of queries and views, both under the Closed and the Open World Assumption (Abiteboul & Duschka, 1998; Grahne & Meldelzon, 1999; Calvanese et al., 2000a, 2000b).

In general, query answering techniques are preferable in contexts where exact answers are rather unlikely to be obtained (e.g., integration of heterogeneous data sources, like websites), and when requirements on response times are not very stringent. On the other hand, as noted in Grahne & Meldelzon (1999), query answering methods can be extremely inefficient, because it is difficult or even impossible to process only the "useful" views and to apply optimization techniques such as pushing selections and joins. As a consequence, the rewriting approach is more appropriate in contexts like OLAP systems, where the amount of data is very large and fast response times are required (Goldstein & Larson, 2001), and for query optimization, where different query plans need to be maintained in main memory and efficiently compared (Albrecht et al., 2000; Afrati, Li, & Ullman, 2001).

Rewriting Aggregate Queries

As it was shown, a typical elementary multidimensional query is described by an aggregate group by query, applied on the join of the fact table with two or more dimension tables. This motivates the interest of many researchers on the rewriting of this form of queries and views.

In Gupta, Harinarayan, & Quass (1995), an algorithm is proposed to rewrite conjunctive queries with aggregations using views of the same form. The technique is based on the concept of generalized projection (GP) and on some transformation rules which can be used by an optimizer and which enable the query and the views to be put in a particular *normal form*, based on GPSJ (Generalized Projection/ Selection/Join) expressions. The query and the views are analyzed in terms of their query tree, i.e., the tree representing the way to compute them by applying selections, joins, and generalized projections on the base relations. By using the transformation rules, the algorithm tries to produce a match between one or more view trees and subtrees of the query tree (and consequently to replace the computations with accesses to the corresponding materialized views). The results are extended to NGPSJ (Nested GPSJ) expressions in Golfarelli & Rizzi (2000a, 2000b).

In Srivastava et al. (1996), an algorithm is proposed to rewrite a single block (conjunctive) SQL query with GROUP BY and aggregations using some views of the same form. The considered aggregate functions are MIN, MAX, COUNT, and SUM. The algorithm is based on the detection of homomorphisms from the view to the query, as in the non-aggregate context (Levy et al., 1995). It is shown, however, that more restrictive conditions need to be considered when dealing with aggregates, because the view has to produce not only the right tuples, but also the correct multiplicities of those tuples.

In Cohen, Nutt, & Serebrenik (1999, 2000), a rather different approach is proposed: the original query to be rewritten, the usable views, and the rewritten query are all expressed by using an extension of Datalog with aggregate functions (again COUNT, SUM, MIN, and MAX) as query language. Queries and views are assumed to be conjunctive. Several rewriting candidates of particular forms are considered, and for each rewriting candidate, the views in its body are unfolded (i.e., replaced by their body in the view definition). Finally, the unfolded candidate is compared with the original query to verify equivalence by using known equivalence criteria for aggregate queries, particularly those proposed in Nutt, Sagiv, & Shurin (1998) for COUNT, SUM, MIN, and MAX queries. The technique can be extended by using the equivalence criteria for AVG queries presented in Grumbach, Rafanelli, & Tininini (1999) and based on the syntactic notion of *isomorphism modulo a product*.

An important issue in query rewriting is that of identifying the views that may be actually useful in the rewriting process: this is often referred to as the *view usability problem*. In the non-aggregate context, it is shown (Levy et al., 1995) that a conjunctive view is usable to produce a conjunctive rewritten query if a homomorphism exists from the body of the view to the body of the query. In Grumbach, Rafanelli, & Tininini (1999), it is shown that more restrictive (necessary and sufficient) conditions are needed for the usability of conjunctive count views to rewrite conjunctive count queries, based on the concept of *sound* homomorphisms. It is also shown that in the presence of aggregations, considering rewriting candidates of conjunctive form is not sufficient, and more complex forms of rewritings may be required, particularly those based on the concept of isomorphism modulo a product.

All rewriting algorithms proposed in the literature are based on trying to obtain a rewritten query having a particular form by using (possibly only) the available views. An interesting question is: "Can I rewrite more by considering rewritten queries of more complex form?" and the even more ambitious one "Given a collection of views, is the information they provide sufficient to rewrite a query?" In Grumbach & Tininini (2000a), the problem is investigated in a general framework based on the concept of *query subsumption*. Basically, the information content of one or more queries is characterized by their distinguishing power, i.e., by their capability to determine that two database instances are different. Hence a collection of views subsumes a query if it is able to distinguish any couple of instances that is distinguishable by the query, and it is shown that a rewriting of a query using some views exists if the views subsume the query. In the particular case of count and sum queries defined over the same fact table, an algorithm is proposed which is demonstrated to be *complete*. In other words, even if the algorithm (as any algorithm of practical use) considers rewritten queries of particular forms, it is shown that no improvement could be obtained by considering rewritten queries of more complex forms.

Finally, in Grumbach & Tininini (2000b), a completely different approach to the problem of aggregate rewriting is proposed. The technique is based on the idea of formally expressing the relationships (metadata) existing between raw and aggregate data, and also among aggregate data of different type and/or level of detail. The data are stored in standard relations, while the metadata are represented by *numerical dependencies*, namely Horn clauses expressing in formal manner the semantics of the aggregate attributes. The mechanism is tested by transforming the numerical dependencies into Prolog rules and then exploiting the Prolog inference engine to produce the rewriting.

(Efficiently) Computing Cubes and Materialized Views

In this section we briefly analyze the techniques proposed to compute cubes and more generally materialized views containing aggregates. We only focus on the problem of computing the views from scratch, since the problem of maintaining a collection of views in the presence of insertions, deletions, and updates has been covered in Chapter 8.

As already pointed out, a typical multidimensional query is constituted by an aggregate group by query applied on the join of the fact table with two or more dimension tables, and consequently has the form of an aggregate conjunctive query, e.g.:

```
SELECT D.dim1, D.dim2, AGG(F.measure)
FROM fact_table F, dim_table1 D1, dim_table2 D2
WHERE F.dimKey1 = D1.dimKey1
AND F.dimKey2 = D2.dimKey2
GROUP BY D.dim1, D.dim2
```

where AGG is an aggregation function, like SUM, MIN, AVG., etc.

Traditional query processing systems perform first all joins expressed in the FROM and WHERE clause, and only afterwards the grouping on the result of the join and the aggregation on each group. The algorithms to produce the groups can be broadly classified as techniques based on a a) sorting on the GROUP BY attributes and b) on hashing tables (see Graefe, 1993). However, there are many common cases where an early evaluation of the GROUP BY is possible, and it can significantly reduce the computation time since: i) it reduces the size of the input of the join (usually very large in the context of multidimensional databases), and ii) it enables the query processing engine to use indexes to perform the (early) GROUP BY on the base tables.

In Chaudhuri & Shim (1995, 1996), some techniques and applicability conditions are proposed for transforming execution plans into equivalent (more efficient) ones. As it is typical in query optimization, the technique is based on *pull-up* transformations, which delay the execution of a costly operation, e.g., a group by on a large dataset, by moving it towards the root of the query tree, and on *push-down* transformations, which are used for example to anticipate an aggregation so decreasing the size of a join.

Several transformations for multidimensional queries are also proposed in Gupta, Harinarayan, & Quass (1995), based on the concept of generalized projection (GP, see above). The transformations enable the optimizer: i) to push a GP down the query tree; ii) to pull a GP up the query tree; iii) to coalesce two GPs into one, or conversely to split up one GP into two. Query tree transformations are also used in the rewriting process.

In Agarwal et al. (1996), some algorithms are proposed and compared to extend the traditional techniques for the computation of GROUP BY queries to the processing of the CUBE operator. The algorithms are applicable in the case of distributive aggregate functions and are based on the property that higher level aggregates can be computed from the lower level ones in case of distributive aggregate functions.

In MOLAP systems, however, the above methods for the computation of the CUBE are inadequate, since they are substantially based on sorting and hashing techniques which cannot be applied to multidimensional arrays. In Zhao, Deshpande, & Naughton (1997), an algorithm is proposed for the computation of CUBEs in a MOLAP environment. It is shown that the computation can be made significantly more efficient (and even more efficient than ROLAP-based computations) by exploiting the inherently compressed representation of data of MOLAP systems. Particularly, the algorithm benefits from the compactness of multidimensional arrays, enabling the query processor to transfer larger "chunks" of data to main memory and to efficiently process them.

In many practical cases GROUP BYs at the finest level of granularity correspond to *sparse* data cubes, i.e., cubes where a high percentage of points correspond to null values. Consider for instance the CUBE corresponding to the query "Number of calls by customer, day, and antenna": it is rather obvious that a considerable number of combinations correspond to zero. Fast techniques to compute sparse data cubes are proposed in Ross & Srivastava (1997). They are based on: i) decomposing the fact table into fragments that can be stored in main memory; ii) computing the data cube in main memory for each fragment; and finally iii) combining the partial results.

Using Indexes to Improve Efficiency

Probably the most common indexes used in traditional DBMSs are B+-trees (Comer, 1979): a particular form of trees labeled using the index key values and having a list of record identifiers on the leaf level, i.e., a list of elements specifying the actual position of each record on the disk. The index key values may be constituted by one or more columns of the indexed table. OLTP queries usually retrieve a very limited number of tuples (or even a single tuple accessed through the primary key index), and in these cases B+-trees have been shown to be particularly efficient.

In contrast, OLAP queries typically involve large groups of tuples to be aggregated, requiring specifically designed indexing structures. Unlike in the OLTP context, there is no "universally good" index for multidimensional queries, but rather a variety of techniques, each of which can perform well for specific type of data and forms of queries, and be instead inappropriate for other ones.

Let us again consider the typical multidimensional query expressed by the SQL query above. The core operations related to its evaluation are: 1) the joins of the fact table with two or more dimension tables; 2) the grouping of tuples according to some dimensional values; 3) the application of an aggregation function on each group of tuples. An interesting type of index, which can be used to efficiently perform operation 1, is the *join index* (Valduriez, 1987): while conventional indexes map column values to records in one table, join indexes map values to records of two (or more) joined tables, thus constituting a particular form of materialized view. Join indexes in their original version are not directly usable to evaluate OLAP queries efficiently, but can be very effective in combination with other indexing techniques, like bitmaps and partitioning.

Bitmap indexes (ON87; O'Neil & Graefe, 1995; Chan & Ioannidis, 1998) are useful to perform the grouping of tuples and some forms of aggregation. In practice these indexes use a bitmap representation for the list of record identifiers in the leaf level of the tree: if the table *t* contains *n* records, then each leaf of the bitmap index (corresponding to a specific value *c* of the indexed column *C*) contains a sequence of *n* bits, where the *i*-th bit is set to 1 if $t_i.C=c$, and to 0 otherwise. Bitmap representations are indicated, when the number of distinct key values is low and when several predicates of the form (Column = value) are to be combined in AND/OR, since the operation can be efficiently performed by AND-ing/OR-ing bit to bit the corresponding bitmap representation, an operation which can be parallelized and performed very efficiently on modern processors. Finally, bitmap indexes can be used for a fast computation of count queries, since counting can be performed directly on the bitmap representation, without even accessing the selected tuples.

In *projection indexes* (French, 1995; O'Neil & Quass, 1997), the tree access structure is coupled with a sort of materialized view, representing the projection of the table on the indexed column. The technique has some analogies with vertically partitioned tables and is indicated when the aggregate operations need to be performed on one or more indexed columns.

Bit-sliced indexes (O'Neil & Quass, 1997) can be considered as a combination of the two previous techniques: the values of the projected column are encoded, and for each bit component of the encoding, a bitmap is associated. The technique has some analogies with bit transposed files, which were proposed in Wong et al. (1985, 1986) to compute queries on very large scientific and statistical databases. These indexes reach best performances for SUM and AVG aggregations, but are not well-suited for aggregations involving more than one column.

In Chan & Ioannidis (1998, 1999), some variations on the general idea of bitmap indexes are presented, by studying encoding schemes, time-optimal and space-optimal indexes, as well as trade-off solutions. A comparative study on the use of STR-tree-based indexes (a particular form of spatial index) versus some variations of bitmap indexes in the context of OLAP range queries can be found in Jürgens & Lenz (1999).

CONCLUSIONS

In this chapter we have illustrated the main issues related to the querying of multidimensional data. Querying is strictly connected to the way the data are modeled and particularly to the concepts of data cube, fact table, measure, and dimension. Multidimensional queries have several distinctive features with respect to conventional queries on transactional systems, and in this differentiation aggregation functions and dimension hierarchies play a major role.

We have shown the main proposals to express multidimensional queries, when the data are represented by the relational model (particularly star and snowflake schemes), as well as the approaches based on specifically designed multidimensional data models, usually extensions of the relational one. We have analyzed languages based on an algebra (usually an extension of the relational one), languages based on a calculus (and hence more declarative), and visual languages, usually relying on a collection of algebraic operators. We have shown that two important problems to face, when considering languages based on specific multidimensional models, are the increased complexity of expressing conventional relational operations and the difficulty to symmetrically deal dimensions and measures, i.e., to transform dimensions into measures and vice versa.

We have also focused on the problem of query evaluation, i.e., on how the query expressed in a given language can be translated, using the available materialized views, into an (efficient) evaluation plan which retrieves the necessary information and calculates the required results. This is known in the literature as the query rewriting problem. We have presented some results on aggregate query equivalence and rewriting of aggregate queries that can be used to optimize the evaluation of queries on multidimensional data. Finally, the main indexing techniques for OLAP queries have been illustrated, as well as the contexts of applicability.

REFERENCES

Abiteboul, S., & Duschka, O.M.(1998). Complexity of answering queries using materialized views. *Proceedings of the ACM Symposium on Principles of Database Systems (PODS'98)*, 254-263.

Abiteboul, S., Hull, R., & Vianu, V. (1995). *Foundations of Databases*. City: Addison-Wesley.

Afrati, F.N., Li, C., & Ullman, J.D. (2001). Generating efficient plans for queries using views. *Proceedings of the ACM International Conference on Management of Data (SIGMOD'01)*.

Agarwal, S., Agrawal, R., Deshpande, P., Gupta, A., Naughton, J.F., Ramakrishnan, R., & Sarawagi, S. (1996). On the computation of multidimensional aggregates. *Proceedings of the International Conference on Very Large Data Bases (VLDB'96)*, 506-521.

Agrawal, R., Gupta, A., & Sarawagi, S. (1997). Modeling multidimensional databases. *Proceedings of the International Conference on Data Engineering (ICDE'97)*, 232-243.

Albert, J. (1991). Algebraic properties of bag data types. *Proceedings of the International Conference on Very Large Data Bases (VLDB'91)*, 211-219.

Albrecht, J., Hümmer, W. Lehner, W., & Schlesinger, L. Query Optimization by using derivability in a data warehouse environment. In *ACM International*

Workshop on Data Warehousing and OLAP (DOLAP 2000).

Cabibbo, L., & Torlone, R. (1997). Querying multidimensional databases. *Proceedings of the International Workshop on Database Programming Languages (DBPL'97),* 319-335.

Cabibbo, L., & Torlone, R. (1998a). From a procedural to a visual query language for OLAP. *Proceedings of the International Conference on Scientific and Statistical Database Management (SSDBM'98),* 74-83.

Cabibbo, L., & Torlone, R. (1998b). A logical approach to multidimensional databases. *Proceedings of the International Conference on Extending Database Technology (EDBT'98),* 183-197.

Cabibbo, L., & Torlone, R. (1999). A framework for the investigation of aggregate functions in database queries. *Proceedings of the International Conference on Database Theory (ICDT'99),* 383-397.

Calvanese, D., De Giacomo, G., Lenzerini, M., & Vardi, M.Y. (2000a). Answering regular path queries using views. *Proceedings of the International Conference on Data Engineering (ICDE'00),* 389-398.

Calvanese, D., De Giacomo, G., Lenzerini, M., & Vardi, M.Y. (2000b). View-based query processing for regular path queries with Inverse. *Proceedings of the ACM Symposium on Principles of Database Systems (PODS'00),* 58-66.

Calvanese, D., De Giacomo, G., Lenzerini, M., & Vardi, M.Y. (2000c). View-based query processing and constraint satisfaction. *Proceedings of the IEEE Symposium on Logic in Computer Science (LICS'00),* 361-371.

Calvanese, D., De Giacomo, G., Lenzerini, M., & Vardi, M.Y. (2000d). What is view-based query rewriting? *Proceedings of the International Workshop on Knowledge Representation Meets Databases (KRDB'00),* 17-27.

Chan, C.Y., & Ioannidis, Y.E. (1998). Bitmap index design and evaluation. *Proceedings of the ACM International Conference on Management of Data (SIGMOD'98),* 355-366.

Chan, C.Y., & Ioannidis, Y.E. (1999). An efficient bitmap encoding scheme for selection queries. *Proceedings of the ACM International Conference on Management of Data (SIGMOD'99),* 215-226.

Chan, P., & Shoshani, A. (1981). SUBJECT: A directory-driven system for organizing and accessing large statistical databases. *Proceedings of the International Conference on Very Large Data Bases (VLDB'81),* 553-563.

Chatziantoniou, D., & Ross, K.A. (1996). Querying multiple features of groups in relational databases. *Proceedings of the International Conference on Very Large Data Bases (VLDB'96),* 295-306.

Chaudhuri, S., & Dayal, U. (1997). An overview of data warehousing and OLAP technology. *SIGMOD Record,* 26(1), 65-74.

Chaudhuri, S., & Shim, K. (1995). An overview of cost-based optimization of queries with aggregates. *Data Engineering Bulletin,* 18(3), 3-9.

Chaudhuri, S., & Shim, K. (1996). Optimizing queries with aggregate views. *Proceedings of the International Conference on Extending Database*

Technology (EDBT'96), 167-182.

Codd, E.F., Codd, S.B., & Salley, C.T. (1993). *Providing OLAP (On-Line Analytical Processing) to User Analysts: An IT Mandate.* Technical Report, E.F. Codd & Associates.

Cohen, S., Nutt, W., & Serebrenik, A. (1999). Rewriting aggregate queries using views. *Proceedings of the ACM Symposium on Principles of Database Systems (PODS'99),* 155-166.

Cohen, S., Nutt, W., & Serebrenik, A. (2000). Algorithms for rewriting aggregate queries using views. *Proceedings of the Conference ABDIS-DASFAA'00,* 65-78.

Comer, D. (1979). The ubiquitous B-tree. *ACM Computing Surveys,* 11(2), 121-137.

Duschka, O.M., & Genesereth, M.R. (1997). Answering recursive queries using views. *Proceedings of the ACM Symposium on Principles of Database Systems (PODS'97),* 109-116.

French, C.D. (1995). "One-size-fits-all" database architectures do not work for DDS. *Proceedings of the ACM International Conference on Management of Data (SIGMOD'95),* 449-450.

Ghosh, S.P. (1986). Statistical relational tables for statistical database management. *IEEE Transactions on Software Engineering,* 12(12), 1106-1116.

Goldstein, J., & Larson, P. (2001). Optimizing queries using materialized views: A practical, scalable solution. *Proceedings of the ACM International Conference on Management of Data (SIGMOD'01).*

Golfarelli, M., & Rizzi, S. (2000a). View materialization for nested GPSJ queries. *Proceedings of the Workshop on Design and Management of Data Warehouses (DMDW'00).*

Golfarelli, M., & Rizzi, S. (2000b). Comparing nested GPSJ queries in multidimensional databases. *Proceedings of the Workshop on Data Warehousing and OLAP (DOLAP'00).*

Graefe, G. (1993). Query evaluation techniques for large databases. *ACM Computing Surveys,* 25(2), 73-170.

Grahne, G., & Mendelzon, A.O. (1999). Tableau Techniques for Querying Information Sources through Global Schemas. In *International Conference on Database Theory (ICDT'99)* pp. 332-347).

Gray, J., Bosworth, A., Layman, A., & Pirahesh, H. (1996). Data cube: A relational aggregation operator generalizing group-by, cross-tab, and sub-total. *Proceedings of the International Conference on Data Engineering (ICDE'96),* 152-159.

Grumbach, S., & Milo, T. (1996). Towards tractable algebras for bags. *Journal of Computer and System Sciences,* 52(3), 570-588.

Grumbach, S., Rafanelli, M., & Tininini, L. (1999). Querying aggregate data. *Proceedings of the ACM Symposium on Principles of Database Systems*

(PODS'99), 174-184.

Grumbach, S., & Tininini, L. (2000a). On the content of materialized aggregate views. *Proceedings of the ACM Symposium on Principles of Database Systems (PODS'00)*, 47-57.

Grumbach, S., & Tininini, L. (2000b). Automatic aggregation using explicit metadata. *Proceedings of the International Conference on Scientific and Statistical Database Management (SSDBM'00)*, 85-94.

Gupta, A., Harinarayan, V., & Quass, D. (1995). Aggregate-query processing in data warehousing environments. *Proceedings of the International Conference on Very Large Data Bases (VLDB'95)*, 358-369.

Gyssens, M., & Lakshmanan, L.V.S. (1997). A foundation for multidimensional databases. *Proceedings of the International Conference on Very Large Data Bases (VLDB'97)*, 106-115.

Halevy, A.Y. (2000). Theory of answering queries using views. *SIGMOD Record*, 29(4), 40-47.

Hurtado, C.A., & Mendelzon, A.O. (2001). Reasoning about summarizability in heterogeneous multidimensional schemas. *Proceedings of the International Conference on Database Theory (ICDT'01)*, 375-389.

Inmon, W.H. (1996). *Building the Data Warehouse*. New York: John Wiley & Sons.

Jagadish, H.V., Lakshmanan, L.V.S., & Srivastava, D. (1999). What can hierarchies do for data warehouses? *Proceedings of the International Conference on Very Large Data Bases (VLDB'99)*, 530-541.

Jarke, M., Lenzerini, M., Vassiliou, Y., & Vassiliadis, P. (2000). *Fundamentals of Data Warehouses*. City: Springer Verlag.

Jürgens, M., & Lenz, H.J. (1999). Tree-based indexes vs. bitmap indexes—a performance study. *Proceedings of the International Workshop on Design and Management of Data Warehouses (DMDW'99)*.

Kimball, R. (1996). *The Data Warehouse Toolkit*. New York: John Wiley & Sons.

Klug, A. (1982). Equivalence of relational algebra and relational calculus query languages having aggregate functions. *Journal of ACM*, 29(3), 699-717.

Lehner, W. (1998). Modeling large-scale OLAP scenarios. *Proceedings of the International Conference on Extending Database Technology (EDBT'98)*, 153-167.

Lenz, H.J., & Shoshani, A. (1997). Summarizability in OLAP and statistical data bases. *Proceedings of the International Conference on Scientific and Statistical Database Management (SSDBM'97)*, 132-143.

Levy, A.Y., Mendelzon, A.O., Sagiv, Y., & Srivastava, D. (1995). Answering queries using views. *Proceedings of the ACM Symposium on Principles of Database Systems (PODS'95)*, 95-104.

Li, C., & Wang, X.S. (1996). A data model for supporting On-Line Analytical Processing. *Proceedings of the Conference on Information and Knowl-*

edge Management (CIKM'96), 81-88.

Libkin, L., & Wong, L. (1993). Some properties of query languages for bags. *Proceedings of the International Workshop on Database Programming Languages (DBPL'93)*, 97-114.

Libkin, L., & Wong, L. (1997). Query languages for bags and aggregate functions. *Journal of Computer and System Sciences,* 55(2), 241-272.

Nutt, W., Sagiv, Y., & Shurin, S. (1998). Deciding equivalences among aggregate queries. *Proceedings of the ACM Symposium on Principles of Database Systems (PODS'98)*, 214-223.

O'Neil, P.E. (1987). Model 204 Architecture and Performance. In *International Workshop on High Performance Transaction Systems (HPTS'87)* pp. 214-23).

O'Neil, P.E., & Graefe, G. (1995). Multi-table joins through bitmapped join indices. *SIGMOD Record,* 24(3), 8-11.

O'Neil, P.E., & Quass, D. (1997). Improved query performance with variant indexes. *Proceedings of the ACM International Conference on Management of Data (SIGMOD'97)*, 38-49.

Özsoyoglu, G., Özsoyoglu, Z.M., & Matos, V. (1987). Extending relational algebra and relational calculus with set-valued attributes and aggregate functions. *Transactions on Database Systems,* 12(4), 566-592.

Pourabbas, E., & Rafanelli, M. (2000). Hierarchies and relative operators in the OLAP environment. *SIGMOD Record,* 29(1), 32-37.

Rafanelli, M., Bezenchek, A., & Tininini, L. (1996). The aggregate data problem: A system for their definition and management. *SIGMOD Record,* 25(4), 8-13.

Ross, K.A., & Srivastava, D. (1997). Fast computation of sparse datacubes. *Proceedings of the International Conference on Very Large Data Bases (VLDB'97)*, 116-125.

Ross, K.A., Srivastava, D., & Chatziantoniou, D. (1998). Complex aggregation at multiple granularities. *Proceedings of the International Conference on Extending Database Technology (EDBT'98)*, 263-277.

Shoshani, A., & Wong, H.K.T. (1985). Statistical and scientific database issues. *IEEE Transactions on Software Engineering,* 11(10), 1040-1047.

Srivastava, D., Dar, S., Jagadish, H.V., & Levy, A.Y. (1996). Answering queries with aggregation using views. *Proceedings of the International Conference on Very Large Data Bases (VLDB'96)*, 318-329.

Su, S.Y.W. (1983). SAM*: A semantic association model for corporate and scientific/ statistical databases. *Information Sciences,* 29, 151-199.

Tininini, L., Paolucci, M., Sindoni, G., & De Francisci, S. Spatio-temporal information systems in a statistical context. *Proceedings of the International Conference on Extending Database Technology (EDBT'02)*, 307-316.

Valduriez, P. (1987). Join indices. *ACM Transactions on Database Systems*, 12(2), 218-246.

Widom, J. (1995). Research problems in data warehousing. *Proceedings of the Conference on Information and Knowledge Management (CIKM'95)*, 25-30.

Wong, H.K.T., Liu, H.F., Olken, F., Rotem, D., & Wong, L. Bit transposed files. *Proceedings of the International Conference on Very Large Data Bases (VLDB'85)*, 448-457.

Wong, H.K.T., Li, J., Olken, F., Rotem, D., & Wong, L. (1986). Bit transposition for very large scientific and statistical databases. *Algorithmica* 1(3), 289-309.

Zhao, Y., Deshpande, P., & Naughton, J.F. (1997). An array-based algorithm for simultaneous multidimensional aggregates. *Proceedings of the ACM International Conference on Management of Data (SIGMOD'97)*, 159-170.

Chapter X

Incomplete Information in Multidimensional Databases

Curtis E. Dyreson
Washington State University, USA

Torben Bach Pedersen
Aalborg University, Denmark

Christian S. Jensen
Aalborg University, Denmark

ABSTRACT

While incomplete information is endemic to real-world data, current multidimensional data models are not engineered to manage incomplete information in base data, derived data, and dimensions. This chapter presents several strategies for managing incomplete information in multidimensional databases. Which strategy to use is dependent on the kind of incomplete information present, and also on where it occurs in the multidimensional database. A relatively simple strategy is to replace incomplete information with appropriate, complete information. The advantage of this strategy is that all multidimensional databases can manage complete information. Other strategies require more substantial changes to the multidimensional database. One strategy is to reflect the incompleteness in computed aggregates, which is possible only if the multidimensional database allows incomplete values in its hierarchies. Another strategy is to measure the amount of incompleteness in aggregated values by tallying how much uncertain information went into their production.

INTRODUCTION

Multidimensional databases are a relatively recent and popular phenomenon. A concise description of a multidimensional database is that it is a hierarchy of aggregate values. Values higher in the hierarchy are further aggregations of those lower in the hierarchy. The utility of the hierarchical organization is that the user can easily navigate among high and low precision views of the same aggregate data using *drill-down* and *roll-up* operations. Drill-down increases the precision of aggregate data being viewed while roll-up decreases the precision. For instance, suppose that a grocery store manager uses a multidimensional database to examine monthly sales for apples and notices that sales in January were low. To analyze the poor sales, the manager might drill-down to look at monthly sales by kind of apple, or she might roll-up to view sales for all fruits combined. Several vendors already have multidimensional database products on the market, either as add-ons to existing databases or as stand-alone tools, and a "cube" operator has been proposed for inclusion in future SQL standards (Gray et al., 1996, 1997).

A database model is only a small part of reality, but complete information about even this limited part is rare. For example, suppose that in an on-line grocery store, a customer fills out a web form to make a purchase. Some customers will invariably leave a portion of the form blank. Although a database could reject an incomplete purchase order, this would result in lower sales. A well-designed database should accept and store the data about an incomplete purchase, especially if the incomplete information is not essential to the purchase order. To cope with incomplete information that is present in real data, database management systems have evolved techniques for storing and reasoning about it. The most common technique in relational and object-oriented database management systems uses a *null value* to represent incomplete information. An ANSI report identified 14 possible interpretations of a null value in SQL, ranging from a value that is inapplicable, to one that is known to exist, but has an unknown value (ANSI/X3/SPARC, 1975).

A multidimensional database can also have incomplete information. The *base data* could be incomplete. For example, a multidimensional database for an on-line grocery store could store incomplete purchase orders. Incomplete information can also appear in the *derived data*. The derived data is typically an aggregate or summary view of the base data. For example, a count of incomplete purchase orders might result in an incomplete count. Finally, and somewhat surprisingly, it is not uncommon to have incomplete information in the *dimensions*. This is often the result of an incomplete specification. For example, suppose that a customer leaves the city field in a purchase order form blank because the customer lives on a farm. Customer orders per state will be undercounted if only customers in cities are counted.

The objective of this chapter is to present techniques for managing incomplete information in a multidimensional database. Although incomplete information is endemic to real-world data, this area has received less attention in the research community than other areas (Vassiliadis, 2000), and few commercial products have support for any kind of incompleteness. The lack of research and commercial emphasis on managing uncertainty in a multidimensional database is somewhat surprising since a multidimensional database is a decision support tool, and humans have a long history of making decisions based on approximations and other incomplete summaries of data. There are several simple techniques that can be used to facilitate the management of incomplete information in a multidimensional database. The next section gives a background on incomplete information in database systems, discussing the limited amount of related work. In a multidimensional database, incomplete information can appear in the base data, the measures, the hierarchy, and in the metadata. In this chapter we discuss the problems that arise in each case and present specific techniques for handling the incomplete information. The chapter concludes with a brief examination of the open problems in the management of incomplete information. The presentation throughout the chapter is informal and practical.

BACKGROUND

Research in multidimensional databases has been extensive (Mendelzon, 2001). In this section we first give a brief overview of a multidimensional database and introduce terminology used in the remainder of the chapter. Next, we outline the treatment of incomplete information in other database models, e.g., the relational model. Finally we consider related work in multidimensional databases.

Overview of a Multidimensional Database

A useful way of understanding a multidimensional database is to conceive of it as a hierarchy in a multidimensional space. Each dimension is composed of a set of related *categories*. A category can be thought of as a system of measurement. For example, a Time dimension has categories corresponding to various ways to measure time, such as Months and Quarters, while a Geography dimension could have categories for Countries and Continents. A category is a set of individual measurements called *category units*, which we will shorten in this chapter to *units*. In a Time dimension, the units in the Month category would be values such as "January 2001" and "March 2001." In this chapter we use strings to label individual units.

A node in the hierarchy is a point in multidimensional space. The coordinates of a node are a unit chosen from each dimension. Each node contains one or more values, which are called *measures*. A measure is a quantity being measured. Each measure in a node is the result of an *aggregate* computed on a set of facts, usually obtained from a database relation. The possible aggregates typically include **count**, **max**, **min**, and **sum**. To compute a measure in a node, the facts are grouped by unit and the aggregate is applied to the group.

Figure 1: A Multidimensional Space

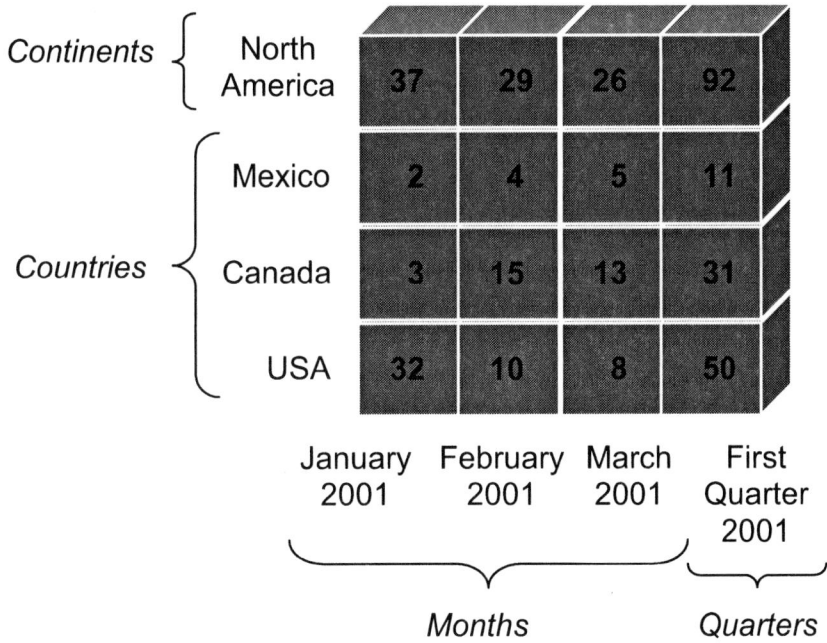

Figure 1 shows an example multidimensional space with two dimensions: Geography and Time. The Time dimension has the categories Months and Quarters. The figure depicts only the units for the first three months in 2001. The Geography dimension has categories for Countries and Continents. The value or measure depicted at each point in the multidimensional space is a **count** of the number of apples sold at a store in that geographic location during that time.

The hierarchy also has edges. An edge represents a data dependency: the data at the "to" node is derived from the data at the "from" node. An edge connects a unit to a less precise version of itself in another category. For example the hierarchy would have an edge from the unit "January 2001" in the category of Months to the unit "First Quarter 2001" in the category of Quarters. *Figure 2* shows the hierarchy for the multidimensional space of *Figure 1*. The hierarchy has several base nodes with no incoming edges. The measures in the base are computed directly from the underlying data. The base nodes are shaded light gray in *Figure 2*. The non-base nodes are called *derived* nodes. The measures in each derived node are the result of an aggregate applied to the measures at nodes on incoming edges. In the example of apple count data, the **sum** aggregate is used is to add the counts.

For the derived nodes to be computable from other nodes, the relationship between a pair of categories must be *strict, covering,* and *onto* (Pedersen et al., 2001). Consider the relationship between a fine category and a coarse category. By *strict,* we mean that the relationship is many-to-one, i.e., a unit in the fine category maps to at most one unit in the coarse category. By *covering,* we mean that there is an edge from every unit in the fine category to some unit in the coarse category. Said differently, the fine category totally participates in the relationship. Finally, by *onto* we mean that every unit in the coarse category has some incoming edge from a unit in the fine category. Said differently, the coarse category participates totally in the relationship. The categories, units, and the relationships among them are the *metadata* in a multidimensional database. Although the cube is a multidimensional space, each dimension is specified in isolation.

A multidimensional database can be implemented using a *lazy, eager,* or *semi-eager* strategy (Widom, 1995). The eager strategy materializes *every* aggregate value in the hierarchy. The advantage of the eager strategy is that values can be quickly fetched from the cube during a query. The primary disadvantage is high storage cost (Shukla et al., 1996, present algorithms for estimating cube size). For many applications, eager cubes are just too big, often 200-500 times the size of the

Figure 2: A Hierarchy in the Multidimensional Space

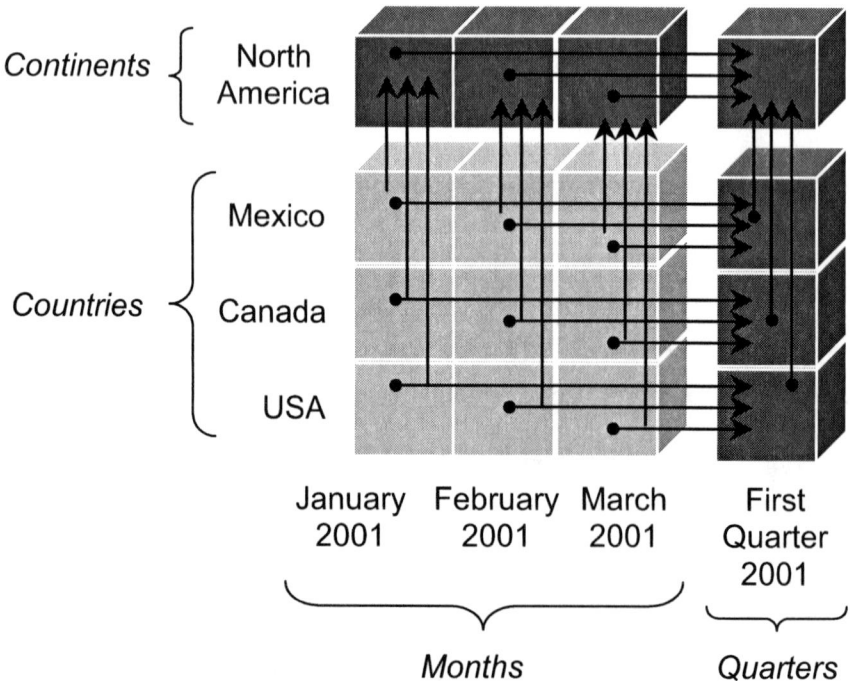

base data (OLAP Report, 2001). A lazy implementation strategy does *not* materialize values. Instead, the values in the hierarchy are computed from the underlying base data during a query. The disadvantage of the lazy strategy is slower query evaluation (faster evaluation algorithms are being researched, e.g., Agrawal et al., 1996). The semi-eager strategy eagerly materializes only *some* regions in the multidimensional database, and lazily computes others when needed during a query (Harinarayan et al., 1996), providing a good trade-off between storage use and query response time.

INCOMPLETE INFORMATION IN DATABASE MANAGEMENT SYSTEMS

There has been extensive research on incomplete information in the context of relational, deductive, and object-oriented databases (Dyreson, 1997a). This research has covered incompleteness at the level of attribute values, tuples, and in the schema. Many kinds of incomplete information have been identified. The term "incomplete" covers fuzzy, imprecise, indeterminate, indefinite, missing, partial, possible, probabilistic, unknown, uncertain, and vague information. In the following sections, the most common kinds of incomplete information are discussed.

What Do We Mean By Incomplete?

The definition of the term "incomplete" given in *The Concise Oxford English Dictionary* is perhaps too concise: incomplete is defined as "not complete." The definition of "complete" is of more help: complete means "entire" or "whole." An object, then, is incomplete with respect to an object that is entire or whole. The incomplete object is missing something from its more complete partner. For example, an ancient Greek statue that is missing an arm is incomplete with respect to the statue with that arm. If the statue's arm is unearthed and reaffixed, the statue could be made complete.

It is important to note that information is incomplete only with respect to more complete information (which in turn could be incomplete with respect to still more complete information). The difference between complete and incomplete information is often a matter of *precision*. For instance, the fact that the temperature is 56 degrees is less precise than that the temperature is 56.4 degrees. Both statements about the temperature are, however, more precise than the assertion that the temperature is "in the fifties."

Unknown Values

An *unknown value* is a value that is known to exist, but the actual value is unknown (Codd, 1979). The unknown value is assumed to belong to some specified domain of values. Unknown values are a very common kind of incomplete information; a null value in SQL is often interpreted to mean an unknown value. For example, in a grocery database, every product must have a calorie count, but the number of calories in a pack of gum could be recorded as unknown. The unknown value indicates that the gum has some non-negative number of calories (i.e., the number of calories is not "purple" or "-3"). An unknown value has various names in the literature, including *missing null* (Goldstein, 1981) and *existential null* (Biskup, 1984; Minker, 1982).

An unknown value is often semantically overloaded to mean either an unknown value or an inapplicable value (Codd, 1979), in which case it is a *no information null*. Problems that result from such an overloading have been identified (Zaniolo, 1984). The use of four-valued logics has been advocated to untangle the semantic confusion (Gessert, 1991).

Imprecise Values

A generalization of an unknown null is an *imprecise value*. An imprecise value is an attribute value that is known to exist and known to be a single value from a *sequence* of the attribute domain. For example, in a grocery database, the number of grams of protein in an apple is recorded as between 1 and 3 (the sequence 1, 2, 3), but the precise amount is not known since it varies from apple to apple. The amount of protein is said to be imprecise. A special case of an imprecise value is an unknown value (the subset is the domain itself). Another special case is a precise or complete value (the subset is a singleton set). *Partially known values* (Grant, 1979) and *set nulls* (Keller & Wilkins, 1985) are imprecise values.

Disjunctive Values

Another generalization of an unknown null is a *disjunctive value* (Grant & Minker, 1986), also known as an *indefinite value* (Liu & Sunderraman, 1990, 1991). A disjunctive value is a non-empty subset of the attribute domain. For example, an apple might be a McIntosh, a Granny Smith, or a Fuji apple. The disjunction can be *exclusive* (Ola, 1992) or *inclusive* (Homenda, 1991). If it is an exclusive disjunction, one and only one disjunct is the true value. If it is inclusive, then more than one disjunct can be the true value.

Exotic Incomplete Values

A *maybe value* is an attribute value that might or might not exist (Gessert, 1991). If it does exist, the value is known. For instance, we could store in our grocery database that the price of a packet of coffee creamer is maybe 10 cents. Coffee creamer packets are usually given away freely, and therefore the price attribute is inapplicable, but some stores sell the coffee creamer at a cost of 10 cents. A *maybe tuple* is similar to a maybe value, but the entire tuple might not be part of the relation (Liu & Sunderraman, 1991). Maybe tuples are produced when one disjunct of an inclusive disjunctive fact is found to be true. The other disjuncts become maybe tuples.

A combination of inclusive disjunctive and maybe information is *open* information. An open value indicates that an attribute of a particular tuple is under the open world assumption (Gottlob & Zicari, 1988). The attribute value may not exist, could be exactly one value, or could be many values. For example, in the employee database an open value could be used for Jill's previous employment history. This value means that Jill possibly had previous employment (this could be Jill's first job), Jill might have had one previous job, or Jill might have been employed many times previously. The open value covers all these possibilities.

A *no information value* is a combination of an open value and an unknown value (Zaniolo, 1984). The no information value restricts an open value to resemble an unknown value. A no information value might not exist, but if it does, then it is a single value that is unknown, rather than possibly many values.

A generalization of open information is *possible* information (Lipski, 1979). Possible information is an attribute value that has an undetermined existence. If it does exist, it could be multiple values from a *subset* of the attribute domain. For example, in an employee database, an employee's previous employment history could be narrowed to possibly two companies; she could have worked for both companies, only one, or neither. A special case of a possible attribute value is an open value (the subset is the domain itself). Another special case is a maybe value (the subset is a singleton set).

Probabilistic Values

A *probabilistic* value is a generalization of an exclusive disjunctive value. A probabilistic value is a set of alternatives, where each alternative has an associated probability that it is the attribute value (Barbará et al., 1992; Cavallo & Pittarelli, 1987). For example, in a grocery database, assume that we do not know the fat content of Spam exactly, but are 70% sure it is 50 grams and 30% sure that it is 45 grams. The fat content is a probabilistic data value; the value exists, it is a value from a known subset of the attribute domain, it is exactly one value, and we know that some alternatives are more likely than others. In some models, one of the members of the set of alternatives could be an unknown value, in which case the associated probability is distributed uniformly over the elements in the domain (Barbará et al., 1992). In other models the probability is at the tuple-level, indicating the likelihood that the tuple is a member of a relation (Gelenbe & Hebrail, 1986).

Possibilistic Values

Another variety of weighted incomplete information is a *possibilistic* or *fuzzy set* value. A possibilistic value is similar to a possible value. A fuzzy set is a set of possibilities. Each possibility is a maybe value, that is, it may belong to the set or it might not (Zadeh, 1989). The possibility that it does belong is given by a membership function (also known as the *degree* of membership). The degree is a value between 0 and 1 (inclusive). A fuzzy set can be an attribute value.

INCOMPLETE INFORMATION IN MULTIDIMENSIONAL DATABASES

While multidimensional data models that store and manage complete information have been studied extensively, there are only a few papers that address incomplete information for multidimensional databases.

Shoshani (1997) compares multidimensional databases with statistical data models. In the tradition of the influential SUBJECT data model (Chan & Shoshani, 1981), multidimensional data models support only two kinds of nodes: cluster nodes (for units) and cross-product nodes (for combining dimensions). The greater sophistication in describing statistical data found in later statistical data models such as SAM* (Su, 1983) and STORM (Rafanelli & Shoshani, 1990) is absent from most multidimensional data models because of a difference in data modeling requirements. Statistical data models are designed to model complicated, nonstandard, heterogeneous, real-world data sets, whereas multidimensional database models create their own simple, standard, homogeneous statistical data set; consequently a simpler data model suffices. The statistical database aspect of multidimensional databases is best understood as an extension of the work done by Malvestuto (and others) on data integration in statistical databases (Malvestuto, 1991). Data integration creates a unified view of a set of different, but homogeneous, statistical tables.

Dyreson (1996) first wrote about incomplete information in a multidimensional database context. He developed a data cube (a multidimensional database) that can contain regions of *unknown values*. Queries on the unknown regions are either redirected to the nearest complete information regions or computed along with *completeness measures*. The completeness measure is a percentage of how much complete information is used in the evaluation of a query. The key to the incomplete data cube is a high-level specification of which regions in multidimensional space are complete, an idea that first gained acceptance in semantic, statistical data models (Sato, 1991). Efficient algorithms to query, update, and reorganize the cube are given. Dyreson (1997) describes a research prototype that automates the loading of data from log files into an incomplete data cube.

Making use of incomplete regions within a complete cube is also the motivation behind *quasi-cubes* (Barbará & Sullivan, 1997). A quasi-cube trades accuracy for space. In a quasi-cube, regions of an eager (fully materialized) multidimensional database are replaced with a single approximated value. The approximated value is subsequently used in operations to quickly provide an *estimate* of an actual value. Complete, accurate values can be computed from the base data when desired. The approximation is a kind of incomplete information. Quasi-cubes have techniques for differentiating between approximated data and complete data in a query.

Pedersen et al. (1999, 2001) describe a complete multidimensional data model that supports both incomplete data and metadata (among other things). One culprit that leads to uncertainty in aggregate values is an incompletely specified hierarchy in the metadata. Pourabbas & Rafanelli (2000) refer to incomplete hierarchies as partial classification hierarchies. For example, consider the problem of *non-strictness*, that is, many-to-many mappings between categories, in a hierarchy. Aggregating nodes that contribute to more than one unit may lead to problems. If a tomato is categorized as both a fruit and a vegetable, the computation of an aggregate value for produce (which includes both fruits and vegetables) will be incorrect because tomatoes will have been considered twice. In some sense the problem is caused by an incomplete, but very human, specification of the hierarchy. A complete specification would have a category for vegetable-fruits that would contribute exactly once to each of the produce, fruit, and vegetable categories. Research in this area has proposed efficient techniques to automatically translate among specifications.

Jagadish et al. (1999) also discuss techniques for handling problems in the specification of metadata and how to aggregate with imprecise values in the grouping attributes (those that map to somewhere above the base of the hierarchy).

TECHNIQUES FOR HANDLING INCOMPLETENESS

In this section we present techniques for handling incompleteness in the base data, derived data, and metadata.

In the base data, a measure could be incomplete. The basic problem posed by incomplete measures is arithmetic. For instance, what is the result of an aggregate, such as **max**, computed on a set of values if some of those values are unknown or otherwise incomplete? The two basic strategies are to replace the incomplete measure with a complete value and then compute the aggregate, or to manufacture an incomplete result and insert it into the hierarchy. The chief advantage of a replacement strategy is that it is easy to implement since multidimensional databases are designed to handle complete data. The disadvantage is that no matter what replacement value is used, some useful information is lost.

The base data could also have incomplete "grouping" attributes. This will result in facts that group at more than one base node. For example, if a multidimensional database is constructed to count how many apples are sold each day and the database records a sale on an unknown date, to which base node should the sale belong? We discuss techniques for calculating "bounds" on groups and for modifying the hierarchy to introduce new nodes for incompletely specified groups.

The derived data (i.e., the values in the hierarchy) could be incomplete. The chief source of incomplete derived data is that techniques for repairing incomplete base data often generate incomplete values. The basic problem and solution is similar to that for incomplete measures. While there are many similarities with incomplete base data, additional techniques can utilize complete values that are "nearby" in the hierarchy.

Finally, the metadata can be incomplete. We present several examples of incomplete metadata and discuss techniques for completing incomplete metadata specifications.

INCOMPLETE MEASURES

A measure is a value in the base data that the multidimensional database will aggregate at each point in the hierarchy. In this section we outline the problems posed by three of the kinds of incomplete information that could appear in a measure attribute: unknown values, imprecise values, and probabilistic values. The techniques proposed to manage the incomplete information are adaptable to other kinds of incompleteness.

Unknown Measure Values

Unknown values will be the most common kind of incomplete information in a multidimensional database. The reason is that the base data is often stored in an SQL-compliant database, and only one kind of incomplete information is supported in SQL: *null values*. A null value in SQL usually represents either an unknown value or an inapplicable value (Melton & Simon, 1993). Many real-world datasets have null values. SQL is not very adept at dealing with null values in aggregate operations; it discards null values in aggregates (in effect treating them as inapplicable values).

To illustrate techniques for dealing with unknown values in the base data for a multidimensional database, assume that the protein attribute of a Nutrition tuple for Fuji apples is an unknown value. The incomplete information indicates that Fuji apples have some amount of protein, but it is unknown exactly how much. The problem is how to best utilize the unknown value, while remaining sensitive to the limited knowledge provided by the value.

There are several strategies for computing the maximum protein and counting the overall amount of protein for apples with an unknown amount of protein. The strategies are discussed below and illustrated in *Figure 3*. In each part of the figure, two base data facts and a single base node are shown. The base node measures the maximum amount of protein among products sold.

Discard the unknown value: This strategy is to throw out the incomplete information rather than using it, which is the default SQL behavior for aggregates. The advantage of this technique is that it is extremely simple and efficient. All multidimensional database products can implement the strategy with no loss of efficiency. *Figure 3(a)* shows the idea. The unknown value is represented by a '@' character.

One disadvantage of discarding unknown values is that the technique will likely produce semantically *incorrect* information. The unknown value that is discarded could be the maximum or could be a non-zero quantity in a summation. Another problem occurs when all the base data is unknown. Consider the case of computing the maximum protein for Fuji apples shown in *Figure 3(b)*. The base node needs to store the **max** protein for Fuji apples. Unfortunately, there is no such value since the amount of protein is unknown for all the base data. Some number, however, must be inserted, so a reasonable default value is used. The obvious candidate is a negative number. The negative number is a good choice since complete information about an amount of protein will always be zero or greater. So when points higher in the hierarchy are needed, e.g., get the maximum protein over all fruits, the negative values (incomplete information) will be discarded in favor of non-negative ones (complete information). For **sum** aggregates, a default value of zero would be used. A user will have to correctly interpret any spurious replacement values that appear in a result, but the technique is simple to implement.

Replace with an *imputed* value: Imputation is the process of using other information to estimate an unknown value. Techniques for imputing missing values are common in statistics. A standard method is to use a mean or median value. For example, the mean amount of protein in other kinds of apples is 3.2 grams of protein, so 3.2 could be imputed for Fuji apples that have unknown values. *Figure 3(c)* shows an example of replacing the unknown with the mean.

Imputing a value could be a more complex computation that depends on the values of other attributes and additional metadata about the unknown value. For example, assume that apples have a color attribute, which can be the value red, green, or yellow. Fuji apples are green. Available metadata about green apples states that they lack a color-dependent gene found in red apples that produces protein. Other metadata about apples includes the knowledge that the amount of protein is within 20% of the amount of vitamin C in an apple. So the amount of protein in Fuji apples is computed as an estimate of the lower bound of the amount in red apples, scaled by a factor of the amount of vitamin C. The primary advantage of imputation is that it produces a statistically meaningful replacement value, as opposed to a non-statistical default value. A second advantage is that in an eager implementation, the cost of imputing a value is incurred at most once.

One disadvantage of imputation is that an imputed value is typically an average, mean, median, or expectation rather than an extreme value, so it would have no effect on a **max/min** aggregate. A second disadvantage is that in spite of the availability of metadata for imputing values in many applications, especially in the health-related industries, no database product that we are aware of offers any support for using metadata to impute values. Such support would further degrade the performance of lazy multidimensional databases since the impu-tation is done during query evaluation. A third disadvantage is that the imputed value is given the same status within the hierarchy as a non-imputed value (both are just numbers), but in reality we have less confidence in the imputed value. This can be remedied by storing the method of imputation and a value representing the degree of confidence in the result along with the imputed value. But the multidimensional database must then have some technique for handling confidence values in the hierarchy. Finally, the use of imputation will often mean that the *variance* of the data will be artificially small, since the imputed values most often do not contribute to the variance. This can be solved by using *multiple imputation* (Rubin, 1987), where imputation is performed in a number of separate passes, the results of which are then combined into the overall result.

Generate an unknown result: An unknown value is used as the result. Computing a **max** on an unknown value would result in an unknown answer as shown in *Figure 3(d)*. This pushes the problem up the hierarchy.

Measure the completeness: The ratio of unknown-to-known values is used as a measure of the *completeness* of the result. Different notions of completeness have been advanced, but a common notion is that the completeness is a *percentage* of the known values used to compute an aggregate. An aggregate computed entirely on known values has 100% completeness, while one computed half of unknown values has a completeness of 50%. An alternative method of computing completeness is to weight the alternatives. Note that the completeness can be added to the multidimensional database as a measure; it is a count of known values and unknown values. *Figure 3(e)* shows the computation of a completeness measure for a **max** aggregate. The complete-ness is represented in parentheses next to the value as a pair of the count of known values and all values. If the multidimensional database treats the completeness as a measure, it can be computed easily for derived nodes. However, the measure should be displayed to the user as a percentage.

Imprecise Measure Values

The handling of imprecise values differs slightly from unknown values. An imprecise value is a sequence of potential values. In the sense that it represents a set

Figure 3: Options for Handling an Unknown Value (in a Max Aggregate)

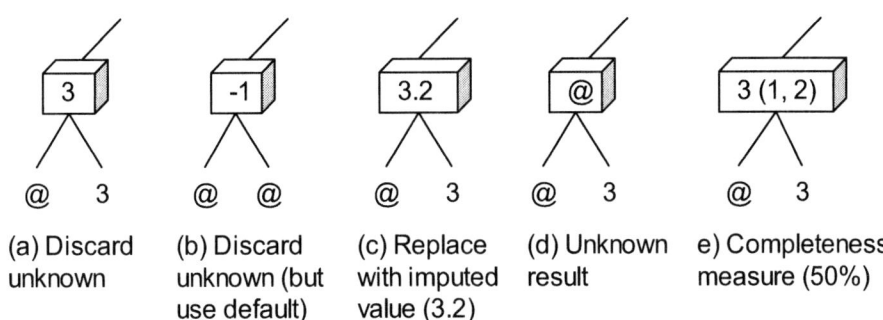

| (a) Discard unknown | (b) Discard unknown (but use default) | (c) Replace with imputed value (3.2) | (d) Unknown result | e) Completeness measure (50%) |

of potential values, one of which is the actual value, it is similar to an unknown value. The key difference is that the imprecise value may possibly map to somewhere below *top* in the hierarchy in each dimension, since the hierarchy is effectively a precision hierarchy. So there is often a point in the hierarchy above at which the imprecise value is completely known. For example, a date of July 1, 2001, is imprecise with respect to a category of hours since it spans 24 potential hours. But the date is complete information at the category of days, since the exact day is known.

One source of imprecise values is a data warehouse that integrates several data sources. Normally, the data warehouse cleanses or scrubs the data prior to insertion. One aspect of cleansing is to remove category mismatches among the data sources by mapping the data to the coarsest category. For example, if the time of sale from one source is given in hours, while that of a second source is given in days, then the hourly data is scrubbed by mapping each hour to the day that contains it. The scrubbing makes "time of sale" data uniform across the data sources, so that the data can be inserted into the data warehouse. The drawback is that some information is discarded during the data scrubbing. An alternative strategy is to record imprecise values. Each day can be represented as an imprecise value of 24 hours.

The following techniques can be used to handle imprecise values. They are similar to those for unknown values, but include extra conditions to manage the mapping of the imprecise value to some precise location in the hierarchy. *Figure 4* illustrates the techniques. The figure assumes that a **max** aggregate is being performed on an imprecisely known amount of protein in two varieties of apple.

Discard the imprecise value: Same as for unknown values. This alternative is not a very good one for imprecise values because an imprecise value is more complete than an unknown value. *Figure 4(a)* shows the strategy. The imprecise value is discarded, along with the information that the amount of protein might be larger than 3.2.

Replace with an imputed value, default value, or a randomly sampled value: Imputation can be more accurate for imprecise values because the set of potential values is smaller. Imputation metadata for an imprecise value would typically include a probability distribution over the range of potential values. The probability distribution could be used to impute the expected value. In *Figure 4(b)* the expected value, 3.5, is used to replace the imprecise value. It happens to be the maximum. Expected values are unappealing for extreme value aggregates like **max** and **min**. An alternative is to use random sampling. In random sampling, the probability distribution is used to select a potential value. The randomly chosen value is used instead of the expected value. In aggregation, for large numbers of imprecise values, random sampling has the same desirable effect as imputing the expected value, since the values will average out to the expected value. But random sampling will produce extreme values that are absent when imputing the expected value. Since values are chosen randomly, however, the same aggregate computed on the same data twice in succession may yield different results.

Replace with a "bound": An imprecise value is a range of potential values, from a least potential value to a greatest potential value. For **max** and **sum** aggregates, a "pessimistic" choice would be to replace the imprecise value with the least potential value. An "optimistic" choice would be to use the greatest potential value. For a **min** aggregate the choices would be reversed. In *Figure 4(c)* the greatest potential value is chosen, and it turns out to be the maximum.

The advantage of all the replacement strategies is that the replacement value is a complete value, so the implementation is simple and the rest of the multidimensional database does not change. The disadvantage is that no matter what choice is made, some useful information is lost because the underlying data value is imprecise.

Figure 4: Options for Handling an Imprecise Value (in a Max Aggregate)

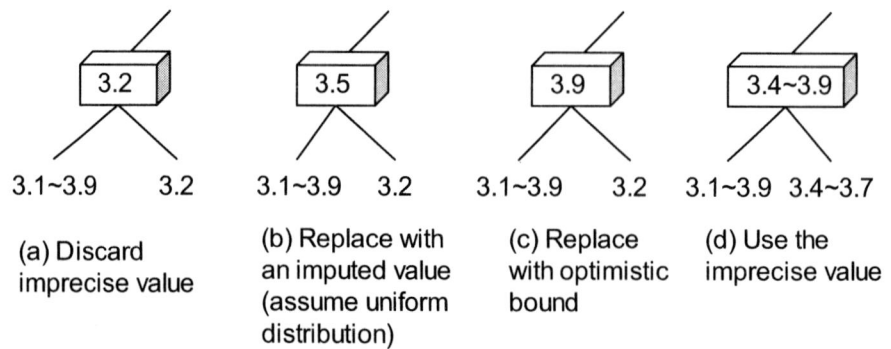

(a) Discard imprecise value

(b) Replace with an imputed value (assume uniform distribution)

(c) Replace with optimistic bound

(d) Use the imprecise value

Use the imprecise value: The aggregate uses the imprecise value to compute an imprecise result. The bounds are computed by interval arithmetic. As the category becomes coarser, the imprecision will gradually reduce until it eventually disappears. *Figure 4(d)* depicts an example of computing the bounds. Two imprecise values are compared, and the maximum of the lower and upper bounds are propagated into the result.

Probabilistic (and Possibilistic) Values in the Data

A probabilistic value in the data is a weighted, exclusive, disjunctive value. The weights provide additional information about which potential values are more likely, but change the management of the incomplete information only slightly from that for unweighted values.

Replace with an imputed value: As discussed in the handling of imprecise values, it is natural to use the probabilities to impute an expected value or to use random sampling to select a potential value. In aggregation, for large numbers of probabilistic values, random sampling has the same desirable effect as imputing the expected value, since the values will average out to the expected value. But random sampling will produce extreme values that are absent when imputing the expected value.

Use a probabilistic value: Bayesian probability theory has sound formulas for computing the **count**, **max**, **min**, and **sum** of probabilistic values. The drawback is that the formulas have relatively high space and time complexity. For a set of n probabilistic values, each with m potential values, the worst-case time and space cost to compute an aggregate is $O(m^n)$.

The handling of possibilistic values, which are weighted inclusive disjunctive values, is similar to that of probabilistic values.

INCOMPLETE GROUPING ATTRIBUTES

Incompleteness in a grouping attribute is a different problem than incompleteness in a measure. In this section we focus on problems and techniques for grouping of exclusive and inclusive disjunctive values only. The techniques that are presented can be adapted to cover other kinds of incomplete information in grouping attributes.

Exclusive Disjunctive Values in the Data

An exclusive disjunctive value is similar to an imprecise value, but the potential values do not have to be in a sequence. The primary problem in handling an exclusive disjunctive value is ascertaining the group it belongs to since it potentially could belong

298 Dyreson, Pedersen, and Jensen

to several base nodes. The techniques discussed below are in the context of a **count** aggregate that is counting the number of apples sold at a store. There are many different kinds of apples, and apples are just one kind of product sold at the store.

Discard the unknown value: This alternative is unattractive because an exclusive disjunctive value could have a very limited number of alternatives. *Figure 5(a)* shows a fragment of a multidimensional database for counting the number of apples sold. One McIntosh, one Fuji, and one additional apple are sold, but the cashier could not decide if the last apple sold was a McIntosh or a Fuji. The incomplete sale information is represented by the exclusive disjunctive value {McIntosh, Fuji}. In *Figure 5(a)* the incomplete information is discarded during the computation of the count aggregate, resulting in an undercount of the number of apples sold. This technique has a straightforward and easy implementation, but it at best underestimates the data.

Maintain min-max bounds: An exclusive disjunctive value potentially belongs in several groups at base nodes. With respect to a base node, the maximum bound assumes that the value is in the group for that node. The minimum bound, on the other hand, assumes that the value is in some other group. The aggregate for a base node is computed with respect to both the minimum bound group and the maximum bound group. *Figure 5(b)* illustrates the technique. The base nodes contain a min-max pair. The exclusive disjunctive value could belong to the McIntosh group or the Fuji group. So the maximum bound is two in each base node, whereas the minimum bound is one. Note that this technique introduces an imprecise value into the hierarchy.

Introduce unknown nodes: Min-max bounds maintain the limits on the possible undercount and overcount in a multidimensional database. But the limits become artificially exaggerated higher in the hierarchy. The distance between the min-max bounds in the base nodes is at most n, where n is the number of exclusive disjunctive values. But at the next level higher up in the hierarchy, the distance becomes $n*g$, where g is the number of base nodes. A refinement is to introduce a new base node to count the number of exclusive disjunctive values. An example is shown in *Figure 5(c)*. The additional base node is labeled *"unknown."* It records the number of apples that could not be precisely identified, i.e., it counts the number of exclusive disjunctive values. When the count of all apples is needed, the unknown base node adds its contribution to the overall count.

Introduce XOR nodes: The main problem with introducing an unknown node is that it discards some information. The exclusive disjunctive value is almost always just a small number of potential values, one of which is the actual value.

Pedersen et al. (1999) develop a clever strategy that loses less information. The technique is to add a node for each unique exclusive disjunctive value rather than a single unknown node. The additional node "fixes" the problem of overcounting and undercounting at the next higher level in the hierarchy. We will call the new node an XOR node (an exclusive-or node). *Figure 5(d)* illustrates the method. An XOR node labeled "McIntosh or Fuji" is introduced to count the number of {McIntosh, Fuji} exclusive disjunctive values. XOR nodes are hidden from the user since they are introduced by the system. In *Figure 5(d)* the node is shaded gray to indicate that it is hidden.

Exclusive disjunctive values are grouped only at XOR base nodes for computing aggregates. Only complete information is used to compute non-XOR base nodes. A further modification of the hierarchy allows min-max bounds, or other derived incomplete information, to be kept. Each non-XOR base node is split

Figure 5: Options for Handling an Exclusive Disjunctive Value (in a Max Aggregate)

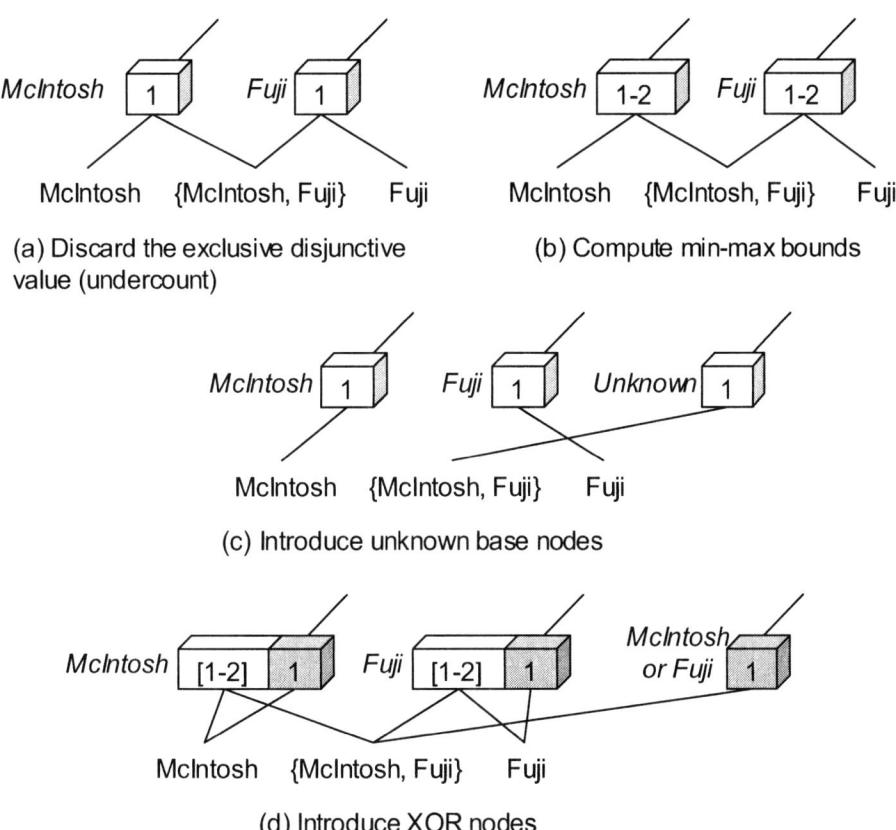

into a hidden and a visible component. The hidden component stores the count of complete information. The visible component stores the min-max computations (they become *terminal* measures that do not contribute up the hierarchy). *Figure 5(d)* illustrates the method. The base nodes for McIntosh and Fuji apples are split; the hidden component of each node is shaded in gray. The exclusive disjunctive value contributes to the visible component, but only the complete data values contribute to the hidden component. The hidden components are used for aggregating up the hierarchy.

The chief disadvantage of this strategy is the space cost. Potentially one new unit is added for each exclusive disjunctive value in a dimension. This increases the size of the dimension, which in turn makes the multidimensional space larger since the number of points in the multidimensional space is a product of the size of each dimension. Correctly managing incomplete information can be expensive.

Inclusive Disjunctive Values in the Data

An inclusive disjunctive value represents a potential set of values, more than one of which could be the actual value. For example, assume that a color scheme for apples is introduced. Experts judge apples by a "primary" color. Newton apples are a mix of red and green, and either color may predominate in the apple. So both red and green are possible colors of a Newton apple. The dilemma comes about when we want to count the number of red (or green) apples. Should Newton apples be counted? If so, then if we similarly count green apples and combine the count of red and green apples to get an idea of how many apples we have, Newton apples will be counted twice. If, however, Newton apples are not counted, then there is an undercount of the number of apples. Techniques for handling inclusive disjunctive values are similar to those for managing exclusive disjunctive values. Min-max bounds, however, tend to be even larger. As in the previous section, a good technique is to gerrymander the metadata to avoid the both overcounting and undercounting, but the solution has high space cost.

INCOMPLETE INFORMATION IN THE HIERARCHY

Some of the techniques for managing incomplete information in the base data presented earlier involve inserting incompleteness into the derived data. Unknown and imprecise values are the most common kinds of incomplete derived data. The techniques to handle incomplete derived data are essentially the same as those for handling incomplete base data, but some additional strategies exist for improving the responsiveness of queries on the incomplete data.

Suggest an alternative, complete query: A query on incomplete derived data can be automatically redirected to the "nearest" complete derived data (Dyreson, 1996; Pedersen et al., 2001). *Figure 6* shows a fragment of a hierarchy for maintaining the count of the number of food items sold in a grocery store application. In the example, three apples are sold, but it is unknown how many Fuji and McIntosh apples are sold because the variety of one of the apples is incompletely known (the value of the variety attribute is the imprecise value "Apple," so the fact is effectively grouped higher in the hierarchy (Pedersen et al., 1999; Jagadish et al., 1999; Pourabbas & Rafanelli, 2000)).

To manage the incomplete base data, the base nodes are split into visible and hidden components. The hidden component is shaded in gray. It accurately counts the complete information for the base node. The visible component is shaded in white. Each visible component contains an unknown value indicating that some number of apples were sold, but it is unknown how many. A query for the number of Fuji apples cannot be satisfied since the number is unknown. The query should be redirected to the closest complete information. The closest complete information can be found by moving *up* the hierarchy, since the incompleteness inherent in imprecise and disjunctive values disappear at some coarse, imprecise category (sometimes only at the *top*). In the hierarchy shown in *Figure 6*, the closest complete information is found immediately above, at the "Apples" unit. The user is advised to ask that query to obtain a definite answer.

Computing min-max bounds: Incomplete derived data is often bounded by complete derived data above and below it in the hierarchy. For a **count** aggregate, if the information in a derived (or base) node is incomplete, the lower bound is the closest complete information below the node, whereas the upper bound is the closest complete information above it. In the hierarchy shown in *Figure 6*, the min-max bounds for Fuji apples are [1-3]. The hidden component

Figure 6: Suggesting an Alternative, Complete Query

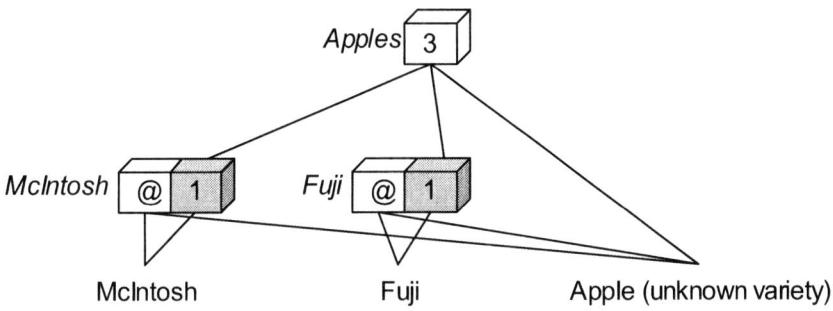

shows that there is at least one Fuji apple. Above it in the hierarchy is the information that there are at most three apples. This is a crude upper bound that could be made tighter by discounting the known contribution of other apple varieties, e.g., since there is at least one McIntosh apple, the upper bound on Fuji apples is two (Dyreson, 1996).

INCOMPLETE METADATA

Less common is incompleteness in the *metadata*. The metadata in a multidimensional database are the dimensions, categories, and units. Often, the metadata can be stored as a set of *mapping* tables, which encodes the relationship between a finer and a coarser category. The set of mapping tables for a dimension specifies the hierarchy for that dimension. A mapping table has two columns, labeled Fine and Coarse as shown in *Figure 7*. The mapping in the table is from individual kinds of Items sold in a store to Food Groups. In the mapping table, each unit in the Fine category is associated with some unit in the Coarse category. As already discussed, in order to be summarizable, all such relationships must be totaled on both sides, and many-to-one from the finer to the coarser category (Lenz & Shoshani, 1997). The example table shown in *Figure 7* has several problems (Pedersen et al., 1999, 2000).

- *Charcoal is not a food.* The null value in the Food Group associated with charcoal indicates that there is no Food Group to which charcoal belongs. The mapping is said to be *non-covering*.
- *Tomatoes are both a vegetable and a fruit.* Technically, a tomato is a fruit, but most people think of tomatoes as vegetables, so the table specification indicates that they are both (the idea is to permit such specifications and let the multidimensional database solve the dilemma). The mapping is a many-to-many relationship, no longer many-to-one, and is referred to as *non-strict*.
- *Monsteros are difficult to identify.* A monstero is a food that is familiar only in the tropics; it is about the size of a cob of corn, and is covered with dark green, hexagonal scales. The store has imported some monsteros from Australia, but the metadata specification is not quite clear on whether monsteros are fruits or vegetables. The mapping is *non-strict*.
- *Grass is a food group, and some Item sold by the store fits into the group, but it is not clear which Item.* The mapping is *non-onto* since the Fine category has partial participation.

The problem of incomplete information in metadata is that it makes the hierarchy non-summarizable. Proper management of the incomplete information can restore the hierarchy. Each case is considered below in detail.

Non-covering hierarchies: *Figure 8(a)* shows an example of a non-covering hierarchy. The problem is subtle. There is no Food Group unit to which

Figure 7: A Mapping Table from Items to Food Groups

Fine (Item)	Coarse (Food Group)
Orange	Fruit
Charcoal	@
Tomato	Vegetable
Tomato	Fruit
Monstero	Fruit or Vegetable
@	Grass

"Charcoal" belongs; but, skipping a level, "Charcoal" does belong to the "Picnic Supplies" unit in the Store Aisles category. The non-covering hierarchy is a problem in a semi-eager multidimensional model that materializes Food Groups. Even though Food Groups is a finer category than Store Aisles, the materialized values for Food Groups cannot be used to compute Store Aisles since "Picnic Supplies" will be undercounted. The solution is to complete the non-covering mapping by adding a hidden unit as shown in *Figure 8(b)*. The unit "Hidden PS" is added to the Food Groups category. The unit is only seen by the system, that is, it is hidden from a user who queries for Food Group measures. At most one hidden unit is added for each non-covering violation.

Figure 8: Repairing a Non-Covering Mapping

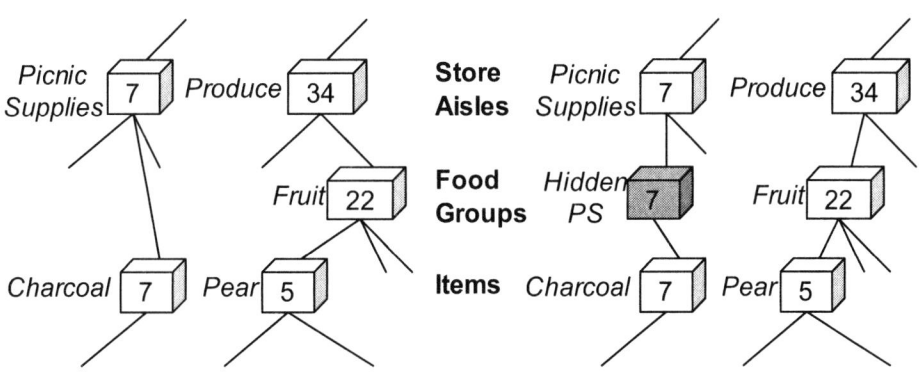

(a) A non-covering mapping from Items to Food Groups

(b) A hidden granule is added to complete the incomplete mapping

Non-onto hierarchies: Techniques for repairing non-onto hierarchies (where the coarse category does not totally participate) are similar to those for completing non-covering hierarchies. Here, the hierarchy is padded with finer, hidden values below the existing values, e.g., a hidden "Grass" value is inserted into the Item level and mapped to "Grass" in the Food Groups level to make the hierarchy onto.

Non-strict hierarchies: A non-strict hierarchy is a many-to-many mapping between two categories. In general, non-strict mappings can be handled by introducing new units in the hierarchy. The example presented above had two separate cases of non-strict mappings. The first case is that tomatoes are considered to be both a fruit and a vegetable. This could lead to overcounting, as shown in *Figure 9(a)*. "Tomato" contributes to both "Fruit" and "Vegetable," and so is counted twice at "Produce." The second case is that a "Monstero" could be considered a "Fruit" or a "Vegetable." This could lead to overcounting (both mappings are possible) or undercounting (neither mapping is definite). The technique to repair the non-strictness in both cases is the same.

To repair a non-strict hierarchy, two goals must be simultaneously attained. First, the counting problems must be fixed. In *Figure 9(a)* the counting problem manifests itself as an overcounting. The count at the "Produce" unit should be two rather than three since the produce is one "Tomato" and one "Apple." The second goal to achieve is the correct display of values at a coarse category.

Figure 9: Repairing a Non-Strict Hierarchy

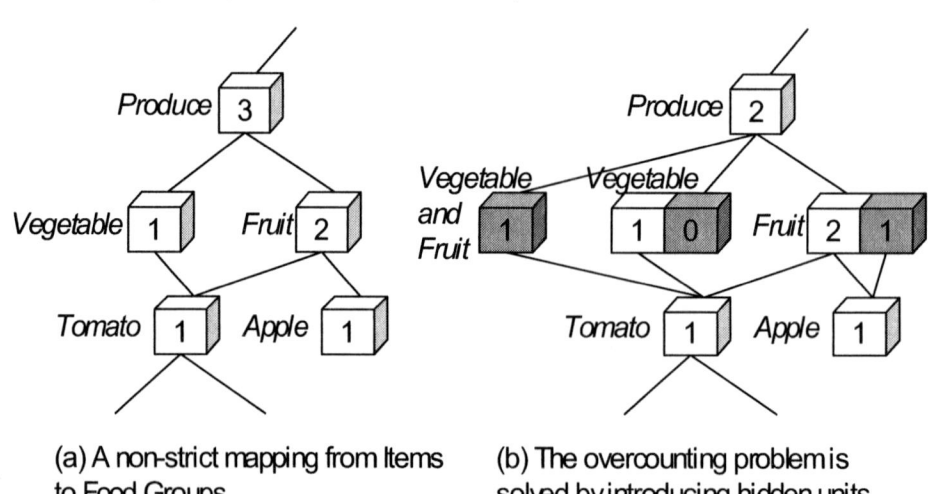

(a) A non-strict mapping from Items to Food Groups

(b) The overcounting problem is solved by introducing hidden units.

When data at a coarse category is viewed, it must include the multiple mapped unit(s) at the finer category. In the example, if the Food Group units are viewed then the count for "Vegetable" should be one, and for "Fruit," it should be two since a "Tomato" is both a fruit and a vegetable.

To solve the counting problem, a new (hidden) unit is added for each non-strict mapping. In *Figure 9(b)* the overcounting has been corrected by adding a "Vegetable and Fruit" unit. "Tomato" maps to the new unit. The solution to the second problem is to split each unit into a visible unit and a hidden unit. The hidden units are used to aggregate higher up the hierarchy. In *Figure 9(b)* the hidden units are shaded in gray. Note that "Tomato" does not map to the hidden "Vegetable" or "Fruit" units. The visible units are used for display purposes only. They are the original non-strict mappings. The visible units are shaded white in the Food Groups category of *Figure 9(b)*. The relationship between the visible units in Food Groups and the units in Items is the same in both *Figure 9(a)* and *Figure 9(b)*.

FUTURE TRENDS

The management of incomplete information in multidimensional databases is a young field with more challenges than solutions. The most important open challenge is perhaps implementation. Many of the techniques outlined in this chapter appear to be easy to implement at low cost, and will add value to a multidimensional database product. But implementation in this area lags behind the research. Some research prototypes are available. In particular, the solutions we outlined for handling incomplete metadata (i.e., partial classification hierarchies) can alleviate a frustrating problem for many users (Pedersen et al., 2000).

Multidimensional databases are complex systems, and introducing a technique to handle incomplete information will impact other areas. For example, solving the exclusive disjunctive value problem by adding hidden units changes the storage size estimation for eager implementations (Deshpande et al., 1997). For semi-eager multidimensional databases, allowing imprecise values in the hierarchy means that new strategies must be devised to determine which points in the multidimensional space should be materialized, i.e., should incomplete values be materialized (Yang et al., 1997; Baralis et al., 1997)? Another area of open research is in the design of visualization tools to judge the amount of incomplete information in a multidimensional database and to direct queries to (more) complete areas. Finally, multidimensional databases have evolved with a limited set of aggregates, drawn from aggregates in relational databases. In contrast, the field of statistics is rife with techniques for aggregating and reasoning with incomplete data. Borrowing more heavily from the field of statistics would enhance multidimensional database research and practice.

CONCLUSION

Incomplete information is endemic to real-world data collections. Often, there are values that are unknown or imprecise, but other kinds of incompleteness, such as probabilistic data, could also be present. Multidimensional databases are not currently engineered to manage incomplete information in base data, derived data, and dimensions. Different techniques are needed to correctly aggregate incomplete information during multidimensional operations such as drill-down and roll-up.

In this chapter we presented several strategies for managing incomplete information in a multidimensional database. The strategy to use depends upon the kind of incomplete information and also on where it occurs in the multidimensional database. A common technique is to replace incomplete information with complete information. The advantage of this technique is that all multidimensional databases can manage complete information. The trick is in replacing the incompleteness with an appropriate, complete value. Other strategies for managing incomplete information require more substantial changes to a multidimensional database. One strategy is to include the incompleteness in a computed aggregate, but this is possible only if the multidimensional database can have incomplete values in the hierarchy. Another technique is to measure the amount of incompleteness in an aggregated value by tallying how much uncertain information was used in its production. The final general strategy is to gerrymander the hierarchy to accommodate incomplete information.

REFERENCES

Agrawal, S., Agrawal, R., Deshpande, P.M., Gupta, A., Naughton, J., Ramakrishnan, R., & Sarawagi, S. (1996). On the computation of multidimensional aggregates. *Proceedings of the 22nd International Conference on Very Large Databases*, 506-521,

ANSI/X3/SPARC. (1975). Interim report of the Study Group on Database Management Systems. *FDT (ACM SIGMOD Bulletin)*, 7(2).

Baralis, E., Paraboschi, S., & Teniente, E. (1997). Materialized view selection in a multidimensional database. *Proceedings of the 23rd International Conference on Very Large Databases*, 156-165.

Barbará, D., Garcia-Molina, H., & Porter, D. (1992). The management of probabilistic data. *IEEE Transactions on Knowledge and Data Engineering*, 4(5), 487-502.

Barbará, D., & Sullivan, M. (1997). Quasi-cubes: Exploiting approximations in multidimensional databases. *ACM SIGMOD Record*, 26(3), 12-17.

Biskup, J. (1984). Extending the relational algebra for relations with maybe tuples and existential and universal null values. *Fundamenta Informaticæ*, VII(1), 129-150.

Cavallo, R., & Pittarelli, M. (1987). The theory of probabilistic databases. *Proceed-*

ings of the 13th International Conference on Very Large Databases, 71-81.

Chan, P., & Shoshani, A. (1981). SUBJECT: A directory-driven system for organizing and accessing large statistical databases. *Proceedings of the Seventh International Conference on Very Large Databases*, 553-563.

Codd, E.F. (1979). Extending the database relational model to capture more meaning. *ACM Transactions on Database Systems*, 4(4), 397-434.

Deshpande, P.M., Naughton, J.F., Ramasamy, K., Shukla, A., Tufte, K., & Zhao, Y. (1997). Cubing algorithms, storage estimation, and storage and processing alternatives for OLAP. *IEEE Data Engineering Bulletin*, 20(1), 3-11.

Dyreson, C. (1996). Information retrieval from an incomplete data cube. *Proceedings of the 22nd International Conference on Very Large Databases*, 532-543.

Dyreson, C. (1997). Using an incomplete data cube as a summary data sieve. *IEEE Data Engineering Bulletin*, 20(1), 19-26.

Dyreson, C. (1997a). A bibliography on uncertainty management in information systems. *Uncertainty Management in Information Systems: From Needs to Solutions*. City: Kluwer Academic Publishers, 415-458.

Gelenbe, E., & Hebrail, G. (1986). A probability model of uncertainty in databases. *Proceedings of the 2nd International Conference on Data Engineering*, 328-333.

Gessert, G.H. (1991). Handling missing data by using stored truth values. *ACM SIGMOD Record*, 20(3), 30-42.

Goldstein, B. (1981). Constraints on null values in relational databases. *Proceedings of the 7th International Conference on Very Large Databases*, 101-110.

Gottlob, G., & Zicari, R. (1988). Closed world databases opened through null values. *Proceedings of the 14th International Conference on Very Large Databases*, 50-61.

Grant, J. (1979). Partial values in a tabular database model. *Information Processing Letters*, 9(2), 97-99.

Grant, J., & Minker, J. (1986). Answering queries in indefinite databases and the null value problem. In Kanellakis, P. (Ed.), *Advances in Computing Research* (Vol. 3). London: JAI Press, 247-267.

Gray, J., Bosworth, A., Layman, A., & Pirahesh, H. (1996). Data cube: A relational aggregation operator generalizing group-by, cross-tab, and sub-totals. *Proceedings of the 12th International Conference on Data Engineering*, 152-159.

Gray, J., Chaudhuri, S., Bosworth, A., Layman, A., Reichart, D., Venkatrao, M., Pellow, F., & Pirahesh, H. (1997). Data cube: A relational aggregation operator generalizing group-by, cross-tab and sub-totals. *Data Mining and Knowledge Discovery*, 1(1), 29-54.

Harinarayan, V., Rajaraman, A., & Ullman, J. (1996). Implementing data cubes efficiently. *Proceedings of the ACM SIGMOD International Conference on Management of Data*, 205-216,.

Homenda, W. (1991). Databases with alternative information. *IEEE Transactions on Knowledge and Data Engineering*, 3(3), 384-386.

Jagadish, H.V., Lakshmanan, L.V.S., & Srivastava, D. (1999). What can hierarchies do for data warehouses? *Proceedings of the 25th International Conference on Very Large Databases*, 530-541.

Keller, A.M., & Wilkins, M.W. (1985). On the use of an extended relational model to handle changing incomplete information. *IEEE Transactions on Software Engineering*, 11(7), 620-633.

Lenz, H., & Shoshani, A. (1997). Summarizability in OLAP and statistical databases. *Proceedings of the 9th International Conference on Statistical and Scientific Database Management*, 39-48.

Lipski Jr., W. (1979). On semantic issues connected with incomplete information databases. *ACM Transactions on Database Systems*, 4(3), 262-296.

Liu, K.C., & Sunderraman, R. (1990). Indefinite and maybe information in relational databases. *ACM Transactions on Database Systems*, 15(1), 1-39.

Liu, K.C., & Sunderraman, R. (1991). A generalized relational model for indefinite and maybe information. *IEEE Transactions on Knowledge and Data Engineering*, 3(1), 65-76.

Malvestuto, F. (1991). Data integration in statistical databases. In Michalewicz, Z. (Ed.), *Statistical and Scientific Databases.* City: Ellis Horwood, 201-232.

Melton, J., & Simon, A.R. (1993). *Understanding the New SQL: A Complete Guide*. San Mateo, CA: Morgan Kaufmann.

Mendelzon, A. (2001). *Data Warehousing and OLAP: A Research-Oriented Bibliography (in progress)*. Available on-line at: http://www.cs.toronto.edu/~mendel/dwbib.html. Current as of November 2001.

Minker, J. (1982). On indefinite databases and the closed-world assumption. *Proceedings of the 6th Conference on Automated Deduction*. Lecture Notes in Computer Science, Vol. 138. City: Springer-Verlag, 292-308.

Ola, A. (1992). Relational databases with exclusive disjunctions. *Proceedings of the 8th International Conference on Data Engineering*, 328-336.

The OLAP Report. (2001). *Database Explosion*. Available on-line at: http://www.olapreport.com/Database-Explosion.htm. Current as of December 2001.

Pedersen, T.B., Jensen, C.S., & Dyreson, C.E. (1999). Extending practical pre-aggregation in On-Line Analytical Processing. *Proceedings of the 25th International Conference on Very Large Databases*, 663-674.

Pedersen, T.B., Jensen, C.S., & Dyreson, C.E. (2000). The TreeScape System: Reuse of precomputed aggregates over irregular OLAP hierarchies. *Proceedings of the 26th International Conference on Very Large Databases*, demonstration track.

Pedersen, T.B., Jensen, C.S., & Dyreson, C.E. (2001). A foundation for capturing and querying complex multidimensional data. *Information Systems* 26(5), 383-423.

Pourabbas, E., & Rafanelli, M. (2000). Hierarchies and relative operators in the OLAP environment. *ACM SIGMOD Record*, 29(1).

Rafanelli, M., & Shoshani, A. (1990). STORM: A statistical object representation model. *Proceedings of the 5th International Conference on Statistical and Scientific Database Management*, 14-29.

Rubin, D.B. (1987) *Multiple Imputation for Nonresponse in Surveys*. New York: John Wiley & Sons.

Sato, H. (1991). Statistical data models: From a statistical table to a conceptual approach. In Michalewicz, Z. (Ed.), *Statistical and Scientific Databases*. City: Ellis Horwood, 167-200.

Shoshani, A. (1997). OLAP and statistical databases: Similarities and differences. *Proceedings of the 16th ACM Symposium on Principles of Databases Systems*, 185-196.

Shukla, A., Deshpande, P.M., Naughton, J., & Ramasamy, K. (1996). Storage estimation for multidimensional aggregates in the presence of hierarchies. *Proceedings of the 22nd International Conference on Very Large Databases*, 522-531.

Su, S. (1983). SAM*: A semantic association model for corporate and scientific/statistical databases. *Information Sciences*, 29(1), 151-199.

Vassiliadis, P. (2000). Gulliver in the land of data warehousing: practical experiences and observations of a researcher. *Proceedings of the 2nd International Workshop on the Design and Management of Data Warehouses*, December 1-16.

Widom, J. (1995). Research problems in data warehousing. *Proceedings of the 4th International Conference on Information and Knowledge Management*, 25-30.

Yang, J., Karlapalem, K., & Li, Q. (1997). Algorithms for materialized view design in a data warehousing environment. *Proceedings of the 23rd International Conference on Very Large Databases*, 136-145.

Zadeh, L. (1989). Knowledge representation in fuzzy logic. *IEEE Transactions on Knowledge and Data Engineering*, 1(1), 89-100.

Zaniolo, C. (1984). Database relations with null values. *Journal of Computer and System Sciences*, 28, 142-166.

Chapter XI

Privacy in Multidimensional Databases

Francesco Malvestuto
University of Rome–La Sapienza, Italy

Marina Moscarini
University of Rome–La Sapienza, Italy

ABSTRACT

When answering queries that ask for summary statistics, the query-system of a multidimensional database should guard confidential data, that is, it should avoid revealing (directly or indirectly) individual data, which could be exactly calculated or accurately estimated from the values of answered queries. In order to prevent the disclosure of confidential data, the query-system should be provided with an auditing procedure which, each time a new query is processed, checks that its answer does not allow a (knowledgeable) user to disclose any sensitive data. A promising approach consists in keeping track of (or auditing) answered queries by means a dynamic graphical data structure, here called the answer map, whose size increases with the number of answered queries and with the number of dimensions of the database, so that the problem of the existence of an efficient auditing procedure naturally arises. This chapter reviews recent results on this problem for "additive" queries (such as COUNT and SUM queries) by listing some polynomially solvable problems as well as some hard problems, and suggests directions for future work.

INTRODUCTION

The explosive increase in access to "statistical databases" has aroused concerns on the compromise of individual privacy. A *statistical database* (Denning, 1982; Date, 1983; Ullman, 1983; Michalewicz, 1991) is a database which contains information about individuals (companies, organizations, etc.) and whose users are only allowed to access summary statistics, but "sensitive" ones, that is, summary statistics that could lead to the exact or approximate disclosure of confidential data of single individuals. A naïve policy for guarding the confidentiality of individual data consists in leaving unanswered queries asking for sensitive summary statistics. This policy is not satisfactory because, given a set of non-sensitive summary statistics, there is a more-or-less large set of summary statistics that are implicitly released in that they can be computed in an exact or approximate way so that, if one of such summary statistics were sensitive, then the confidentiality of some individual data would be compromised. To guarantee privacy of individual records, a mechanism of inference control must be embodied in the statistical database interface according to some security policy, which can be said to be effective only if sensitive summary statistics are neither explicitly nor implicitly released.

We can imagine how an ideal control method should work. If Q is the set of previously answered queries and Q is a new query, then the value of Q is initially "locked." Next, Q is tested for sensitivity; if the test is positive, the answer to Q will be denied; otherwise, the sensitivity test is applied to each of the summary statistics that would be implicitly revealed if Q were answered; if none of these summary statistics comes out to be sensitive, then and only then the value of Q is "unlocked" and Q is answered. Such a security procedure raises the computational problem of finding out the summary statistics that are implicitly revealed from the values of a given set of answered queries. This problem, sometimes called the "inference problem," proves to be hard in the general case and has been solved in an efficient way only under certain restrictive assumptions. The complexity of the inference problem essentially depends on the query type and on the sensitivity criterion in use. The query type is defined by the following three parameters: the aggregation function (count, sum, average, max, min, etc.), the data type (real, nonnegative real, integer, nonnegative integer, etc.) and the data structure (*simple* if the answer to the query consists of a scalar, and *complex* if the answer consists of a structured data such as a table). Note that a complex query can be always thought of as a set of simple queries, and that average queries can be thought of being couples of queries. As to the sensitivity criteria, several proposals exist (Adam & Wortmann, 1989; Adam, Gangopadyay & Holowczak, 1999).

In this chapter, we review some recent results on the computational complexity of the inference problem for simple count and sum queries, which will be referred to as simple "additive queries," with data of real and nonnegative real type.

The inference problem for additive queries with data of integer and nonnegative integer type is far more complicated, and some results can be found in Chin (1986) and Kleinberg, Papadimitriou, & Raghavan (2000), where max and min queries are also discussed. It is worth mentioning that a similar inference problem arises in the publication of statistical two- and three-dimensional tables; it is then called the problem of the "statistical disclosure control" (Cox, 1980, 1981; Kelly, Golden & Assad, 1992; Carvalho, Dellaert, & Osorio, 1994; Irvin & Jerrum, 1994), and surveys can be found in Schackis (1993), Cox & Zayatz (1995), Willenborg & de Waal (1996, 2000), and Roehrig (2001). Although most results of the theory of statistical disclosure control apply to complex additive queries on a statistical database, there are two main aspects of the inference problem in statistical databases that make it harder: one is the on-line characteristic of statistical query processing, and the other is that the statistical disclosure control of a table aims at protecting sensitive cells only, and not general forms of sensitive information which might consist of sensitive sets of cells.

To give a taste of the security issues connected with statistical data, we now present some examples of statistical disclosure control on tables which serve to introduce basic notions. Let us suppose that the economic division of a census bureau has collected a wide range of data and wishes to publish these data without violating confidentiality laws. Normally, economic data is published by geography and standard industrial classification (SIC) codes. For example, Table 1 shows state-level data for various types of food stores.

This table shows that only one establishment reported candy store sales for this state. If this table were published, any data user would know the candy establishment's precise sales value. Also, this table shows only two establishments reporting fruit store sales. Either of these two establishments, knowing their own sales figure, could calculate the other establishment's precise sales value. Thus, publishing this table would result in a violation of the privacy right. Cell values such as these are considered sensitive, and must not be published. A way to prevent the disclosure of sensitive values is to simply not publish the values. When this table is published, the sensitive cell values are suppressed or "obscured" (*primary suppression*). Table 2 shows an incomplete table where the two sensitive cell values have been suppressed.

Table 1

SIC		Number of Establishments	Value of Sales ($)
54	All Food Stores	347	200,900
541	Grocery	333	196,000
542	Meat and Fish	11	1,500
543	Fruit Stores	2	2,400
544	Candy	1	1,000

Table 2

SIC		Number of Establishments	Value of Sales ($)
54	All Food Stores	347	200,900
541	Grocery	333	196,000
542	Meat and Fish	11	1,500
543	Fruit Stores	2	
544	Candy	1	

Generally speaking, if only sensitive cells of a table are suppressed, users could derive their contents from non-sensitive data owing to the data additivity. Therefore, to fully protect the contents of suppressed sensitive cells, additional cells must be suppressed (*complementary suppression*). Of course, one must be certain that no suppressed cell values can be derived exactly. It is rarely sufficient to merely look at a table and determine that the complementary suppression scheme fully protects all sensitive cell values. Often a two-dimensional table seems to have an adequate number of complementary suppressions, but mathematical manipulations may reveal a suppressed cell value. For example, consider Table 3 and assume that cells (2, 1), (3, 2), (3, 3), and (4, 3) are sensitive.

Table 4 is obtained from Table 3 by suppressing the four sensitive cells (primary suppressions), and cells (1, 2), (1, 4), (2, 3), (3, 4), and (4, 1) (complementary suppressions).

Table 3

	j=1	j=2	j=3	j=4	
i=1	25	70	40	20	T(1,+) = 155
i=2	0	20	75	30	T(2,+) = 125
i=3	30	0	40	80	T(3,+) = 150
i=4	45	10	5	20	T(4,+) = 80

T(+,1) = 100   T(+,2) = 100   T(+,3) = 160   T(+,4) = 150

Table 4

	j=1	j=2	j=3	j=4	
i=1	25		40		T(1,+) = 155
i=2		20		30	T(2,+) = 125
i=3	30				T(3,+) = 150
i=4		10		20	T(4,+) = 180

T(+,1) = 100   T(+,2) = 100   T(+,3) = 160   T(+,4) =150

At first, it could seem that there is a sufficient number of suppressions to protect the values of all sensitive cells. However, the value $T(3, 3)$ of the sensitive cell $(3, 3)$ can be easily determined as follows:

$$
\begin{aligned}
T(3, 3) \ &= [T(1, 2) + T(1, 4)] + [T(3, 2) + T(3, 3) + T(3, 4)] - \\
&\quad - [T(1, 2) + T(3, 2)] - [T(1, 4) + T(3, 4)] = \\
&= [T(1, +) - T(1, 1) - T(1, 3)] + [T(3, +) - T(3, 1)] - \\
&\quad - [T(+, 2) - T(2, 2) - T(4, 2)] - [T(+, 4) - T(2, 4) - T(4, 4)] = \\
&= (155{-}25{-}40) + (150{-}30) - (100{-}20{-}10) - (150{-}30{-}20) = \\
&= 90 + 120 - 70 - 100 = 40.
\end{aligned}
$$

Moreover, even when no sensitive cell value can be determined, one can always estimate every suppressed value if the data is of nonnegative type. For example, consider Table 5 where all interior cells are obscured.

Table 5

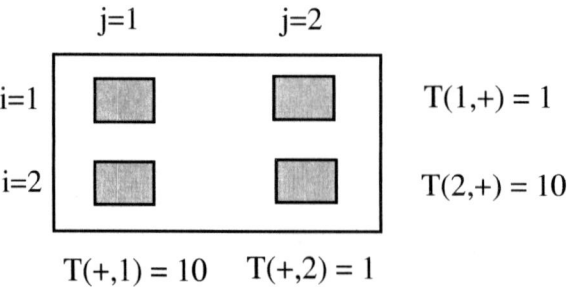

	j=1	j=2	
i=1			T(1,+) = 1
i=2			T(2,+) = 10

T(+,1) = 10   T(+,2) = 1

If the data is of real type, then the feasible values of each suppressed cell range from $-\infty$ to $+\infty$. But, if the data is of nonnegative real type, then with some linear programming, a data user can find out that the feasible values for each suppressed cell range in a bounded interval; for example, the feasible values for the cell value $T(1, 1)$, $T(1, 2)$, and $T(2, 2)$ span the interval $[0, 1]$ and the feasible values for the cell value $T(2, 1)$ span the interval $[9, 10]$. Worst of all, if the data is of nonnegative integer type, there are exactly two feasible values for each cell.

Table 1 showed an obvious type of sensitive cell: only one or two firms contributed at a cell. A not-so-obvious sensitivity occurs when more than two firms contribute to a cell, but one firm is able to estimate the data for another firm very closely. With magnitude data, if the data is of nonnegative type, a stronger sensitivity criterion is commonly used by census bureaus, namely, the $(n, k\%)$ *dominance rule*, where n is a positive integer and k is a percentage. According to the $(n, k\%)$ dominance rule, a cell (i, j) is considered sensitive if the sum of n or fewer individual contributions to the cell value $T(i, j)$ accounts for more than $k\%$ of $T(i, j)$. If this is the case, a coalition of at most $n-1$ individuals in cell (i, j) can get an accurate estimate of the largest contribution to $T(i, j)$.

For frequency tables, that is, with count data, the n *threshold rule* is commonly used; accordingly, a cell (i, j) is considered sensitive if $T(i, j) \leq n$.

To sum up, a security policy is defined by two parameters which are kept secret to the data users: the sensitivity criterion (e.g., the dominance criterion, the threshold criterion, etc.) and the protection type (from exact or approximate inference).

In the next sections, we discuss the computational issues connected with the implementation of a security policy in an on-line database environment. The chapter is organized as follows. First we introduce some basic definitions on additive queries. The next section contains the formal framework for the inference problem and the notion of an "answer map," which is the tool of a database user to disclose sensitive information. We then deal with "graphical" answer maps, and efficient algorithms are given. A few notions of safety are introduced, and the computational scheme of an auditing procedure is presented. The next section deals with two-dimensional tables, and the final sections contain future direction and conclusions.

ADDITIVE QUERIES

Consider a population I of individuals and a database $\{r_i : i \in I\}$ of individual records which contain *identifying* (or key) attributes (e.g., **name**, **SSN**, **tax-code**), and *descriptive* attributes, which may be both qualitative (e.g., **sex**) and quantitative (e.g., **salary**). A *category predicate* is an arbitrary condition on descriptive attributes built up with the logical connectives (conjunction, disjunction and negation).

A *count query* Q is a question such as:

What is the number of individuals whose descriptive attributes satisfy condition *P*?

where *P* is a category predicate. If by *I*[Q] we denote the subset of *I* selected by *P*, the *value* of Q is the number of individuals in *I*[Q]; that is, |*I*[Q]|.

A *sum query* Q on a quantitative descriptive attribute **a** is a question such as:

What is the sum of **a** over the individuals whose descriptive attributes satisfy condition *P*?

where *P* is a category predicate involving descriptive attributes other than **a**. The attribute **a** is called the *summary attribute* of the query Q. If by *I*[Q] we denote the subset of *I* selected by *P*, the *value* of Q, denoted by *q*, is the sum of the values a_i of **a** for all individuals *i* in *I*[Q]; that is,

$$q = \Sigma_{i \in I[Q]} \, a_i$$

Note that, if the category predicate of an additive query Q is logically inconsistent, then *I*[Q] = Ø and the value of Q is taken to be zero. Such as query will be referred to as a *null* query.

Example 1: Consider the following database

name	sex	age	salary
John	M	25	2.0
Andrea	M	30	3.0
Mike	M	30	2.5
Mary	F	35	3.7
Helen	F	40	3.0
Anna	F	55	3.8

The following are four examples of sum queries on **salary**:

Q1: What is the sum of the salaries of individuals with (**sex** = M)?
   Answer: 7.5

Q2: What is the sum of the salaries of individuals with (**sex** = F)?
   Answer: 10.5

Q3: What is the sum of the salaries of individuals with (**age** < 50)?
   Answer: 14.2

Q4: What is the sum of the salaries of individuals with (**age** > 50)?
   Answer: 3.8

Suppose that our database contains a descriptive attribute **a** which is confidential. We saw in the previous section some sensitivity criteria; we now apply some of them to state precisely what queries are considered sensitive. A count query Q whose category predicate contains **a** is considered sensitive according to the *n* threshold rule if $|I[Q]| \leq n$. Consider now a sum query Q with summary attribute **a** which is of numeric type. Let us distinguish two cases depending on whether the domain of **a** is the set of reals or integers, or is the set of nonnegative reals or integers. In the former case, we apply the n threshold criterion so that Q is sensitive if $|I[Q]| \leq n$. In the latter case, we can apply either the n threshold criterion or the $(n, k\%)$ dominance rule. The security problem for additive queries can be stated as follows: Given a sensitivity criterion, what measures suffice to avoid disclosing values of sensitive queries? The following two examples on sum queries show that memory-less control methods (which only refuse to answer sensitive queries) are not secure.

Example 1 (continued): Assume that **salary** is a confidential attribute and that the sensitivity criterion in use is the n threshold criterion with $n = 2$. Thus, a sum query Q on **salary** is considered sensitive if $|I[Q]| \leq 2$. Consider now the above-mentioned queries Q1, Q2, Q3 and Q4; then Q1, Q2 and Q3 are not sensitive, and Q4 is sensitive. Suppose that Q1, Q2 and Q3 were answered and Q4 was left unanswered; nevertheless, the value of Q4 can be simply computed from the values of Q1, Q2 and Q3 by subtracting the answer (14.2) to Q3 from the sum (18.0) of the answers to Q1 and Q2.

Example 2: Consider the following database

name	department	salary
e1	Direction	10.05
e2	Direction	3.0
e3	Direction	1.95
e4	Administration	4.0
e5	Administration	2.55
e6	Administration	2.45
e7	Services	3.0
e8	Services	2.0
e9	Services	1.0
e10	Marketing	1.5

Again assume that **salary** is confidential and that the (2, 80%) dominance criterion is in use. Consider now the following four sum queries on **salary**:

Q1: What is the sum of the salaries of employees with (**department** = Direction or **department** = Administration)?
Answer: 24.0

Q2: What is the sum of the salaries of employees with (**department** = Administration or **department** = Services)?
Answer: 15.0

Q3: What is the sum of the salaries of employees with (**department** = Direction or **department** = Services)?
Answer: 21.0

Q4:   What is the sum of the salaries of employees with (**department** = Direction)?
Answer: 15.0.

The queries Q_1, Q_2 and Q_3 are not sensitive, and Q_4 is sensitive. Suppose that Q_1, Q_2 and Q_3 have been answered; then, even if Q_4 was left unanswered, its value is uniquely determined from the values of Q_1, Q_2 and Q_3 since it can be computed as

$$q_4 = (q_1 - q_2 + q_3)/2$$
where q_v is the value of Q_v, $1 \leq v \leq 4$.

From the foregoing it follows that an effective control method should also take into account previously answered queries before deciding whether a new query can or cannot be answered. Such a control method is usually referred to as "auditing" (Chin & Ozsoyoglu, 1982; Malvestuto & Moscarini, 1999). In view of getting a secure control method, we need to identify those queries which, given the values of a set of answered queries, are implicitly answered in an exact or approximate way. To achieve this, in the next section we present a formal model of inference.

THE USER'S INFERENCE MODEL

Suppose that our user is interested in the value of a sensitive query Q, which henceforth will be referred to as the *target-query*. If he asks Q, the query-answering system of the statistical database will refuse to answer Q. But, as shown in the previous examples, the user could exactly compute or estimate the value of Q from the values of his (non-sensitive) queries that have been previously answered. We now introduce a formal method which allows the user to decide if the target-query is indirectly answered in an exact or approximate way given the set of previously answered queries. In what follows, we limit our considerations to the case that all queries posed by the user are either count queries or sum queries sharing the summary attribute, so that queries are fully specified by their category predicates; moreover, we assume that the values of each descriptive attribute are exhaustive and mutually exclusive. By $\mathbf{b} = (\mathbf{b}', \mathbf{b}'', \ldots)$ we denote the multidimensional attribute whose components are the descriptive attributes of the records in the database; we shall refer to $\mathbf{b}$ as the *category variable*. Let $\mathbf{B}$ be the domain of $\mathbf{b}$; that is, $\mathbf{B}$ is the Cartesian product of the domains of $\mathbf{b}'$, $\mathbf{b}''$, ... A query is called *disjunctive* if its category predicate is given by

$$\vee_{b \in B} (\mathbf{b} = b) \tag{1}$$

where B is a nonempty subset of $\mathbf{B}$; such a query will be denoted by $Q(B)$. Note that the number of possible distinct disjunctive queries is $2^{|\mathbf{B}|} - 1$. Owing to the exhaustive and mutually exclusive nature of values of the descriptive attributes, every nonnull query is equivalent to a disjunctive query (Malvestuto & Moscarini,

1990) since its category predicate is logically equivalent to a formula such as (1). Therefore, without loss of generality, we shall consider disjunctive queries only, and in what follows the adjective "disjunctive" is omitted.

Suppose that, before asking the target-query, the user posed n queries $Q_1, \ldots, Q_n$ which were all answered. Let $Q = \{Q_1, \ldots, Q_n\}$, let $Q_v = Q(B_v)$ where B_v is a nonempty subset of **B**, and let q_v be the value of Q_v, $1 \leq v \leq n$. The set

$$\cup_{v=1,\ldots,n} B_v$$

will be referred to as the *active domain* of the category variable (with respect to Q). By the *characteristic partition* of the active domain of **b**, we mean its coarsest partition $\gamma = \{C_1, \ldots, C_m\}$ such that the sets B_v belong to the set field generated by γ, that is, each B_v can be obtained as the union of one or more classes of γ. Note that m is not greater than the size of the active domain of **b**. Let $V = \{1, \ldots, n\}$ and $E = \{e_1, \ldots, e_m\}$ where

$$e_k = \{v \in V: C_k \subseteq B_v\} \qquad\qquad (1 \leq k \leq m)$$

The sets V and E will be respectively viewed as being the vertex set and the edge set of a hypergraph $H = (V, E)$, which will be referred to as the *hypergraph associated* with Q. Note that, if each edge of H contains less than three vertices, then H can be pictured as an ordinary graph and the query set Q is called *graphical*.

Example 2 (continued): Consider again the three sum queries Q_1, Q_2, and Q_3, and let $Q = \{Q_1, Q_2, Q_3\}$. Then, with our notation we have
$\quad B_1 = \{\text{Direction}, \text{Administration}\},$
$\quad B_2 = \{\text{Administration}, \text{Services}\},$
$\quad B_3 = \{\text{Direction}, \text{Services}\}.$

The active domain of **department** is $\{\text{Direction}, \text{Administration}, \text{Services}\}$, and its characteristic partition is the point partition; that is, $\gamma = \{C_1, C_2, C_3\}$ where
$\quad C_1 = \{\text{Direction}\},$
$\quad C_2 = \{\text{Administration}\}$
$\quad C_3 = \{\text{Services}\}.$

Thus, $V = \{1, 2, 3\}$ and $E = \{e_1, e_2, e_3\}$ with $e_1 = \{1, 3\}$, $e_2 = \{1, 2\}$, and $e_3 = \{2, 3\}$. The hypergraph associated with Q is the graph shown in Figure 1 where, for each vertex v, the value of Q_v is also reported.

Let **D** be the data type of the values q_v. Thus, if the queries Q_v are count queries, then **D** is the set Z_0^+ of nonnegative integers, and if the queries Q_v are sum queries,

Figure 1: The Graph Associated with Three Queries

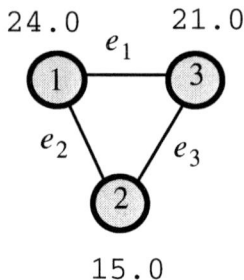

then **D** is the domain of the summary attribute (typically, **D** is the set $\Re$ of real numbers, or the set $\Re_0^+$ of nonnegative real numbers, or the set Z_0^+ of nonnegative integers). The *constraint system associated* with **Q** is the equation system

$$\Sigma_{e\in E(v)}\, x(e) = q_v \qquad\qquad (v \in V),$$

where $E(v)$ is the set of edges of H containing vertex v, with the addition of the *m* type constraints

$$x(e) \in \mathbf{D} \qquad\qquad (e \in E).$$

Note that, if $H = (h_v)_{v\in V}$ is the incidence matrix of the hypergraph H and $q = (q_v)_{v\in V}$, then the equation system above can be written in matrix notation as

$$H\,x = q \qquad\qquad (2)$$

The constraint system associated with **Q** can be summarized in the vertex-weighted hypergraph $(H, \mathbf{q})$, which will be referred to as the *answer map* of **Q**. We shall show how it allows the user to decide if the target-query Q is indirectly answered given **Q**, by which we mean that either the value of Q can be exactly computed from **Q** or the range of the feasible values of Q given **Q** is bounded. Before giving the decision algorithms for these two cases, we need to introduce some further definitions.

Given a nonempty set F of edges of H, let

$$C[F] = \cup k: e_k \in F\ C_k.$$

By the *query corresponding* to F, we mean the query $Q(C[F])$, that is, the query with category predicate

$$\vee b \in C[F]\ (\mathbf{b} = b).$$

Note that, if $F = E(v)$ for some vertex v of H, then $C[F] = B_v$ and the query corresponding to F is exactly Q_v. Moreover, if F is a singleton (that is, if F contains exactly one edge), then the query corresponding to F is called an *elementary query*.

The *closure* of **Q** is the set of the queries of the type $Q(C[F])$ such that the sum expression $\Sigma_e \in F\, x(e)$ is a **D**-*invariant* in the answer map $(H, \mathbf{q})$, by which we mean that, for every two solutions x_1 and x_2 of the constraint system associated with **Q**, one has $\Sigma_e \in F\, x_1(e) = \Sigma_e \in F\, x_2(e)$. Thus, if the target-query Q belongs to the closure of Q, then the value of Q can be exactly computed from Q, that is, Q is implicitly answered given Q.

Example 2 (continued): Assume that the domain of **salary** is . Then, the system of linear constraints associated with Q reads

$$\left|\begin{array}{l} x(e_1) + x(e_2) = 24.0 \\ x(e_1) + x(e_3) = 15.0 \\ x(e_1) + x(e_3) = 21.0 \\ x(e_1)\, x(e_2)\, x(e_3) \in \mathfrak{R}_0^+ \end{array}\right.$$

Since there is exactly one solution with
$$x(e_1) = 15.0, \quad x(e_2) = 9.0, \qquad\qquad x(e_3) = 6.0,$$
every sum query on **salary** whose category predicate contains only values of**department** from {Direction, Administration, Services} belongs to the closure of Q. It follows that the sensitive query Q4 is implicitly answered given Q.

Consider now the case that the target-query Q does not belong to the closure of Q but is *covered* by Q in the sense that there exists a nonempty set F of edges of H such that $Q = Q(C[F])$. Then the user can determine the tightest lower bound, say α_1, and the tightest upper bound, say α_2, on the value of Q by solving the following two problems:

minimize $\sum_{e \in F} x(e)$ subject to the constraint system associated with Q

maximize $\sum_{e \in F} x(e)$ subject to the constraint system associated with Q.

If $\alpha_1 \neq -\infty$ and $\alpha_2 \neq +\infty$, then the target-query Q is indirectly answered in an approximate way and the intersection of **D** with the real interval $[\alpha_1, \alpha_2]$ is called the *feasibility range* for the value of Q.

Example 3: Let $Q = \{Q_1, Q_2, Q_3\}$ be a set of additive queries with values from **D** whose answer map is the vertex-weighted graph shown in Figure 2.

Thus, the equation system in the constraint system associated with Q contains the following three equations

Figure 2

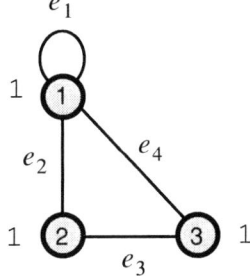

$$\begin{cases} x(e_1) + x(e_2) + x(e_4) = 1 \\ \quad x(e_2) + x(e_3) = 1 \\ \quad x(e_3) + x(e_4) = 1 \end{cases}$$

and the parametric expression of the general solution of this equation system is

$x(e_1) = 1 - 2\rho \qquad x(e_2) = \rho \qquad x(e_3) = 1 - \rho \quad x(e_4) = \rho$

Consider now the type constraints

$x(e_1), x(e_2), x(e_3), x(e_4) \in \mathbf{D}$.

We now separately discuss the three cases $\mathbf{D} = \Re$, $\mathbf{D} = \Re_0^+$, and $\mathbf{D} = Z_0^+$.

If $\mathbf{D} = \Re$, then the parameter ρ ranges from $-\infty$ to $+\infty$. In this case, it is easy to see that the closure of Q only contains the queries corresponding to the edge sets $E(1)$, $E(2)$, and $E(3)$ and, hence, it coincides with Q. Moreover, for any other nonempty edge set, the tightest lower and upper bounds on the value of the corresponding query are $-\infty$ to $+\infty$, respectively.

If $\mathbf{D} = \Re_0^+$, then the parameter ρ ranges from 0 to 1/2. Again, the closure of Q coincides with Q. But, for each nonempty edge set F, the feasibility range for the value of the query corresponding to F is bounded. For example, the feasibility range for the value of the query corresponding to $F = \{e_1, e_3\}$ is the real interval $[1/2, 2]$. To sum up, every query covered by Q and not in Q is indirectly answered in an approximate way.

If $\mathbf{D} = Z_0^+$, then $\rho = 0$ and, hence, the closure of Q coincides with the set of queries covered by Q.

In the rest of this section, we discuss the computational aspects of the problem of testing the membership of the target-query Q in the closure of Q. Let $Q = Q(B)$ where B is a nonempty subset of the domain $\mathbf{B}$ of $\mathbf{b}$. The following procedure allows the user to decide if Q is covered by Q:

COVERING TEST

Step 1. Set $F := \emptyset$.
Step 2. For $k = 1, ..., m$ do
                    if Ck is a subset of B then set $F := F \cup \{ek\}$ and $B := B - Ck$.
Step 3. If $B = \emptyset$ then and only then conclude that Q is covered by Q.

Assume now that the target-query Q is covered by Q and that $B = C(F)$ where F is the output of Covering Test. Then Q belongs to the closure of Q if and only if the sum expression $\sum_{e \in F} x(e)$ is a **D**-invariant of the answer map (H, q). We now examine the three cases $\mathbf{D} = \Re$, $\mathbf{D} = \Re_0^+$, and $\mathbf{D} = Z_0^+$.

The Case $\mathbf{D} = \Re$

To decide if the sum expression $\sum_{e \in F} x(e)$ is an $\Re$-invariant of (H, q), it is sufficient to check that the incidence vector f of F in H is a linear combination of the rows h_v of the incidence matrix H of H, that is, that the equation system

$$H^T y = f \tag{3}$$

is consistent (Malvestuto & Moscarini, 1990). Note that if the sum expression $\sum_{e \in F} x(e)$ is an $\Re$-invariant of (H, q) and y is a solution of equation system (3), then the target-query belongs to the closure of Q and the value of Q is given by

$$\sum_{v \in V} y_v \, q_v.$$

It should be noted that if the sum expression $\sum_{e \in F} x(e)$ is not an $\Re$-invariant of (H, q), then the tightest lower and upper bounds on the value of the target-query are $-\infty$ to $+\infty$, respectively.

The Case $\mathbf{D} = \Re_0^+$

One can apply linear programming methods to compute the tightest lower bound α_1 and the tightest upper bound α_2 on the value of the sum expression $\sum_{e \in F} x(e)$, and conclude that it is an $\Re_0^+$-invariant of (H, q) if and only if $\alpha_1 = \alpha_2$. We now present an alternative method. Let $kernel(H)$ denote the set of edges e of H such that $x(e) = 0$ for every solution x of the constraint system associated with Q. The set $kernel(H)$ can be determined by solving for each edge e of H the following linear programming problem:

find the maximum value $\alpha(e)$ of the variable $x(e)$ subject to the constraint system associated with Q.

Then, $kernel(H)$ is given by the set of edges e of H for which $\alpha(e) = 0$. Let K be the incidence matrix of the hypergraph K obtained from H by deleting all the edges in $kernel(H)$, and let g be the incidence vector of the edge set $F-kernel(H)$ in K. Then, the sum expression $\sum_{e \in F} x(e)$ is an $\Re_0^+$-invariant of (H, q) if and only if the equation system

$$K^T y = g \tag{4}$$

is consistent (Malvestuto, 1993).

The Case $\mathbf{D} = Z_0^+$

One can apply integer linear programming methods to compute the tightest lower bound α_1 and the tightest upper bound α_2 on the value of the sum expression $\sum_{e \in F} x(e)$, and conclude that it is a Z_0^+-invariant of (H, q) if and only if $\alpha_1 = \alpha_2$. Note that if β_1 and β_2 are respectively the tightest lower and upper bounds on the value of the sum expression $\sum_{e \in F} x(e)$ under the assumption $\mathbf{D} = Z_0^+$, then

$$\alpha_1 \geq \lceil \beta_1' \rceil \qquad \text{and} \qquad \alpha_2 \geq \lfloor \beta_2 \vee \rfloor .$$

It is also worth mentioning that if the incidence matrix of H is "totally unimodular" (Garfinkel & Nemhauser, 1972)—and this is the case if H is a bipartite graph—then relaxing the integrality constraints does not affect the values of the tightest lower and upper bounds on the value of the sum expression $\sum_{e \in F} x(e)$, that is, $\alpha_1 = \beta_1$ and $\alpha_2 = \beta_2$.

GRAPHICAL QUERY SETS

If the set Q of previously answered queries is graphical (that is, if the answer map of Q is a graph), then the procedure for testing the membership of the target-query Q in the closure of Q can be efficiently implemented in the two cases $\mathbf{D} = \mathfrak{R}$ and $\mathbf{D} = \mathfrak{R}_0^+$; moreover, if Q is an elementary query, then the tightest lower and upper bounds on the value of Q can be efficiently computed with flow-network methods in the case $\mathbf{D} = \mathfrak{R}_0^+$. Before showing that, in the next subsection we introduce some basic definitions from the theory of graphs (Bondy & Murty, 1976) and of flow networks (Ahuja, Magnanti & Orlin, 1993).

Graphs and Networks

An edge e is *incident* to vertex v if v belongs to e. An edge incident to exactly one vertex is called a *self-loop,* and an edge that is not a self-loop is called a *link.* Two vertices are *adjacent* if they belong to some edge, of which they are the *endpoints*. Let $H = (V, E)$ be a graph. Given a subset E' of E, the subgraph of H *induced* by E' is the graph (V, E'); moreover, the subgraph of H induced by $E-E'$ is denoted by $H-E'$. Given a subset V' of V, the subgraph of H *induced* by V' is the graph (V', E') where $E' = \{\{u, v\} \in E : u$ and v are both in $V'\}$; moreover, the subgraph of H induced by $V-V'$ is denoted by $H-V'$. A *path* in H is a sequence $(v_1, ..., v_k)$ of distinct vertices such that, for all $h < k$, (v_h, v_{h+1}) is an edge of H; then, the vertices v_1 and v_k are called the *endpoints* of the path. If $(v_1, ..., v_k)$ is a path in H, and the pair $\{v_1, v_k\}$ is an edge of H, then the sequence $(v_1, ..., v_k, v_1)$ is called a *cycle*, which is said to be *even* (or *odd*) if k is even (respectively, odd). H is *bipartite* if it contains no odd cycles. Two vertices of H are *connected* if they are the endpoints of some path in H, and H is connected if every two vertices of H are connected. The subgraphs of H induced by its maximal sets of pairwise connected vertices are called the *connected components* of *H*. The *bipartite components* of *H* are the connected components of *H* that are bipartite. Most of the results stated in this section rely on the following fact (Dantzig, 1963; Conforti & Rao, 1987):

The rank of the incidence matrix of *H* is equal to $|V|-r$ where *r* is the number of bipartite components of *H*.

For subsets *V'* and *V''* of *V*, we denote by $[V', V'']_H$ the (possibly empty) set of edges of *H* with one end in *V'* and the other in *V''*. An *edge cut* of *H* is a nonempty subset of *E* of the form $[V', V-V']_H$ where *V'* is a proper subset of *V*. If *H* is a bipartite connected graph with at least one edge, then there exists a bipartition $\{V_1, V_2\}$ of *V* such that the edge set *E* of *H* coincides with the edge cut $[V_1, V_2]_H$; then the sets V_1 and V_2 are called the *sides* of *H*. A *bond* of *H* is a minimal (with respect to set-inclusion) edge cut of *H*. (Note that every edge cut of *H* is the disjoint union of bonds of *H*.) An edge *e* of *H* is a *cut edge* (or a "bridge") if $\{e\}$ is a bond of *H*. A bond *F* of a bipartite graph *H* is *simple* (Malvestuto, 1993) or "bipartite" (Kao, 1997) if, for each connected component *H'* of *H–F*, *F* is incident to at most one side of *H'* (see Figure 3).

A mixed graph is a generalization of an ordinary graph. A *mixed* graph is a pair $H = (V, E \cup E)$ where *E* is a collection of subsets of *V* of cardinality 1 or 2, and *E* is a set of ordered pairs of elements of *V*. The elements of *E* are called *undirected edges* of *H*, and the elements of are called *directed edges* of *H*. If the set of undirected edges of *H* is empty, then *H* is called a *digraph*. A *directed path* in *H* from vertex *u* to vertex *u'* is a sequence $(v_1, \ldots, v_k)$ of distinct vertices such that $u = v_1, u' = v_k$ and for all $h < k$, $\{v_h, v_{h+1}\}$ is in *E* or $<v_h, v_{h+1}>$ is in . Two vertices *u* and *u'* are *strongly connected* if there exist a directed path from *u* to *u'* and a directed path from *u'* to *u*. A mixed graph is *strongly connected* if every two vertices are strongly connected. The *strongly connected components* of *H* are the subgraphs of *H* induced by its maximal sets of pairwise strongly connected vertices. The *cross-component edges* of *H* are the directed edges joining two strongly connected components of *H*.

A *network* N is a digraph (V,E) with two distinguished vertices, called the *source* and the *sink* of N respectively, where each directed edge *e* has associated to it a nonnegative quantity (possibly, $+\infty$), which is called the *capacity* of *e* and

Figure 3: A Simple Bond of a Bipartite Graph

denoted by $c(e)$. Let s and t be the source and the sink of N, respectively. A *flow* in N is a real-valued function φ defined on such that

$$0 \le \varphi(e) \le c(e) \qquad\qquad (\varepsilon \in E)$$

and

$$\Sigma_v \ \varphi(<v, u>) = \Sigma_v \ \varphi(<u, v>) \qquad\qquad (u \in V-\{s, t\})$$

the summations being extended over all v such that $<v, u> \in E$ and $<u, v> \in E$, respectively. A *maximum flow* in N is a flow in N that maximizes the quantity

$$\Phi(s{\to}t) = \Sigma_v \ \varphi(<s, v>)$$

the summation being extended over all v such that $<s, v> \in E$. Maximum flows in N can be computed in $O(|V|^3)$ time.

Example 4: Consider the network N shown in Figure 4.

A maximum flow in N is shown in Figure 5.

Testing Invariance

Let Q be a graphical query set, and let (H, q) be the answer map of Q. Without loss of generality, in what follows we assume that H is a connected graph with m edges and n vertices. We separately discuss the problem of testing the sum expression $\Sigma_{e \in F} \ x(e)$ for **D**-invariance in the cases $\mathbf{D} = \Re$, $\mathbf{D} = \Re_0^+$.

Figure 4

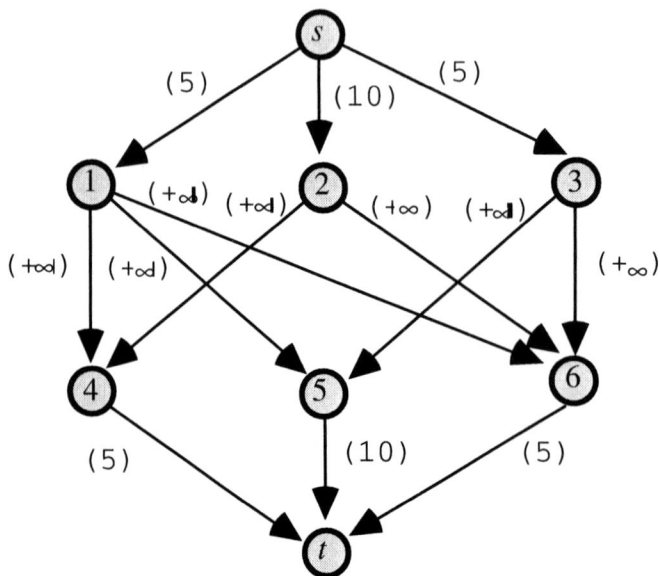

Figure 5

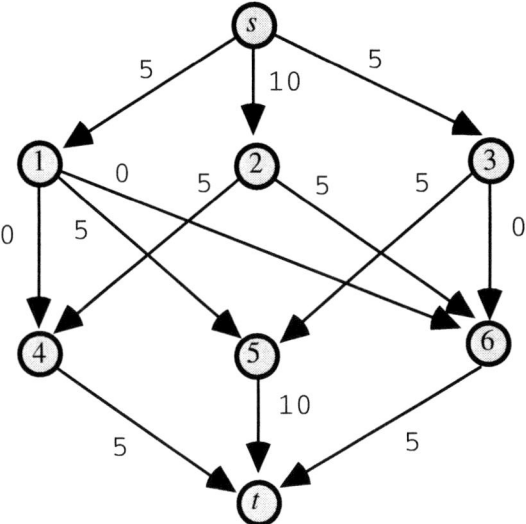

The Case **D** = $\Re$

We saw that deciding if the sum expression $\sum_{e \in F} x(e)$ is an $\Re$-invariant of $(H,$ $q)$ is equivalent to testing the consistency of equation system (3). We now show that this test can be carried out in linear time (Malvestuto & Moscarini, 1999). Since the rank of the incidence matrix of H is equal to $n-1$ if H is bipartite, and is equal to n otherwise, equation system (3) has at most ∞^1 solutions and can be transformed to an equivalent system Λ of $m-n+1$ linear equations in one unknown, denoted by λ. It consists of choosing an arbitrary vertex r of H, and performing a depth-first search (DFS) traversal of H with start-vertex r. During the traversal of H, each vertex v, when v is first visited, will be labeled by a linear expression $\varepsilon_v(\lambda)$ in λ and, for each edge not in the traversal tree of H, a linear equation in λ is added to Λ as is stated by the following algorithm whose output variable *inv* is True if and only if the sum expression $\sum_{e \in F} x(e)$ is an $\Re$-invariant of (H, q).

Algorithm = $\Re$ -INVARIANCE

Input: A connected graph H, the incidence vector f of F, and a vertex r of H.
Step 1. *inv* := False; $\varepsilon_r(\lambda) := \lambda$; $\Lambda := \emptyset$.
Step 2. Start a DFS traversal of H at vertex r. During the traversal of H, when
vertex v is reached using edge e, then
— if e is a link, say $e = \{u, v\}$, then
if v is first visited then set

$$\varepsilon_v(\lambda) := -\varepsilon_u(\lambda) + f(e);$$

otherwise, add to Λ the equation

$$\varepsilon_u(\lambda) + \varepsilon_v(\lambda) = f(e);$$

— if e is a self-loop, say $e = \{v\}$, then add to Λ the equation

$$\varepsilon_v(\lambda) = f(e).$$

Step 3. Test Λ for consistency. If Λ turns out to be consistent, then set $inv :=$ True.

It is clear that equation system (3) is equivalent to equation system Λ. Moreover, an equation in Λ of the form $\varepsilon_u(\lambda) + \varepsilon_v(\lambda) = f(e)$ can have no solutions (i.e., the equation is impossible), one solution (the equation is determined), or ∞ s□olutions (the equation is an identity), and an equation in Λ of the form $\varepsilon_v(\lambda) = f(e)$ has always one solution. Note that an equation in Λ is determined if and only if it corresponds to a non-tree edge of H whose addition to the traversal tree creates an odd cycle. Since Λ is consistent if and only if no equation in Λ is impossible and the determined equations in Λ (if any) have all the same solution, it is easy during the traversal of H to check the consistency of Λ so that one can decide in linear time whether equation system Λ and, hence, equation system (3) is consistent and, if this is the case and λ_0 is a solution of equation system Λ, one can determine a solution y of equation system (3) by taking $y_v = \varepsilon_v(\lambda_0)$.

Example 5: Consider again the query set $Q = \{Q_1, Q_2, Q_3\}$ of Example 3, whose answer map is shown in Figure 2, and assume that $\mathbf{D} =$. Suppose that $F = \{e_1, e_3\}$. Using algorithm $\mathfrak{R}$-INVARIANCE, we can label the vertices as shown in Figure 6 (we are assuming that the DFS traversal starts at vertex 1 and reaches vertices 2 and 3 using edges e_2 and e_3).

Accordingly, the equation system Λ is the pair of the following two equations corresponding to e_1 and e_3 respectively:

Figure 6

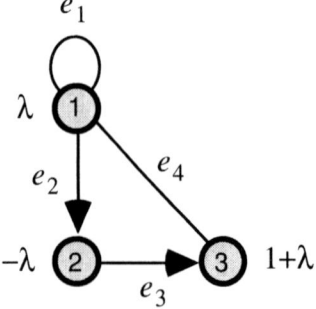

$$\lambda = 1 \qquad \text{and} \qquad 1 + 2\lambda = 0$$

and turns out to be inconsistent. Therefore, we can conclude that the sum expression $\sum_{e \in F} x(e)$ is not an $\Re$-invariant of (H, q).

Using algorithm $\Re$-INVARIANCE, one can prove that, if H is bipartite, then the sum expression $\sum_{e \in F} x(e)$ is an $\Re$-invariant of (H, q) if and only if F is the disjoint union of simple bonds (Malvestuto, 1993; Kao, 1997); as a consequence, if the target-query is an elementary query, that is, $F = \{e\}$ for some edge e of H, then Q belongs to the closure of $\boldsymbol{Q}$ if and only if e is a cut edge of H.

The Case $\mathbf{D} = \Re_0^+$

We now prove that the procedure for deciding if the sum expression $\sum_{e \in F} x(e)$ is an $\Re_0^+$-invariant of $(H, \boldsymbol{q})$ can be implemented in $O(n^3)$ time using network computation. We distinguish two cases depending on whether H is or is not bipartite.

Assume that $H = (V, E)$ is a bipartite graph with sides V_1 and V_2. Let $Net(H, \boldsymbol{q})$ be the network $(V \cup \{s, t\})$ with

$$E = \{<s, v> : v \in V_1\} \cup \{<u, v> : \{u, v\} \in E, u \in V_1 \text{ and } v \in V_2\} \cup \{<v, t> : v \in V_2\},$$

where the capacities of the directed edges of $Net(H, \boldsymbol{q})$ are defined as follows

$$c(\vec{e}) = \begin{cases} q_v & if\ \vec{e} = \langle s, v \rangle\ or\ \vec{e} = \langle v, t \rangle \\ +\infty & else \end{cases}$$

Gusfield (1988) proved that:

(i) The solutions of the constraint system associated with $\boldsymbol{Q}$ coincide with the restrictions to E of maximum flows in $Net(H, \boldsymbol{q})$.

(ii) Given a maximum flow φ in $Net(H, \boldsymbol{q})$, let $H(\varphi)$ be the mixed graph obtained from H by directing each edge $\{u, v\}$, $u \in V_1$, and $v \in V_2$, with $\varphi(<u, v>) = 0$ from V_1 to V_2; then the set of cross-component edges of $H(\varphi)$ equals *kernel(H)*.

Therefore, the following algorithm correctly decides if the sum expression $\sum_{e \in F} x(e)$ is an $\Re_0^+$-invariant of $(H, \boldsymbol{q})$.

Algorithm $\Re_0^+$-INVARIANCE

Step 1. Construct the network $Net(H, \boldsymbol{q})$ and find a maximum flow φ in $Net(H, \boldsymbol{q})$.

Step 2. Let w be the restriction of φ to E. Construct the mixed graph $H(\varphi)$ obtained from H by directing from V_1 to V_2 each edge $\{u, v\}$, $u \in V1$, and $v \in V2$, with $j(<u, v>) = 0$.

Step 3. Set *kernel(H)* to the set of cross-component edges of $H(w)$.

Step 4. For each connected component $H' = (V', E')$ of the graph $H - kernel(H)$ with $F \cap E' \neq \emptyset$, apply algorithm $\Re$-INVARIANCE with input H' and the incidence vector of $F \cap E'$. If each time the algorithm terminates with *inv* =

Figure 7

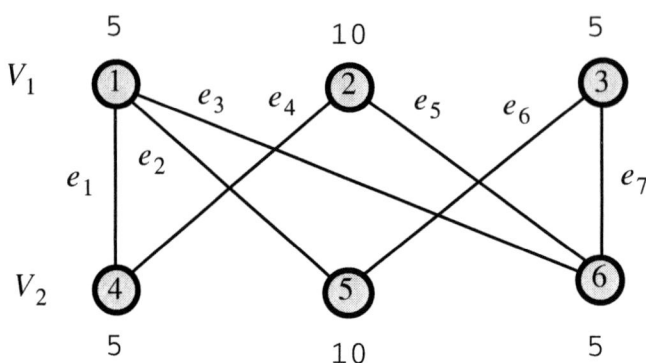

True, then and only then conclude that the sum expression $\sum_{e\in F} x(e)$ is an $\mathfrak{R}_0^+$-invariant of (H, q).

Since a maximum flow in $Net(H, q)$ can be computed in $O(n^3)$ time, algorithm $\mathfrak{R}_0^+$-INVARIANCE can be implemented in $O(n^3)$ time.

Example 6: Let $Q = \{Q_1, Q_2, Q_3, Q_4, Q_5, Q_6\}$ be a set of additive queries with values in $\mathfrak{R}_0^+$, whose answer map is shown in Figure 7.

Then, $Net(H, q)$ is the network of Figure 4 and a maximum flow $\varphi\in$ in $Net(H, q)$ is shown in Figure 5. The mixed graph $H(\varphi)$ is shown in Figure 8.

Figure 8

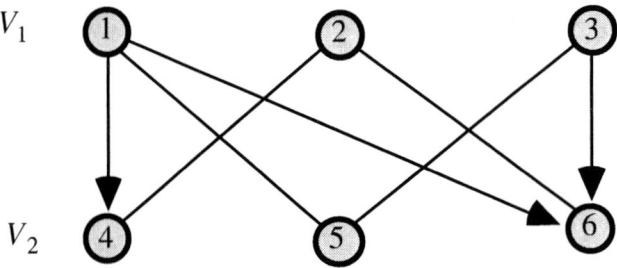

Figure 9

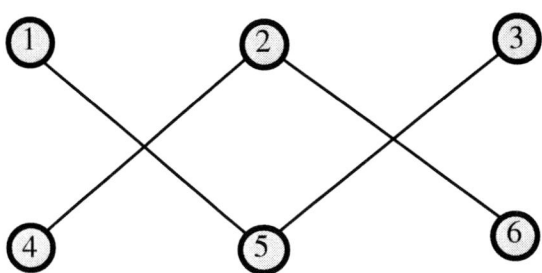

The cross-component edges of $H(\varphi)$ are $<1, 4>$, $<1, 6>$, and $<3, 6>$. Therefore, *kernel(H)* is formed by the three edges e_1, e_3, and e_7. Of course, the elementary query corresponding to each edge in *kernel(H)* belongs to the closure of Q. The graph *H–kernel(H)* is shown in Figure 9.

Here, each edge is a cut edge of a bipartite graph and, hence, the corresponding elementary query belongs to the closure of Q. To sum up, every elementary query (and, hence, the target-query) belongs to the closure of Q.

We now discuss the case that H is not bipartite. We associate with H a bipartite, connected graph $H^* = (V^*, E^*)$ which contains $2n$ vertices and $2m–l$ edges, where l is the number of self-loops of H. The graph H^* is constructed as follows. Let G be a bipartite partial graph of H obtained from a spanning tree of H by adding the nontree edges that create even cycles. Let V' be a "copy" of V, that is, $V' \cap V = \emptyset$ and $|V'| = |V|$. If v is a vertex of H, then by v' we denote the copy of v; moreover, if $e = \{u, v\}$ is a link of H, then by e' we denote the set $\{u', v'\}$. The vertex set V^* of H^* is taken to be the union of V and V', that is, $V^* = V \cup V'$. The edge set E^* of H^* is taken to be

$E^* = \cup_{e \in E} \beta[e]$

where $\beta[e]$ is defined as follows:

— if e is an edge of G, then $b[e] = \{e, e'\}$;
— if $e = \{u, v\}$ is a link of H but not an edge of G, then $b[e] = \{\{u, v'\}, \{u', v\}\}$;
— if e is a self-loop of H, say $\{v\}$, then $\beta[e] = \{\{v, v'\}\}$.

Note that since H is a nonbipartite, connected graph, H^* is bipartite and connected. Furthermore, if V_1 and V_2 are the sides of G, then the sides of H^* are $V^*_1 = V_1 \cup V_2'$ and $V^*_2 = V_1' \cup V_2$, where $V_i' = \{v': v \in V_i\}$, $i = 1, 2$. For each v in V set $q^*_v := q_v$, and for each v' in V', set $q^*_{v'} := q_v$.

Example 7: Consider again the query set $Q = \{Q_1, Q_2, Q_3\}$ of Example 3, whose answer map was shown in Figure 2, and assume that $\mathbf{D} = \mathfrak{R}^+_0$. By choosing as G the graph shown in Figure 10, we obtain the bipartite vertex-weighted graph H^* shown in Figure 11.

Figure 10   *Figure 11*

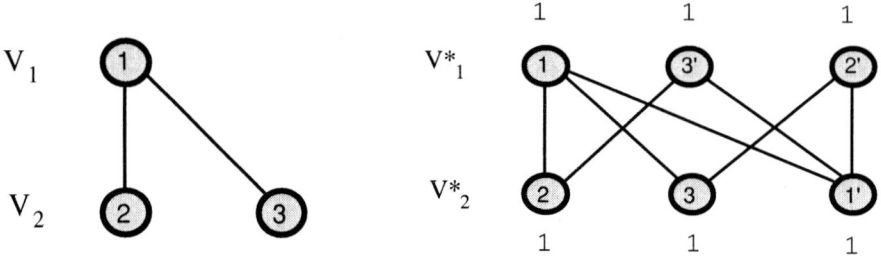

Let Γ^* be the system of linear constraints given by the equation system

$$H^* x^* = q^* ,$$

where H^* is the incidence matrix of H^*, with the addition of the nonnegativity constraints

$$x^*(e^*) \in \mathfrak{R}_0^+ .   (e^* \in E^*)$$

Malvestuto & Mezzini (2000) proved that:

(i)   If w is a solution of G, then the vector w^* with

$$w^*(e^*) = w(e) \text{ for } e^* \in \beta[e]$$

is a solution of Γ^*, and if w^* is a solution of Γ^*, then the vector w with

$$w(e) = (1/|\beta[e]|)\sum_{e^* \in \beta[e]} w * (e^*)$$

is a solution of Γ.

(ii) $kernel(H) = \{e \in E: \beta[e] \subseteq kernel(H^*)\}$.

By combining these results with Gusfield's results, one has also in the case that H is not bipartite; the procedure for testing invariance can be implemented in $O(n^3)$ time.

Example 7 (continued): The constraint system Γ^* reads

$$
\begin{aligned}
x * (\{1,2\}) + x * (\{1,3\}) + x * (\{1,1'\}) &= 1 \\
x * (\{2,3'\}) + x * (\{1',3'\}) &= 1 \\
x * (\{2',3\}) + x * (\{1',2'\}) &= 1 \\
x * (\{1,2\}) + x * (\{2,3'\}) &= 1 \\
x * (\{1,3\}) + x * (\{2',3\}) &= 1 \\
x * (\{1,1'\}) + x * (\{1',3'\}) + x * (\{1',2'\}) &= 1 \\
x * (\{1,2\}), x * (\{1,3\}), x * (\{1',2'\}), x * (\{1',3'\}) &\in \mathfrak{R}_0^+ \\
x * (\{1,1'\}, x * (\{2,3'\}), x * (\{2',3\}) &\in \mathfrak{R}_0^+
\end{aligned}
$$

Figure 12

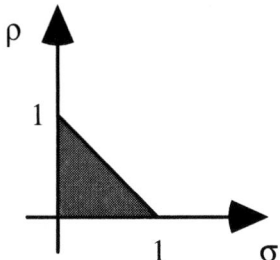

The general solution of Γ^* is

$x^*(\{1, 2\}) = \rho$ $\qquad\qquad x^*(\{1', 2'\}) = \sigma$

$x^*(\{1, 3\}) = \sigma$ $\qquad\qquad x^*(\{1', 3'\}) = \rho$

$x^*(\{1, 1'\}) = 1 - \rho - \sigma$

$x^*(\{2, 3'\}) = 1 - \rho$ $\qquad\qquad x^*(\{2', 3\}) = 1 - \sigma$

where ρ an σ are bounded as shown in Figure 12.

We now show how *kernel*(*H*) can be graphically obtained. Figure 13 shows *Net*(*H**, ***q***)and Figure 14 shows a maximum flow φ^* in *Net*(*H**, ***q***).

Figure 13

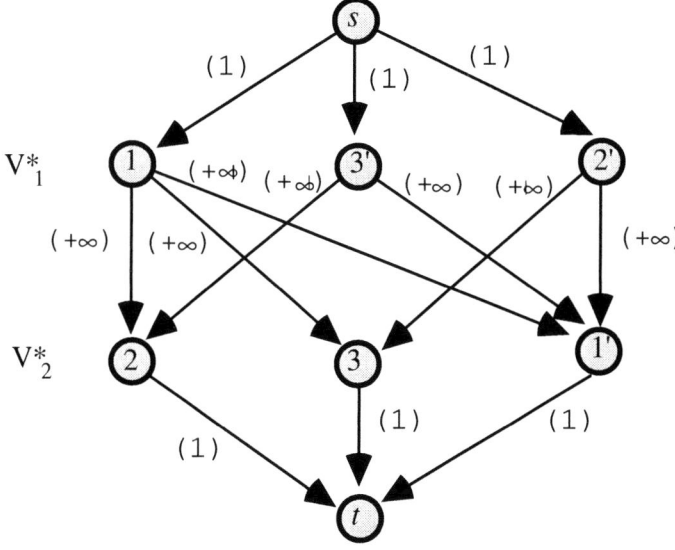

Figure 14

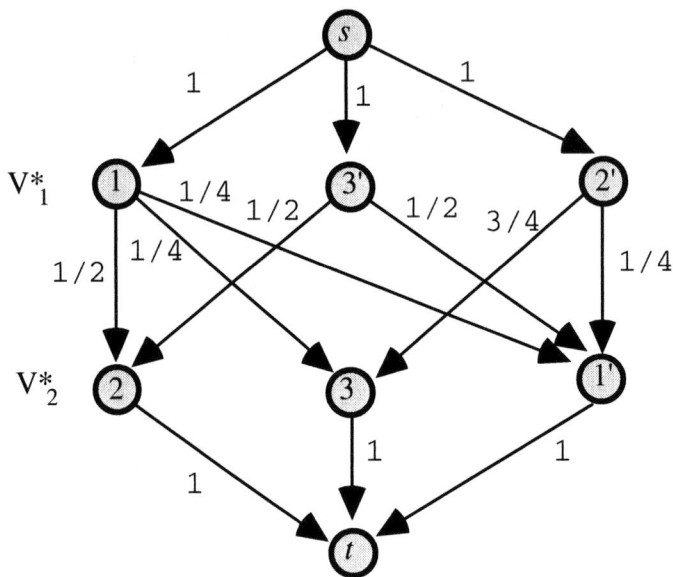

The mixed graph $H^*(\varphi^*)$ coincides with H^* which is strongly connected. So, $kernel(H^*)$ is empty, and $kernel(H)$ is empty too.

Computing the Feasibility Range

Assume that the target-query Q is an elementary query and that e is the edge corresponding to Q. Let q' and q'' be the tightest lower and upper bounds on the value of Q. Here we show how q' and q'' can be efficiently computed in the case $\mathbf{D} = \mathfrak{R}_0^+$.

We distinguish two cases depending on whether H is or is not bipartite, and we shall solve the problem of computing q' and q'' with the method given by Gusfield (1988) in the former case, and by Malvestuto & Mezzini (2002) in the latter case.

Assume that H is a bipartite graph with sides V_1 and V_2, and let $e = \{u_0, v_0\}$ with $u_0 \in V_1$ and $v_0 \in V_2$. We first find a maximum flow φ in the network $Net(H, q)$. Given φ, let us construct the digraph N with vertex set V and directed-edge set

$$\cup_{(u,v) \in E} \{<u, v>, <v, u>\}.$$

For each directed edge $<u, v>$ of N, take its capacity to be

$$c(\langle v,u \rangle) = \begin{cases} \varphi\langle u, v \rangle & \text{if } v \in V_2 \\ M & \text{else} \end{cases}$$

where M is a finite number larger than the largest q_v for v in V_1. Let N_1 be the network with underlying digraph N having source u_0 and sink v_0, and let φ_1 be a maximum flow in N_1. Analogously, let N_2 be the network with underlying digraph N having source v_0 and sink u_0, and let φ_2 be a maximum flow in N_2. If $\Phi_1(u_0 \rightarrow v_0)$ and $\Phi_2(v_0 \rightarrow u_0)$ are respectively the values of φ_1 and φ_2, then q' and q'' are given by

$$q' = \max(0, \varphi(<u_0, v_0>) - \Phi_1(u_0 \to v_0) + M) \qquad (5)$$

and

$$q'' = \Phi_2(v_0 \to u_0). \qquad (6)$$

Moreover, if H is a complete bipartite graph, then α_1 and α_2 are simply given by

$$q' = \max(0, q_{u_0} + q_{v_0} - \sum_{v \in \square V_1} q_v$$

and

$$q'' = \min(q_{u_0}, q_{v_0}).$$

Therefore, the tightest lower and upper bounds on the value of Q can be found in $O(n^3)$ time.

Example 8: Let $Q = \{Q_1, Q_2, Q_3, Q_4, Q_5, Q_6\}$ be a set of additive queries with values in , whose answer map is shown in Figure 15.

Figure 15

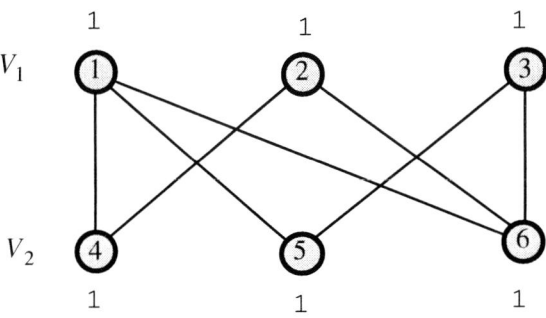

Figure 16

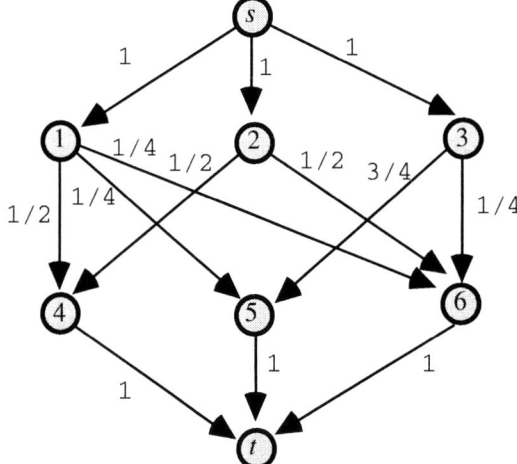

Figure 16 shows a maximum flow φ in *Net(H, q)*.

Suppose that the target-query Q is the query corresponding to the edge {1, 6}. In order to compute the tightest lower bound *q'* and the tightest upper bound *q"* on the value of Q, we construct the digraph N, which is shown in Figure 17.

Next, consider the network N_1 on N with source vertex 1 and sink vertex 6, and with capacities defined as follows:

Figure 17

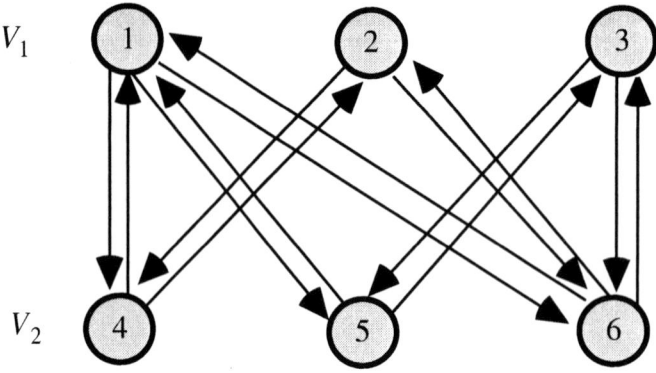

A maximum flow $φ_1$ in N_1 is shown in Figure 18.

The value of $φ_1$ is $Φ_1(1→6) = 11/4$ so that, by formula (5), one obtains
$q' = \max(0, φ(<1, 6>) - Φ_1(1→6) + 2) = \max(0, 1/4 - 11/4 + 2) = 0$.
Consider now the network N_2 which differs from N_1 only in that the source and the sink are vertices 6 and 1, respectively. A maximum flow $φ_2$ in N_2 is shown in Figure 19.

The value of $φ_2$ is $Φ_2(6→1) = 1$ so that, by formula (6), one obtains
$q" = Φ_2(6→1) = 1$.
Consider now the case that H is not bipartite. Let H* be a bipartite transform of H and Γ* the constraint system as stated in the "Graphs and Networks" subsection above. Let β[e] be the image of e in H*. If e is a self-loop and β[e] = {e*}, then the tightest lower and upper bounds on the value of Q are given by

$$q' = \min x^*(e^*) \qquad\qquad \text{subject to } Γ^*$$
$$\text{and}$$

Figure 18

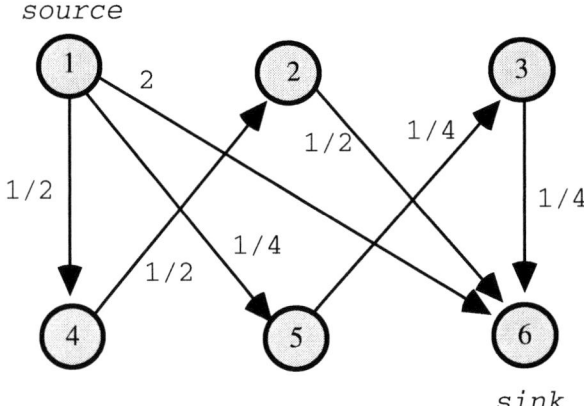

Figure 19

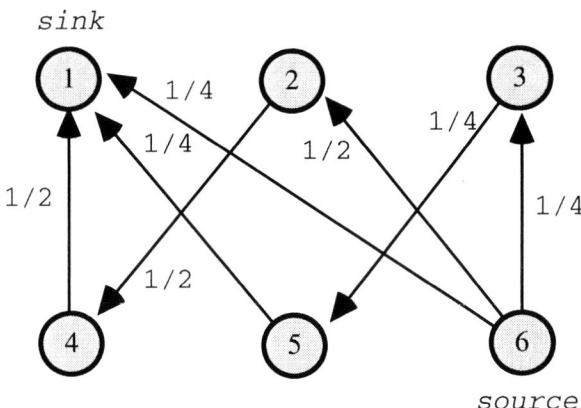

$$q'' = \max x^*(e^*) \qquad\qquad \text{subject to } \Gamma^*$$

so that q' and q'' can be determined in $O(n^3)$ time using Gusfield's method applied to H^*. Otherwise, that is, if e is a link and $\beta[e] = \{e^*_1, e^*_2\}$, let

$$\alpha_1 = \min \left[x^*(e^*_1) + x^*(e^*_2) \right] \quad \text{subject to } \Gamma^*$$

and

$$\alpha_2 = \max \left[x^*(e^*_1) + x^*(e^*_2) \right] \quad \text{subject to } \Gamma^*.$$

Then, the tightest lower and upper bounds on the value of Q are given by

$$q' = (1/2)\,\alpha_1 \qquad\qquad \text{and} \qquad\qquad q'' = (1/2)\,\alpha_2 \qquad\qquad (7)$$

With an example we show that α_1 and α_2 (and hence q' and q'') can be computed in $O(n^3)$ time.

Example 7 (continued): Suppose that the target-query Q is the elementary query corresponding to the self-loop {1}. By formula (5), the tightest lower and upper bounds on the value of Q are given by

$$q' = \min x^*(\{1, 1'\}) \qquad \text{and} \qquad q'' = \max x^*(\{1, 1'\}).$$

To compute them, we can proceed as in Example 8 when $\min x(\{1, 6\})$ and $\max x(\{1, 6\})$ were computed. So, one has

$$q' = \min x^*(\{1, 1'\}) = 0 \text{ and} \qquad q'' = \max x^*(\{1, 1'\}) = 1.$$

Suppose now that the target-query Q is the elementary query corresponding to the edge {2, 3}. The tightest lower and upper bounds on the value of Q are respectively given by formula (7) where

$$\alpha_1 = \min [x^*(\{2, 3'\}) + x^*(\{2', 3\})] \qquad \text{subject to } \Gamma^*$$

$$\text{and}$$

$$\alpha_2 = \max [x^*(\{2, 3'\}) + x^*(\{2', 3\})] \qquad \text{subject to } \Gamma^*.$$

We show how to compute α_1 and α_2 in cubic time. For both of them, we make use of a maximum flow in $Net(H^*, q^*)$. Let φ be the maximum flow in $Net(H^*, q^*)$ shown in Figure 20.

Figure 20

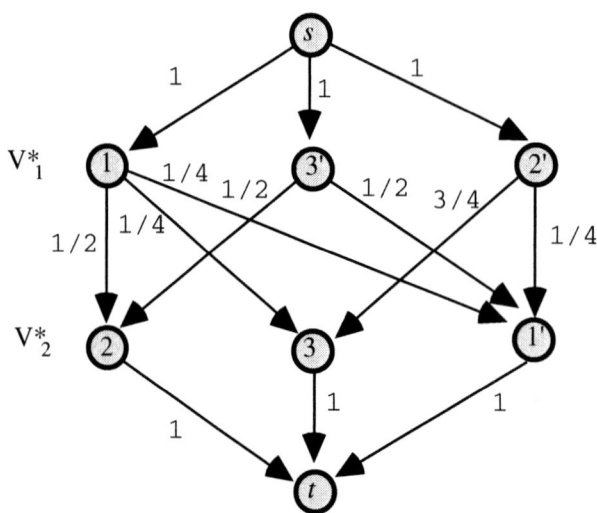

Let us begin with the computation of α_1. First of all, we compute $\min x^*(\{2,3'\})$ subject to Γ^* using the above procedure for bipartite graphs. More precisely, we construct the network N_1 on the digraph with source vertex 3' and sink vertex 2, and with capacities defined as follows:

$$c^*(\langle v^*, u^* \rangle) = \begin{cases} \varphi \langle u^*, v^* \rangle & if\ v^* \in V^*_2 \\ 2 & else \end{cases}$$

A maximum flow φ_1 in N_1 is shown in Figure 22.

Figure 21

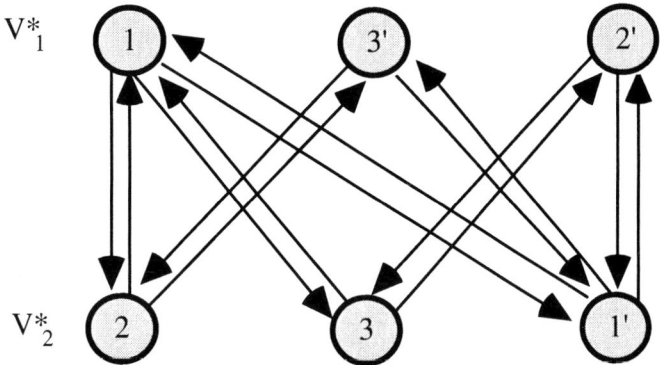

Figure 22

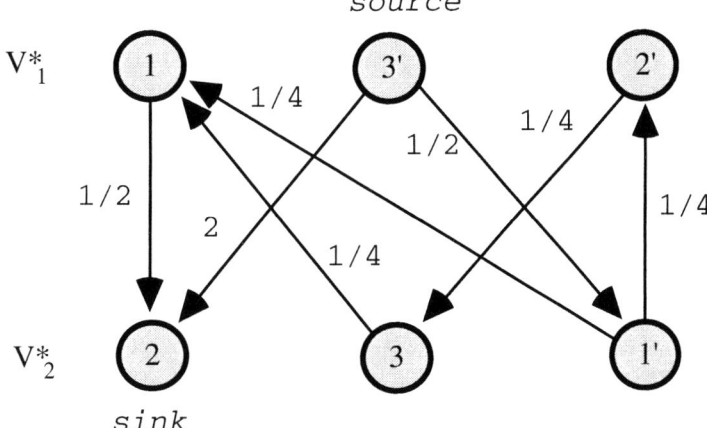

Figure 23

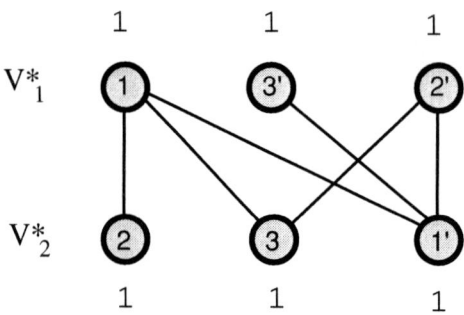

The value of φ_1 is $\Phi_1(3'{\to}2) = 5/2$ so that, by formula (5), one has

$$\min x^*(\{2, 3'\}) = \max\,(0,\,\varphi(<3', 2>) - \Phi_1(3'{\to}2) + 2) = \max\,(0,\,1/2 - 5/2 + 2) = 0.$$

After computing $\min x^*(\{2, 3'\})$ ($= 0$), we calculate α_1 as $\min x^*(\{2, 3'\})$ plus the quantity

$$\min x^*(\{2', 3\}) \qquad \text{subject to } \Gamma^* \text{ and } x^*(\{2, 3'\}) = 0.$$

To achieve this, let $(H^*{-}\{2, 3'\}, q^*_1)$ be the vertex-weighted graph obtained from (H^*, q^*) by deleting the edge $\{2, 3'\}$ and subtracting $\min x^*(\{2, 3'\})$ ($= 0$) from the weights of the endpoints of the edge $\{2, 3'\}$ (see Figure 23).

Then, the quantity

$$\min x^*(\{2', 3\}) \qquad\qquad \text{subject to } \Gamma^* \text{ and } x^*(\{2, 3'\}) = 0$$

equals $\min x^*(\{2', 3\})$ in $(H^*{-}\{2, 3'\}, q^*_1)$, which is found with the same technique employed to find $\min x^*(\{2, 3'\})$ in (H^*, q^*). That is, we first find a maximum flow χ in $Net(H^*{-}\{2, 3'\}, q^*_1)$. Let χ be the maximum flow in $Net(H^*{-}\{2, 3'\})$ shown in Figure 24.

Next, let D be the digraph (see Figure 25) with vertex set V^* and directed-edge set

$$\cup\;(u^*,v^*) {\in} E^*{-}\{\{2, 3\}\}\;\{<u, v>,\, <v, u>\}.$$

Then, we construct the network M_1 on D with source vertex 2' and sink vertex 3, and capacities defined as follows:

$$c_1(\langle v^*, u^* \rangle) = \begin{cases} \chi\langle u^*, v^* \rangle & \text{if } v^* \in V^*_2 \\ 2 & \text{else} \end{cases}$$

Figure 24

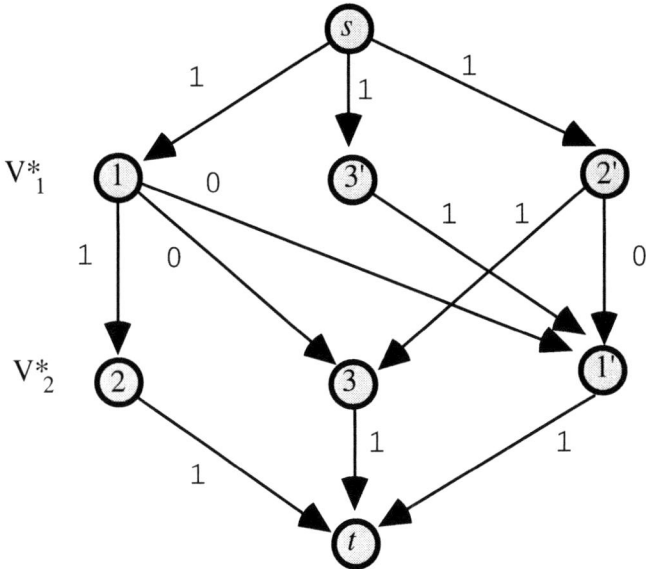

Figure 25

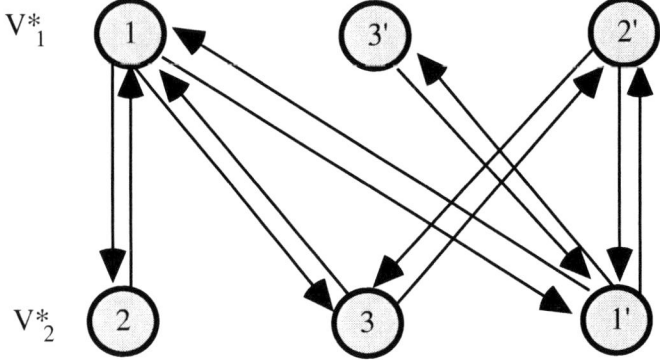

and compute a maximum flow in M_1. Let χ_1 be the maximum flow in M_1 shown in Figure 26.

The value of χ_1 is $X_1(2'{\to}3) = 2$ so that, by formula (5), one has that min $x^*(\{2',$ $3\})$ in $(H^* - \{2, 3'\}, q^*_1)$ is equal to

$$\max (0, \chi(<2', 3>) - X_1(2'{\to}3) + 2) = \max (0, 1 - 2 + 2) = 1 .$$

Figure 26

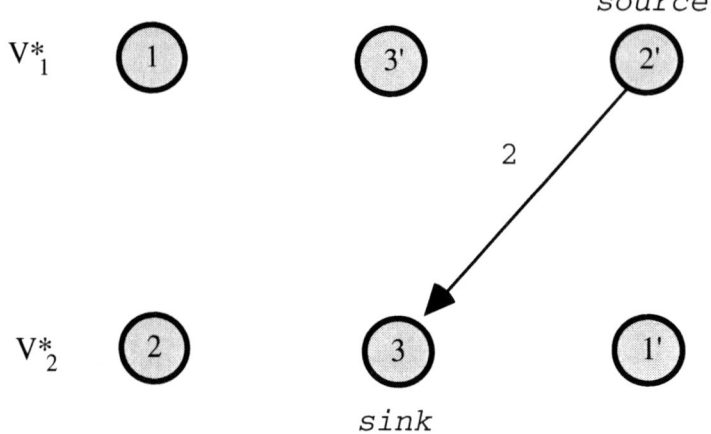

Finally, we obtain
$$\alpha_1 = 0 + 1 = 1$$
and, by formula (7),
$$q' = (1/2)\,\alpha_1 = 1/2.$$

As to α_2, we proceed as follows. First of all, we compute $\max x^*(\{2,3'\})$ subject to Γ^*. To this end, we construct the network N_2 which differs from N_1 only in the source and the sink which are not set to vertex 3' and vertex 2, respectively. A maximum flow φ_2 in N_2 is shown in Figure 27.

Figure 27

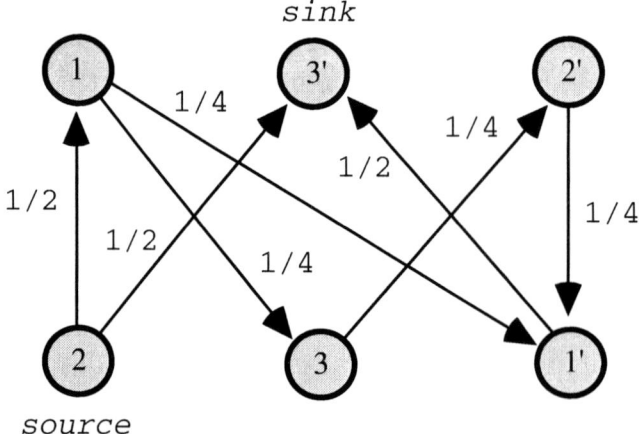

The value of φ_2 is $\Phi_2(2 \rightarrow 3') = 1$ so that, by formula (6), one has

$$\max x^*(\{2, 3'\}) = \Phi_2(2 \rightarrow 3') = 1.$$

After computing $\max x^*(\{2, 3'\})$, we calculate α_2 as $\max x^*(\{2, 3'\})$ plus the quantity

$\quad\quad \max x^*(\{2', 3\})$ $\quad\quad\quad\quad$ subject to Γ^* and $x^*(\{2, 3'\}) = 1$.

To achieve this, let $(H-\{2,3'\}, q^*_2)$ be the vertex-weighted graph obtained from H^* by deleting edge $\{2, 3'\}$ and subtracting $\max x^*(\{2, 3'\})$ $(= 1)$ from the weights of the endpoints of edge $\{2, 3'\}$ (see Figure 28).
Then, the quantity

$\quad\quad \max x^*(\{2', 3\})$ $\quad\quad\quad\quad$ subject to Γ^* and $x^*(\{2, 3'\}) = 1$

equals $\max x^*(\{2', 3\})$ in $(H^*-\{2, 3'\}, q^*_2)$, which is found with the same technique employed to find $\max x^*(\{2, 3'\})$ in (H^*, q^*). That is, we first find a maximum flow in $Net(H^*-\{2,3'\}, q^*_2)$. Let ψ the maximum flow in $Net(H^*-\{2, 3'\}, q^*_2)$ shown in Figure 29.

Figure 28

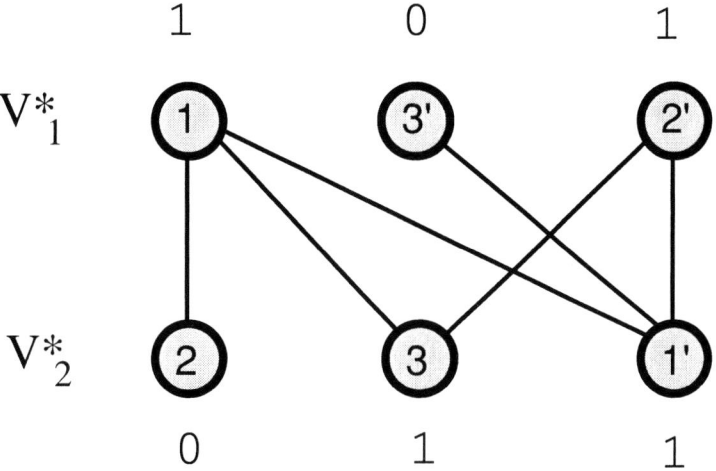

Figure 29

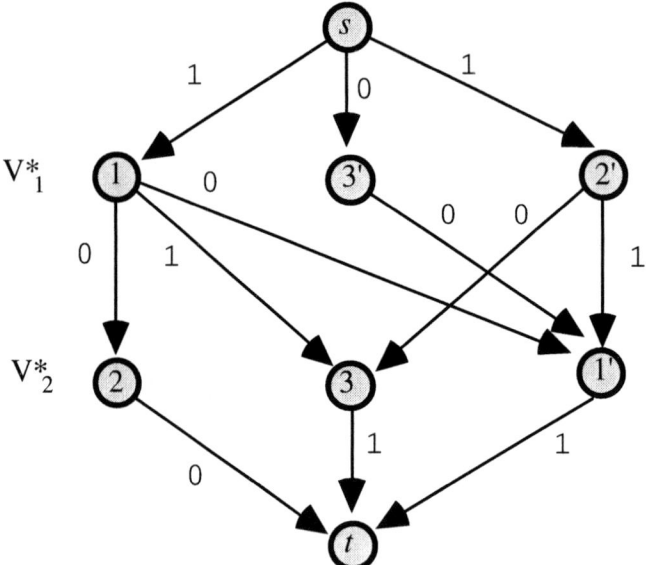

Then, we construct the network M_2 on D with source vertex 3 and sink vertex 2', and capacities defined as follows:

$$c_2(\langle v^*, u^* \rangle) = \begin{cases} \psi\langle u^*, v^* \rangle & \text{if } v^* \in V^*_2 \\ 2 & \text{else} \end{cases}$$

and compute a maximum flow in M_2. Let ψ_2 be the maximum flow in M_2 shown in Figure 30.

The value of ψ_2 is $\Psi_2(3 \rightarrow 2') = 1$ so that, by formula (6), one has
  $\max x^*(\{3, 2'\}) = \Psi_2(3 \rightarrow 2') = 1$.
Finally, we obtain
  $\alpha_2 = 1 + 1 = 2$
and
  $q'' = (1/2)\, \alpha_2 = 1$.

Before closing this section, we note that the existence of an efficient membership test, and of an efficient procedure for computing the tightest lower and upper bounds on the value of an elementary query in the case $\mathbf{D} = Z^+_0$, is an open problem; however, if H is bipartite, then the total unimodularity of its incidence matrix allows us to apply the procedures above stated for the case $\mathbf{D} = \Re^+_0$.

Figure 30

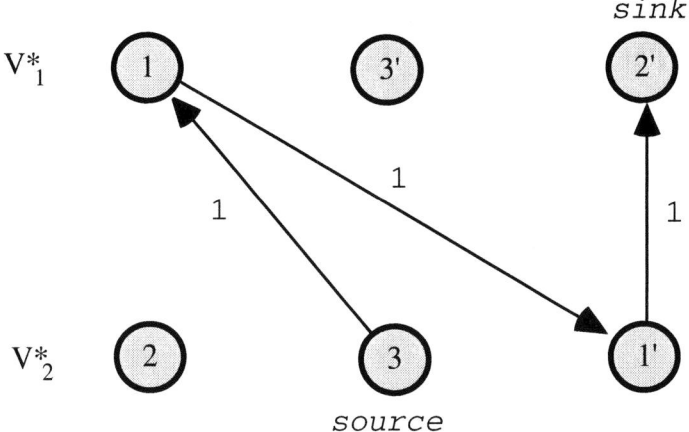

SAFETY

Let Q be a set of queries that are all either count queries or sum queries with the same summary attribute. Loosely speaking, given a sensitivity criterion, Q is considered "safe" by the query-answering system if every sensitive information contained in Q is protected. We now make this notion precise. First of all, it should be noted that most sensitivity criteria—including the n threshold rule and the $(n, k\%)$ dominance rule—can be expressed by means of a linear subadditive function defined on the powerset of the underlying population I and called a *sensitivity function* (Cox, 1981), and a query Q is sensitive according to the to the criterion specified by the sensitivity function σ if $\sigma(I[Q]) > 0$. For example, the $(n, k\%)$ dominance rule for sum queries with summary attribute **a** is specified by the sensitivity function σ which in correspondence of a subset J of I takes the value

$$\sigma(J) = \sum_{i \in J_n} a_i - (k/100) \times \sum_{i \in J} a_i$$

where a_i is the value of **a** for individual i, and $J_n = J$ if $|J| \leq n$ and J_n is the set of the n individuals with largest values of **a** otherwise.

By the subadditivity of the function σ, if J and J' are two subsets of I, then

$$\sigma(J \cup J') \leq \sigma(J) + \sigma(J').$$

Remark 1: Let Q, Q', and Q" be queries such that $I[Q] = I[Q'] \cup I[Q"]$. Then, if neither Q' nor Q" is sensitive, then Q is not sensitive; on the other hand, if Q is sensitive, then Q' or Q" is sensitive.

Consider now a set of queries Q and let (H, q) be the answer map of Q. Let Q^* be the closure of Q, and let Q_σ be the set of sensitive queries $Q(C[F])$ for some subset F of E; that is,

$$Q_\sigma = \{Q(C[F]): F \subseteq E \text{ and } \sigma(I[Q(C[F])]) > 0\}.$$

We say that Q is *strongly safe* if

$$Q_\sigma \cap Q^* = \emptyset.$$

We now introduce a subset of Q^* which is sufficient to assure strong safety. Recall that the queries in Q^* correspond to the sets F of edges of H for which the sum expression $\sum_{e \in F} x(e)$ is a **D**-invariant. Let F* be the family of such subsets of E and let F+ be the subfamily of F* formed by minimal (with respect to set-inclusion) nonempty members of F*. From the very definition of a **D**-invariant sum expression, it follows that F* is closed under disjoint union and proper difference, which implies that every nonempty member of F* is the disjoint union of one or more members of F+. Let Q^+ be the set of queries in Q^* that correspond to edge sets in F+. Of course, if Q is strongly safe, then $Q_\sigma \cap Q^+ = \emptyset$. On the other hand, if Q is not strongly safe, then there exists a query $Q(C[F])$ in Q_σ such that F belongs to F*. Since F cannot be empty, there is a subset F' of F that belongs to F+ and, by Remark 1, is such that $Q(C[F'])$ is in Q_σ which proves that $Q_\sigma \cap Q^+ \neq \emptyset$. To sum up, Q is strongly safe if and only if $Q_\sigma \cap Q^+ = \emptyset$.

The following is a weaker condition than strong safety which is often used in the security analysis of statistical tables. A query set Q is *weakly safe* if no elementary query in Q^* is sensitive. Of course, if Q is strongly safe, then Q is also weakly safe; but, the converse need not hold as shown by the following example.

Example 9: Let the graph shown in Figure 31 be the hypergraph associated with a set Q of four additive real-valued queries.

Q is weakly safe since no elementary query belongs to the closure of Q. However, if the query Q corresponding to the pair F of self-loops $\{1\}$ and $\{2\}$ were sensitive, then Q would be not strongly safe because Q belongs to the closure of Q

Figure 31

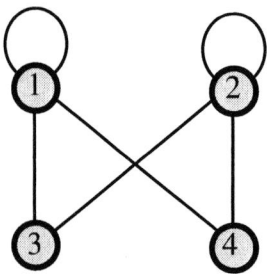

(the incidence vector of F can be written as a linear combination of the rows of the incidence matrix of the graph).

The previous two notions of safety are related to exact inference. We now introduce their counterparts when approximate inference is taken into account. To this end, consider any query Q corresponding to some subset of E, and let q be the value of Q. By the *protection interval* for Q, we mean the intersection of **D** with the real interval

$$[q(1 - k/100), q(1 + k/100)]$$

for a fixed percentage k.

A query set **Q** is *strongly safe with respect to the k% rule* (*strongly k-safe*, for short) if, for each sensitive query Q corresponding to a set of edges of H, the feasibility range for the value of Q strictly contains the protection interval for Q. Note that **Q** is strongly safe if and only if **Q** is strongly 0-safe; furthermore, when $k > 0$, if **Q** is strongly k-safe, then **Q** is strongly safe, but the converse need not hold.

A query set **Q** is *weakly safe with respect to the k% rule* (*weakly k-safe*, for short) if, for each sensitive elementary query Q, the feasibility range for the value of Q strictly contains the protection interval for Q. Note that **Q** is weakly safe if and only if **Q** is weakly 0-safe; furthermore, when $k > 0$, if **Q** is weakly k-safe, then **Q** is weakly safe, but the converse need not hold.

At this point, the following question naturally arises: how can the query-answering system test **Q** for safety? In the next two subsections, we answer this question by considering the four above safety criteria in the cases that the queries in **Q** are additive queries with values in and in . The safety tests we present make use of an edge-weighted hypergraph (H, w), we call the *data map* of **Q**, where H is the hypergraph associated with **Q** and, for each edge e of H, the weight $w(e)$ of e is given by the value of the elementary query corresponding to e. Moreover, each edge e of H is labeled by the corresponding class of the characteristic partition of the active domain of the category variable. Finally, an edge e of H is marked (sensitive)

Figure 32: A Data Map

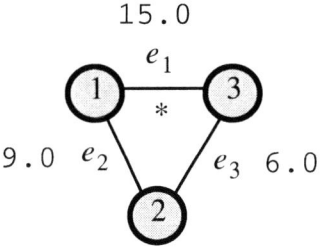

if and only if the corresponding elementary query is sensitive. Note that, by Remark 1, every edge set corresponding to a sensitive query must contain at least one marked edge.

Example 2 (continued): The data map (H, w) of $Q = \{Q_1, Q_2, Q_3\}$ is shown in Figure 32. The weights of the edges of H are $w(e_1) = 15.0$, $w(e_2) = 9.0$, and $w(e_3) = 6.0$; moreover, the labels of $e_1, e_2,$ and e_3 are $\{\texttt{Direction}\}, \{\texttt{Administration}\}$, and $\{\texttt{Services}\}$, respectively. Finally, only the edge e_1 is marked ($*$).

The Case **D** = $\mathfrak{R}$

Testing Q for weak safety. Let $Inv_H(\mathfrak{R})$ be the set of edges of H corresponding to elementary queries in the closure of Q. The test for weak safety consists in determining the set $Inv_H(\mathfrak{R})$ and, if no edge in $Inv_H(\mathfrak{R})$ is marked, then and only then is Q recognized to be weakly safe.

The set $Inv_H(\mathfrak{R})$ can be determined by checking, for each edge e of H, the consistency of equation system (3) where f is taken to be the incidence vector of the singleton $\{e\}$. However, if Q is graphical, there is a linear algorithm for determining $Inv_H(\mathfrak{R})$ (Malvestuto & Mezzini, 2002), which is based on the following property: an edge of H belongs to $Inv(\mathfrak{R})$ if and only if its removal increases the number of bipartite components. It is easily seen that the removal of an edge e from a connected graph H increases the number of bipartite components if and only if either e participates in all odd cycles, or e is a cut edge and at least one of the two subgraphs of H separated by e is bipartite. (Note that, if H is bipartite, then $Inv(\mathfrak{R})$ coincides with the set of cut edges.) Since both the set of edges of H that participate in all odd cycles and the set of cut edges separating bipartite subgraphs of H can be determined in linear time (Malvestuto & Mezzini, 2002), the problem of testing Q for weak safety can be solved in linear time.

Testing Q for strong safety. The family of edge sets in F^+ is first determined. If for each set F in F^+, the query corresponding to F is not sensitive, then and only then is Q recognized to be strongly safe.

An open problem is the existence of an effective procedure for determining F^+ without examining all sets of edges of H. What is known is only a graphical characterization of the sets in F^+ when Q is graphical (Malvestuto & Mezzini, 2001), which entails that if H is bipartite, then F^+ consists of all and the only simple bonds of H (see Figure 3). Unfortunately, as shown below, the number of simple bonds of a bipartite graph can be exponential in the numbers of its edges, which entails that the size of F^+ may be exponential in the size of H.

Example 10: Let H be the bipartite (connected) graph shown in Figure 33. Note that H contains $O(k^2)$ vertices and $O(k^2)$ edges.

Figure 33   *Figure 34*

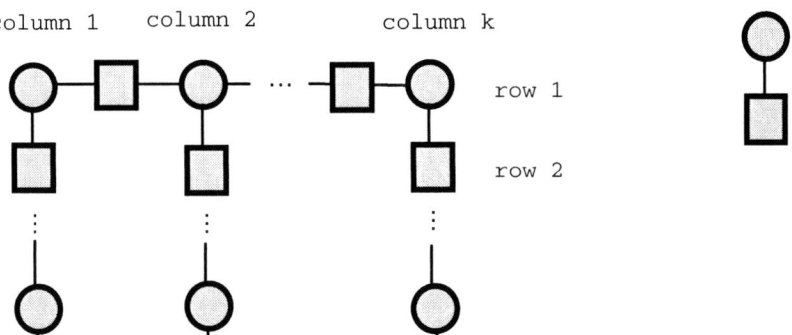

Let *F* be any set of *k* edges of the form shown in Figure 34

one for each of the *k* columns. It is easily seen that *F* is a simple bond of *H*, and that the number of such simple bonds is equal to k^k.

The Case $\mathbf{D} = \mathfrak{R}_0^+$

Testing Q for weak safety. Let $Inv_H(\mathfrak{R}_0^+)$ be the set of edges of *H* corresponding to elementary queries in the closure of Q. The test for weak safety consists in determining the set $Inv_H(\mathfrak{R}_0^+)$, and if no edge in $Inv_H(\mathfrak{R}_0^+)$ is marked, then and only then is Q recognized to be weakly safe. Since $Inv_H(\mathfrak{R}_0^+) = kernel(H)$ $\cup Inv_{H'}(\mathfrak{R}_0^+)$, where $H' = H - kernel(H)$, if Q is graphical then both $kernel(H)$ and $Inv_{H'}(\mathfrak{R}_0^+)$ can be determined in linear time from the data map of Q and, hence, the problem of testing Q for weak safety can be solved in linear time.

Testing Q for strong safety. The test of strong safety is the same as in the case $\mathbf{D} = \mathfrak{R}$ after determining $kernel(H)$, which is computed as above.

Testing Q for weak k-safety. For each marked edge *e* of *H*, the feasibility range for the value of the elementary query corresponding to *e* is determined. If for each marked edge *e* of *H*, the feasibility range for the value of the elementary query corresponding to *e* strictly contains the protection interval for the elementary query corresponding to *e*, then and only then is Q recognized to be weakly *k*-safe. We saw that if Q is graphical, then the feasibility range for the value of an elementary query can be determined in cubic time so that testing Q for weak k-safety requires a polynomial time.

Testing Q for strong k-safety. The family F_σ of the edge sets of H to which correspond sensitive queries is first determined. Next, for each F in F_σ, the feasibility range for the value of the query corresponding to F is determined and compared with its protection interval. If, for each F in F_σ, the feasibility range strictly contains the protection interval, then and only then is Q recognized to be strongly k-safe.

The existence of a procedure for determining F_σ without examining all the edge sets in H is an open problem.

AN AUDITING PROCEDURE

In this section we sketch an auditing procedure, which given
— a sensitivity criterion, specified by a "sensitivity test";
— a safety criterion, specified by a "safety test," which given the data map of a query set Q, correctly decides whether Q is or is not safe;
— a safe set of queries $\{Q_1, ..., Q_n\}$, specified by its data map (H, w), where $H = (V, E)$ with $V = \{1, ..., n\}$ and $E = \{e_1, ..., e_m\}$; and
— a new query Q_{n+1} specified by a subset B of the domain of the category variable and with   value Q_{n+1}

computes the value of a Boolean variable *safe*, which will come out to be True if and only if Q_{n+1} can be safely answered. The procedure is as follows.

Step 1. Set *safe* := False.

Step 2. Apply the sensitivity test to Q_{n+1}. If Q_{n+1} turns out to be sensitive, then Exit.

Step 3. If Q_{n+1} belongs to the closure of $\{Q_1, ..., Q_n\}$, then set *safe* := True and Exit.

Step 4. Construct the data map (H', w') of $\{Q_1, ..., Q_n, Q_{n+1}\}$.

Step 5. If no edge of H' is marked, then set *safe* := True and Exit.

Step 6. Apply the safety test to $\{Q_1, ..., Q_n, Q_{n+1}\}$. If $\{Q_1, ..., Q_n, Q_{n+1}\}$ turns out
        to be safe, set *safe* := True.

The implementation of Steps 1, 2, 5, and 6 are straightforward. The problem of computing the closure of a query set was discussed earlier. We now detail Step 4, that is, how the data map (H', w') of $\{Q_1, ..., Q_n, Q_{n+1}\}$ can be constructed from the data map (H, w) of $\{Q_1, ..., Q_n\}$. Let V' and E' be the vertex set and the edge set of H', respectively. Recall that each edge e of H is labeled by a class, denoted by *label*[e], of the characteristic partition of the active domain of the category variable with respect to $\{Q_1, ..., Q_n\}$. For each edge e' of H', by *label'*[e'] we denote the class of the characteristic partition of the active domain of the category variable with respect to $\{Q_1, ..., Q_n, Q_{n+1}\}$.

(4.1) Set $V' := V \cup \{n+1\}$, $E' := E$, and $m' := m$.

(4.2) For $k = 1, ..., m$, if $B \cap label[ek] = \emptyset$ then set $w'(ek) := w(ek)$ and $label'[ek] := label[ek]$;   otherwise, do:

    begin

    if $label[ek] \subseteq B$ then do:

            begin

            $ek := ek \cup \{n+1\}$;

            $B := B - label[ek]$, $qn+1 := qn+1 - w(ek)$

            end;

        otherwise, do:

            begin

            $label'[ek] := label[ek] - B$;

            compute the value of Q($label'[ek]$) and set $w'(ek)$ to it;

            apply the sensitivity test to Q($label'[ek]$) and mark ek if Q($label'[ek]$) turns

            out to be sensitive;

            $m' := m'+1$, $em' := ek \cup \{n+1\}$;

            add em' to E';

            $label'[em'] := B \cap label[ek]$; $B := B - label'[em']$;

            if $B = \emptyset$ then set $w'(em') := an+1$ and Exit;

            otherwise, do:

                begin

                compute the value of Q($label'[em']$) and set $w'(em')$ to it;

                apply the sensitivity test to Q($label'[em']$) and mark em' if Q($label'[em']$) turns out to be sensitive;

                $qn+1 := qn+1 - w'(em')$

                end

            end

    end

(4.3) Set $m' := m'+1$, $em' := \{n+1\}$;

    add em' to E';

    $label'[em'] := B$;

    $w'(em') := qn+1$;

    apply the sensitivity test to Q($label'[em']$) and mark em' if Q($label'[em']$) turns out to be   sensitive.

Example 11: Consider again the sum queries Q_1, Q_2, and Q_3 of Example 2. Suppose we are given the data map of $\{Q_1, Q_2\}$, which is shown in Figure 35.

Figure 35

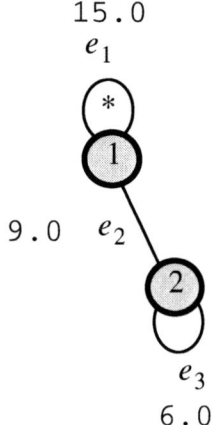

Figure 36

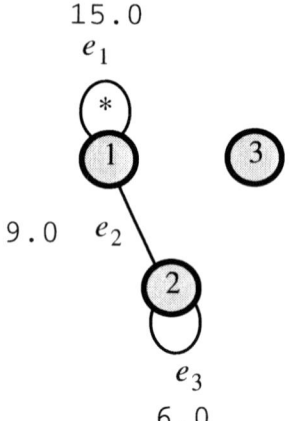

Here, edge e_1 = {1} is marked because the corresponding query is sensitive. The other two edges, e_2 = {1, 2} and e_3 = {2}, are not marked. The labels of these edges are *label*[e_1] = {Direction}, *label*[e_2] = {Administration} and *label*[e_3]={Services}.

We now apply the procedure above to construct the data map of $\{Q_1, Q_2, Q_3\}$ from the data map of $\{Q_1, Q_2\}$. After executing Step 4.1, the current value of (H', w') is as shown in Figure 36.

During the execution of Step 4.2, edge e_2 is not modified, but edge e_1 will contain both vertex 1 and the vertex 3, and edge e_3 will contain both vertex 2 and vertex 3. The data map of $\{Q_1, Q_2, Q_3\}$ is then as shown in Figure 32. Step 4.3 is not executed for $B = \emptyset$.

From a computational point of view, the worst case of Step 4 occurs if, when Step 4.2 is performed, $B \cap label[e_k] \neq \emptyset$ and neither $label[e_k] \not\subset B$ nor $B \not\subset label[e_k]$ for each k, $1 \mathrm{d} \leq k \mathrm{d} \leq m$; then, the procedure has to evaluate $2m$ queries and apply the sensitivity test to each of them. To avoid such an overhead, we may introduce a query-overlap restriction (Malvestuto & Moscarini, 1999), according to which the query Q_{n+1} is left unanswered whenever during the execution of Step 4 an edge with more than two vertices is created. Accordingly, the data map of every set of answered queries is always a graph so that at Steps 3 and 6 we can also use the efficient algorithms for graphical query sets we reported in earlier sections.

Before closing this section, we wish point out a simple technique to reduce the size of a data map (H, w). It consists of deleting each vertex that is contained in exactly one edge of H. If $H' = (V', E')$ is the resulting hypergraph and w' is the restriction of w to E', then it should be clear that, since no edge in $E–E'$ can be marked in H (for, otherwise, a query in Q would be sensitive), the edge-weighted hypergraph (H', w') is equivalent to (H, w) from the point of view of safety, and one can test weak and strong safety using (H', w') instead of (H, w). We call the edge-weighted hypergraph (H', w') the *reduced data map* of Q.

TWO-DIMENSIONAL TABLES

In this section we show how the data-map approach can be used to decide if an incomplete two-dimensional table (i.e., a table with suppressions) is or is not weakly safe. We shall consider the general case that primary and complementary suppressions are not limited to internal entries; thus, entries corresponding to row totals and column totals can also be suppressed (Kelly, Golden & Assad, 1992; Malvestuto & Moscarini, 1996a, 1996b; Chu, 1997). The following example is taken from Malvestuto & Moscarini (1996b), where the problem of testing weak safety is solved with an *ad hoc* method. Consider the following incomplete table and suppose that it was obtained from a complete table by suppressing the following entries:

$T(1, 1) = 0$	$T(1, 2) = 10$	$T(1, 3) = 0$
$T(2, 1) = 2$	$T(2, 2) = 3$	$T(2, 3) = 0$

$$T(3, 3) = 5 \qquad T(3, 4) = 0 \qquad T(3, +) = 35$$
$$T(4, 4) = 5 \qquad T(4, +) = 50$$

$$T(+, 4) = 45$$

We assume that $T(i, j)$ is the value of a summary attribute **a** of nonnegative real type. Moreover, without loss of generality, we can suppose that i and j index the values of two descriptive attributes **b**' and **b**'', respectively with domains **B**' = {b'[i]: i = 1, 2, 3, 4} and **B**'' = {b''[j]: j =1, 2, 3, 4}, which were used to categorize the individuals of the underlying population in the original complete table; thus, the category variable is **b** = (**b**', **b**'') and the domain of **b** is the Cartesian product **B** = **B**' X **B**''. Then, Table 6 can be thought of as being a set Q of 12 sum queries on **a** corresponding to its entries; for example, the category predicates of the sum queries on **a** with values $T(1, 4)$, $T(1, +)$, and $T(+, 1)$ are respectively
b = (b'[1], b''[4]),   D b''∈ **B**'' (**b** = (b'[1], b''),and D b'∈ **B**' (**b** = (b', b''[1]).

Accordingly, the active domain of **b** with respect to Q is given by
{   (b'[1], b''[1]), (b'[1], b''[2]), (b'[1], b''[3]), (b'[1], b''[4]),
          (b'[2], b''[1]), (b'[2], b''[2]), (b'[2], b''[3]), (b'[2], b''[4]),
          (b'[3], b''[1]), (b'[3], b''[2]), (b'[3], b''[3]),
          (b'[4], b''[1]), (b'[4], b''[2]), (b'[4], b''[3])   }

and its characteristic partition g is the point partition. The data map of Q is then a weighted hypergraph (H, w) with 12 vertices (corresponding to the unsuppressed entries in Table 6)

Table 6

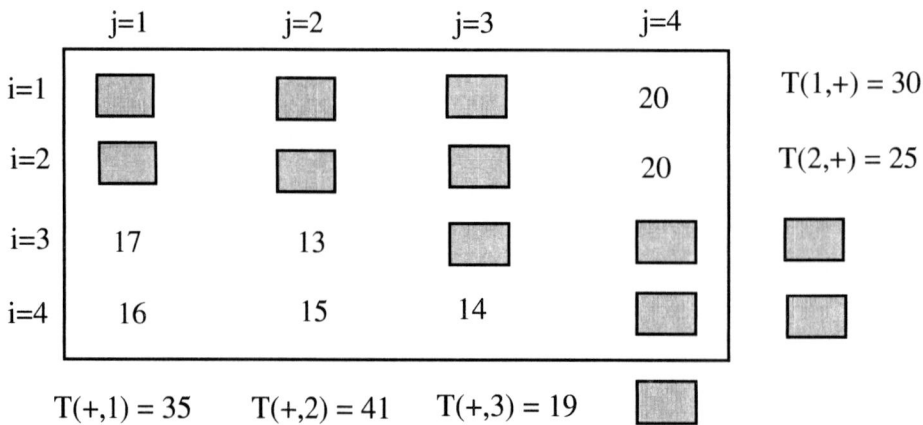

Figure 37

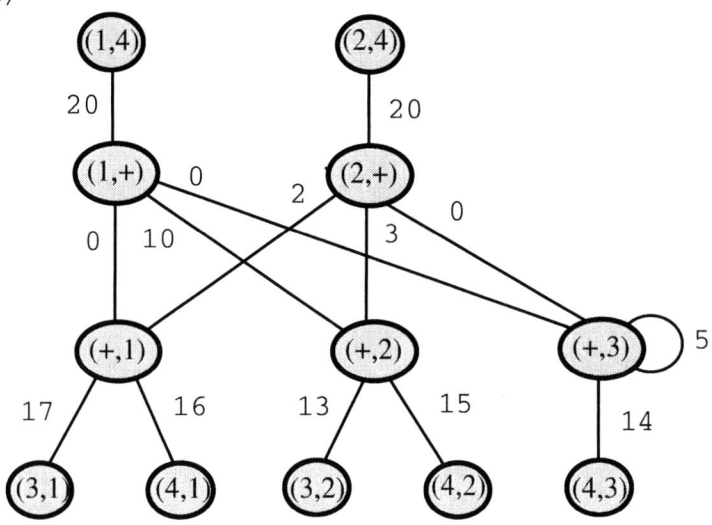

Figure 38

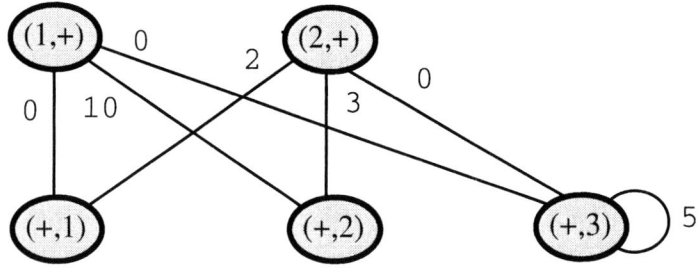

$$(1, 4), (2, 4), (3, 1), (3, 2), (4, 1), (4, 2), (4, 3), (1, +), (2, +), (+, 1), (+, 2), (+, 3)$$

and 14 edges labeled by the classes of γ. The edges of H contain one or two vertices; for example, the edge labeled by the class $\{(b'[3], b''[3])\}$ of γ contains exactly one vertex, the vertex $(+, 3)$, and the edge labeled by the class $\{(b'[1], b''[1])\}$ of γ contains exactly two vertices, the vertices $(1, +)$ and $(+, 1)$. The data map of Q is shown in Figure 37 and the reduced data map of Q is the graph shown in Figure 38.

It should be noted that, generally speaking, the data map of the query set Q associated with an incomplete table need not be a graph, for if for some i and j none of the cells (i, j), $(i, +)$, and $(+, j)$ has been suppressed, then the data map of Q will contain the edge $\{(i,j), (i, +), (+,j)\}$. However, the reduced data map of Q is always a graph.

Before closing this section, we give the formal definition of the data map and the reduced data map of the query set Q associated with a given incomplete version of a complete table T. Let S be the set of suppressed internal cells, I the set of suppressed row totals, and J the set of suppressed column totals. The data map (H, w) of Q is defined as follows. The vertex set of H is taken to be

$$V = \{(i, j): (i, j) \notin S\} \cup \{(i, +): i \notin I\} \cup \{(+, j): j \notin J\}.$$

As for the edges of H, we proceed as follows. Let $S' = \{(i, j) \in S: i \in I$ and $j \in J\}$. Moreover, for each $i \notin I$, let $J_i = \{j \in J: (i, j) \in S\}$; analogously, for each $j \notin J$, let $I_j = \{i \in I: (i, j) \in S\}$. For example, for Table 6 one has $J_1 = J_2 = \emptyset$, and $I_1 = I_2 = \emptyset$ and $I_3 = \{3\}$. Then the active domain of $\mathbf{b}$ with respect to Q is

$$\{(b'[i], b''[j]): (i,j) \notin S'\}$$

and its characteristic partition γ contains

one elementary class $\{(b'[i], b''[j])\}$ for each $(i, j) \notin S$;

one elementary class $\{(b'[i], b''[j])\}$ for each i and j such that $(i, j) \in S$ and $i \notin I$ and $j \notin J$;

one class $\{(b'[i], b''[j]): j \in J_i\}$ for each $i \notin I$ with $J_i \neq \emptyset$;

one class $\{(b'[i], b''[j]): i \in I_j\}$ for each $j \notin J$ with $I_j \neq \emptyset$.

The edge set of H is taken to be

$$
\begin{aligned}
E = \quad & \{\{(i, j), (i, +), (+, j)\}: (i, j) \notin S, i \notin I, j \notin J\} \cup \\
& \cup \{\{(i, j), (i, +)\}: (i, j) \notin S, i \notin I, j \in J\} \cup \\
& \cup \{\{(i, j), (+, j)\}: (i, j) \notin S, i \in I, j \notin J\} \cup \\
& \cup \{\{(i, j)\}: (i, j) \notin S, i \in I, j \in J\} \cup \\
& \cup \{\{(i, +), (+, j)\}: (i, j) \in S, i \notin I, j \notin J\} \cup \\
& \cup \{\{(i, +)\}: i \notin I, J_i \neq \emptyset\} \cup \\
& \cup \{\{(+, j)\}: j \notin J, I_j \neq \emptyset\}.
\end{aligned}
$$

The labels and the weights of the edges e of H are reported below:

e	$label[e]$	$w(e)$
$\{(i,j), (i,+), (+,j)\}$	$\{(b'[i], b''[j])\}$	$T(i,j)$
$\{(i,j), (i,+)\}$	$\{(b'[i], b''[j])\}$	$T(i,j)$
$\{(i,j), (+,j)\}$	$\{(b'[i], b''[j])\}$	$T(i,j)$
$\{(i,j)\}$	$\{(b'[i], b''[j])\}$	$T(i,j)$
$\{(i,+), (+,j)\}$	$\{(b'[i], b''[j])\}$	$T(i,j)$
$\{(i,+)\}$	$\{(b'[i], b''[j]): j \in Ji\}$	$\sum_{j\in J_i} T(i,j)$
$\{(+,j)\}$	$\{(b'[i], b''[j]): i \in Ij\}$	$\sum_{i\in I_j} T(i,j)$

The reduced data map of Q is the graph with vertex set
$$\{(i,+): i \notin I\} \cup \{(+,j): j \notin J\}$$

and edge set
$$\{\{(i,+), (+,j)\}: (i,j) \in S, i \notin I, j \notin J\} \cup$$
$$\cup \{\{(i,+)\}: (i,j) \in S, i \notin I, J_i \cdot \bullet \varnothing\} \cup \{\{(+,j)\}: (i,j) \notin S, i \in I, I_j \neq \varnothing\}$$

FUTURE DIRECTIONS

Future research is called to answer all the complexity questions raised by the inference problem discussed in the previous sections, as well as the case of data of integer and nonnegative integer type. Other lines of research will have to cover multidimensional tables and special classes of hypergraphs for which safety tests can be efficiently worked out. Some results in these directions will profit of results stated in Irving & Jerrum (1994) and in Björner & Karlander (1993), respectively. It is also worth mentioning another research topic given by the so-called "usability" of a statistical database (Horak, Brankovic, & Miller, 1999, 2000).

CONCLUSIONS

We have reported some recent results on the inference problem for additive queries on a statistical database. We met with some computationally hard problems (on strong safety and strong p-safety), which are likely to have no efficient solutions. If so, one should be content with weaker protection levels (such as weak safety and weak p-safety). In any case, we can conclude that much work still remains to be done for the security of statistical databases to be solved in a satisfactory way.

REFERENCES

Adam, N.R., & Wortmann, J.C. (1989). Security control methods for statistical databases: A comparative study. *ACM Computing Surveys, 21*, 515-556.

Adam, N.R., Gangopadyay, A., & Holowczak, R.D. (1999). A survey of research on database protection. *Statistical Data Protection*. EUROSTAT, 29-43.

Ahuja, R.K., Magnanti, T.L., & Orlin, J.B. (1993). *Network Flows*. City, NJ: Prentice-Hall.

Björner, A., & Karlander, J. (1993). The mod *p* rank of incidence matrices for connected uniform hypergraphs. *European Journal of Combinatorics, 14*, 151-155.

Bondy, J.A., & Murty, U.S.R. (1976). *Graph Theory with Applications*. New York: North Holland.

Brankovic, L., Horak, P., & Miller, M. (2000). An optimization problem in statistical databases. *SIAM Journal of Discrete Mathematics, 13*, 346-353.

Carvalho, F.D. de, Dellaert, N.P., & Osorio, M. de S. (1994). Statistical disclosure in two-dimensional tables: general tables. *Journal of the American Statistical Association, 89*, 1547-1557.

Chin, F.Y. (1986). Security problems on inference control for sum, min, and max queries. *Journal of the ACM, 33*, 451-464.

Chin, F.Y., & Ozsoyoglu, G. (1982). Auditing and inference control in statistical databases. *IEEE Transactions on Software Engineering, 8*, 574-582.

Chu, P.C. (1997). Cell suppression methodology: The importance of suppressing marginal totals. *IEEE Transactions on Knowledge and Data Engineering, 9*, 513-523.

Conforti, M., & Rao, M.R. (1987). Some new matroids on graphs: Cutsets and the max cut problem. *Mathematics of Operations Research, 12*, 193-204.

Cox, L.H. (1980). Suppression methodology and statistical disclosure control. *Journal of the American Statistics Association, Theory and Method Section, 75*, 377-385.

Cox, L.H. (1981). Linear sensitivity measures in statistical disclosure control. *Journal of Statistical Planning and Inference, 75*, 153-164.

Cox, L.H., & Zayatz, L.V. (1995). An agenda for research in statistical disclosure limitation. *Journal of Official Statistics, 11*, 205-220.

Dantzig, G.B. (1963). *Linear Programming and Extensions*. Princeton, NJ: Princeton University Press.

Date, C.J. (1983). *An Introduction to Database Systems: Vol. II*. Reading, MA: Addison-Wesley.

Denning, D.E.R. (1982). *Cryptography and Data Security*. Reading, MA: Addison-Wesley.

Garfinkel, R.S., & Nemhauser, G.L. (1972). *Integer Programming*. New York: John Wiley & Sons.

Gusfield, D. (1988), A graph theoretic approach to statistical data security. *SIAM Journal of Computing, 17*, 552-571.

Horak, P., Brankovic, L., & Miller, M. (1999). A combinatorial problem in database security. *Discrete Applied Mathematics, 91*, 119-126.

Irving, R.W., & Jerrum, M.R. (1994). Three-dimensional statistical data security problems. *SIAM Journal of Computing, 23*, 170-184.

Kao, M.-Y. (1997). Efficient detection and protection of information in cross-tabulated tables II: Minimal linear invariants. *Journal of Combinatorial Optimization, 1*, 187-202.

Kelly, J.P., Golden, B.L., & Assad, A.A. (1992). Cell suppression: Disclosure protection for sensitive tabular data. *Networks, 22*, 397-417.

Kleinberg, J.M., Papadimitriou, C.H., & Raghavan, P. (2000). Auditing Boolean attributes. *Proceedings of the 14th ACM Symposium on Principles of Database Systems,* 86-91.

Malvestuto, F.M., & Moscarini, M. (1990). Query evaluability in statistical databases. *IEEE Transactions on Knowledge and Data Engineering, 2*, 425-430.

Malvestuto, F.M. (1993). A universal-scheme approach to statistical databases containing homogeneous summary tables. *ACM Transactions on Database Systems, 18*, 678-708.

Malvestuto, F.M., & Moscarini, M. (1996a). Suppressing marginal totals from a two-dimensional table to protect sensitive information. *Statistics and Computing, 7*, 101-114.

Malvestuto, F.M., & Moscarini, M. (1996b). Censoring statistical tables to protect sensitive information: Easy and hard problems. In Svenson, P., & French, J.C. (Eds.), *Proceedings of the 8th International Conference on Scientific and Statistical Database Management,* 12-21.

Malvestuto, F.M., & Moscarini, M. (1999). An audit expert for large statistical databases. *Statistical Data Protection.* EUROSTAT, 29-43.

Malvestuto, F.M., & Mezzini, M. (2002). A linear algorithm for finding the invariant edges of an edge-weighted graph. *SIAM Journal of Computing* (31, 1438-1455).

Malvestuto, F.M., & Mezzini, M. (2001). *Minimal Linear and Sum Invariants of an Edge-Weighted Graph.* Unpublished Manuscript.

Michalewicz, Z. (1991). Security of a statistical database. Chapter 13 of Michalewicz, Z. (Ed.), *Statistical and Scientific Databases.* New York: Ellis Horwood.

Roehrig, S.F. (2001). An overview of disclosure auditing in categorical databases. *Proceedings of the World Multiconference on Systemics, Cybernetics, and Informatics.* Volume XIV, 510-514.

Schackis D. (1993). Manual on disclosure control methods. EUROSTAT, Technical Report.

Ullman, J.D. (1983). *Principles of Database Systems.* Rockville, MD: Computer Science Press.

Willenborg, L., & de Waal, T. (1996). Statistical disclosure control in practice. *Lecture Notes in Statistics,* Volume 111. New York: Springer-Verlag.

Willenborg, L., & de Waal, T. (2000). Elements of statistical disclosure control. *Lecture Notes in Statistics*, Volume 155. New York: Springer-Verlag.

Chapter XII

Source Integration for Data Warehousing

Andrea Calí
Università di Roma–La Sapienza, Italy

Domenico Lembo
Università di Roma–La Sapienza, Italy

Maurizio Lenzerini
Università di Roma–La Sapienza, Italy

Riccardo Rosati
Università di Roma–La Sapienza, Italy

ABSTRACT

While the main goal of a data warehouse is to provide support for data analysis and management's decisions, a fundamental aspect in design of a data warehouse system is the process of acquiring the raw data from a set of relevant information sources. We will call source integration system the component of a data warehouse system dealing with this process. The main goal of a source integration system is to deal with the transfer of data from the set of sources constituting the application-oriented operational environment, to the data warehouse. Since sources are typically autonomous, distributed, and heterogeneous, this task has to deal with the problem of cleaning, reconciling,

*and integrating data coming from the sources. The design of a source
integration system is a very complex task, which comprises several different
issues. The purpose of this chapter is to discuss the most important problems
arising in the design of a source integration system, with special emphasis on
schema integration, processing queries for data integration, and data cleaning
and reconciliation.*

INTRODUCTION

The typical architecture of a data warehouse system is constituted by two
different components, usually called back-end and front-end, respectively. While the
latter is intended to provide support for the main task of the system, namely, data
analysis and management's decisions, the former is responsible for acquiring the raw
data from a set of relevant information sources. One of the basic assumptions in data
warehouse architectures is that a correct modularization requires these two components
be decoupled. It is the task of the back-end component to free the front-end from the
knowledge on where data are, and how data are structured at the sources.

The goal of this chapter is to discuss the most important issues in the design of
the back-end component of a data warehouse. We will call such component "source
integration system," because its main goal is to deal with the task of transferring data
from the set of sources constituting the application-oriented operational environment
to the data warehouse. When data passes from one environment to the other, possible
inconsistencies and redundancies should be resolved, so that the warehouse is able
to provide an integrated and reconciled view of data of the organization (Inmon,
1996).

The constraints that are typical of data warehouse applications restrict the large
spectrum of approaches that have been proposed for information integration (Hull
& Zhou, 1996; Inmon, 1996; Jarke et al., 1999). In particular, since the data
warehouse should reflect the informational needs of the organization, it should be
based on a unified, corporate view of data, called *global schema*. Without the
definition of a global schema, the risk arises of concentrating on what is in the sources
at the operational level, rather than on what is really needed in order to perform the
required analysis on data (Devlin, 1997).

The architecture of a source integration system is usually described in terms of
two types of modules: wrappers and mediators (Wiederhold, 1994; Ullman, 1997).
The goal of a wrapper is to access a source, extract the relevant data, and present
such data in a specified format. The role of a mediator is to collect, clean, and combine
data produced by different wrappers (or mediators), so as to meet a specific
information need of the data warehouse. The specification and the realization of
mediators is the core problem in the design of a source integration system.

The design of a source integration system is a very complex task, which
comprises several different issues, including the following:

1. Definition of both the global schema, and the relationships between the global schema and the sources.
2. Specification of how to load and refresh data according to the global schema.
3. Cleaning and reconciliation of the data to be transferred from the sources to the data warehouse.

Problem (1) deals with the fact that sources are typically heterogeneous, meaning that they adopt different models and systems for storing data. This poses challenging problems in specifying the global schema. We call *schema integration* the activities to be carried out for this purpose. The goal of schema integration is to design the global schema so as to provide an appropriate abstraction of all the data residing at the sources. However, the specification of the global schema is not the only issue in schema integration. Indeed, another crucial aspect in schema integration is the specification of the relation between the sources and the global schema. To this purpose, two basic approaches have been proposed in the literature. The first approach, called *global-as-view* (or simply GAV), requires that a view over the sources is associated to each element of the global schema, so as to specify the global schema in terms of the data residing at the sources. On the contrary, the second approach, called *local-as-view* (or simply LAV), requires that a view over the global schema is associated to each element of the sources. Thus, in the LAV approach, we specify the meaning of the sources in terms of the elements of the global schema.

In the lifetime of a data warehouse, the explicit representation of relationships between the sources and the materialized data in the data warehouse is useful in several tasks: from the initial loading, where the identification of the relevant data within the sources is critical, to the refreshment process, which may require a dynamic adaptation depending on the availability of the sources, as well as on their reliability and quality that may change over time. Moreover, the extraction of data from a primary data warehouse for data mart applications, where the primary warehouse is now regarded as a data source, can be treated in a similar way. In addition, even though the sources within an enterprise are not as dynamic as in other information integration frameworks, they are nonetheless subject to changes; in particular, creation of new sources and deletion of existing ones must be taken into account. Consequently, the maintenance of the data warehouse requires several upgrades of the data flows towards the data warehouse. In other words, a data warehouse, especially in large organizations, should be regarded as an incremental system, which critically depends upon the relationships between the sources and the data warehouse.

Whereas the construction of the global schema concerns the intentional level of the source integration system, problems (2) and (3) refer to a number of issues arising when considering the integration at the extensional/instance level. Problem (2) is concerned with one of the most important issues in a source integration system, namely, the choice of the method for computing the answer to queries posed in terms of the global schema. Being able to carry out query processing on the global schema is indeed crucial for *data integration*, i.e., both for loading and refreshing the data warehouse according to the global schema. The main issue is that the system should be able to re-express queries expressed on the global schema in terms of a suitable set of queries posed to the sources. In this reformulation process, the crucial step is deciding how to decompose the query on the global schema into a set of subqueries on the sources, based on the meaning of the sources in terms of the concepts in the global schema. The computed subqueries are then shipped to the sources, and the results are assembled into the final answer.

Problem (3) deals with the fact that sources are generally heterogeneous not only at the intentional level, but also at the instance level. In particular, the data returned by various sources during query processing need to be converted/reconciled/combined to provide the data warehouse with high-quality data. The complexity of this reconciliation step is due to several problems, such as possible mismatches between data referring to the same real-world object, possible errors in the data stored in the sources, or possible inconsistencies between values representing the properties of the real-world objects in different sources (Galhardas et al., 1999). The above task is known in the literature as *data cleaning and reconciliation*.

In the rest of this chapter we discuss the above three problems in more detail. After presenting some basic definitions, we provide a survey on the problem of schema integration, starting from the early work found in the database literature more than two decades ago, and pointing out the recent developments in this area. We then deal with data integration, by describing the most advanced approaches to query processing in the context of source integration systems. The final section discusses the most important approaches that have been proposed for carrying out the data cleaning and reconciliation task.

BACKGROUND

In this section we set up a formal framework for source integration in Data Warehousing. In particular, our main goal is to define the notion of source integration systems, which is intended to represent the component of a data warehouse system dealing with the task of integrating the sources of information for the data warehouse system. We characterize a source integration system as constituted by three elements, namely, the global schema, the sources, and the mapping between the two. Finally, we provide the semantics both of the system, and of query answering.

The formal definition of a source integration system is given below.

Definition 1: A *source integration system* I is a triple $<G, S, M>$, where G is the global schema, S is the source schema, and M is the mapping between G and S.

The following comments on the above formal definition are in order.

- The global schema is intended to specify the structure of the information needed in the data warehouse. From a methodological point of view, such a schema is a reconciled view of the information stored in the sources. In what follows, we denote with A_G the finite alphabets for the elements of the global schema. According to Devlin (1997), a conceptual data model, e.g., the entity-relationship model, is generally used for expressing the global schema. However, our formalization is completely independent from the particular data model used.

- The source schema provides the specification of the structure of the various data sources. Such a schema contains the intentional description of all the sources of the data warehouse application. Although in principle the various source schemas may be expressed using different data models and notations, it is common to define suitable wrappers that present all the schemas of the sources in a predefined form, e.g., in terms of the relational model. Therefore, the source schema is usually expressed as a set of relation schemas. In what follows, we denote with A_S the finite alphabets for the elements of the source schema.

- The mapping M establishes a relationship between elements of the global schema G and those of the source schema S. As we already said in the introduction, two basic approaches, namely GAV and LAV, have been proposed for specifying the mapping, and we will distinguish between these two types of mappings when specifying the semantics of a source integration system.

Let us turn our attention to the semantics of a source integration system $I=<G, S, M>$. We assume that the databases involved in our framework (both global databases and source databases) are defined over a fixed (infinite) alphabet Γ of symbols. In order to assign semantics to I, we start by considering a *source database* for I, i.e., a database D for the source schema S. Based on D, we can specify which is the information content of the global schema G at the extensional level. We call *global database* for I any database for G.

Definition 2: Let $I=<G, S, M>$ be a source integration system, and D a source database for I. A global database B for I is said to be *legal for I with respect to D*, if:

- B is legal with respect to G, i.e., B satisfies all the constraints of G;
- B satisfies the mapping M with respect to D.

The notion of B satisfying the mapping M with respect to D depends on the type of the mapping considered, GAV or LAV.

- **GAV mapping.** In the GAV approach, the mapping M associates to each element r in G a view, i.e., a query, over S, denoted by $\rho(r)$. We say that B satisfies M with respect to D if, for each element r of G, the set of tuples r^B that B assigns to r contains the set of tuples $\rho(r)^D$ that satisfy the query $\rho(r)$ in D, i.e.,

$$\rho(r)^D \subseteq r^B$$

  Note that this means that the view associated to r is *sound*: the data provided by the sources satisfy the element of global schema, but are not necessarily complete.

- **LAV mapping.** In the LAV approach, instead, the mapping M associates to each source s in S a view, i.e., a query, over G, denoted by $\rho(s)$. In this case, we say that B satisfies M with respect to D, if for each source s of S, the set of tuples s^D that D assigns to s is contained in the set of tuples $\rho(s)^B$ that satisfy the query $\rho(s)$ in B, i.e.,

$$s^D \subseteq \rho(s)^B$$

  Note that, analogously to the previous case, this means that the view associated to s is *sound*.

Queries posed to a source integration system I are expressed in terms of a query language $\mathcal{L}_Q$ over the alphabet $\mathcal{A}_G$, i.e., over the global schema. In the following, if DB is a database, and q is a query, then q^{DB} denotes the result of evaluating q over the DB.

Definition 3: Let $I=<G, S, M>$ be a source integration system, D a source database for I, and q a query of arity n to I. The answer $q^{I,D}$ to q with respect to D, is the set of tuples $(c_1,...,c_n) \in \Gamma^n$ such that $(c_1,...,c_n) \in q^B$ for each global database B legal for I with respect to D.

Since, in general, several global databases exist that are legal for I with respect to D, in the terminology of source integration, $q^{I,D}$ is often called the set of *certain answers* of q with respect to D.

As we said in the introduction, the main activities that are carried out in the design of a source integration system are: schema integration, data integration, and data cleaning. To relate these activities to the formalization presented in this section, we observe that:

- Schema integration has the goal to provide the specification of the three main components of the system, namely, the global schema, the source schema, and the mapping.
- Data integration aims at defining the correct method for acquiring data from the sources, so as to populate (either virtually or physically) the elements of the

global schema. In other words, the purpose of data integration is to come up with a suitable method for answering queries over the global schema, by accessing the data at the sources.

- The goal of data cleaning is to design the mapping M of the source integration system in such a way that, when acquiring data at the sources, suitable conversion, transformation, and reconciliation actions are performed on these data.

SCHEMA INTEGRATION

Schema integration is the activity of integrating the schemas of the various sources in order to produce a homogeneous description of the data of interest.

The work on schema integration is relevant to any information integration approach and in particular to the context of data warehousing, since, in order to integrate data, schema integration must either implicitly or explicitly be done.

Schema integration is divided into several methodological steps (Batini, Lenzerini, & Navathe, 1986). Such steps are essentially independent of the integration strategy adopted, and aim at relating the different components of schemas, finding and resolving conflicts in the representation of the same data among the different schemas, and eventually merging the conformed schemas into a global one.

In particular, the following methodological steps are singled out:
1. preintegration
2. schema comparison
3. schema conforming
4. schema merging and restructuring
5. schema mapping

Traditionally, schema integration is a "one-shot" activity, resulting in a global schema in which all data are represented uniformly (Batini, Lenzerini, & Navathe, 1986). However, more recently, in order to deal with autonomous and dynamic information sources, an incremental approach is arising (Catarci & Lenzerini, 1993). Such an approach consists of building a collection of independent partial schemas, formalizing the relationships among entities in the partial schemas by means of the so-called *interschema assertions*. In principle, under the assumption that the various information sources remain unchanged, the incremental approach would eventually result in a global schema, similar to those obtained through a traditional one-shot approach, although in practice, due to the dynamics of the sources, such a result is never achieved. Additionally, the integration may be partial, taking into account only certain aspects or components of the sources (Catarci & Lenzerini, 1993).

In the rest of this section, we review recent studies on schema integration, according to the steps they address. Furthermore, we analyze the following key aspects: i) whether a global schema is produced or not; ii) among the ones above mentioned, which is the schema integration methodological step the work refers to; iii) which is the formalism used for representing data schemas.

We refer to Batini, Lenzerini, & Navathe (1986) for a comprehensive survey on previous work in this area.

Preintegration

Preintegration consists of an analysis of the schemas to decide the general integration policy: choosing the schemas to be integrated, deciding the order of integration, and possibly assigning preferences to entire schemas or portions thereof. The choices made in this phase influence the usefulness and relevance of the data corresponding to the global schema. During this phase also additional information relevant to integration is collected, such as assertions or constraints among views in a schema. Such process is sometimes referred to as *semantic enrichment* (García-Solaco, Saltor, & Castellanos, 1995a, 1995b; Blanco, Illarramendi, & Goñi, 1994; Reddy et al., 1994). It is usually performed by translating the source schemas into a richer data model, that allows for representing information about dependencies, null values, and other semantic properties, thus increasing interpretability and believability of the source data.

For example, Johanneson (1994) defines a collection of transformations on schemas represented in a first-order language augmented with rules to express constraints. Such transformations are correct with respect to a given notion of information preservation, and constitute the core of a "standardization" step in a new schema integration methodology. This step is performed before schema comparison and logically subsumes the schema conforming phase, which is not necessary in the new methodology.

In Blanco, Illarramendi, & Goñi (1994), relational schemas are enriched using a class-based logical formalism, a *description logic* (DL), available in the terminological system BACK (Peltason, 1991). Instead García-Solaco, Saltor, & Castellanos (1995a, 1995b) use as a unifying model a specific object-oriented model with different types of specialization and aggregation constructs.

The creation of a knowledge base (terminology) in the preintegration step is proposed in Sheth, Gala, & Navathe (1993). More precisely, a hierarchy of attributes is generated, thus representing the relationship among attributes in different schemas. Then, source schemas are classified: the terminology thus obtained corresponds to a partially integrated schema. Such a terminology is then restructured by using typical reasoning services of class-based logical formalisms. The underlying data model is hence the formalism used for expressing the terminology: more precisely, a description logic (CANDIDE) is used.

Schema Comparison

Schema comparison (also called schema matching) is the phase in which the correlations among concepts of different schemas are determined and possible conflicts are detected. Moreover, interschema properties are typically discovered during this phase.

There has been a considerable amount of research in studying the types of conflicts that arise when comparing source schema components (see, e.g., Batini, Lenzerini, & Navathe, 1986; Krishnamurthy, Litwin, & Kent, 1991; Spaccapietra, Parent, & Dupont, 1992; Ouksel & Naiman, 1994; Reddy et al., 1994), and consensus has arisen on their classification, which can be summarized as follows:

- *Heterogeneity conflicts* arise when different data models are used for the source schemas.
- *Naming conflicts* arise because different schemas may refer to the same data using different terminologies. Typically one distinguishes between *homonyms*, where the same name is used to denote two different concepts, and *synonyms*, where the same concept is denoted by different names.
- *Semantic conflicts* arise due to different choices in the level of abstraction when modeling similar real-world entities.
- *Structural conflicts* arise due to different choices of constructs for representing the same concepts.

In general, this phase requires a strong knowledge of the semantics underlying the concepts represented by the schemas. The more the semantics is represented formally in the schema, the easier can similar concepts in different schemas be automatically detected, possibly with the help of specific CASE tools that support the designer. Traditionally, schema comparison was performed manually (Batini, Lenzerini, & Navathe, 1986). However, recent methodologies and techniques emphasize automatic support to this phase.

For example, Blanco, Illarramendi, & Goñi (1994) exploit the reasoning capabilities of the terminological system to classify relational schema components and derive candidate correspondences between them expressed in the description logic BACK.

In Miller, Yoannidis, & Ramakrishnan (1994), the problem of deciding equivalence and dominance between schemas is analyzed, based on a formal notion of information capacity given in Hull (1996). Specifically, schemas are expressed in a graph-based data model which allows for the representation of inheritance and simple forms of integrity constraints. It is proven that such a problem is undecidable in schemas that occur in practice; moreover, *sufficient* conditions for schema dominance are defined, based on a set of schema transformations that preserve schema dominance. A schema S_1 is dominated by a schema S_2 if there is a sequence of such transformations that converts S1 to S2.

Krishnamurthy, Litwin, & Kent (1991) discuss reconciliation of semantic discrepancies in the relational context due to information represented as data in one database and as meta-data in another. The paper proposes a solution based on reifying relations and databases by transforming them into a structured representation.

An architecture where schema comparison and the subsequent phase of schema conforming are iterated is proposed in Gotthard, Lockemann, & Neufeld (1992). At each cycle, the system proposes correspondences between concepts that can be confirmed or rejected by the designer. The system uses newly established correspondences both to conform the schemas and to guide its proposals in the following cycle. A data model that essentially corresponds to an entity-relationship model extended with complex objects is used to express both the component schema and the resulting global schema.

In Bouzeghoub & Comyn-Wattiau (1990), schema comparison in an extended entity-relationship model is performed by analyzing structural analogies between subschemas through the use of similarity vectors. Subsequent conforming is achieved by transforming the structures into a canonical form.

Palopoli, Saccà, & Ursino (1999) present semi-automatic techniques for detecting synonym and homonym relationships between objects belonging to different entity-relationship schemas. In particular, such techniques are based on algorithms whose input is a set of weighted synonym, homonym, and inclusion relationships between objects. The weight represents the "plausibility factor" for the relationship to hold. Based on such relationships, new weighted relationships are automatically derived, which in turn are expected to hold with a plausibility degree corresponding to the computed weight. The method for deriving new relationships consists of pairwise comparison of schema objects E_1 and E_2, which measures the similarity of all objects related to E_1 and E_2 in the respective schemas. Such techniques have been implemented in the system DIKE (Palopoli, Terracina, & Ursino, 2000).

In Madhavan, Bernstein, & Rahn (2001), an algorithm is presented for the detection of correspondences between schema elements in a very general data model that is able to capture relational, object-oriented, and XML schemas. The algorithm takes into account several aspects of schema elements (names, data types, constraints), integrating linguistic and structural matching techniques. Moreover, the algorithm is able to cope with mappings of shared types and with some forms of schema constraints (e.g., foreign key constraints).

Finally, Bergamaschi et al. (2001) present a schema comparison technique that computes, for each pair of objects A_1, A_2 in the schemas, a value corresponding to the "affinity" between A_1 and A_2. Such an affinity is obtained as the weighted sum of three kinds of affinity: name, data type, and structural affinity. Name affinity is obtained by resorting to thesauri that specify relationships (e.g., synonym, hypernym) between object names; data type affinity is obtained by means of a table that defines compatibilities between data types, while structural affinity is obtained by analyzing the similarity of the relationships the objects participate in, in the respective schemas. In this framework, schemas are represented using description logics. Such techniques have been implemented in ARTEMIS, a tool for schema integration, which is used as a component of the MOMIS system (Beneventano et al., 2000) for integration of relational, object-oriented, and semi-structured source schemas.

Schema Conforming

The phase of schema conforming has the goal of conforming or aligning schemas to make them compatible for integration. *Conflict resolution* is the most challenging aspect of this phase. Typically, semi-automatic solutions to schema conforming are proposed, in which intervention of the designer is requested by the system when conflicts have to be resolved. Recent methodologies and techniques also emphasize the automatic resolution of specific types of conflicts (e.g. structural conflicts). However, a logical reconstruction of conflict resolution is far from being accomplished and is still an active topic of research.

For example, in Vidal & Winslett (1994), a general methodology for schema integration is presented, in which the semantics of updates is preserved during the integration process. Specifically, three steps are defined: combination, restructuring, and optimization. In the first phase, a combined schema is generated, which contains all source schemas and assertions (constraints) expressing relationship among entities in different schemas. The restructuring step is devoted to normalizing (through schema transformations) and merging views, thus obtaining a global schema, which is refined in the optimization phase. Such a methodology is based on a semantic data model which allows for declaring constraints containing indications on what to do when an update violates that constraint. A set of schema transformations is defined which is *update semantics,* preserving in the sense that any update specified against the transformed schema has the same effect as if it was specified against the original schema.

Qian (1996) presents a formal analysis of the problem of establishing correctness of schema transformations. More specifically, schemas are modeled as abstract data types, and schema transformations are expressed in terms of signature interpretations. The notion of schema transformation correctness is based on a refinement of Hull's notion of information capacity (Hull, 1986). In particular, such a refinement allows for a formal study of schema transformations between schemas expressed in different data models.

Schema Merging and Restructuring

During this phase the conformed schemas are superimposed, thus obtaining a (possibly partial) global schema. Such a schema is then tested against qualities such as completeness, correctness, minimality, and understandability. This analysis may give rise to transformations of the schema obtained.

Geller et al. (1992a, 1992b) present an integration technique (*structural integration*) which allows for the integration of entities that have structural similarities, even if they differ semantically. An object-oriented model, called DUAL model, is used, in which structural aspects are represented as object types, and semantic aspects are represented as classes. Two notions of correspondence between classes are defined: full structural correspondence and partial structural correspondence. The (partial) integration of two schemas is then obtained through a generalization of the classes representing the original schemas.

In Spaccapietra, Parent, & Dupont (1992), a methodology for schema integration is presented. Such a methodology allows for automatic resolution of structural conflicts and building of the integrated schema without requiring conforming of the initial schemas. The methodology is applicable to various source data models (relational, entity-relationship, and object-oriented models), and is based on an expressive language to state interschema assertions that may involve constructs of schemas expressed in different models. Data model independent integration rules that correspond to the interschema assertions are defined in the general case and are also specialized to the various classical data models. Quality issues are addressed in an informal way: specifically, correctness is achieved by selecting, in case of conflicts, the constructs with weaker constraints. The methodology includes strategies that avoid introducing redundant constructs in the generated schema. However, completeness is not guaranteed, since the model adopted for the global schema lacks a generalization construct.

Schema Mapping

Once the global schema is generated, the various source schemas are related to the global schema. The way in which the data at the sources are related to elements of the global schema, i.e., the way in which the *mapping* is specified, may assume different forms.

As explained in the previous sections, two basic approaches, GAV and LAV, have been used to specify the mapping between the sources and the global schema (Lenzerini, 2001; Levy, 1999, 2000; Li & Chang, 2000). The GAV approach, also called query-based approach, requires that the global schema is expressed in terms of the data sources. More precisely, to every element of the global schema, a view over the data sources is associated, so that its meaning is specified in terms of the data residing at the sources. The LAV approach, also called source-based approach, requires the global schema to be specified independently from the sources. In turn, the sources are defined as views over the global schema. The relationships between the global schema and the sources are thus established by specifying the information content of every source in terms of a view over the global schema.

Intuitively, the GAV approach provides a method for source integration with a more procedural flavor with respect to the LAV approach. Indeed, whereas in LAV the designer may concentrate on specifying the content of the source in terms of the global schema, in the GAV approach, one is forced to specify how to get the data of the global schema by queries over the sources.

A comparison of the LAV and the GAV approaches is reported in Ullman (1997). It is known that the former approach ensures an easier extensibility of the integration system, and provides a more appropriate setting for its maintenance. For example, adding a new source to the system requires only provision of the definition of the source, and does not necessarily involve changes in the global schema. On the contrary, in the GAV approach, adding a new source may in principle require changing the definition of the concepts in the global schema.

DATA INTEGRATION

Data integration activities concern with the problem of acquiring data from different sources, making them available in an integrated and reconciled form to users' applications. This is a central issue in several contexts: applications requiring accessing or re-engineering legacy systems, information integration on the World Wide Web, data mining or enterprise resource planning, or data warehouse systems.

With regard to the data explicitly managed by the integration systems, two different approaches exist, called *materialized* and *virtual*.

In the materialized approach, data at the sources are replicated in the integration system and can be directly queried by the users. In this setting, the problem arises of refreshing the materialized views maintained in the system in order to cope with updates at the sources. A naïve way to deal with this problem is to recompute views entirely from scratch in the presence of changes at the sources. This is expensive and makes frequent refreshing of views impractical. The study of materialized view management is an active research topic, and it is concerned both with the problem of choosing the views to be materialized into the data warehouse, and reducing the overhead in view recomputation. We do not discuss this further in this chapter and refer the reader to Jarke et al. (1999) for more details on the topic. The materialized approach to data integration is the most closely related to data warehousing. In this context, data integration activities are relevant for the initial loading and for the refreshing of the warehouse according to the global schema produced by the schema integration activities. Notice that, whereas the global schema provides a reconciled *intentional* representation of all source schemas, data integration deals with the problem of computing its instances, thus pertaining to the *extensional* level of the source integration system of the data warehouse.

In the virtual approach to data integration, data residing at the sources are not replicated, and the systems provide the user with a virtual global schema, i.e., a global schema whose extension is not materialized, e.g., in a data warehouse. Hence, sources are accessed each time a user query is posed to the system, rather than in the process of loading the data warehouse. Despite such differences, most of the issues that arise, for the case in which the views are virtual, are also relevant in the case of materialized views, which can be, therefore, conceived as a specialization of the virtual views. Hence, in the following we focus on the problem of processing user requests for new reconciled data, without distinguishing between the cases in which such data have to be materialized or not.

Generally speaking, the main goal of the data integration process is to answer a query expressed over the global schema only on the basis of the data stored at the sources, i.e., to solve the *query processing* problem. Given a user query, the system should be able to decompose it in terms of a suitable set of queries posed to the sources that are considered relevant to the original query, send such queries to the sources, and assemble the results into the final answer. With reference to the formal framework for source integration in data warehousing described earlier, query processing amounts to computing the set of certain answers to the query.

Query processing has been studied under different assumptions on the model adopted to represent the global schema and the sources (conceptual, object-oriented, semistructured, etc.), the language used to express both queries and views (conjunctive queries, datalog queries, etc.), the nature of constraints imposed on the global schema, and the way in which the mapping between data at the sources and the elements of the global schema is specified. As we have already said, two basic approaches exist to specify the mapping. The first approach, called *global-as-view* (GAV), requires that a view over the data sources is associated to every element of the global schema. On the contrary, in the second approach, called *local-as-view* (LAV), the relationships between the global schema and the sources are established by associating a view over the global schema to every element in the sources. Calì et al. (2002b) discussed the relationship between the expressive power of LAV and GAV in a setting where the global schema and the sources are relational, and either queries or views are conjunctive queries. In that paper the authors first show that, when no integrity constraints are allowed in the global schema, the two approaches are incomparable. Then they propose techniques that exploit the presence of suitable constraints in the global schema to transform a system following the LAV (GAV) approach into a query-preserving GAV (respectively, LAV) system, i.e., a system in which processing the same query, or an equivalent suitable transformation of it, produces the same answer.

In the rest of this section we describe several solutions proposed in the literature to address the problem of query processing, either in LAV or in GAV. For the sake of simplicity, we assume to deal with relational databases.

Query Processing in the LAV Approach

Since in LAV the mapping between the sources and the global schema is described as a set of views over the global schema, query processing amounts to finding a way to answer a query posed over a database using a set of views over the same database. This problem, known in the literature as *query answering using views*, is widely studied, since it has applications in many areas. In query optimization (Chaudhuri et al., 1995), the problem is relevant because using materialized views may speed up query processing. In database design, using views provides means for maintaining the physical perspective of the data independent from its logical perspective (Tsatalos, Solomon, & Ioannidis, 1986). It is also relevant for query processing in distributed databases (Keller & Basu, 1996) and in federated databases (Levy, Srivastava, & Kirk, 1995). Finally, since the views provide partial knowledge on the database, answering queries using views can be seen as a special case of query answering with incomplete information (Imielinski & Lipski, 1984; Abiteboul & Duschka, 1998).

Query Rewriting

The most common approach proposed in the literature to deal with the problem of query answering using views is by means of *query rewriting*. In query rewriting, a query over the global schema, and a set of views defining the sources, expressed in turn over the global schema, are provided. The goal is to reformulate the query into an expression, the *rewriting*, which refers only to the views and supplies the answer to the query. Query answering via query rewriting is divided into two steps: (1) reformulating the query in terms of a given query language over the alphabet of the view names, i.e., the alphabet of the sources, and (2) evaluating the rewriting over the view extensions, i.e., the data at the sources.

In order to be able to compare different reformulations of queries, we first introduce the notion of containment between queries. Given two queries q_1 and q_2, we say that q_1 is *contained* in q_2 if for all databases $\mathcal{DB}$ we have that $q_1^{\mathcal{DB}} \subseteq q_2^{\mathcal{DB}}$. We say that q_1 and q_2 are *equivalent* if q_1 is contained in q_2 and q_2 is contained q_1. The problem of query containment has been studied in various settings. In particular, the containment of conjunctive queries is established to be an NP-complete problem in Chandra & Merlin (1977), and in the presence of inequalities, it is proved to be a Π^p_2-complete problem in Klug (1988) and van der Meyden (1992).

Let us turn our attention to the problem of rewriting. Formally, given a query q and a set of views $\mathcal{V}$, both expressed over the global schema, the query q_r is a *rewriting* of q using $\mathcal{V}$ if: *(i)* q_r refers only to the alphabet of the sources, and *(ii)* the *expansion* of q_r, i.e., the query obtained by replacing the views in it with their definitions in terms of the global schema is contained in q.

In general, the set of sources available in a source integration system may not store all the data needed to answer a query, and therefore the goal is to find a query expression that provides all the answers that can be obtained from the views. So, whereas in different contexts, e.g., query optimization or maintaining physical data independence, the focus is on finding rewritings that are logically equivalent to the original query, in data integration one is interested in *maximally contained rewritings* (Halevy, 2000). Formally, given a query q, a set of views $\mathcal{V}$, and a query language $\mathcal{L}_Q$, the query q_m is a *maximally contained rewriting* of q using $\mathcal{V}$ with respect to $\mathcal{L}_Q$ if: *(i)* q_m is a rewriting of q using $\mathcal{V}$, and *(ii)* there is no q' rewriting of q using $\mathcal{V}$ in $\mathcal{L}_Q$ such that $q' \neq q_m$ and q_m is contained in q'.

An important theoretical result concerning the problem of query rewriting, in the case that views and queries are conjunctive queries, is presented in Levy et al. (1995). In that paper the authors demonstrate that, when a query q is a union of conjunctive queries, if an equivalent conjunctive rewriting of q exists, then such a rewriting has at most as many atoms as q. Such a result leads immediately to nondeterministic polynomial-time algorithms to find either equivalent conjunctive rewritings or maximally contained rewritings that are the union of conjunctive queries. In both cases it is sufficient to consider each possible conjunction of views that produces a *candidate rewriting* whose size is less or equal to the size of the original query, and then check the correctness of the rewriting. Note that the number of candidate rewritings is exponential in the size of the query.

In order to compute all the rewritings that are contained in (and not necessarily equivalent to) the original query, the *bucket algorithm*, presented in Levy, Rajaraman, & Ordille (1996a), improves the technique described above since it exploits a suitable heuristic that enables one to prune the space of candidate rewritings. The algorithm was proposed in the context of the Information Manifold (IM) system (Levy, Srivastava, & Kirk, 1995), a project developed at AT&T. To compute the rewriting of a query q, the bucket algorithm proceeds in two steps: *(i)* for each atom g in q, it creates a bucket that contains the views from which tuples of g can be retrieved; then *(ii)* it considers as a candidate rewriting each conjunctive query obtained by combining one view from each bucket, and checks by means of a containment algorithm whether such a query is contained in q. If so, the candidate rewriting is added to the answer. If the candidate rewriting is not contained in q, before discarding it, the algorithm checks if it can be modified by adding comparison predicates in such a way that it is contained in q.

The proof that the bucket algorithm generates the maximally contained rewriting, when the query language is a union of conjunctive queries, is given in Grahne & Mendelzon (1999). Note that, on the basis of the results in Levy et al. (1995), the algorithm considers only rewritings that have at most the same number of atoms as the original query. As shown in the same paper and in Rajaraman, Sagiv, & Ullman (1995), the given bound on the size of the rewriting does not hold in the presence of arithmetic comparison predicates, functional dependencies expressed in the global schema, or limitations in accessing the sources. In such cases we have to consider rewritings that are longer than the original query. According to Levy, Rajaraman, & Ordille (1996b), the bucket algorithm, in practice, does not miss solutions because of the length of the rewritings it considers; but other results (Duschka, Genesereth, & Levy, 2000; Gryz, 1998a, 1998b) show that in the presence of functional dependencies and limitations in accessing the sources, the union of conjunctive queries does not suffice to obtain the maximal contained rewritings, and it is necessary to make use of recursive rewritings.

Two improved versions of the bucket algorithm are the *MiniCon algorithm* (Pottinger & Levy, 2000) and the *shared-variable bucket algorithm* (Mitra, 1999). In both the algorithms the basic idea is to examine the interaction among variables of the original query and variables in the views, in order to reduce the number of views inserted in the buckets, and hence the number of candidate rewritings to be considered. Experimental results related to the performance of MiniCon and shared-variable bucket show that these algorithms scale out and outperform the bucket algorithm.

Differently from the algorithms described above, the *inverse rules algorithm* (Duschka & Genesereth, 1997), developed in the context of the Infomaster system, an information integration tool of Stanford University, does not search the space of candidate rewritings, but generates a rewriting in time that is polynomial in the size of the query, and does not require any containment test. The algorithm constructs a set of rules that "invert" the view definitions and show how to obtain the instances of the global schema from the data stored at the sources. Given a query q and a set of views $\mathcal{V}$, the rewriting is the Datalog program consisting of both the query and the inverse rules obtained from $\mathcal{V}$. The inverse rules algorithm is proved to return a maximally contained rewriting with respect to union of conjunctive queries, in a time that is polynomial in the size of the query. Furthermore, Duschka, Genesereth, & Levy (2000) show that the algorithm can also handle recursive datalog queries, the presence of functional dependencies in the global schema, or the presence of limitations in accessing the sources, by extending the obtained query plan with other specific rules.

Finally, we cite the *unification-join algorithm* (Qian, 1996), an exponential-time query answering algorithm based on a skolemization phase and on the representation of conjunctive queries by means of hypergraphs. Qian (1996) also describes a polynomial time version of the algorithm developed for the case in which queries are acyclic conjunctive queries. The unification join algorithm is extended in Gryz (1998a, 1998b) in order to deal with inclusion and functional dependencies expressed over the global schema.

Other studies are concerned with the problem of query rewriting using views, under the different assumptions that queries are recursive queries (Afrati, Gergatsoulis, & Kavalieros, 1999), description logics queries (Beeri, Levy, & Rousset, 1997; Calvanese et al., 2001), queries for semi-structured data (Calvanese et al., 1999; Papakonstantinou & Vassalos, 1999), queries with aggregates (Grumbach, Rafanelli, & Tininini, 1999; Cohen, Nutt, & Serebrenik, 1999), or in presence of limitations in accessing the views (Rajaraman, Sagiv, & Ullman, 1995; Kwok & Weld, 1996; Levy, Rajaraman, & Ullman, 1996; Florescu et al., 1999).

Query Answering

In the query rewriting approach, the query reformulation step is carried out, not taking into account the source database, i.e., the extensions of the views. Only in a second step the rewritten query is evaluated over such a database. A more general approach, simply called *query answering* (Abiteboul & Duschka, 1998; Grahne & Mendelzon, 1999; Calvanese et al., 2001), is to consider, besides the query and the view definitions, the extensions of the views. In such a case we do not pose any limit to query processing, and the only goal is to compute the answer to the query by exploiting all available information, in particular the view extensions.

The complexity of answering queries using views for different languages (both for the query and for the views) is studied in Abiteboul & Duschka (1998). In that paper, the authors consider two different assumptions, called *open-world assumption* and *closed-world assumption*, respectively. In the first case, each view stores only, but not necessarily all, the tuples that satisfy the view definition, whereas in the second case, each view stores exactly the tuples that satisfy the view definition. In the literature such views are also called *sound* and *exact*, respectively. Under the *open-world assumption*, Abiteboul & Duschka show practical cases in which the complexity of query answering using views is polynomial, and prove that the problem becomes coNP-hard already in the case where the views are conjunctive queries and the query is a conjunctive query with inequality, or where the query is a conjunctive query and the views are positive queries. Actually, as shown in Calvanese et al. (2000b), the complexity in the case in which the views are sound is coNP-hard also for very simple query languages containing union. Under the *closed-world assumption,* Abiteboul & Duschka prove that the problem is coNP-hard already in the case in which views and queries are conjunctive queries.

The problem of query answering using views in the context where constraints over the global schema are expressed in terms of description logics, and queries and views are unions of conjunctive queries, is addressed in Calvanese, De Giacomo, & Lenzerini (1998, 2000). Finally, the case of queries for semi-structured data is dealt with in Calvanese et al. (2000b, 2000c).

Query Processing in the GAV Approach

When the mapping is specified following the GAV approach, each relation of the global schema is expressed in terms of a view over the sources. Hence, it is in general assumed that, in order to answer a query posed over the global schema on the basis of data stored at the sources, it is sufficient to *unfold* each atom of the original query with the corresponding view (Ullman, 1997). It is worthwhile to stress that, even if the process of unfolding is a simple mechanism, defining the views associated with the global elements implies precisely understanding the relationships among the sources, that is in general a non-trivial task. We can say that, in the GAV approach, query processing by unfolding is realized at design time, and the main issue turns out to be the specification of mediators that synthesize the definition of views associated to global elements.

The reason why simple unfolding is assumed sufficient to process a query is that the GAV mapping, which has been commonly considered so far in the literature, essentially specifies a single database satisfying the global schema, i.e., the database obtained by populating the global schema according to the view definitions. We call this database the *retrieved database*. Evaluating the query over the retrieved database is equivalent to evaluating its unfolding over the sources.

This is, for example, the case of TSIMMIS (The Stanford-IBM Manager of Multiple Information Sources) (Chawathe et al., 1994), a joint project of Stanford University and IBM's Almaden Research Center, based on an architecture that presents a hierarchy of wrappers and mediators. In TSIMMIS the global schema is simply the collection of objects exported by the mediators, and no rules or integrity constraints are expressed among them, in such a way that the mediators only realize an explicit layer between the user' applications and the data sources. Hence, user' queries are posed in terms of objects synthesized at a mediator or directly exported by a wrapper, and are processed by suitable modules, called Mediator Specification Interpreters (MSIs) (Papakonstantinou, Garcia-Molina, & Ullman, 1996; Yerneni et al., 1999), each performing integration independently.

Differently from TSIMMIS, the Garlic system (Carey et al., 1995), developed at IBM Almaden Research Center, provides the user with an integrated data perspective by means of an architecture comprising a middleware layer for query processing and data access software. The middleware layer presents an object-oriented data model based on the ODMG standard (Cattell & Barry, 1997) that allows data from various information sources to be represented uniformly. Given a query over the middleware layer, its execution plan is simply the set of sub-queries produced by expanding the objects in the original query with their definitions provided by suitable wrappers. Each wrapper translates the sub-queries into the source native query language, taking into account the query processing power of the source. Note that many of the tasks done by the mediators in TSIMMIS, such as the integration of objects from different sources, in Garlic are performed by wrappers, and the notion of mediator is actually missing.

In MOMIS (Bergamaschi et al., 2001), a system jointly developed at the University of Milano and the University of Modena and Reggio Emilia, query processing and optimization are the tasks of the *query manager*, a component which is part of the mediator. The query manager exploits extensional relationships to first identify all sources whose data are needed to answer a user query posed over the global schema. Then it reformulates the original query into queries to the single sources, and sends the obtained sub-queries to the wrappers, which execute them and report the results to the query manager. Finally, it combines the single results to provide the answer to the original query. During the process, query optimization is performed taking into account intentional intra- and inter-source relationships, both for the original query and the local queries to the sources.

The *Squirrel* project, developed at the University of Colorado, provides a framework for data integration that supports either the virtual or the materialized approaches to data integration. The framework is based on the *integration mediators*, special mediators originally constructed to support only fully materialized views (Zhou et al., 1995a, 1995b), and then generalized in order to support a hybrid of virtual and materialized views (Zhou et al., 1996; Hull & Zhou, 1996). The architecture of a general Squirrel mediator consists of components that deal with the problem of refreshing the materialized portion of the supported view, and components related to the problem of query processing, namely the query processor (QP) and the virtual attribute processor (VAP). With regard to the two last components, the QP provides the integrated view interface to the user. When it receives a user query, it tries to answer the query on the basis of the materialized portion of the view maintained in a local store. If the QP needs virtual data to answer the query, the VAP constructs temporary relations containing the relevant data. To obtain temporary relations, the VAP uses information provided by a virtual decomposition plan (VDP) maintained in the local store. The notion of VDP is analogous to that of query decomposition plan in query optimization. More specifically, the VDP specifies the global relations (materialized, virtual, or hybrid) that the mediator maintains, and provides the basic structure for retrieving data from the sources or supporting incremental maintenance.

As we already said at the beginning of this section, all the systems described above assume that the GAV mapping specifies a single database for the global schema. Furthermore, during query processing, they do not take into account constraints imposed by the data model used to specify the global schema. Actually, as shown in Lenzerini (2001), when the global schema contains integrity constraints, its semantics is best described in terms of a set of databases rather than a single one, and this implies that, as in the LAV approach, query processing in GAV is intimately connected to the notion of querying incomplete databases (van der Meyden, 1998). Hence, to process a query in GAV, unfolding is in general not sufficient, and it is necessary to make use of reasoning mechanisms. A technique to effectively process queries when both global schema and sources are relational, and key and foreign key constraints are expressed in the global schema, is described in Calì et al. (2002a), where it is shown that, starting from the retrieved database, it is possible to build a *canonical* database that has the property of faithfully representing all the databases that satisfy the global schema. Furthermore, in that paper an algorithm is given that makes it possible to find the answers to a query over the given canonical database without actually building it. The algorithm constructs a suitable reformulation of the query, called *query expansion*, such that the answer to the expansion computed over the retrieved database is equal to the answer to the original query over the canonical database. In Calì et al. (2001), the same technique is used to deal with the mandatory participation and functional attribute constraints imposed by a conceptual model adopted to represent the global schema.

Other studies deal with the problem of query processing in GAV in the presence of limitations on how sources can be accessed (Li & Chang, 2001). To answer queries over such sources, one generally needs to start from a set of constants (provided, e.g., by the user filling in a form, or taken from a source without access limitations) to bind attributes. Such bindings are used to access sources and there obtain new constants which in turn can be used for new accesses. Hence, query processing in GAV in the presence of access limitations in general requires the evaluation of a recursive query plan. It is worth noticing that also in this case, unfolding is not sufficient to answer the query.

DATA CLEANING AND RECONCILIATION

The quality of data is very important for company business (Stoker, 2000). The ability of a company to make critical decisions in a complex business environment largely depends on how the company uses information. If the quality of company data is poor, also the information produced by data mining tools is not correct, due to the well-known "garbage in, garbage out" principle.

Usually, a large part of the data of a company are stored in legacy databases, which complicate the task of building a data warehouse. When information coming from heterogeneous sources is gathered and integrated, it is likely that data quality problems are present. This is because there may be inconsistencies due to the fact that different sources contain redundant data with different representations. Moreover, problems may arise within a single source: possible problems are missing information, misspellings, or any kind of invalid data.

Data cleaning (Bouzeghoub & Comyn-Wattiau, 1990; Rahm & Do, 2000; Quass, 1999) consists in detecting and removing inconsistencies from data, in order to improve data quality. It constitutes the most important step of the so-called ETL (extraction, transformation, loading) process. The task of data cleaning is orthogonal to that of semantic data integration. Even if query answering problems are solved in a data warehouse, retrieved data may still be dirty and inconsistent.

The process of data cleaning is disregarded by many IT managers (Maydanchick, 2000). One of the reasons of this is the high and unpredictable cost of this operation. Moreover, it is very difficult to plan a data cleaning process, due to the overwhelming complexity of business rules in a modern data warehouse; without such a plan, not only is the data cleaning procedure unlikely to end on time, but measuring the success of the operation becomes impossible.

Henceforth, we will assume that data are represented in the relational model. There are many reasons for data to be dirty; Rahm & Do (2000) and Quass (1999) present a number of them. We group data problems in two main categories: invalid data and differences in representation of the same data.

- Invalid data can be caused by extracting data from multiple sources, or they can exist in a single source, due to incorrect data entries. For example, a "City" field may contain the value "Italy," or a simple misspelling may occur, e.g., the field "Country" contains the value "Brazik" instead of "Brazil." A slightly more complex problem is due to inconsistencies among different fields of the same record; for example, a record regarding a person may have the value "12 December 1973" for the date of birth and the value "12" for the age. Violation of functional dependencies within a table is another typical example of such inconsistencies. In general, having a global schema for a data warehouse means also having integrity constraints expressed in the schema itself. Such constraints are not related to the underlying data sources; instead they are derived from the semantics of the global schema, or, in other words, from the real world. We cannot expect independent data sources to produce data which respect these constraints. Therefore, a data warehouse has to cope with many violations of constraints over the global schema (also called "business rules"). An effective strategy for dealing with violation of foreign key constraints is presented in Calì et al. (2001).

- When gathering information from different data sources, in order to build a data warehouse, it is likely that the same information is represented in different ways in different sources. Rahm & Do (2000) identify three main categories of such conflicts. *Naming conflicts* arise when the same name is used for different objects, or when different names are used for the same object. *Structural conflicts* are more general, and occur when the difference in the representation lies in the structure of the representation itself. A simple case is that of the format of a field: for example, in a source the gender of a customer is denoted with "M" or "F" while another source uses the values "0" and "1". Also the format of dates is a common problem. Moreover, data that are represented by more than one field in a source may be represented in a single field in another source; Borkar, Deshmukh, & Sarawagi (2000) present an approach to give a field structure to free-text addresses. Other structural conflicts, in the relational model, are due to different relational modeling in different sources, e.g., attribute vs. relation representation, different data types, different integrity constraints, etc. Finally, *data conflicts* appear only at the instance level, and are mainly related to different value representations. For example, the currency may be expressed in Japanese Yen in one source and in German Marks in another. Another typical case is the different representation of the same value; for instance, "Simon M. Sze" versus "Sze, S.M."

A major problem in data cleaning is that of overlapping data (Hernàndez & Stolfo, 1995; Monge & Elkan, 1996, 1997; Roychoudhury, Ramakrishnan, & Swift, 1997; Monge & Elkan, 2000), also referred as *duplicate elimination problem* or *merge/purge problem*. This problem arises when different records of data representing the same real-world entity are gathered in a data warehouse. In this case, duplicates have to be detected and merged. Most efforts in data cleaning have been devoted to the solution of the duplicate detection problem, which proves to be a central issue. In fact, in order to be able to detect different records related to the same real-world entity, a system has to resolve all conflicts relating that entity, due to different representations in different sources. In the following, we will focus our attention on the duplicate elimination problem.

Before starting the actual data cleaning process, a *pre-processing* is usually performed. Pre-processing consists in standardization operations (domain-dependent standardization techniques are presented, e.g., in Borkar, Deshmukh, & Sarawagi (2000) and Roychoudhury, Ramakrishnan, & Swift (1997), for example simple format conversions, or discovering abbreviations. The aim of this preliminary step is to make candidate duplicate records more similar, in order to make their identification easier during the duplicate detection step.

Two metrics (Low, Lee, & Ling, 2001) are usually defined to measure the effectiveness of a data cleaning system: *recall* and *false-positive error*. Recall is also known as *percentage hits*, *true positives*, or *true merges*, and it is defined as the percentage of the duplicate records which are correctly identified, out of all duplicate records. The false-positive error is the percentage of records which are wrongly identified as duplicates, out of all identified records. This metrics, also known as *false merges*, is related to the *precision* metrics, being precision=100%−falsepositive errors.

Determining exact duplicates is an easy task (Bitton & DeWitt, 1983), and can be done by sorting all records and checking whether there are identical neighbors. Detecting *inexact* duplicates relating to the same real-world entity is a quite complex task instead. Once we have a technique to compare two records, if we have N records, the naïve approach would require one to make N^2 comparisons, which is obviously inefficient.

In Hernàndez & Stolfo (1995), a more efficient technique is presented, called *sorted neighborhood method*. This method consists of three steps:

- First, a key is computed for every set of homogeneous records in the database (or data warehouse); the key selection process is strongly domain-dependent, and it is crucial for the effectiveness of the technique. This highly knowledge-intensive step is to be performed by a specialist who has very deep knowledge of the characteristics of the data, and also of possible errors in the data.
- The second step is a simple sorting of the records, according to the key computed in the previous step.
- The final step consists of merging the duplicate records. A fixed size, sliding window is moved along the ordered list of records. At each step, the window, which contains w records, goes one record ahead, and the newly entered record is compared with the other w-1 records in the window. When two records are found to have the same key, then they are merged into a single record. The scan is performed in $O(wN)$ steps.

The main drawback of the sorted neighborhood method lies in the fact that the merging heavily depends on the sorting, and therefore on the key. If two inexact duplicates fall far apart after sorting, they will never be recognized as duplicate. This problem can be partly solved by enlarging the size of the window, thus improving the recall; on the other hand, this increases the computational complexity of the merging phase.

As the choice of the key is crucial in sorted neighborhood method, in order to increase recall, a good strategy is that of performing multiple passes (Monge & Elkan, 1997; Hernàndez & Stolfo, 1995), each one with a different key, and computing the *transitive closure* of the relations computed at each step. That is, if record A matches with record B in a scan, and record B matches with record C in another, then we deduce that A matches with C. This approach solves the problem of the window size, as multiple passes with a small window size have shown to perform better, and to be more efficient, than a single scan with a large window size. On the other hand, introducing transitivity increases the false-positive errors, lowering the overall precision.

An alternative to window scan consists of partitioning the records in *clusters*, where in each cluster records are stored that match each other according to a certain criterion. Duplicate elimination can be performed on each cluster separately, and in parallel. This approach suffers from the same drawbacks as the usual sorted neighborhood method; moreover, if the database is small, many very small (or even singleton) clusters are likely to occur.

A clustering technique, presented in Monge & Elkan (1997), makes use of a *priority queue*, containing a set of clusters. Records not already appearing in the clusters are compared with representative records of each cluster: if the match succeeds, then the record is added to the corresponding cluster. If the record cannot be put in any cluster, then a new cluster is created, containing only that record. As the priority queue could increase significantly, a maximum number of clusters is fixed; if the limit is exceeded, then the cluster with lowest priority is removed. At any time, the cluster that was most recently created has the highest priority. In this technique, the choice of the representative records is crucial, and it strongly influences the results. Heuristics have to be developed in order to optimize such choice.

Finally, *knowledge-based approaches* have been used in data cleaning. In Galhardas et al. (2001), an attempt of separating the logical and the physical level is done: the logical level supports the design of the data transformation procedures, while the physical level supports the implementation and optimization of such procedures. In this framework a declarative language for data cleaning transformations is presented. Also, a declarative specification of user interaction and a declarative way to specify properties of the matching operations are introduced. These techniques have been implemented in an actual data cleaning system.

A framework for knowledge-based data cleaning, based on a set of rules, has been proposed in Low, Lee, & Ling (2001). In the approach presented in this paper, the processing stage is performed according to a set of rules: duplicate identification rules, merge/purge rules, update rules, alert rules (specifying some interaction with the user). Also, a pre-processing step is defined, in which data standardization is executed on the data. At the end of the data cleaning process, a framework for human validation and verification of the results is proposed.

In conclusion, the problem of cleaning data in the context of data warehousing is still far from having a solution. The identification of inexact duplicates is inherently characterized by a certain degree of uncertainty, and this means that human validation is needed in order to evaluate the performance of a data cleaning system. On the other hand, human inspection is infeasible when the size of the data is very large, a case that is quite common in practice. Another serious problem is that of the dependence on the domain: an algorithm performing very well on a database may behave badly on another. Declarative approaches seem to introduce some flexibility, which may produce good results on several domains. Anyway, the results in this case are again strongly dependent on the knowledge provided to the system, which in turn is produced by human reasoning.

CONCLUSIONS

In this chapter we have introduced the main aspects of data source integration, which is the process of transferring data from a set of heterogeneous data sources into a data warehouse. Source integration comprises several complex aspects, which make it a very difficult task. Schema integration consists of integrating the schemas of the sources in order to obtain a homogeneous representation of the whole set of data, called global schema, and in specifying the mapping between the global schema and the sources. We have shown the various steps of the schema integration process, comparing the different approaches that have been adopted in the literature. Data integration consists of making data sources available through a set of materialized views, retrieving the data from the sources themselves. We have illustrated several techniques and mechanisms proposed in the literature to deal with this task, classified according to the approach used to specify the mapping (LAV vs. GAV). Data cleaning and reconciliation consists of removing inconsistencies in the data retrieved from the sources, which are due to errors in the data or differences in the representations of the data themselves in different sources. After having shown the main causes of such "dirt," we have presented the main approaches to the duplicate detection problem, which is a major task in data cleaning; it consists of the identification of distinct records relating the same real-world entity.

A lot of work is still to be done in the field of source integration. This problem is very hard, and all its aspects need further investigation both from the theoretical and practical point of view. The latest approaches aim to provide automatic reasoning services; yet, some source integration tasks inherently depend on the designer's knowledge.

REFERENCES

Abiteboul, S., & Duschka, O. (1998). Complexity of answering queries using materialized views. *Proceedings of the 17th ACM SIGACT SIGMOD SIGART Symposium on Principles of Database Systems (PODS'98)*, 254-265.

Afrati, F.N., Gergatsoulis, M., & Kavalieros, T. (1999). Answering queries using materialized views with disjunction. *Proceedings of the 7th International Conference on Database Theory (ICDT'99)*. Lecture Notes in Computer Science, No. 1540. City: Springer-Verlag, 435-452.

Batini, C., Lenzerini, M., & Navathe, S.B. (1986). A comparative analysis of methodologies for database schema integration. *ACM Computing Surveys*, 18(4), 323-364.

Beeri, C., Levy, A.Y., & Rousset, M.-C. (1997). Rewriting queries using views in description logics. *Proceedings of the 16th ACM SIGACT SIGMOD SIGART Symposium on Principles of Database Systems (PODS'97)*, 99-108.

Beneventano, D., Bergamaschi, S., Castano, S., Corni, A., Guidetti, R., Malvezzi, G., Melchiori, M., & Vincini, M. Information integration: The MOMIS project demonstration. *Proceedings of the 26th International Conference on Very Large Data Bases (VLDB'00).*

Bergamaschi, S., Castano, S., Vincini, M., & Beneventano, D. (2001). Semantic integration of heterogeneous information sources. *Data and Knowledge Engineering*, 36(3), 215-249.

Bitton, D., & DeWitt, D.J. (1983). Duplicate record elimination in large data files. *Database Systems*, 8(2), 255-265.

Blanco, J.L., Illarramendi, A., & Goñi, A. (1994). Building a federated relational database system: An approach using a knowledge-based system. *Journal of Intelligent and Cooperative Information Systems*, 3(4), 415-455.

Borkar, V.R., Deshmukh, K., & Sarawagi, S. (2000). Automatically extracting structure from free text addresses. *IEEE Data Engineering Bulletin*, 23(4), 27-32.

Bouzeghoub, M., & Comyn-Wattiau, I. (1990). View integration by semantic unification and transformation of data structures. *Proceedings of the 9th International Conference on the Entity-Relationship Approach (ER'90)*, 413-430.

Bouzeghoub, M., & Lenzerini, M. (Eds.). (2001). Article Title. *Information Systems, Special Issue on Data Extraction, Cleaning and Reconciliation*, 26(8).

Calì, A., Calvanese, D., De Giacomo, G., & Lenzerini, M. (2001). Accessing data integration systems through conceptual schemas. *Proceedings of the 20th International Conference on Conceptual Modeling (ER'01).*

Calì, A., Calvanese, D., De Giacomo, G., & Lenzerini, M. (2002a). On the expressive power of data integration systems. *Proceedings of the 21st International Conference on Conceptual Modeling (ER'02).*

Calì, A., Calvanese, D., De Giacomo, G., & Lenzerini, M. (2002b). Data integration under integrity constraints. *Proceedings of the 14th Conference on Advanced Information Systems Engineering (CAiSE, 2002).*

Calvanese, D., De Giacomo, G., & Lenzerini, M. (1998). What can knowledge representation do for semi-structured data? *Proceedings of the 15th National Conference on Artificial Intelligence (AAAI'98)*, 205-210.

Calvanese, D., De Giacomo, G., & Lenzerini, M. (2000). Answering queries using views over description logics knowledge bases. *Proceedings of the 17th National Conference on Artificial Intelligence (AAAI 2000)*, 386-391.

Calvanese, D., De Giacomo, G., Lenzerini, M., Nardi, D., & Rosati, R. (2001). Data integration in data warehousing. *International Journal of Cooperative Information Systems*, 10(3), 237-271.

Calvanese, D., De Giacomo, G., Lenzerini, M., & Vardi, M.Y. (1999). Rewriting of regular expressions and regular path queries. *Proceedings of the 18th ACM*

SIGACT SIGMOD SIGART Symposium on Principles of Database Systems (PODS'99), 194-204.

Calvanese, D., De Giacomo, G., Lenzerini, M., & Vardi, M.Y. (2000a). Answering regular path queries using views. *Proceedings of the 16th IEEE International Conference on Data Engineering (ICDE'00)*, 389-398.

Calvanese, D., De Giacomo, G., Lenzerini, M., & Vardi, M.Y. (2000b). What is query rewriting? *Proceedings of the 7th International Workshop on Knowledge Representation Meets Databases (KRDB'00)*, 17-27. CEUR Electronic Workshop Proceedings, http://ceur-ws.org/Vol-29/.

Calvanese, D., De Giacomo, G., Lenzerini, M., & Vardi, M.Y. (2000c). Query processing using views for regular path queries with inverse. *Proceedings of the 19th ACM SIGACT SIGMOD SIGART Symposium on Principles of Database Systems (PODS'00)*, 58-66.

Carey, M.J., Haas, L. M., Schwarz, P. M., Arya, M., Cody, W. F., Fagin, R., Flickner, M., Luniewski, A., Niblack, W., Petkovic, D., Thomas, J., Williams, J.H., & Wimmers, E.L. (1995). Towards heterogeneous multimedia information systems: The Garlic approach. *Proceedings of the 5th International Workshop on Research Issues in Data Engineering-Distributed Object Management (RIDE-DOM'95)*. City: IEEE Computer Society Press, 124-131.

Catarci, T., & Lenzerini, M. (1993). Representing and using interschema knowledge in cooperative information systems. *Journal of Intelligent and Cooperative Information Systems*, 2(4), 375-398.

Cattell, R.G.G., & Barry, D.K. (Eds.). (1997). *The Object Database Standard: ODMG 2.0*. Los Altos, CA: Morgan Kaufmann.

Chandra, A.K., & Merlin, P.M. (1977). Optimal implementation of conjunctive queries in relational data bases. *Proceedings of the 9th ACM Symposium on Theory of Computing (STOC'77)*, 77-90.

Chaudhuri, S., Krishnamurthy, S., Potarnianos, S., & Shim, K. (1995). Optimizing queries with materialized views. *Proceedings of the 11th IEEE International Conference on Data Engineering (ICDE'95)*, Taipei, Taiwan.

Chawathe, S.S., Garcia-Molina, H., Hammer, J., Ireland, K., Papakonstantinou, Y., Ullman, J.D., & Widom, J. (1994). The TSIMMIS project: Integration of heterogeneous information sources. *Proceedings of the 10th Meeting of the Information Processing Society of Japan (IPSJ'94)*, 7-18.

Cohen, S., Nutt, W., & Serebrenik, A. (1999). Rewriting aggregate queries using views. *Proceedings of the 18th ACM SIGACT SIGMOD SIGART Symposium on Principles of Database Systems (PODS'99)*, 155-166.

Devlin, B. (1997). *Data warehouse: From Architecture to Implementation*. Reading, MA: Addison Wesley.

Duschka, O.M., & Genesereth, M.R. (1997). Query planning in Infomaster. *Proceedings of the ACM Symposium on Applied Computing*, San Jose, California.

Duschka, O.M., & Genesereth, M.R., & Levy, A.Y. (2000). Recursive query plans for data integration. *Journal of Logic Programming*, 43(1), 49-73.

Florescu, D., Levy, A.Y., Manolescu, I., & Suciu, D. (1999). Query optimization in the presence of limited access patterns. *Proceedings of the ACM SIGMOD International Conference on Management of Data*, 311-322.

Galhardas, H., Florescu, D., Shasha, D., & Simon, E. (1999). *An Extensible Framework for Data Cleaning.* Technical Report 3742, INRIA, Rocquencourt.

Galhardas, H., Florescu, D., Shasha, D., Simon, E., & Saita, C. (2001). *Declarative Data Cleaning: Language, Model and Algorithms.* Technical Report 4149, INRIA.

García-Solaco, M., Saltor, F., & Castellanos, M. (1995a). A structure-based schema integration methodology. *Proceedings of the 11th IEEE International Conference on Data Engineering (ICDE'95)*, 505-512.

García-Solaco, M., Saltor, F., & Castellanos, M. (1995b). A semantic-discriminated approach to integration in federated databases. *Proceedings of the 3rd International Conference on Cooperative Information Systems (CoopIS'95)*, 19-31.

Geller, J., Perl, Y., Cannata, P., Sheth, A.P., & Neuhold, E. (1992a). A case study of structural integration. *Proceedings of the International Conference on Information and Knowledge Management (CIKM'92)*, 102-111.

Geller, J., Perl, Y., Neuhold, E., & Sheth, A.P. (1992b). Structural schema integration with full and partial correspondence using the dual model. *Information Systems*, 17(6), 443-464.

Gotthard, W., Lockemann, P.C., & Neufeld, A. (1992). System-guided view integration for object-oriented databases. *IEEE Transactions on Knowledge and Data Engineering*, 4(1), 1-22.

Grahne, G., & Mendelzon, A.O. (1999). Tableau techniques for querying information sources through global schemas. *Proceedings of the 7th International Conference on Database Theory (ICDT'99)*. Lecture Notes in Computer Science, No. 1540. City: Springer-Verlag, 332-347.

Grumbach, S., Rafanelli, M., & Tininini, L. (1999). Querying aggregate data. *Proceedings of the 18th ACM SIGACT SIGMOD SIGART Symposium on Principles of Database Systems (PODS'99)*, 174-184.

Gryz, J. (1998a). An algorithm for query folding with functional dependencies. *Proceedings of the 7th International Symposium on Intelligent Information Systems*, 7-16.

Gryz, J. (1998b). Query folding with inclusion dependencies. *Proceedings of the 14th IEEE International Conference on Data Engineering (ICDE'98)*, 126-133.

Halevy, A.Y. (2000). Theory of answering queries using views. *SIGMOD Record*, 29(4), 40-47.

Hernàndez, M.A., & Stolfo, S.J. (1995). The merge/purge problem for large databases. *Proceedings of the ACM SIGMOD International Conference on Management of Data (SIGMOD'95).*

Hull, R. (1986). Relative information capacity of simple relational database schemas. *SIAM Journal on Computing*, 15(3), 856-886.

Hull, R., & Zhou, G. (1996). A framework for supporting data integration using the materialized and virtual approaches. *Proceedings of the ACM SIGMOD International Conference on Management of Data*, 481-492.

Imielinski, T., & Lipski Jr., W. (1984). Incomplete information in relational databases. *Journal of the ACM*, 31(4), 761-791.

Inmon, W.H. (1996). *Building the Data Warehouse* (2nd Ed.). New York: John Wiley & Sons.

Jarke, M., Lenzerini, M., Vassiliou, Y., & Vassiliadis, P. (Eds.). (1999). *Fundamentals of Data Warehouses*. City: Springer-Verlag.

Johanneson, P. (1994). Schema standardization as an aid in view integration. *Information Systems*, 19(3), 275-290.

Keller, A.M., & Basu, J. (1996). A predicate-based caching scheme for client-server database architectures. *Very Large Database Journal*, 5(1), 35-47.

Klug, A.C. (1998). On conjunctive queries containing inequalities. *Journal of the ACM*, 35(1), 146-160.

Krishnamurthy, R., Litwin, W., & Kent, W. (1991). Language features for interoperability of databases with schematic discrepancies. *Proceedings of the ACM SIGMOD International Conference on Management of Data*, 40-49.

Kwok, C.T., & Weld, D. (1996). Planning to gather information. *Proceedings of the 13th Nat. Conference on Artificial Intelligence (AAAI'96)*, 32-39.

Lenzerini, M. (2001). Data integration is harder than you thought. *Proceedings of the 9th International Conference on Cooperative Information Systems (CoopIS'01)*, 22-26.

Levy, A.Y. (1999). *Answering Queries Using Views: A Survey*. Technical Report, University of Washington.

Levy, A.Y. (2000). Logic-based techniques in data integration. In Minker, J. (Ed.), *Logic-Based Artificial Intelligence*. City: Kluwer Academic Publisher.

Levy, A.Y., Mendelzon, A.O., Sagiv, Y., & Srivastava, D. (1995). Answering queries using views. *Proceedings of the 14th ACM SIGACT SIGMOD SIGART Symposium on Principles of Database Systems (PODS'95)*, 95-104.

Levy, A.Y., Rajaraman, A., & Ordille, J.J. (1996a). Querying heterogeneous information sources using source descriptions. *Proceedings of the 22nd International Conference on Very Large Data Bases (VLDB'96).*

Levy, A.Y., Rajaraman, A., & Ordille, J.J. (1996b). Query answering algorithms for information agents. *Proceedings of the 13th National Conference on Artificial Intelligence (AAAI'96)*, 40-47.

Levy, A.Y., Rajaraman, A., & Ullman, J.D. (1996). Answering queries using limited external query processors. *Proceedings of the 15th ACM SIGACT SIGMOD SIGART Symposium on Principles of Database Systems (PODS'96)*, 227-237.

Levy, A.Y., Srivastava, D., & Kirk, T. (1995). Data model and query evaluation in global information systems. *Journal of Intelligent Information Systems*, 5, 121-143.

Li, C., & Chang, E. (2000). Query planning with limited source capabilities. *Proceedings of the 16th IEEE International Conference on Data Engineering (ICDE'00)*, 401-412.

Li C., & Chang, E. (2001). On answering queries in the presence of limited access patterns. *Proceedings of the 8th International Conference on Database Theory (ICDT'01)*, 219-233.

Low, W.L., Lee, M.L., & Ling, T.W. (2001). A knowledge-based approach for duplicate elimination in data cleaning. *Information Systems, Special Issue on Data Extraction, Cleaning and Reconciliation*, 26(8).

Madhavan, J., Bernstein, P.A., & Rahn, E. (2001). Generic schema matching with Cupid. *Proceedings of the 27th International Conference on Very Large Data Bases (VLDB'01)*.

Maydanchick, A. (2000). Challenges of efficient data cleaning. *DM Direct*, (September).

Miller, R.J., Yoannidis, Y.E., & Ramakrishnan, R. (1994). Schema equivalence in heterogeneous systems: Bridging theory and practice. *Information Systems*, 19(1), 3-31.

Mitra, P. (1999). *An Algorithm for Answering Queries Efficiently Using Views*. Technical Report, University of Southern California, Information Science Institute, Stanford, California. Available on-line at: http://dbpubs.stanford.edu/pub/1999-46.

Monge, A.E., & Elkan, C.P. (1996). The field matching problem: Algorithms and applications. *International Conference on Practical Applications of Prolog (PAP'97)*.

Monge, A.E., & Elkan, C.P. (1997). An efficient domain-independent algorithm for detecting approximately duplicate database records. *Proceedings of the ACM-SIGMOD Workshop on Research Issues on Knowledge Discovery and Data Mining*.

Monge, A.E., & Elkan, C.P. (2000). Matching algorithms within a duplicate detection system. *IEEE Data Engineering Bulletin*, 23(4), 14-20.

Ouksel, A.M., & Naiman, C.F. (1994). Coordinating context building in heterogeneous information systems. *Journal of Intelligent Information Systems*, 3, 151-183.

Palopoli, L., Saccà, D., & Ursino, D. (1999). Semi-automatic techniques for deriving interscheme properties from database schemes. *Data and Knowledge Engineering*, 30(3), 239-273.

Palopoli, L., Terracina, G., & Ursino, D. (2000). The system DIKE: Towards the semi-automatic synthesis of cooperative information systems and data warehouses. *Proceedings of Symposium on Advances in Databases and Information Systems (ADBIS-DASFAA'00)*, 108-117.

Papakonstantinou, Y., Garcia-Molina, H., & Ullman, J.D. (1996). MedMaker: A mediation system based on declarative specifications. In Su, S.Y.W. (Ed.), *Proceedings of the 12th IEEE International Conference on Data Engineering (ICDE'96)*, 132-141.

Papakonstantinou, Y., & Vassalos, V. (1999). Query rewriting using semistructured views. *Proceedings of the ACM SIGMOD International Conference on Management of Data.*

Peltason, C. (1991). The BACK system—An overview. *SIGART Bulletin,* 2(3), 114-119.

Pottinger, R., & Levy, A.Y. (2000). A scalable algorithm for answering queries using views. *Proceedings of the 26th International Conference on Very Large Data Bases (VLDB'00)*, 484-495.

Qian, X. (1996). Correct schema transformations. In Apers, P., Bouzeghoub, M., & Gardarin, G. (Eds.), *Proceedings of the 5th International Conference on Extending Database Technology (EDBT'96).* Lecture Notes in Computer Science, No. 1057. City: Springer-Verlag, 114-128.

Qian, X. (1996). Query folding. *Proceedings of the 12th IEEE International Conference on Data Engineering (ICDE'96)*, 48-55.

Quass, D. (1999). A Framework for Research in Data Cleaning. Unpublished draft paper.

Rahm, E., & Do, H.H. (2000). Data cleaning: Problems and current approaches. *IEEE Data Engineering Bulletin,* 23(4), 3-13.

Rajaraman, A., Sagiv, Y., & Ullman, J.D. (1995). Answering queries using templates with binding patterns. *Proceedings of the 14th ACM SIGACT SIGMOD SIGART Symposium on Principles of Database Systems (PODS'95).*

Reddy, M.P., Prasad, B.E., Reddy, P.G., & Gupta, A. (1994). A methodology for integration of heterogeneous databases. *IEEE Transactions on Knowledge and Data Engineering,* 6(6), 920-933.

Roychoudhury, A., Ramakrishnan, I.V., & Swift, T. (1997). A rule-based data standardizer for enterprise data bases. *Proceedings of the International Conference on Practical Applications of Prolog (PAP'97)*, 271-289.

Sheth, A.P., Gala, S.K., & Navathe, S.B. (1993). On automatic reasoning for schema integration. *Journal of Intelligent and Cooperative Information Systems,* 2(1), 23-50.

Spaccapietra, S., Parent, C., & Dupont, Y. (1992). Model independent assertions for integration of heterogeneous schemas. *Very Large Database Journal,* 1, 81-126.

Stoker, S. (2000). Good data housekeeping. *DM Direct*, (September).

Tsatalos, O.G., Solomon, M.H., & Ioannidis, Y.E. (1996). The GMAP: A versatile tool for physical data independence. *Very Large Database Journal*, 5(2), 101-118.

Ullman, J.D. (1997). Information integration using logical views. *Proceedings of the 6th International Conference on Database Theory (ICDT'97).* Lecture Notes in Computer Science, No. 1186. City: Springer-Verlag, 19-40.

van der Meyden, R. (1992). *The Complexity of Querying Indefinite Information.* PhD Thesis, Rutgers University, City, New Jersey.

van der Meyden, R. (1998). Logical approaches to incomplete information. In Chomicki, J., & Saake, G. (Eds.), *Logics for Databases and Information Systems.* City: Kluwer Academic Publisher, 307-356.

Vidal, V.M., & Winslett, M. (1994). Preserving update semantics in schema integration. *Proceedings of the 3rd International Conference on Information and Knowledge Management (CIKM'94)*, 263-271.

Wiederhold, G. (1992). Mediators in the architecture of future information systems. *IEEE Computer*, 25(3), 38-49.

Yerneni, R., Li, C., Garcia-Molina, H., & Ullman, J.D. (1999). Computing capabilities of mediators. *Proceedings of the ACM SIGMOD International Conference on Management of Data*, 443-454.

Zhou, G., Hull, R., & King, R. (1996). Generating data integration mediators that use materializations. *Journal of Intelligent Information Systems*, 6, 199-221.

Zhou, G., Hull, R., King, R., & Franchitti, J.-C. (1995a). Data integration and warehousing using H20. *IEEE Bulletin of the Technical Committee on Data Engineering*, 18(2), 29-40.

Zhou, G., Hull, R., King, R., & Franchitti, J.-C. (1995b). Using object matching and materialization to integrate heterogeneous databases. *Proceedings of the 3rd International Conference on Cooperative Information Systems (CoopIS'95)*, 4-18.

Chapter XIII

Cooperation with Geographic Databases

Elaheh Pourabbas
Istituto di Analisi dei Sistemi ed Informatica, C.N.R., Italy

ABSTRACT

The purpose of this chapter is to create cooperation between geographic databases (GDBs) and multidimensional databases (MDDBs), which are considered as the most promising and efficient information technologies for supporting decision making. We focus on the common key elements between geographic and multidimensional data which allow effective support in data cooperating. These elements are basically time *and* space, *which are present implicitly or explicitly in MDDB and are modeled on the dimensions,* Time *and* Location. *Thus, because GDBs are primarily concerned with geographic data, we will focus on* space *as a bridge element for cooperating MDDBs and GDBs. We propose an approach that extends the geographic data structure through special attributes, called* binding attributes, *in order to describe all phenomena represented by MDDBs. This extension will make it possible to answer more specific "OLAP-based" queries within GDBs without modifying the physical organization of data in both environments.*

INTRODUCTION

In recent years, the enormous increase in data and its sources, due to the growing number of independent databases widely accessible through computer networks, has created a new challenge, which is to find a way of sharing data and programs across different databases. It has motivated cooperation between database systems, creating systems that are sometimes referred to as multi-database or federated database systems. They support collection of cooperating but independent database systems, without requiring data to be physically moved and without incurring significant complications in the functioning of the individual systems. The great advantage that a large community of users has when connecting many data sources nowadays is the tremendous increase in the amount of available data, and this allows enterprises to become more competitive. Unusual trends in particular applications can be identified through the analysis of a huge amount of data, creating opportunities for new business or for forecasting production needs. In this field, decision support systems that treat data in very large databases have recently attracted research attention. These databases may represent business information (such as transaction data), medical information (such as patient treatment and results), scientific data (such as large sets of experimental measurements), or spatial information (such as geographic data and its visualization as maps).

Currently, in the research community, multidimensional databases (MDDBs) and geographic information systems (GISs) are seen as the most promising and efficient information technologies for supporting decision making. Multidimensional databases, through On-Line-Analytical-Processing (OLAP) techniques, provide business dataset handling and summarization over multiple dimensions (OLAP Council, 1997).

Geographic information systems, which are geographic data base (GDB)-dependent, through graphic display functionalities and complex data structures, facilitate the storage and manipulation of either geographic data and its related attributes or data which refers to the phenomena of interest.

A feature notably lacking in most GDBs is the capability of accessing and manipulating data stored in MDDBs on which analyses of transaction-based business data (OLAP) are carried out. Therefore, a new challenge would be to create cooperation between or a federation of these sophisticated technologies in order to introduce a new feature into decision-making support.

In this chapter, we focus on the common key elements between geographic and multidimensional data which allow effective support in data cooperating. These elements are basically *time* and *space,* which are present implicitly or explicitly in MDDBs and are modeled on the dimensions, *Time* and *Location.* Cooperation between the above-mentioned environments based on each of these elements needs an appropriate approach and data modeling. Thus, because GDBs are primarily concerned with geographic data, we will focus on *space* as a bridge element for cooperating MDDBs and GDBs. In this way, the location dimension, which is defined as a "linguistic category corresponding to different ways of looking at the geographic information" in MDDBs, will be viewed as a more complex concept in GDBs.

In order to clarify the need for such cooperation, let us consider the case of a GDB user wishing to display his query result on a map with different combinations of geographic and non-geographic data. Let the query be as follows:

> *Query 1: "Find all the Italian regions which are adjacent to the Tuscany region, in which the number of cars sold in 1990, in the case of <Corolla>, was greater than 10,000."*

It is necessary not only to retrieve the adjacent regions, but also to perform some OLAP operations on the "time" and "product" dimensions of the "Car_Sales" cube. The former analyzes the topological spatial relationship (i.e., adjacency) and can be performed only in GDBs.

The solution of such queries depends essentially on a data model that enables the component databases to cooperate but remain independent. The main idea is to explore the additional input that can come to GDBs from MDDBs, and also to look at what geographic data can add to OLAP applications.

In this chapter, we give some basic definitions for describing geographic data to support multidimensional data. Then, we propose an approach that extends the geographic data structure through special attributes, called *binding attributes*, in order to describe all phenomena represented by MDDBs. This extension will make it possible to answer more specific "OLAP-based" queries within GDBs without modifying the physical organization of data in both environments.

This chapter is composed of six main sections following this introduction. The next section examines the fundamental characteristics of GDBs. This is followed by the section on geographic and multidimensional data modeling, briefly reviewing their basic concepts. Then, an overview on the related works are given. The sections on cooperation by binding elements and inferring binding attributes are the most substantial and describe the main idea of heterogeneous data cooperation. The final section draws some conclusions and proposes some future lines of research.

FUNDAMENTAL CHARACTERISTICS OF GEOGRAPHIC DATABASES

Geographic data is characterized mainly by its four components: geographic position or coordinates; attribute values; topological relationships, and time (Aronof, 1991). The modeling of geographic data is based on *field-based* and *object-based* approaches.

The field-based approach treats a spatial object as a continuous surface, and data is stored and processed as a collection of regular grid cells with a given measure and area. The object-based approach treats a spatial object as an identifiable entity with which key or object-id are associated. Through this model, data is stored and processed as a list of coordinates.

These different models have their advantages and disadvantages. For a detailed discussion about their trade-offs, see Aronof (1991), Maguire, Goodchild, & Rhind (1991), and Laurini & Thompson (1992). Since data analysis and storing, and data querying are performed on object-based spatial data, the vector format is the most spatial data model used in GIS.

A fundamental requirement for geographic database design which uses the vector format is the ability to model spatial properties, i.e., to associate part of space with an attribute. Parts of space are usually represented by points, lines, and regions and are known as *geometric features*. Basically, the geographic data models are based on geographic objects and geographic classes (or themes or layers). The geographic objects correspond to the individual data items of the real world. They are defined by a description component, which is a set of descriptive or alphanumeric attributes (e.g., the name and population of a country), and a spatial component representing the object location in the underlying geographic space and shape. The geographic classes correspond to a collection of geographic objects having the same structure or type.

The spatial structure of a geographic object cannot be modeled by any built-in data types, such as integer, string, etc., in a computer environment. In order to overcome this lack of modeling power, abstract data types (ADTs) were introduced. The main idea of this approach is to hide the structure of the data types from users and allow them to access the data types only through a set of operations. The spatial types for representing the geometric features are *Type point, Type polyline,* and *Type region.* For each of them a set of operations is defined and the result of spatial operations should be one of the existing data type. They can be either unary or binary operations with Boolean, scalar, or spatial results. Many of these operations concern the spatial topological relationships among objects.

The description of spatial relationships between spatial objects and the definition of an appropriate terminology for these relationships and their semantics has been dealt with by using theoretical models (Egenhofer & Franzosa, 1991, 1995). These studies concern the definition of the fundamental properties of the spatial "regions" and the formalization of a minimal set of spatial topological relationships using point-set theory. This is based upon the intersection of the *boundary* and *interior* of two objects, named A and B, to be compared, and distinguishes only "empty" and "non-empty" intersections. A 2×2 matrix, called the 4-intersection, represents these criteria as follows:

$$\begin{pmatrix} \partial A \cap \partial B & \partial A \cap B^{\circ} \\ A^{\circ} \cap \partial B & A^{\circ} \cap B^{\circ} \end{pmatrix}$$

By considering the above-mentioned empty and non-empty values, 16 binary topological relations can be distinguished. Among them, only eight can be realized for two regions with connected boundaries if the objects are embedded in R^2. They are called *disjoint, meet, equal, inside, contains, covers, covered by*, and *overlap* (see Egenhofer & Herring, 1990). This set of eight relations provides a complete coverage, and they are mutually exclusive so that exactly one of these topological relations holds true between any two regions (Egenhofer & Franzosa, 1991).

Concerning data modeling, there are two classic data models that are mostly used to define and manage GDBs: the relational and object-oriented data models. The relational models aim at distributing the attributes in various relations so that some rules on database design known as normal forms hold. The relation, as the single construct of relational models, is limited to modeling complex spatial entities. Therefore, there have been several proposals to extend the original relational model with spatial abstract data types (ADTs) (Stonebraker, Runestein, & Guttman, 1983; Stonebraker & Rowe, 1986; Stonebraker, 1986, Gardarin et al., 1989; Gargano, Nardelli, & Talamo, 1991). Even though relational data models are powerful because of their theoretical validity and are easy to implement, they are not completely suitable for the manipulation of complex data, such as spatial data. The object-oriented paradigm (see Hughes, 1991; Ullman, 1988;. Kim, 1989), with its well-known common characteristics such as the object identifier, data abstraction, inheritance, and encapsulation, is recognized as the most advanced approach for modeling spatial data. It was promoted originally to overcome the limitations of relational data models.

In an object-oriented (O-O) model, each instance of an entity is modeled as an object which includes a behavior description. Objects comprise both attributes and methods. Methods are also considered as procedure-valued attributes. This model provides the notion of class. Several spatial data models, which benefit the power of such a paradigm, have been proposed (see Worboys, Hearnshaw, & Maguire, 1990; Gunther & Riekert, 1993; Milne, Milton, & Smith, 1993; Leung, Leung, & He, 1999; Rigaux, Scholl, & Voisard, 2001).

Spatial Partition Hierarchies

The best-known metaphor in a GDB is that of a *map*. A map is a generalized, simplified abstraction of reality. It consists of a set of topographic data displayed in visual form providing a frame of reference (e.g., the location data or position data).

Maps are the most natural way to convey geographical information, and they are excellent support for visualizing analytical data about phenomena that have a geographical extent. They are also as faithful as possible to the real-world location and shape.

A central element of maps is the concept of *partition*. A partition is a subdivision of the 2-D plane into pairs of disjoint regions where each region is associated with an attribute which can have a simple or complex structure. There are many examples of partitions in the real world, like the subdivision of a given territory into administrative boundaries such as countries, states, and counties, or in the classification of land according to soil type, etc.

Partitions are identified as an important spatial concept, and they are widely used in the generation of many scale-dependent maps from a single database. This is called *generalization* (Brassel & Weibel, 1988). This process is used to convert spatial data from one scale-dependent representation into another by calculating the geometry of a more abstract object through the union of the geometries of lower level objects. It is referred to as *abstract generalization* and is concerned basically in changing the object representation according to the level of abstraction at which data is represented. The main goal of abstract generalization is spatial analysis. A given set of geographic objects may have distinct representations depending on the level of abstraction. At a more abstract level, we obtain a more simplified representation of objects. This provides a hierarchy of partitions over geographic objects where each level corresponds to a given representation (Rigaux & Scholl, 1994; Volta & Egenhofer, 1993; Frank, Volta, & McGranaghan, 1997). Therefore, the hierarchy of partitions of a single 2-D space can be sketched as follows (see Rigaux & Scholl, 1995):

Definition 1: Let S be a subset of plane ($S \subset \Re^2$) and $G \in 2^S$ be a partition of S, such that $\cup_G = S$ and $\forall g, g' \in G$ $g \cap g' = \varnothing$. Let $G_1, ..., G_n$ be the set of partitions of S and $\preceq$ be the partial order defined as follows:

$$G_i \preceq G_j \text{ iff } \forall g_i \in G_i, g_j \in G_j, g_i \cap g_j = g_i.$$

Definition 2: Let A_{g1} and A_{g2} represent the geometric attributes of objects belonging to two distinct classes, the domains of which are defined by $D(A_{g1}) = G_1 \in 2^S$, $D(A_{g2}) = G_2 \in 2^S$. The partial order $\preceq$ among spatial partitions induces partial order, renamed *Contains* relationship, on geographic classes and objects.

Contains (or *space inclusion*) is the most common hierarchical relationship among geographic classes and objects, the inverse of which is the well-known "is-in" relationship. For instance, in the case of administrative subdivision, a country contains several states, and states contain counties. Furthermore, any two spatial entities related by inclusion relationship satisfy some constraints on their common attributes. In other words, the value of some numeric non-spatial or *aspatial* attributes of an object belonging to a given level is the aggregation of the corresponding attribute values of objects belonging to the lower level. For example, the population of a given state is the sum of the population of its counties. With regard to the geometry, aggregation implies the spatial union of the geometry of objects belonging to the lower level.

BASIC NOTIONS

In this section, we describe some basic notions of multidimensional data and geographic data needed to define the cooperation approach. They are based on the common concepts discussed extensively in several proposals.

Geographic Data

The geographical data model under consideration is based on the existence of sets and definitions discussed by Pourabbas & Rafanelli (2000) as follow:

- a finite set of domains $D = \{D_1, ..., D_n\}$

- an infinite numerical set $A = \{A_1, ..., A_n\}$ of symbols called *instance variables*

- an infinite numerical set of *identifiers*

- set C of all the classes defined in the database

- set O of all the objects defined in the database

- set M of all the methods defined in the database.

A geographical class is a set of elements called geographical objects $go_1, go_2, ..., go_n$, which have the same properties $A_1, A_2, ..., A_n$ defined over the set of not necessarily distinct domains $D_1, D_2, ..., D_n$ and a set of methods which describe the behavior of the objects.

Definition 3: a *geographic class* (gc) is a quadruple:

$< n, sc, P, M >$ where:

- n is the name of gc;

- sc is the class-parent of gc;

- P is the set of attributes which define the geographic entity properties enclosing the set of attributes inherited from the relative super-class sc. It consists of:

  – a *geometric attribute* A_g, which is the type of geometric structure that refers to the geographic class. It can assume only one value chosen in its domain definition: {*null, point, polyline, region*};

  – a set of *alphanumeric attributes* $A_{\bar{g}} = A_1, ..., A_m$ defined on domains $D_1, ..., D_m$ which are the non-geometric properties of the gc instances; each of them consists of a couple (attribute name, data type);

  – M is the non empty set of methods such that:

  $M = \{inherited\ methods\ from\ sc\} \cup \{methods\ defined\ on\ gc\} \cup \{overridden\ methods\}$

Example 1: The geographic class "province" is defined as:
<province, administrative subdivision, {{(name, string), (surface, integer), (population, integer),.....}, region }, {Get_name, . . .}>

Definition 4: A *geographic object* go instance of a geographic class gc represents a geographic entity of the real world. It consists of a quadruple:

$$< id, gc, V, T >$$ where:

- id is the identifier of a geographic class instance; it is unique.
- gc is the class to which it belongs; it is also unique.
- V represents the ordered set of the alphanumeric attribute values.
- T is the ordered set of couples (latitude, longitude) which specifies the object position with regard to the geographic coordinates.

Example 2: Let us consider an instance of the geographic class "province":
<001, province, {(name, Roma), (surface, 1507.60 km^2), (population, 2915000),....}....{$(x_1, y_1), ..., (x_n, y_n)$}>

Definition 5: The *root class* (rt) is the superclass of all the geographic classes. It is a quadruple $< n, sc, P, M >$, where $sc = \Phi$ (void class) because its parent does not exist.

If the geometric attribute value is null, the geographic class is the *null geographic class* and the quadruple is defined as $< n, rt, P, \Phi >$, where $n =$ "void class," $P =< \Phi, null >$, where Φ represents the set of alphanumeric attributes which defines the non-geometric characteristics of the gc instances, and $null$ represents the type of geometric structure of the instances.

Definition 6: A *method* is a triple
$< nm, f : U \rightarrow R, s >$, where:
- nm is the name of the method.
- $f : U \rightarrow R$ is the method's interface in the form of: $f : U_1 \times U_2 \times ... \times U_n \rightarrow R$ where $U_1, U_2, ... U_n$ are the domains of the arguments, R is the domain of the result, and $U_1 = ist(gc)$, where $ist(gc)$ represents the class instances.
- s is the set of instructions in a given language which determines the semantics of the function f (method implementation).

Definition 7: A *schema of a geographic database* consists of a set of geographic classes; a *geographic database* consists of a set of objects, each of which is an instance of a schema class.

As we mentioned earlier, the *Contains* relationship between a pair of geographic classes, named gc_1 and gc_2, indicates that the instances of the class gc_1 are included in the instances of gc_2. This implies that the geo-feature of gc_1 is internal to or on the boundary of the geo-feature of gc_2; for instance, at the intentional level "REGION Contains MUNICIPALITY," and at the extensional level "Tuscany *Contains* Florence" and "Latium *Contains* Rome."

A variant of *Contains* is the *Full-Contains* relationship (see Ferri et al., 2000). The *Full-Contains* relationship between a pair of geographic classes gc_1 and gc_2 exists so that gc_2 *Contains* gc_1 if the union of class gc_2 instances constitutes a complete coverage of the union of class gc_1 instances; in each class, the instances are disjoined. The following definition illustrates the condition for the existence of the containment relationship: *Full-Contains*.

In the rest of the chapter, we will refer to only three spatial relationships named *Geo-Disjunction, Geo-Union,* and *Geo-Touching* which are, respectively, equivalent to disjoint, union, and meet operators defined by Egenhofer & Franzosa (1991).

Definition 8: Let gc_α and gc_β be two geographic classes. gc_α *Full-Contains* gc_β if:

- $\forall\ go_j: go_j$ *is-an-instance-of* $gc_\beta \Rightarrow \exists!\ go_i: (go_i$ *is-an-instance-of* gc_α) and (go_i *Contains* go_j);

- $\forall\ go_i,\ go_{j1},\ go_{j2}: go_i$ *is-an-instance-of* gc_α, go_{j1} *is-an-instance-of* gc_β, go_{j2} *is-an-instance-of* gc_β, $go_{j1} \neq go_{j2}$, go_i *Contains* go_{j1}, go_i *Contains* $go_{j2} \Rightarrow go_{j1}$ *Geo-Disjunction* go_{j2};

- $\forall\ go_i: go_i$ *is-an-instance-of* $gc_\alpha \Rightarrow \exists\{go_{j1}, ..., go_{jn}\}: go_{j1}$ *is-an-instance-of* gc_β, ..., go_{jn} *is-an-instance-of* gc_β, go_i *Contains* go_{j1}, go_i *Contains* $go_{jn}: go_i \equiv go_{j1}$ *Geo-Union* go_{j2} .... *Geo-Union* go_{jn}.

The above definition shows that *Full-Contains* induces a partially ordered relationship among geographic classes and objects. It satisfies the conditions of the summarizability (disjointness and completeness) of statistical databases discussed by Lenz & Shoshani (1997) from a topological point of view for the GDB environment. The description of these conditions will be given in the next sub-section.

Example 3: Let the geographic classes MUNICIPALITY, PROVINCE, and REGION in the GDB be subclasses of the geographic class Geo Entity defined by the following schema:

```
CLASS = Geo Entity
    attributes
            name: string
            geom: geo-region
            surface: real
```

In Figure 1, the class-subclass hierarchy defined by the *ISA* relationship is shown, as well as the *Full-Contains* relationships between the geographic classes representing the administrative subdivision of the Italian territory.

MULTIDIMENSIONAL DATA

Multidimensional data refer either to statistical data (Chan & Shoshani, 1981; Rafanelli & Shoshani, 1990), which mostly represent applications in the socio-economic area, or OLAP data (Gray et al., 1996; Gyssens & Lakshmanan, 1997; Shoshani, 1997), which emphasize business applications. In the OLAP area the conceptual representation of multidimensionality by *cube* was proposed (see OLAP Council,. 1997; Agrawal, Gupta, & Sarawagi, 1997). According to this concept, a cube is "a group of data cells arranged by the dimensions of the data." A *dimension* is "a structural attribute of a cube, that is, a list of members, all of which are of a similar type in the user's perception of the data." The set of cube dimensions represents the relative data multidimensionality.

Dimensions have often been associated with different hierarchically organized *levels*. The levels correspond to different granularities of viewing data. The name of each cube dimension corresponds to the name of a dimension hierarchy level. We can write the following definition for a dimension:

Definition 9: A dimension D is a tuple $<L, \preceq >$ where L is a set of levels, and $\preceq$ is a partial order over L's elements such that $\forall l_i, l_j \in L,\ l_i \preceq l_j$ if $dom(l_i) \leq dom(l_j)$.

Each level belongs to one and only one dimension D; it means that for any two levels $l' \in D'$ and $l'' \in D''$, $l' \cap l'' = \varnothing$. Given a dimension, the shift from a lower (more detailed) level to a higher (more aggregated) level is carried out by a mapping. This mapping is full, if:

a)   each level instance of a lower level corresponds to only one level instance of a higher level;

b)   each level instance of a higher level corresponds to at least one level instance of a lower level.

For the sake of simplicity, such a mapping will be called *Complete containment function*. It stresses that no cell contains the "Not available" value and there is no missing value.

A *measure* is a particular dimension of the cube as defined by Agrawal, Gupta, & Sarawagi (1997) which represents the extensional fashion of the phenomenon described by the cube, and which is in general a numeric value. The measure is obtained by a *mapping* from the assignment of a value to each dimension of a cube.

A cube represents factual data (for instance, sales, production, etc.) through a set of levels where each level belongs to a dimension. It can be represented simply by a triple $C = < C_{name}, L, f_c >$ where C_{name} denotes the name of the cube, L is a set of levels, and $f_c : L_T \rightarrow M$ is a function by which a measure (M) is associated with the tuples defined over the combinations of levels (denoted by L_T).

Example 4: The above-mentioned concepts are presented through a cube shown in Figure 2. It represents a nationwide car company that owns chain stores

Figure 1: A GDB Schema with Full-Contains Relationships

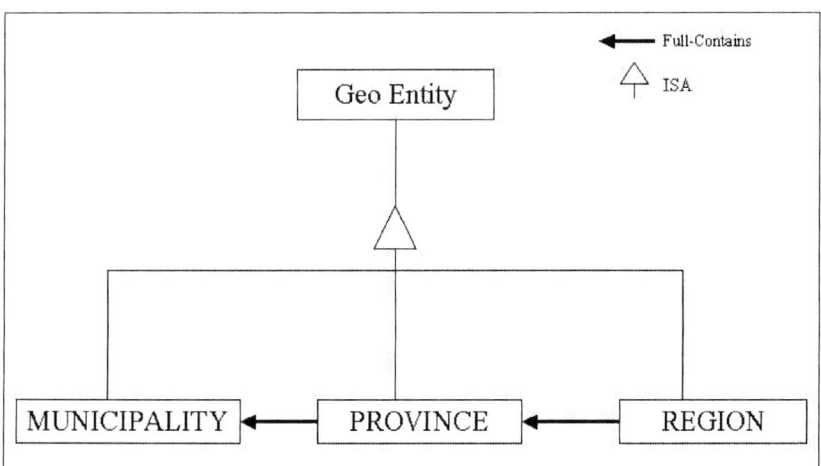

located in all the cities. In this example, we refer to the Italian administrative territory subdivision represented in *Location* hierarchy.

This schema represents the multidimensional data for the sales of cars, organized at the level Month of the dimension *time*, the level Municipality of the dimension *Location*, and the level Model of the dimension *Product*.

The most well-known OLAP operators are *roll-up*, *slice*, and *dice*. The *roll-up* operator decreases the detail of the measure, aggregating it along the dimension hierarchy, which is summarizable. Summarizability is a condition upon which we can correctly obtain from a cube defined at level l_j of a given hierarchy, another cube defined at the higher level l_i of the same hierarchy by using the roll-up function.

The summarizability or correctness of aggregations in OLAP is discussed in Lenz & Shoshani (1997). The authors give three necessary conditions for summarizability, and assume that these conditions are sufficient. The conditions are: disjointness of levels (or category attributes) in hierarchies; completeness in hierarchies; correct use of measure (summary attributes) with statistical functions. Disjointness implies that instances of category attributes in dimensions from disjoint subsets of the elements of a level. Completeness in hierarchies means that all the elements occur in one of the dimensions and every element is assigned to some category on the level above it in the hierarchy. Correct use of measures with statistical functions depends on the type of the measure and the statistical function.

Depending on the summarizability conditions, a *default aggregate function* f_{agg} associated with the measure **M** must be distributive. A distributive aggregate function can be computed on a set by partitioning the set itself into disjoint subsets, aggregating each separately, and then computing the aggregation of these partial results with another aggregate function. In fact, among the SQL aggregate functions, COUNT, SUM, MIN, and MAX are distributive.

Let us consider any two levels l_1 and l_2 of a given hierarchy. The aggregation of measure is represented by *roll-up*$_{l_1 \rightarrow l_2}^{f_{agg}(f_c)}(C)$, and it is defined by the following steps:

- level changing $l_1 \rightarrow l_2$ will yield a new set of tuples over these more aggregated levels denoted by L'_T;

- the application of the aggregate function $f_c : L'_T \rightarrow M$;

- the resulting cube is defined by $C = <C_{name}, L', f_c : L'_T \rightarrow M>$.

Example 5: Let us consider the cube represented in Example 4. If we obtain the total number of cars sold in all provinces, then we have to perform *roll-*$up_{Municipality \rightarrow Province}^{SUM(f_{Car\_Sales})}$ *(Car-Sales)* where f_{Car_Sales} is a function defined as follows:

Figure 2: Example of Cube "Car_Sales"

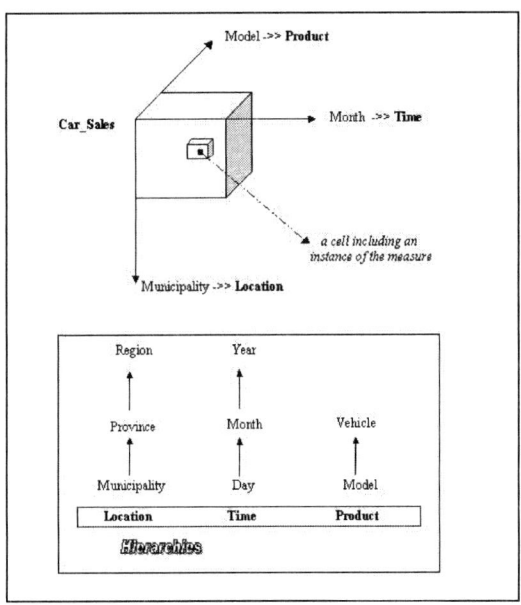

f_{Car_Sales} : {*model* : *Model*, *province* : *Province*, *month* : *Month*}$_T$ → *numeric*

; the resulting cube is defined as:

< *Car_Sales*, {*model*, *province*, *month*}, f_{Car_Sales} >

The *slice* operator omits one dimension of the cube and the result is a cube defined by the remaining dimensions. Let L_1, L_2, and L_3 be the elements of L. The slice on L_2 and L_3 that is represented symbolically by $slice_{L_2,L_3}^{f_{agg}(f_c))}$ (C) consists of the next steps:

- it omits all the levels of L_1 by projecting out L_2 and L_3 which yields L' = L - L_1;
- the application of the aggregate function $f_c : L'_T \to M$;
- the resulting cube is defined by $C = <C_{name}, L', f_c : L'_T \to M>$.

Example 6: The number of cars sold by model and month is obtained by:

$slice_{model,\,month}^{SUM(f_{Car\_Sales})}$ *(Car_Sales)* , where:

f_{Car_Sales} : *{model : Model,month : Month}*$_T$ $\rightarrow$ *numeric* , and the resulting cube is defined as:

$$< Car\_Sales, \{model, month\}, f_{Car\_Sales} > .$$

The *dice* operator restricts the dimension value domain of the cube according to the built-in predicate and it is represented by $dice_\theta(C)$.

Example 7: The number of cars sold in "*Rome*" is obtained by

$dice_{municipality=Rome}(Car\_Sales)$, where:

f_{Car_Sales} : *{model : Model, municipality =" Rome" , month : Month}*$_T$ $\rightarrow$ *numeric*,
and the cube is

$$< Car\_Sales, \{model, municipality =" Rome" , month\}, f_{Car\_Sales} > .$$

RELATED WORKS

In the database community, the cooperation between GDBs and MDDBs is indicated by taking into account the notion of multiple representations of space or map generalization. Generalization has been the subject of research by the cartographic and GIS communities. These studies refer to the *geometric* and *modeling* aspects of the generalization process. Geometric generalization processes are used solely for graphic display, and they are intended for modeling spatial data for visual interaction through map scaling. Modeling generalizations consider the impact of scale and resolution on spatial data modeling and querying (see Muller, Lagrange, & Weibel, 1995). In particular, some attempts have been made to look for a standard set of multidimensional (or statistical) operators based on aggregation/disaggregation. The case of spatial partitioning played a central role in finding such a set of operators. For instance, Gargano et al. (1991) have presented an extension of the relational model by ADTs in order to deal with spatial data and complex aggregated data. They extended the relational algebra essentially by defining two algebraic operators that are able to manipulate either the spatial extension of geographic data or summary (statistical) data. They are named G-Compose and G-Decompose. The first operator is denoted by $G - Compose_X(F_y;Y)$, where X and Y are two non-intersecting subsets of attributes of a relation R. It "merges" all tuples of R which are already projected on Y in a single one whose Y -value is generated by the application of the fusion function. This function, which is represented by F_y, takes a subset of elements of a given type and returns a single element of the same type. In the case of summary data, F_y aggregates the numeric values of Y attributes. The effect of G-Decompose is that all tuples of R projected on Y are "decomposed."

The effect of G-Compose is analogous of the operator "Aggregate Format" proposed by Özsoyoglu, Özsoyoglu, & Victor (1987) for statistical data. Contrary to the "Aggregate Format" operator where the aggregation function is unique, F_y in G-Compose can be a collection of different fusion functions. For a detailed description of these operators, see Gargano, Nardelli, & Talamo (1991).

In the case of summary data, G-Compose is equivalent to the slice operator in OLAP databases or summarization in statistical databases (see Rafanelli & Ricci, 1993). In the proposal by Gargano, Nardelli, & Talamo (1991), the fundamental issues of hierarchies and data aggregation for either spatial or summary data have been omitted. These issues are discussed later in an approach proposed by Rigaux & Sholl (1995).

In this work, the authors make the bridge between the geographic and statistical disciplines by defining an aggregation technique over a hierarchy of space partitions. Their model is based on the concept of "partition" which is used for partitioning either geometric space or other sets (e.g., a set of people). Such a set (geometry or people) is called, for the sake of simplicity, a *population*. The set of partitions on a generic population E is represented by $P(E)$. It can be the domain of a generic attribute A_g. They have introduced the concept of *cover*, which is essentially a relation defined by the schema $O = \{A_1, \ldots, A_n, A_g\}$ such that $\pi_{A_g}(O)$ is a partition in $P(E)$, and there is a biunivocal functional dependency between the attributes $A_1, \ldots, A_n$ and A_g. They defined the *geometric projection* operator on a subset of attributes $S = \{A_1, \ldots, A_q\}$ as follows:

$$apply_{\Sigma_{Geo}} \ (nest_S (\pi_{S,A_g}(O)))$$

where $\pi_{A_g}(O)$ is the N1NF grouping operation (see Abiteboul & Bidoit, 1986) on S and $\Sigma_{Geo}: \{A_g\} \rightarrow A_g$ performs the geometric aggregation function. The operation $(nest_S (\pi_{S,A_g} O))$ gives the result with the schema $\{A_1, \ldots, A_q, B\}$, and $\sum_{Geo}$ performs the union aggregation function on attribute $B = set(A_g)$.

For representing summary data, they use the notion of cover, but each descriptive attribute A can be defined on a hierarchical domain. The same operator is redefined as below:

$$apply_{\Sigma_{Geo}} \ (nest_S (gen_{A:A'}(O)))$$

where before applying the nest operator, the abstraction level of hierarchy to which the attribute A belongs is changed. The effect of this operator, which is indicated by $gen_{A:A'}(O)$, is the same as the *roll-up* operator defined by Cabibbo & Torlone (1998) where in each tuple, attribute A value is replaced by its ancestor (A') belonging to the hierarchy. The result of such an operator is a relation that is no longer a cover, since there are several tuples with the same value for A. Note that in this case, $\sum_{Geo}$ performs the numeric aggregation function SUM.

The model proposed by Rigaux & Scholl (1995) is addressed to generate maps in multiple representations of data using the hierarchy of space partitions and the hierarchy induced by a partial order relationship in the domain of an attribute. In this proposal only one location dimension hierarchy for summary data is considered.

The issue of aggregation of spatial data has also been considered by Shakhar et al. (1999) from a different point of view with regard to the previous proposals. The authors extend the concept of the data cube introduced by Gray et al. (1997) to the spatial domain (called *spatial data cube*) by proposing the *map cube* operator. This operator takes as its arguments: a base map, a base table, a geographic hierarchy, and a set of cartographic preferences. It adds cartographic visualization to the spatial data cube. Such an operator is aimed at generating a collection of maps corresponding to the power sets of all possible spatial and non-spatial aggregation, which can be browsed using OLAP operators.

While the above models give a formal definition for the cooperation between spatial and multidimensional environments, some other works consider the architectural aspects of an integration system. For instance, Kouba, Matousek & Miksovsky (2000) tried to identify some requirements for the correct and consistent functionality of system interconnection. They proposed an integration module which has two different roles: one is the transformation of data from external data sources, and the other refers to the integration of GIS and data warehouse through their common components. The integration module coordinates the actions carried out by the GIS system and data warehouse. The GIS under consideration is based on an object-oriented model which identifies the basic GIS elements that are objects and classes. In the GIS system, the structure of the geographical class hierarchy is stored in a metadata object for accessing directly from the integration module.

In this work, cooperation is carried out by the common elements which are the data warehouse "location" dimension aggregation levels and the GIS objects' taxonomy. The task of the integration module is to provide the following three types of mapping:

- *Class correspondence* maps particular GIS taxonomical levels on the corresponding location dimension and vice versa.
- *Instance correspondence* maps particular instances of aggregation levels on the instances of the geographic classes and vice versa.

- *Action correspondence* consists of the processing of queries in one environment which require information stored in another environment. It guarantees navigation consistency, and provides the information to be modified in the integration module and propagated to the data warehouse and GIS.

Moreover, the system is integrated by a front-end module able to display the results in output. Furthermore, the implementation aspects of the integration of a data warehouse that is the Microsoft SQL Server 7 and ArcView GIS System is also discussed.

Paolucci et al. (2000) considered the integration of several spatio-temporal data collections of the Italian National Statistics Institute. The integration system, called SIT-IN, is defined mainly by a historical database containing the temporal variation of territorial administrative partitions; a statistical data warehouse providing statistical data from a number of different surveys; and a GIS providing the cartography of the Italian territory up to census tract level. The implemented cooperative systems manage the maps of the temporal evolution of a certain number of administrative regions and link to these maps the content of the above-mentioned statistical database of a given year.

The approach that will be discussed in this chapter shares a number of characteristics and goals with the above-mentioned works. Like the proposals of Gargano et al. (1991) and Rigaux & Scholl (1995), the main goal is to provide a formal approach for cooperative query answering. The previous approaches aimed at defining a set of operators applicable to either spatial or summary data without dealing with the "logical organization" of databases at all. Consequently, it is not possible to handle summary queries in the context of GDBs. Conversely, our approach relies on a formal logical model that provides a solid basis for the study of summary data manipulation in GDBs.

Unlike our approach, there are some multidimensional issues that are not considered explicitly in the above-mentioned models like the notion of multiple location dimension hierarchies. Their models are based on multidimensional data formed by solely one location dimension, whereas in our approach we also consider data defined by more than one location dimension and we analyze their effect on data modeling and query answering.

The works by Kouba et al. (2000) and Paolucci et al. (2000) are aimed mainly at technical realization and discuss some interesting issues related to the integrating system implementation. However, the focus of their papers is on the development of querying rather than logical data modeling.

COOPERATION BY BINDING ELEMENTS

As we have seen earlier in this chapter, geographic databases are characterized by entities or classes which are organized through hierarchies whose common semantics are space inclusion. On the other hand, the model of MDDBs is mainly represented by the structure of its dimensions, hierarchies over dimensions, facts, and measures. MDDBs, besides the phenomenon of interest "what" and the temporal "when," support that of the spatial "where," often among several dimensions. Thus, the *location* dimension becomes the binding element between GDBs and MDDBs. The task of cooperation is based on two kinds of correspondences:

- *Level correspondence* maps a given geographic class of a geographic taxonomy to the corresponding level belonging to the location dimension and vice versa.
- *Level instance correspondence* results from the previous step and maps a generic instance of a geographic class to its corresponding instance of location dimension level and vice versa. In this step, for a correct correspondence, any instance changes should be propagated in both environments.

For a clear description of the above correspondences, we introduce some simple denotations. Let l_g be a level of a location dimension whose domain is represented by $dom(l_g)$ and let I_g be its own instance. Let gc be a geographic class which belongs to the geographic taxonomy, and let the set of all its instances be represented by O_{gc}. Then, the above correspondences are achieved by a bijective function λ, where

- $\lambda : gc \rightarrow l_g$ for levels

- $\lambda : O_{gc} \rightarrow dom(l_g)$ for levels instances

The function λ, Full-Contains and Complete containment functions play a central role either in linking the heterogeneous environments or in providing data which lack in one environment by knowing data and their relationships in another. The following theorem shows how the location hierarchy of MDDBs can help to complete our partial knowledge about its equivalent hierarchy in GDBs.

Theorem 1: Let gc' and gc'' be two geographic classes in a GDB such that gc' *Full-Contains* gc''. Let l_g' and l_g'' be two geographic levels in an MDDB. If gc' (gc'') is mapped to l_g' (l_g'') by the function λ, then the *complete containment function* from l_g' to l_g'' exists, and it is unique.

Proof. Let go' and go'' be, respectively, an instance of gc' and gc'', such that between them the *Full-Contains* relationship exists. Applying the bijective function λ to these two instances, i.e., $\lambda(go') = I_g'$ and $\lambda(go'') = I_g''$, we obtain a generic pair $< I'_g, I''_g >$. The relationship between each pair of such geographic levels instances, obtained by λ onto the instances of gc' *Full-Contains* gc'', is a total and surjective function, i.e., it has the same properties as the complete containment function.

We prove that the complete containment function is unique and show the contrapositive.

Let the relationship $< I'_g, I''_g >$ represent complete containment function. We assume there exists another complete containment function that maps another geographic level instance I'''_g (instead of I'_g) to I'_g. It indicates that the geographic instance corresponding to I'''_g, and obtained by the function $\lambda^{-1}(I'''_g) = go'''$ is mapped to $\lambda^{-1}(I'_g) = go'$, and between them as well as between go' and go'' *Full-Contains* relationships exist. It means that to different pairs $< I'_g, I''_g >$, $< I'_g, I'''_g >$ with different complete containment functions, different instances $< go', go'' >$, $< go', go''' >$ with the same *Full-Contain* relationship correspond. This contradicts the *Full-Contains* definition.

Example 8: Let us consider two geographic classes MUNICIPALITY and PROVINCE in the GDB and let between them the *Full-Contains* relationship hold. Let us suppose only the geographic variable "Province" are defined in the MDDB. The geographic variable "Municipality" can be generated thanks to the above theorem by which the complete containment function is defined from "Province" to "Municipality."

Contrary to the above theorem and following the assumption made before about the equivalency of the geographic hierarchies, in both databases we can calculate the geometry of geographic objects. This is described in the following definition and example:

Definition 10: Let l'_g and l''_g be two geographic levels between which the complete containment function from l''_g to l'_g exists. Let gc' be the geographic class corresponding to the geographic variable l'_g by λ. This mapping allows the generation of a geographic class gc'' such that gc'' *Full-Contains* gc'. The geometric attribute value of any instance of the class $gc'' = \lambda^{-1}(I''_g)$ is obtained by

$$Geo - Union_{go_k \in I} \, geom(go_k) \quad \text{where}$$

$$I = \{go_k \mid go_k \; is\_an\_instance\_of \; gc' \; \wedge \; go'' \; Full \text{-} Contains \; go_k\}$$

Example 9: Let us suppose that in the MDDB two geographic variables "Municipality" and "District" are defined and that a complete containment function exists between them. We suppose that in the GDB only the geographic class "MUNICIPALITY" is defined. The mapping between the geographic class "MUNICIPALITY" and the geographic variable "Municipality" implies that for each instance of the variable "Municipality," one instance of the geographic class "MUNICIPALITY" exists and vice versa. In the GDB, the geographic class "DISTRICT" can be generated thanks to the DISTRICT *Full-Contains* MUNICIPALITY relationship. Since *Full-Contains* corresponds to the complete containment function, the "Geo-Union" of the geometric attributes of the class "MUNICIPALITY" instances will generate the geometric attribute of an instance of "DISTRICT."

Definition of Binding Attributes

The actual cooperation is provided by linking dynamically the common geographic elements between GDBs and MDDBs. To achieve such a link, we extend the geographic class schema through some special attributes that describe all the phenomena of interests represented by cubes. For this reason, let us recall the definitions of geographic class schema and cube. According to this formulation, the set of attributes of geographic class can be subdivided into spatial and aspatial. The link that we mentioned before consists of adding new attributes, called *binding attributes*, to the set of aspatial attributes. A binding attribute of a geographic class is a cube derived from a given cube in a multidimensional database whose geographic level corresponds to the same class name. From now on, we will call such a cube a *sub-cube*. The binding attributes, in other words, represent all the phenomena (e.g., Car_Sales by model, month, municipality) modeled by cubes in MDDBs in the spatial context of geographic classes and objects.

The schema of a binding attribute is identical to the schema of the cube introduced in the previous section. However, the set of its levels is defined by all the levels of the cube except one, the geographic level, which is kept implicit by being equal to the class name (e.g., MUNICIPALITY). These are called *local levels*. Note that every binding attribute represents one and only one cube. For a clear description, we give the following example, and then we formalize the above concepts.

Example 10: Figure 3 shows graphically the binding attributes in the schema of the geographic class MUNICIPALITY.

Figure 3: Example of Binding Attributes in the Geographic Class Schema
MUNICIPALITY

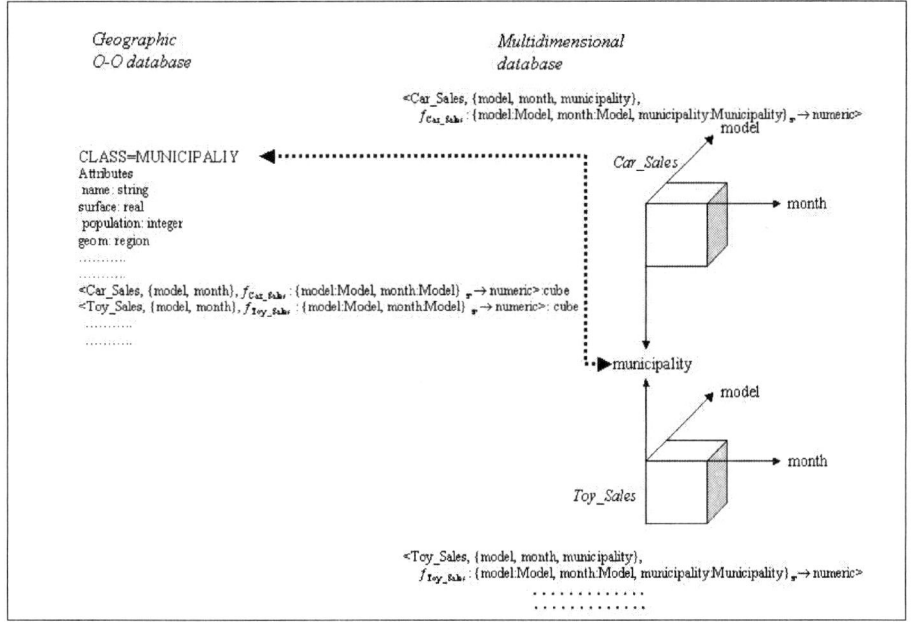

The following definition highlights the fact that the set of attributes in the schema of geographic class is enlarged by new binding attributes corresponding to cubes formed by only one location dimension.

Definition 11: Let C be a cube with the schema $<C_{name}, L, f_c : L_T \to M_T>$, where L is formed solely by one geographic level l_g. Let gc be a geographic class with the schema $< n, sc, P, M >$. If $gc = \lambda^{-1}(l_g)$, then the binding attribute A_B of gc can be defined by the following schema $<C_{name}, L - l_g, f_c : \{L - l_g\}_T \to M>$ and the set of attributes of g_c is $P \cup A_B$.

The value of the binding attribute for an instance go of gc is defined by

$$slice_{L-l_g}^{f_{agg}(f_c)}(dice_{l_g = \lambda(go)}(C)),$$ where I_g represents an instance of a geographic level l_g.

Example 11: In Figure 4 the binding attributes of the instance (Rome) of the geographical class MUNICIPALITY refer to the corresponding instance (Rome) of the geographic level Municipality.   ―

Figure 4: Instances of the Binding Attributes, "Car_Sales" and "Toy_Sales"

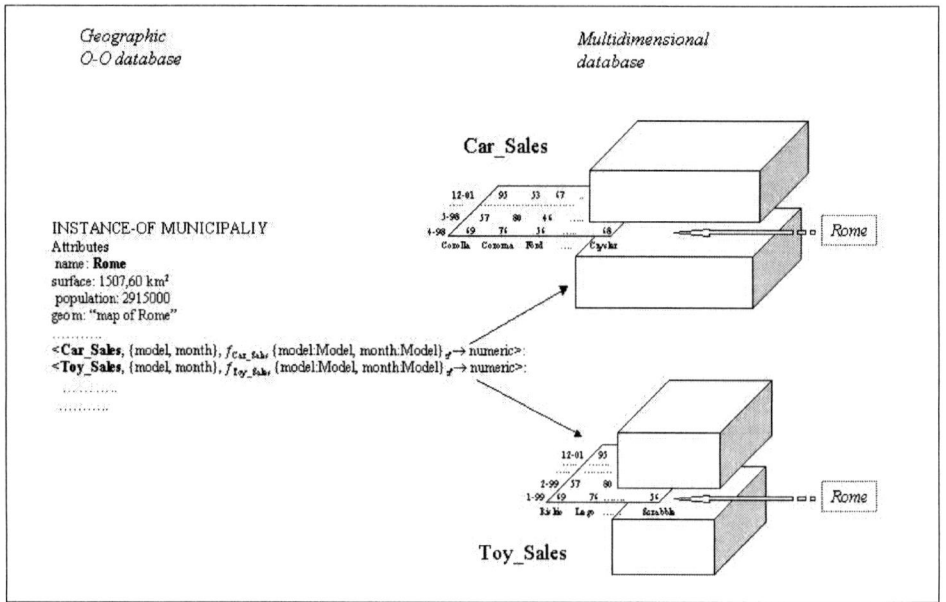

Generally speaking, a cube can be defined with more geographic levels. The next definition describes the binding attributes when the number of geographic levels of the cube under consideration is equal or more than two.

Definition 12: Let the set of levels L of a given cube $C = <C_{name}, L, f_c : \{L\}_T \to M>$ divided into geographic (L_g) and non-geographic ($\overline{L}_g$) levels. Let l'_g and l''_g be any two geographic levels belonging to L_g. Let gc' and gc'' be two geographic classes, which are obtained by $gc' = \lambda^{-1}(l'_g)$ and $gc'' = \lambda^{-1}(l''_g)$. The schemas of these last classes are, respectively, $gc' =< n', sc', P' \cup A'_B, M' >$ and $gc'' =< n'', sc'', P'' \cup A''_B, M'' >$ where their binding attributes are defined as follows:

$$A'_B = <C_{name}, \overline{L}_g \cup l''_g, f_c : \{\overline{L}_g \cup l_g\}_T \to M>$$
$$A''_B = <C_{name}, \overline{L}_g \cup l'_g, f_c : \{\overline{L}_g \cup l'_g\}_T \to M>$$

The different geographic levels of a cube reflect the cases where each of them is a specialization of a given level of possibly the same location hierarchy. For instance, the level "Province of residence" is a specialization of the level "Province" in the location dimension hierarchy: *Municipality $\to$ Province $\to$ Region*.

Consequently, its correspondence obtained by the function λ would also have been a specialization of the geographic class "PROVINCE" named "PROVINCE OF RESIDENCE."

While the process of the binding attributes generation for a "specialized" geographic class is the same as a "non-specialized" geographic class, the meaning of the attributes in the latter is intuitively more complex than those in the former. This is due to the specific organization of each geographic level along the relative dimension hierarchy, which will be discussed further through some illustrative examples.

Generally, in the case of cubes with more geographic levels, two different cases are distinguished as follows:

Case 1. There is no hierarchical relationship among the geographic levels of the cube.

***Example 12*:** Let us consider a nationwide shoe company that owns chain stores located in all provinces. We assume that this company delivers monthly to its stores in various provinces a certain number of shoes from its production departments located in provinces. Then, the delivering of items is modeled by the following cube which has the "Province of Production," "Province of Distribution" as the geographic levels, and "month" as one non-geographic level.

$< Shoe \_ Sales, \{Province\ of\ production, Province\ of\ distributi\ on, Month\},$
$\quad f_{Shoe\_Sales} : \{Province\ of\ production : Province, Province\ of\ distributi\ on : Province, month : Month\}_T \rightarrow integer >$

As we can see, the above geographic levels are specializations of a single level "Province." In addition, the relative mapping by the function λ^{-1} will yield two geographic classes with the same name which are sub-classes of the geographic class "PROVINCE."

CLASS = PROVINCE OF PRODUCTION
attributes
name: string
....

$<$Shoe_Sales, {Province of   distribution, month}, f_{Shoe_Sales} : {Province of distribution: Province, month:Month} $_T$ $>$: cube

and

CLASS = PROVINCE OF DISTRIBUTION
attributes
name: string
....

< Shoe_Sales, {Province of   production, month},

f_{Shoe_Sales} : {Province of production: Province, month:Month} $_T$ >:cube

The above example shows that in the geographic database, we have the generation of the geographic classes "PROVINCE OF PRODUCTION" and "PROVINCE OF DISTRIBUTION" which are specializations of the geographic class "PROVINCE."

We note the appearance of the geographic level "Province of distribution" (similarly Province of production) in the local levels of the binding attribute of the class "PROVINCE OF PRODUCTION" (similarly PROVINCE OF DISTRIBUTION).

Case 2. There is a hierarchical relationship among the geographic levels of the cube.

Example 13: Let us consider the same nationwide company of the above example that owns chain stores located in all provinces. This company delivers monthly a certain number of items to its stores from the production department located in various regions. It is represented by the following cube:

< Shoe _ Sales, {Province of distribution, Region of production, Month},

f_{Shoe_Sales} : {Province of production : Province, Region of distribution : Region, month : Month} $_T$ → integer >

CLASS = REGION OF PRODUCTION
attributes
name: string
....

<Shoe_Sales, {Province of distribution, month}, f_{Shoe_Sales} : {Province of

distribution: Province, month:Month} $_T$ >: cube

and

CLASS = PROVINCE OF DISTRIBUTION
attributes
name: string
....

< Shoe_Sales, {Region of   production, month}, f_{Shoe_Sales} : {Region of

production: Region, month:Month} $_T$ >: cube

The instances of the above geographic classes will assume different semantics. For example, a given instance of the first geographic class represents a region of shoe production in which the items are distributed over several provinces and months, while in the second case an instance represents the province of the distribution of shoes which are made in several regions and months.

A Query Example

In this section, we show the solution of *"Query 1"* through the approach discussed above. Let us recall the query:

Query 1: "Find all the regions adjacent to the Tuscany region, in which the number of cars sold in 1990, for the <Corolla> product, was greater than 10,000."

The procedure for solving this query is formed by the following steps:
1) identification of the geographic class and geographic object/s involved by the geographic operator in the query;

   In the case of our query, they are "Region" and "Tuscany" and the geographic operator is "Touching" (for a detailed description of this operator, see Ferri et al., 1999).

2) resolving the above Step 1;
   The result of the geographic operator *Geo-Touching* is formed by the instances: "Liguria, Emilia-Romagna, Umbria, and Latium."

3) identification of the binding attribute in the geographic class which is the "target" of the query;
   In our case, "Car_Sales" in "regions."

4) selection of the cube identified by the binding attribute in the MDDB;
   In our query, select the cube "Car_Sales."

5) satisfying the constraints on the non- geographic dimension of the cube;
   In our case according to the specific predicates, *dice* operator is applied on the values of Time and Model dimensions.

6) satisfying the constraints on the measure;
   In our query, it will involve all the instances of the measure.

7) selection of the values of the measure which satisfy the previous constraints;
   In our query, they are the regions with the number "greater than 10,000" of " Corolla " cars sold in "1990."

8) Pass the list of the geographic variable instances that satisfy the query to the GDB. This means identifying the corresponding geographic objects.

In the extended geographic environment, the above query may be expressed by the following SQL-like expression:

SELECT   R1.name
FROM   REGION R1,R2
WHERE
   R1.Car_Sales(model=Corolla,
   year=1990)>10,000
AND
   R1.geom *Geo-Touching* R2.geom
AND
   R2.name=Tuscany

Let us recall that the binding attribute is obtained from the formula as follows:

$$roll-up_{Municipality \to Region}^{sum(Car\_Sales)}(Car\_Sales)$$

where the containment function Municipality $\to$ Region is the concatenation of the two containment functions Province $\to$ Region and Municipality $\to$ Province. This means that the above formula is equivalent to the application of the following double roll-up operator on "Car_Sales":

$$roll-up_{Province \to Region}^{sum(Car\_Sales)}(roll-up_{Municipality \to Province}^{sum(Car\_Sales)}(Car\_Sales))$$

To summarize, the query is solved through the following steps:

- in the GDB:

$$Region\ Geo\text{-}Touching\ Region="Tuscany"$$

- and in the MDDB:

$$dice_{Car\_Sales>10,000}(dice_{model=Corolla}(dice_{year=1990}(roll-up_{Month \to Year}^{sum(Car\_Sales)}(Car\_Sales))))$$

Note that all the geographic operations are performed entirely in the GDB, the results of which are automatically reported in the MDDB in order to apply OLAP operators. Thanks to the equivalence properties between these environments, the so-called OLAP results are indicated as the extended part of the geographic objects involved in the query.

INFERRING BINDING ATTRIBUTES

By inferring cubes, we mean the maximum number of cubes which can be inferred from a given binding attribute. These types of cubes are not uncommon, and they indicate essentially the set of possible answers to a class of queries which can be formulated on the basic binding attributes. This fact highlights the usability of binding attributes.

In database community literature, usability is defined as the ratio of the maximum number of answerable queries to the total number of possible queries, which means any query that can be expressed by use of a formula. In this field, many studies have been made to discover the relationship between usability and security of databases. For instance, details of usability of secure statistical databases and SUM range queries is extensively discussed in Brankovic, Miller, & Siran (1996), Chin & Ozsoyoglu (1982), and Brankovic et al. (1997).

In this chapter, the topic of usability is addressed for inferring cubes which are the answers of aggregate COUNT and SUM queries. For a clear explanation, we classify the inferred cubes from binding attributes into two categories named: covering and deduced. The remainder of this section explains each of them, using some appropriate examples. Then, the total number of sub-cubes which can be achieved from basic, covering, and deduced binding attributes will be discussed.

Covering Attributes

These types of attributes are based on a process which is able to generate new binding attributes for the parent class based on those of its subclasses along the hierarchy of geographic classes. Note that the hierarchy to which we are referring is defined by class/sub-classes that are related to each other by the *Full-Contains* relationship. This relationship basically concerns the geometric property of the geographic classes and objects. Therefore to assimilate the power of the well-known *ISA* hierarchy for inferring the covering binding attributes, we introduce the *is-derived* relationship between any two geographic classes or objects. In short, it is denoted by *ISD* and it is represented graphically by a doubled hatch arrow. For a clear description, let gc_1 and gc_2 be two geographic classes. Let gc_2 *Full-Contains* gc_1 and the binding attributes of class gc_1 be known. We define the concept of the *ISD* relationship in the following:

Definition 14: An *ISD* relationship between two geographic classes is a triple gc_2 *ISD* gc_1 where *ISD* indicates that the binding attributes of gc_1 are also the binding attributes of gc_2.

The covering attribute is obtained across the hierarchy of space partitions in a bottom-up fashion, which ensures that a binding attribute produced at a lower level can participate in a later binding attribute at a higher level.

Obviously, the binding attributes of a geographic class are inherited from its superclasses along the *ISA* hierarchy. In fact, if a binding attribute of a class gc'' "is derived" from class gc' (denoted by A''_B), and class gc''' is a subclass of gc'' (see Figure 5), then A''_B is included in the set of binding attributes of gc'''.

Figure 5: The Inheritance of Binding Attribute

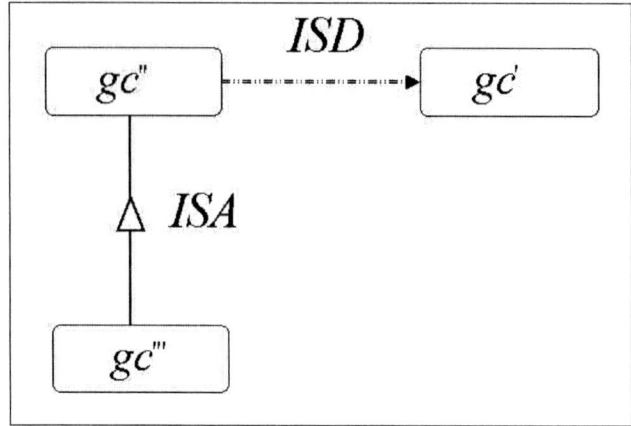

Property: The *ISD* is a partially ordered relation, i.e., it satisfies the following properties.

- *reflexivity:* Let gc be a geographic class, gc *ISD* gc.

- *antisymmetry*: Let gc_i, gc_j be geographic classes where $i \neq j$. If gc_i *ISD* gc_j, then gc_j *ISD* gc_i is not satisfied.

- *transitivity:* Let gc_i, gc_j, gc_k be geographic classes where $i \neq j$ and $j \neq k$. If gc_i *ISD* gc_j and gc_j *ISD* gc_k, then gc_i *ISD* gc_k.

The third property of this relationship indicates that the binding attributes of gc_k will be included in the set of binding attributes of gc_i. They are calculated by a roll-up operator composed of a concatenation of complete containment functions along the geographic dimension. Therefore, each covering attribute captures an orthogonal type of inference.

Definition 15: Let $C = <C_{name}, \mathrm{L}, f_c: \mathrm{L_T} \to \mathrm{M}>$ be a cube, and let $l'_g \in \mathrm{L}$ be a geographic level to which the geographic class $gc' = \lambda^{-1}(l'_g)$ corresponds. The binding attribute of such a class has the schema as follows: $A'_B = <C_{name}, \mathrm{L} - l'_g, f_c: \mathrm{L} - l'_{g_\mathrm{T}} \to \mathrm{M}>$. Let $l''_g \in \mathrm{L}$ be a geographic level that satisfies $l'_g \leq l''_g$. Then, the *Full-Contains* relationship between $gc'' = \lambda^{-1}(l''_g)$ and $gc' = \lambda^{-1}(l'_g)$ exists. The binding attribute of the geographic class gc'' by Definition 14 is $A''_B = <C_{name}, \mathrm{L} - l''_g,$

$f_c: L - l'_g\}_T \rightarrow M>$. It is obtained from $roll-up_{l'_g \rightarrow l''_g}^{f_{agg}(f_c)}$ (C), and it is called *covering*

binding attribute.

Example 14: Let the geographic classes PROVINCE and MUNICIPALITY be related by the *Full-Contains* relationship. As a consequence of this relationship, the binding attribute of the PROVINCE can be inferred from the binding attribute of the MUNICIPALITY (see Figure 6), and each instance of the latter is defined as follows:

For each instance of PROVINCE, for example "Milan," this attribute is calculated by the formula:

$$slice_{model,\ month}\ (dice_{province\ =\ Milan}\ (roll-up_{Municipali\ ty \rightarrow Province}^{sum\ (Car\ \_\ Sales\ )}\ (Car\ \_\ Sales\ )))$$

A similar line of reasoning can be followed for geographic classes. Indeed, we have the generation of more binding attributes corresponding to different specialized geographic classes that are interrelated by the *ISD* relationship. This is described by the next steps:

1. the automatic generation of the *ISD* relationship between each pair of superclasses which are already related by the *Full-Contains* relationship;

2. for each geographic level $l_g \in L$ of $C =< C_{name}$, L, $f_c : \{L\}_T \rightarrow M>$, identify the corresponding geographic class by the function λ in the GDB;

Figure 6: Example of Inferring Binding Attribute

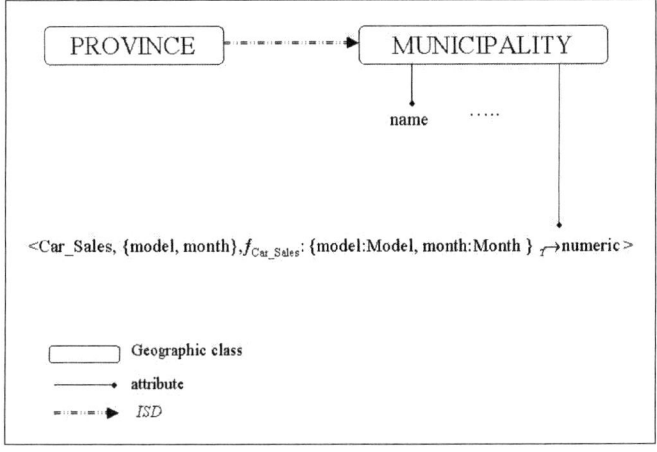

3. if $\lambda^{-1}(l_g)$ is a specialization of the geographic class gc, then create the geographic class s - gc and make it a sub-class of gc;

4. generate the binding attribute of s - gc with the schema $A_B = < C_{name}, L - l_g, f_c : \{L - l_g\}_T \to M >$;

5. for each geographic class belonging to the higher level of geographic taxonomy wrt gc, create its specialization;

6. because of Step 1 between any pair of such specialized geographic classes, the *ISD* relationship exists;

7. the binding attribute of any specialized class s - gc', that is next to s - gc in the geographic taxonomy, is derived from the s - gc's binding attribute through the roll-up operation.

Example 15: We extend the schema of the GDB, which is shown in Figure 1, with the cube illustrated in Example 12. For a clear description, we assume that the production departments are located in municipalities.

$< Shoe\_Sales, \{Province\ of\ distribution, Municiplaity\ of\ production, Month\},$

$\quad f_{Shoe\_Sales} : \{Province\ of\ distribution : Province, Municiplaity\ of\ production : Municipality, month : Month\}_T \to integer >$

Note that in the initial phase of correspondence between the MDDB and the GDB, the classes MUNICIPALITY, PROVINCE, and REGION become superclasses.

Then, we have:

1. the automatic generation of the *ISD* relationship between each pair of superclasses MUNICIPALITY, PROVINCE, and REGION which are already related by the *Full-Contains* relationship;

2. the generation of the specialized geographic class PROVINCE OF DISTRIBUTION with the binding attribute

$< Shoe\_Sales, \{Municiplaity\ of\ production, Month\},$

$\quad f_{Shoe\_Sales} : \{Municiplaity\ of\ production : Municipality, month : Month\}_T \to integer >$

3. the generation of the specialized geographic class REGION OF DISTRIBUTION and the *ISD* relationship between this class and the class defined in Step 2;

4. the generation of the geographic class MUNICIPALITY OF PRODUCTION with the binding attribute

$<: Shoe\text{-}Sales, \{Province\ of\ distribution, Month\},$

$\quad f_{Shoe\text{-}Sales} : \{Province\ of\ distribution : Province, month : Month\}_T \to integer >$

5. the generation of the geographic classes PROVINCE OF PRODUCTION and REGION OF PRODUCTION and *ISD* relationship between them.

Figure 7: Example of Different Steps for the Generation of Specialized Geographic Classes

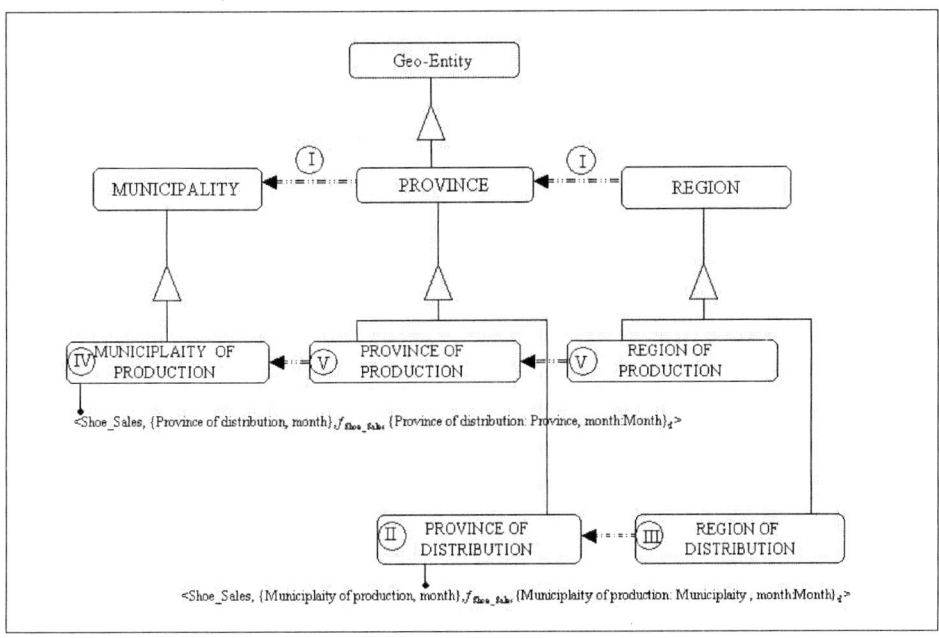

In Figure 7, the result of the above process is shown. Note, each previous labeled step is represented with the same label in the figure.

Analogously in the above discussion, the covering binding attributes are not created in the geographic classes which are at a higher, more aggregate level in the hierarchy.

Deduced Attribute

Deduced attributes may result from the aggregation over the local levels of a given basic binding attribute through roll-up and slice operators as follows:

Definition 16: Let $C = < C_{name}, L, f_c : L_T \to M >$ be a cube, and let $l_g \in L$ be a geographic level to which the geographic class $gc = \lambda^{-1}(l_g)$ corresponds. Starting from $A_B = < C_{name}, L - l_g, f_c : \{L - l_g\}_T \to M >$ of gc, the sub-cubes which are obtained by the application of the following operators:

- $slice_{L'}^{f_{agg}(f_c)}(A_B)$, where $L' \subseteq L - l_g$;

- $roll\text{-}up_{l' \to l''}^{f_{agg}(f_c)}(A_B)$ where $l', l'' \in L$ and they belong to dimension

$D = <L, \preceq>$, such that $l' \leq l''$ and f_{agg} is SUM

are called *deduced* binding attributes.

Example 16: From $A_B = <Car\_Sales, \{model, month\}, f_{Car\_Sales} : \{model:Model, month:Month\}_T \to M>$, the deduced binding attributes, by applying slice operator are, for instance: $< Car\_Sales, \{model\}, f_{Car\_sales} : \{model : Model\}_T \to M>$, $< Car\_Sales, \{month\}, f_{Car\_Sales} : \{month : Month\}_T \to M>$. Applying the roll-up operator, they are: $< Car\_Sales, \{model, year\}, f_{Car\_Sales} : \{model : Model, year : Year\}_T \to M>$, $< Car\_Sales, \{vehicle, year\}, f_{Car\_Sales} : \{vehicle : Vehicle, year : Year\}_T \to M>$.

To compute the deduced binding attributes, we defined a set of methods in the root class Geo-Entity, which are able to invoke a set of OLAP-like operators in the context of geographic classes. They are potentially equivalent to the OLAP operators of MDDBs, and are called G-Slice, G-roll-up, and G-dice. They are inherited by each

---

ALGORITHM	DEDUCED BINDING ATTRIBUTES

Input: a cube with the schema $C = <C_{name}, L, f_c : \{L\}_T \to M>$, a binding attribute $A_B = <C_{name}, \bar{L}, f_c : \{\bar{L}\}_T \to M>$, where $\bar{L} = L - l_g$. I_g is an instance of l_g, F is a predicate, $Op(X)$ (Op = G-slice, G-roll-up, G-dice) is a collection of operations, root class $<rt, \Phi, P, M>$, where A_B belongs to P, gc subclass of the root class

Output: new binding attribute A'_B
case of $Op = G$ - slice :
    for each $k \in \bar{L}$
    then return $A'_B = G\text{-}slice_{\bar{L}\text{-}k}(A_B) \in P'$ and
    $A'_B = <C_{name}, \bar{L}\text{-}k, f_c : \{\bar{L}\text{-}k\}_T \to M>$
    for each $go \in gc$
    then return $G\text{-}slice_{\bar{L}\text{-}k}(A_B) \equiv slice_k(slice_{\bar{L}}(dice_{I_g = \lambda(go)}(C)))$
case of $Op = G$ - roll - up :
    for each $k, k' \in \bar{L}$ if $k \leq k'$
    then return $A'_B = G - roll - up_{k \to k'}^{f_{agg}}(A_B)$ and
    $A'_B = <C_{name}, (\bar{L} - k) \cup k', f_c : \{(\bar{L} - k) \cup k'\}_T \to M>$
    for each $go \in gc$
    then return $G - roll - up_{k \to k'}^{f_{agg}}(A_B) \equiv roll - up_{k \to k'}^{f_{agg}}(slice_{\bar{L}}(dice_{I_g = \lambda(go)}(C)))$

case of $Op = G$ - dice :
    for F **then return** $A'_B = G\text{-}dice_F(A_B)$ and
    $A'_B = <C_{name}, \bar{L}, f_c : \{\bar{L}\}_T \to M>$
    for each $go \in gc$
    then return $G\text{-}dice_F(A_B) \equiv dice_F(slice_{\bar{L}}(dice_{I_g = \lambda(go)}(C)))$
endbegin/*all attempts to generate functional attributes failed*/
return a failure message

---

sub-class of the root class, and each of these operations can be invoked at run time by the substitution of the sub-class parameters. The main functions of such an algorithm is to keep the applications of OLAP operators isomorphic, both to binding attributes in the GDBs and to the cube from which such a binding attribute is calculated in the MDDB. In this way, the processing of a joint query formalized on both databases cannot be perceived by the user.

Depending on each type of operator, the following algorithm shows the schema of resulting cubes or attributes and the relative computation in the context of MDDB through OLAP operators.

Number of Binding Attributes

One interesting issue to take into consideration is related to the combinatorial number of cubes which can result from different types of binding attributes. In order to obtain the total number of the basic, covering, and deduced binding attributes, we first give some definitions which will be used in the theorems introduced afterwards.

Let us consider the following symbols: K is the number of different cubes in an MDDB; N_i represents the total number of levels (geographic and non-geographic) of the i-th cube; G_i represents the number of geographic levels of the i-th cube; $l_{i,h}$ represents the number of basic and covering binding attributes generated by a geographic level ($1 \leq h \leq L_G$) of the cube; $f_{i,k}$ is the number of binding attributes generated by applying roll-up and slice operators to any local level $k \geq 1$ of the i-th base cube.

Any geographic level of each independent cube generates a basic binding attribute: in a generic cube G_i geographic levels exist. Therefore, we have the following definition.

Definition 17: Given a set of n cubes. The number of basic binding attributes of each cube $i = 1,\ldots,n$, denoted by $A_{i_{(b)}}$, is G_i , and the number of basic and covering binding attributes is: $A_{i_{(b,d)}} = \sum_{h=1}^{G_i} l_{i,h}$.

We can calculate the total number of the attributes generated by a cube in a geographic class by the next theorem.

Theorem 2: The total number of basic, covering, and deduced binding attributes of the i-th cube is given as follows:

$$A_{i(Tot)} = \prod_{k=1}^{N_i}(f_{i,k}+1)\sum_{h=1}^{G_i}(l_{i,h}/l_{i,h}+1)$$

Proof: For each base and covering binding attribute, we can apply roll-up and slice operators to the local levels for generating deduced binding attributes. Let $P_{i,h}$ be the number of local levels of a generic binding attribute and let $Q_{i,h,q}$, $q \geq 1$ be the number of deduced binding attributes generated by the application of the roll-up and slice operators to a generic local level. Then, the number of the deduced binding attributes for all local levels is: $\displaystyle\prod_{q=1}^{P_{i,h}}(Q_{i,h,q}+1)$.

From a generic geographic level h, we obtain $l_{i,h}$ basic and covering binding attributes; then, the number of deduced binding attributes generated by such a geographic level of the given cube is $\displaystyle l_{i,h}\prod_{q=1}^{P_{i,h}}(Q_{i,h,q}+1)$.

Hence, the number of geographic levels is G_i , then the total number of basic, covering, and deduced binding attributes generated by i-th cube is $\displaystyle\sum_{h=1}^{G_i} l_{i,h}\prod_{q=1}^{P_{i,h}}(Q_{i,h,q}+1)$.

As we can observe, the geographic levels of the cube appear twice as $l_{i,h}$ and $Q_{i,h,q}$.

$$... + l_{i,h}\,(l_{i,h'}+1)\prod_{q=1}^{P_{i,h}-1}(Q_{i,h,q}+1) + l_{i,h'}\,(l_{i,h}+1)\prod_{q=1}^{P_{i,h}-1}(Q_{i,h,q}+1) + ... =$$

$$= ...+\frac{l_{i,h}}{l_{i,h}+1}(l_{i,h}+1)(l_{i,h'}+1)\prod_{q=1}^{P_{i,h}-1}(Q_{i,h,q}+1) + \frac{l_{i,h'}}{l_{i,h'}+1}(l_{i,h}+1)(l_{i,h'}+1)\prod_{q=1}^{P_{i,h}-1}(Q_{i,h,q}+1) + ...$$

where the term $\displaystyle\prod_{q=1}^{P_{i,h}-1}(Q_{i,h,q}+1)$ refers to non-geographic levels.

Therefore, for all levels N_i we have:

$$A_{i_{Tot}} = \sum_{h=1}^{G_i} l_{i,h}\prod_{q=1}^{P_{i,h}}(Q_{i,h,q}+1) = \prod_{k=1}^{N_i}(f_{i,k}+1)\sum_{h=1}^{G_i}\frac{l_{i,h}}{l_{i,h}+1}$$

Theorem 3: The number of basic binding attributes generated by K independent cubes is: $A_b = \sum_{k=1}^{K} N_k$.

The number of base and covering binding attributes generated by K independent cubes is:

$$A_{b,d_{(Tot)}} = \sum_{i=1}^{K} \sum_{h=1}^{G_i} l_{i,h} \ .$$

Then, the total number of the basic, covering, and deduced functional attributes generated by K cubes is:

$$A_{Tot} = \sum_{i=1}^{K} \left(\prod_{k=1}^{N_i} (f_{i,k}+1) \sum_{h=1}^{G_i} \frac{l_{i,h}}{l_{i,h}+1} \right).$$

Proof: The proof is trivial because the K cubes are independent, then all the results of Theorem 2 are satisfied by them.

To emphasize and clarify the problem of binding attributes blowing up, and to make it easier to understand the general concepts defined in the previous theorems, we give the following examples.

Example 17: Let us consider that an MDDB is composed of the following cubes ($K = 3$) where facts/measures are recorded for Italy only and the hierarchies are those indicated in Figure 1:

$$< Car\_Sales, \ \{city, model\}, f_{Car\_Sales} \ : \{day : Day, model : Model\}_T \ \rightarrow numeric >$$

$< Shoe\_Sales, \{Region \ of \ production, Region \ of \ distribution, Month\},$

$\qquad f_{Shoe\_Sales} \ : \{Region \ of \ production : Region, \ Region \ of \ distribution : Region, month : Month\}_T \rightarrow numeric >$

$<\# \ of \ patients, \{Municipality \ of \ residence, Province \ of \ hospitalization, Day\},$

$\qquad f_{\# of \ patients} \ : \{Municipality \ of \ residence : Municipality, Province \ of \ hospitalization : Province, day : Day\}_T \rightarrow numeric >$

Table 1: Number of Binding Attributes

	Basic	Basic + Covering	Basic + Covering + Deduced
Car _ Sales	1	3	36
Shoe _ Sales	2	2	12
# of patients	2	5	68
All the cubes	5	10	116

The number of base, covering, and deduced binding attributes are indicated in Table 1.

This total number has been given against three simple cubes defined by one and two geographic levels between each pair of which there can be a hierarchical or a non-hierarchical relationship. These numbers are not intended to give a comprehensive evaluation of performance. Rather, they demonstrate the other possible answers to queries which can be driven from binding attributes.

CONCLUSIONS

We have presented a logical data model, based on the extension of the geographical data model, which provides the capability of manipulating complex entities and answering queries by the cooperation with MDDBs. The cooperation of spatio-aggregate data is obtained by introducing new attributes into the geographic data structure. These are called *binding* attributes. The main idea behind them is to retrieve information which is stored in MDDBs and can be invoked from queries formulated in GDBs.

We have shown that our approach to the common "spatial" feature provides the interoperability of spatio-aggregate databases and guarantees the autonomy of data sources. The features of our approach that distinguish it from similar works include:

- uniform treatment of spatio-aggregate data;
- a comprehensive aggregation facility of either geographic or summary data;
- a non explicit different use of a specific set of operators in each database; and
- a specific design for interoperability.

The discussion up to this point has focused upon the *space* feature for realizing the interoperability. However, the role of the *Time* feature in such a cooperation forms an interesting issue to be considered in future work.

An important assumption made in this chapter concerns the summarizability of a data cube, which means no cell contains a "not available" value, and there is no missing value. This restriction can be of practical importance in some cases of data warehousing. In many other cases the partial availability of data is realistic and consequently the comprehensive aggregation of summary data is compromised. The answering of joint aggregate queries formulated on cubes with missing and/or non-available data values in the context of the interoperability of heterogeneous data sources can create another open problem.

We discussed the problem of answering spatio-summary queries in the context of GDBs. Research that considers this question in the special light of MDDBs could raise another interesting issue to be examined.

REFERENCES

Abiteboul, S., & Bidoit, N. (1986). Non first normal form relations: An algebra allowing data restructuring. *Journal of Computer and System Sciences*, 33(3), 361-393.

Agrawal, R., Gupta, A., & Sarawagi, S. (1997). Modeling Multidimensional Databases. *Proceedings of the 13th International Conference on Data Engineering (ICDE'97)*, April 7-11, Birmingham, UK, 232-243.

Aronof, S. (1991). *Geographic Information Systems: A Management Perspective*. Canada: WDL Publications.

Brankovic, L., Miller, M., & Siran. J. (1996). Graphs, 0-1 matrices, and usability of statistical databases. *Congressus Numerantium*, 12, 196-182.

Brankovic, L., Horak, P., Miller, M., & Wrightson, G. (1997). Usability of compromise-free statistical databases for range sum queries. In Ioannidis, Y.E., & Hansen D.M., (Eds.), *Proceedings of Ninth International Conference on Scientific and Statistical Database Management (SSDBM'97)*, August 11-13, Olympia, Washington. Silver Spring, MD: IEEE Computer Society Press, 144-154.

Brassel, K.E., & Weibel, R. (1988). A review and conceptual framework of automated map generalization. *International Journal of Geographical Information Systems*, 2(3), 229-244.

Cabibbo, L., & Torlone, R. (1998). A logical approach to multidimensional databases. In Schek, H.J., Saltor, F., Ramos, I., & Alonso, G. (Eds.), *Proceedings of the 6th International Conference on Extending Database Technology (EDBT'98)*, March 23-27, Valencia, Spain. Lecture Notes in Computer Science, No. 1377. Berlin: Springer-Verlag, 183-197.

Chan, P., & Shoshani, A. (1981). SUBJECT: A directory-driven system for organizing and accessing large statistical databases. *Proceedings of Very Large Data Bases, 7th International Conference*, September 9-11, Cannes, France. Silver Spring, MD: IEEE Computer Society Press, 553-563.

Chin, F.Y., & Ozsoyoglu, G. (1982). Auditing and inference control in statistical databases. *IEEE Transactions on Software Engineering*, SE-8(6), 574-582.

Egenhofer, M.J., & Franzosa, R. (1991). Point-set topological spatial relations. *International Journal of Geographic Information Systems*, 5(2), 161-174.

Egenhofer, M.J., & Franzosa, R. (1995). On the equivalence of topological relations. *International Journal of Geographical Information Systems,* 9(2), 133-152.

Egenhofer, M.J., & Herring, J. (1990). A mathematical framework for the definition of topological relationships. *Proceedings of the Fourth International Symposium on Spatial Data Handling.* Columbus, OH: International Geographical Union, 803-813.

Ferri, F., Pourabbas, E., Rafanelli, M., & Ricci F.L. (2001). Extending geographic databases for a query language to support queries involving statistical data. In

Günther, O., & Lenz H.-J. (Eds.), *Proceedings of 12th International Conference on Scientific and Statistical Database Management (SSDBM)*, July 26-29, Berlin, Germany. Silver Spring, MD: IEEE Computer Society Press, 220-230.

Frank, A.U., Volta, G.S., & McGranaghan, M. (1997). Formalization of families of categorical coverages. *International Journal of Geographical Information Science*, 11(3), 215-231.

Gardarin, G., Cheiney, J.P., Kiernan, G., Pastre, D., & Stora, H. (1989). Managing complex objects in an extensible relational DBMS. *Proceedings of the Conference on Very Large Databases (VLDB)*, pp. 155-165.

Gargano, M., Nardelli, E., & Talamo, M. (1991). Abstract data types for the logical modeling of complex data. *Information Systems*, 16(5), 565-583.

Gray, J., Bosworth, A., Layman, A., & Pirahesh, H. (1996). Data cube: A relational aggregation operator generalizing group-by, cross-tab and sub-total. *Proceedings of the 12th IEEE International Conference on Data Engineering*, February, New Orleans, Louisiana, 152-159.

Gunther, O., & Riekert, W.-F. (1993). The design of GODOT: An object-oriented geographic information system. *IEEE Data Engineering Bulletin*, 4-9.

Gyssens, M., & Lakshmanan, L. (1997). A foundation for multi-dimensional databases. *Proceedings of 23rd International Conference on Very Large Data Bases (VLDB'97)*, August 25-29, Athens, Greece, 1-10.

Hughes, J.G. (1991). Object-oriented databases. *Prentice Hall International Series in Computer Science*. London: Prentice Hall.

Kim, W., & Lochovsky, F.H. (1989). Object-oriented concepts, databases, and applications. New York: ACM Press.

Kouba, Z., Matousek, K., & Miksovsky, P. (2000). On data warehouse and GIS integration. In Ibrahim, M.T., Küng, J., & Revell, N. (Eds.), *Proceedings of the 11th International Conference on Database and Expert Systems Applications (DEXA)*, September 4-8, London, UK. Lecture Notes in Computer Science, No. 1873. Berlin: Springer-Verlag, 604-613.

Laurini, R., & Thompson, D. (1992). *Fundamentals of Spatial Information Systems*. New York: Academic Press.

Lenz, H.-J., & Shoshani, A. (1997). Summarizability in OLAP and statistical data bases. In Ioannidis, Y.E., & Hansen, D.M. (Eds.), *Proceedings of Ninth International Conference on Scientific and Statistical Database Management (SSDBM)*, August 11-13, Olympia, Washington. Silver Spring, MD: IEEE Computer Society Press, 132-143.

Leung, Y., Leung, K.S., & He, J.Z. (1999). A generic concept-based object-oriented geographical information system. *International Journal of Geographical Information Science*, 13(5), 475-498.

Maguire, D., Goodchild, M., & Rhind, D. (1991). *Geographical Information Systems: Principles and Applications*. New York: John Wiley & Sons.

Milne, P., Milton, S., & Smith, J. (1993). Geographical object-oriented databases: A case study. *International Journal of Geographical Information Systems*, 7(1), 39-56.

Muller, J.C., Lagrange, J.P., & Weibel, R. (Eds.). (1995). *GISs and Generalization.* London: Taylor & Francis.

OLAP Council. (1997). *The OLAP Glossary.* Available on-line at: http://www.olapcouncil.org.

Özsoyoglu, G., Özsoyoglu Z.M., & Victor, M. (1987). Extending relational algebra and relational calculus with set-valued attributes and aggregate functions. *ACM Transactions on Database Systems*, 12, 566-592.

Paolucci, M., Sindoni, G., DeFrancisci, S., & Tininini, L. (2000). SIT-IN on heterogeneous data with Java, HTTP, and Relations. *Proceedings of the International Workshop on Java and Databases: Persistence Options* (in NetObject.Days).

Pourabbas, E., & Rafanelli, M. (2000). *A Pictorial Query Language Extended with Cardinal and Positional Operators for Querying Geographic Databases.* Technical Report No. 530, Istituto di Analisi dei Sistemi ed Informatica-CNR, September.

Rafanelli, M., & Ricci, F.L. (1993). Mefisto: A functional model for statistical entities. *IEEE Transactions on Knowledge and Data Engineering*, 5(4), 670-681.

Rafanelli, M., & Shoshani, A. (1990). STORM: A statistical object representation model. In Michalewicz, Z. (Ed.), *Proceedings of the Fifth International Conference on Statistical and Scientific Database Management (SSDBM'90),* April 3-5, Charlotte, North Carolina. Lecture Notes in Computer Science, No. 420. Berlin: Springer-Verlag, 14-29.

Rigaux, P., & Scholl, M. (1994). Multiple representation modeling and querying. In Nievergelt, J., Roos, T., Schek, H.-J., & Widmayer, P. (Eds.), *Proceedings of the International Workshop on Advanced Research in Geographic Information Systems*, February-March, Monte Verità, Ascona, Switzerland. Lecture Notes in Computer Science, No. 884. Berlin: Springer-Verlag, 59-69.

Rigaux, P., & Scholl, M. (1995). Multi-scale partitions: Application to spatial and statistical databases. In Egenhofer, M., & Herrings, J. (Eds.), *Advances in Spatial Databases (SSD'95).* Lecture Notes in Computer Science, No. 951. Berlin: Springer-Verlag, 170-184.

Rigaux, P., Scholl, M., & Voisard, A. (2001). *Spatial Databases with Applications to GIS.* London: Morgan Kaufmann.

Shekhar, S., Lu, C., Tan, X., Chawla, S., & Vatsavai, R. (1999). Mapcubes: A visualization tool for spatial data warehouses. In Miller, H., & Han, J. (Eds.), *Geographic Data Mining and Knowledge Discovery (GKD).*

Shoshani, A. (1997). OLAP and statistical databases: Similarities and differences. *Proceedings of the Sixteenth ACM SIGACT-SIMOD-SIGART Symposium* on *Principles of Database Systems (PODS'97),* May 12-14, Tucson, Arizona, 185-196.

Stonebraker, M. (1986). Inclusion of new types in relational database systems. *Proceedings of the International Conference on Data Engineering,* February 5-7, Los Angeles, CA, 262-269.

Stonebraker, M., & Rowe, L.A. (1986). The design of POSTGRES. *Proceedings of ACM SIGACT-SIGMOD,* June, 340-355.

Stonebraker, M., Runestein, B., & Guttman, A. (1983). Application and abstract data types and abstract indices to CAD databases. *SIGMOD 1983: Engineering Design Applications.* Silver Spring, MD: IEEE Computer Society Press, 107-113.

Ullman J.D. (1988). *Principles of Database and Knowledge-Based Systems.* Volume I. City: Computer Science Press.

Volta, G.S., & Egenhofer, M. (1993). Interaction with GIS attribute data based on categorical coverages. In Frank, A.U., & Campari, I. (Eds.), *Spatial Information Theory: A Theoretical Basis for GIS, International Conference (COSIT'93),* September 19-22, Marciana Marina, Elba Island, Italy. Lecture Notes in Computer Science, No. 716. Berlin: Springer-Verlag, 215-233.

Worboys, M.F., Hearnshaw, H.M., & Maguire, D.J. (1990). Object-oriented modeling for spatial databases. *International Journal of Geographical Information Systems,* 4(4), 369-383.

About the Authors

Maurizio Rafanelli is senior scientist at the Istituto di Analisi dei Sistemi ed Informatica "A. Ruberti" and in charge of the research area, "Information Systems and Knowledge Bases for Complex Information Structures." He is author and co-author of many international publications (journals, conferences, books, etc.). He was a part-time professor at the University of Roma-La Sapienza (Roma-1) and at the University of Roma-Tor Vergata (Roma-2) from 1981 to 1997. He was general chairman at the 4th International Conference on Statistical and Scientific Database Management (*Proceedings*, Lecture Notes in Computer Science, No. 339, Springer-Verlag) in 1988 and general chairman at the 10th International Conference on Scientific and Statistical Database Management (*Proceedings*, IEEE Publications) in 1998. He has been a program committee member of various international conferences and cooperates with many journals as a reviewer. He is author of the entry, "Data Models in Statistical and Scientific Databases," in the *Encyclopedia of Computer Science and Technology* (Kent & Williams, Exec. Eds.).

*   *   *   *   *

Amr El Abbadi received his PhD in Computer Science from Cornell University. Since 1987 he is a professor in the Department of Computer Science at the University of California, Santa Barbara. Between 1990 and 1999 he held visiting positions at various institutions, including the University of Campinas in Brazil, IBM Almaden Research Center, the Swedish Institute of Computer Science in Stockholm, and at IRISA at the University of Rennes in France. He is currently the area editor for *Information Systems: An International Journal*, an editor of *Information Processing Letters (IPL)*, and associate editor of the *Bulletin of the Technical Committee on Data Engineering*. He was vice chair of the 1999 International Conference on Distributed Computing Systems, vice chair for the International Conference on Data Engineering 2002, and the Americas program chair for the 2000 International Conference on Very Large Data Bases. Dr. El Abbadi's main research interests are in the area of distributed information management systems, including databases, digital libraries, and data management of moving objects.

Divyakant Agrawal is a professor of Computer Science at the University of California, Santa Barbara. His research interests include distributed systems, computer networks, databases, large-scale information systems and digital libraries, and data management of moving objects. He received his BE from Birla Institute of Technology and Science, Pilani, India, and his MS and PhD in Computer Science from the State University of New York at Stony Brook. Professor Agrawal is the editor for *Distributed and Parallel Databases, An International Journal* and was a member of the program committee for various international conferences, including ACM SIGMOD International Conference on Management of Data, ACM Symposium on Principles of Database Systems (PODS), and International Conference on Very Large Data Bases (VLDB). He is also a member of the ACM and the IEEE Computer Society.

Elena Baralis is an associate professor at Politecnico di Torino. She obtained a PhD in Information and Systems Engineering from Politecnico di Torino in 1994. Her research activity is focused on active database systems, data warehousing, and data mining. She teaches courses on database design and technology.

Andrea Calí received a Laurea (MS) in Electrical Engineering from the University of Roma-La Sapienza, and now he is a third-year PhD student in Computer Engineering at the Department of Computer and System Sciences of the University of Roma-La Sapienza. His research interests include information integration, data warehousing, database models, knowledge representation. He is involved in several national and international research projects. His latest publications deal with investigating the expressiveness of data integration systems and managing integrity constraints in information integration.

Curtis E. Dyreson is an Assistant Professor of Computer Science at Washington State University. Prior to working in academia, he consulted in industry for Citibank and Chemical Bank, UK. He has held faculty positions in Australia at James Cook University and Bond University, and was a visiting associate professor at Aalborg University, Denmark. He is an author or coauthor of numerous papers on temporal and multidimensional databases. Currently, he is the ACM TODS information director.

Christian S. Jensen is a professor of Computer Science at Aalborg University, Denmark, and an honorary professor at Cardiff University. He has authored or coauthored numerous scientific papers on database semantics, modeling, and performance. He is on the editorial boards of *ACM TODS* and *IEEE Data*

Engineering Bulletin. He was recently co-program committee chair for the Workshop on Spatio-Temporal Database Management, held with VLDB'99, and for the 8th International Symposium on Spatial and Temporal Databases; also he was program committee chair for the 2002 EDBT Conference. He serves on the boards of directors for a small number of companies.

Domenico Lembo is a second-year PhD student at the Department of Computer and System Sciences of University of Rome-La Sapienza, Italy. His research concerns database models, information integration and data warehousing, logic programming, and knowledge representation. He is currently involved in various Italian and European research projects that aim at providing innovative methodologies and advanced reasoning capabilities for information integration and data warehousing. Recently, he co-authored several papers on modeling and semantics of data integration systems.

Maurizio Lenzerini, born in Pavia, has served as full professor in Computer Science and Engineering since 1990. He is the author of several academic books on fundamentals of computer science, software engineering, and database design. Since 1983, he has been carrying out his research activity at the University of Rome-La Sapienza, where he is leading a research group on databases and artificial intelligence. His main research interests are oriented toward conceptual and semantic data modeling, data integration, data warehousing, semistructured data management, knowledge representation and reasoning, and object-oriented methodologies. He is currently involved in national and international research projects on data integration, data warehousing, and semi-structured data. He is the author of more than 200 publications in international conferences and journals, including the most prestigious ones in the above-mentioned areas, such as *Journal of Computer and System Science, Information and Computation, Artificial Intelligence, Information Systems, IEEE Data and Knowledge Engineering, ACM-PODS, ACM-SIGMOD, IEEE-ICDE, VLDB, ICDT, IJCAI, AAAI, KR,* and *CoopIS.* He is the editor of several international books, including a recent one on *Data Warehouse Quality.* He is regularly a member of the program committee of the most important international conferences in the above areas, including IJCAI, AAAI, EDBT, PODS, KR, CoopIS, ER, and ICDT. He organized several international conferences and workshops. He is a member of the editorial board of various international journals, and is the editor of *Information Systems: An International Journal,* for the area of data modeling, knowledge representation, and reasoning. He was program co-chair of the 4th International Conference on Cooperative Information Systems, held in Edinburgh in 1999. He was the conference chair of the International Conference on Conceptual Modeling, held in Paris in 1999. He will serve as the program chair of ICDT 2003.

Francesco M. Malvestuto was born in Sulmona, Italy. He earned his degrees from the University of Rome-La Sapienza, including one in Physics in 1974, a master's in Informatics in 1978, and a master's in System Engineering in 1980. His research interests evolved from statistical databases to uncertainty management in artificial intelligence, and axiomatizion of the probability-theoretic notions of independence and irrelevance. His current research fields include uncertainty management, database theory, computational statistics, and graph theory. He is a professor of Computer Science at the University of Rome-La Sapienza, and is a member of the IASC and SIS.

Alberto Mendelzon was born in Buenos Aires, and received his MA, MSE, and PhD degrees from Princeton University. He spent a post-doctoral year at the IBM T.J. Watson Research Center in 1979-1980 and has been with the University of Toronto since 1980. He has been a visiting scientist at the IBM Centre for Advanced Studies, AT&T Bell Laboratories, NTT Basic Research Labs in Musashino, Japan, and the IASI in Rome. His research interests are in databases and knowledge bases, including database design theory, query languages, database visualization, query processing, Web-based information systems, and data warehousing. He has been associate and acting chair of the Computer Systems Research Institute and chaired or co-chaired the program committees of the major database conferences. He was a guest editor for the *Journal of Computer and System Sciences* and the *VLDB Journal,* and is on the Editorial Board of the *Journal of Digital Libraries, World Wide Web Journal, ACM Digital Reviews,* and *Theory and Practice of Object Systems.* He is the current information director for the ACM Special Interest Group on Management of Data (SIGMOD).

Marina Moscarini was born in Roma. She earned her degree in Mathematics from the University of Rome-La Sapienza in 1973. Since 1993 she is a professor of Computer Science at the University of Rome-La Sapienza, where she currently heads the Computer Science Department. Prior to joining the University of Rome-La Sapienza, she was a full-time researcher at IASI-CNR at Rome until 1989, and from 1989 to 1993 she was professor at the University of Rome-Tor Vergata. Her research interests evolved from the study of classical difficult (NP-complete) graph theoretic problems on particular classes of graphs, to the study of problems related to database design, query optimization, and statistical database security.

Stefano Paraboschi is an associate professor at the Dipartimento di Elettronica e Informazione of Politecnico di Milano. He received the Laurea Degree in Ingegneria Elettronica in 1990, and a PhD in Ingegneria Informatica in 1994, both from Politecnico di Milano. His main research interests are in the area of databases, with a focus on active databases, data warehouses, and the construction of data-intensive websites. He is the author, together with Paolo Atzeni, Stefano Ceri, and Riccardo

Torlone, of the book, *Database Systems: Concepts, Languages and Architectures* (McGraw-Hill, 1999).

Torben B. Pedersen is an associate professor of Computer Science at Aalborg University, Denmark. Previously, he has worked as a visiting scientist at Lawrence Berkeley National Laboratory and as a database specialist at the largest Danish software company, Kommunedata. He has authored or co-authored papers on multidimensional data modeling and semantics, OLAP performance issues, OLAP data integration, and clickstream analysis. He is on the program committees for the EDBT 2002 conference and the DMDW 2002 workshop, and has reviewed papers for SIGMOD, VLDB, ICDE, CAiSE, SSDBM, and KAIS. He has extensive industry collaboration within the business intelligence industry.

Elaheh Pourabbas received her master's degree in Engineering from the University of Roma-La Sapienza in 1992, and her PhD in Bioengineering from the University of Bologna in 1997. She is currently a researcher at the Istituto di Analisi dei Sistemi ed Informatica "Antonio Ruberti" of the Italian National Research Council. Dr. Pourabbas' interests include database theory, data models, data warehousing and OLAP, spatio-temporal databases, and database systems.

Mirek Riedewald is a PhD candidate in the Computer Science Department at the University of California, Santa Barbara. His dissertation research focuses on supporting aggregation and summarization in data warehouses and digital libraries. He was due to graduate in the summer of 2002 and join Cornell University as a research associate. His research interests include database and information systems, especially On-Line Analytical Processing (OLAP), digital libraries, data streams, and distributed systems. His work has been published in the proceedings of prestigious scientific conferences such as the International Conference on Very Large Data Bases (VLDB) and the ACM SIGMOD International Conference on Management of Data.

Riccardo Rosati is assistant professor in Computer Science and Engineering at Università di Roma-La Sapienza, Italy. His main research interests are oriented towards artificial intelligence and databases, in particular knowledge representation and reasoning, planning and cognitive robotics, information integration, and data warehousing. He is currently involved in national and international research projects on information integration. He is the author of more than 50 publications in international conferences and journals, such as *Artificial Intelligence, ACM Transactions on Computational Logic, JAIR, IJCAI, AAAI, KR,* and *CoopIS.*

Arie Shoshani is a senior staff scientist at Lawrence Berkeley National Laboratory; he joined LBNL in 1976. He heads the Scientific Data Management Group. He received his PhD from Princeton University in 1969, and from 1969 to 1976, he was a researcher at System Development Corporation, where he worked on the Network Control Program for the ARPAnet, distributed databases, database conversion, and natural language interfaces for data management. His current areas of work include data models, query languages, temporal data, statistical and scientific database management, storage management on tertiary storage, and grid storage middleware. Dr. Shoshani is also the director of a Scientific Data Management (SDM) Integrated Software Infrastructure Center (ISIC), one of seven centers (budget, $3 million) selected by the SciDAC program at DOE in 2001. In this capacity, he is coordinating the work of collaborators from four DOE laboratories and four universities (see http://sdmcenter.lbl.gov). Dr. Shoshani has published more than 60 technical papers in refereed journals and conferences; chaired several workshops, conferences, and panels in database management; and served on numerous program committees for various database conferences. He also served as an associate editor for the *ACM Transactions on Database Systems.* He was elected a member of the VLDB Endowment Board, served as the publication board chairperson for the *VLDB Journal,* and as the vice-president of the VLDB Endowment. His homepage is http://www.lbl.gov/~arie.

Giuseppe Sindoni has been an applicative researcher at ISTAT for four years. He holds a PhD in Information Systems Engineering and an Honours Degree in Electronic Engineering. He also served as associate professor of Informatics at the Faculty of Economics, Roma Tre University for the 2001-2002 academic year. Before joining ISTAT, he was a scientific collaborator and assistant professor at the Department of Computer Science and Automation, Roma Tre University, and scientific collaborator of the Istituto per l'Analisi dei Sistemi e l'Informatica (IASI) of the Italian National Research Council (CNR). His expertise covers the following areas: integration of (statistical) information systems; Web data quality; databases; (statistical) databases and the Web, and geographical information systems. He has published more than 20 papers in these fields, mainly in refereed journals and international conference proceedings. Additionally, he has 15 years' experience as a freelance consultant to several national and international organizations.

Ernest Teniente is an associate professor in the Software Department at the Universitat Politècnica de Catalunya in Barcelona. He teaches courses on software engineering and database technology. His research interests include data warehousing, database updating and conceptual modeling of information systems. His e-mail address is teniente@lsi.upc.es.

Leonardo Tininini is a researcher at the Italian National Institute of Statistics (ISTAT). He has written several scientific papers on statistical databases, aggregate data, query languages, spatio-temporal databases, and been referee for important international conferences and journals. He is lecturer at the University of Rome-Campus Bio-Medico. He also collaborates with the Institute of Analysis on Systems and Computer Science of the Italian Research National Council, and with the French research institute INRIA. He is leading a project for the dissemination of the Italian Census 2001 data through a Web-based data warehouse.

Riccardo Torlone is a professor of Computer Science at the Università Roma Tre, Italy. In the past, he had teaching and research appointments at IASI-CNR and at University of Roma-La Sapienza. He has authored or co-authored numerous scientific papers on database theory, database design, active databases, object-oriented query languages, and data warehousing. He has also co-authored the book, *Database Systems: Concepts, Languages and Architectures* (McGraw Hill). He served on the program committee of several international conferences on database technology and information systems.

Alejandro Vaisman was born in Buenos Aires. He is a civil engineer and computer scientist, and holds a PhD in Computer Science from the University of Buenos Aires. He was a visiting researcher at the University of Toronto, Canada, and invited lecturer at the Polytechnic University of Madrid. He has authored and co-authored several scientific papers presented in major database conferences like ICDE and VLDB. His research interests are in relational and deductive databases, OLAP and data warehousing, temporal databases, data mining, and Web-based information systems. He has worked in design and operation of database systems, and he is currently vice-dean at the University of Belgrano, Argentina.

Index

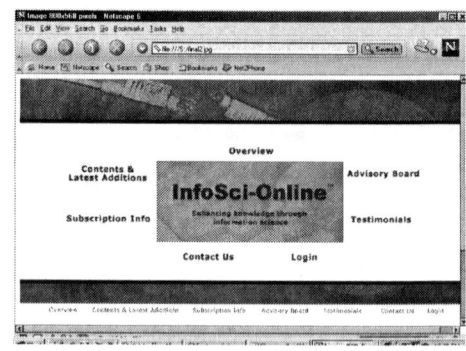